Stimuli-Responsive Materials

From Molecules to Nature Mimicking Materials Design

Stimuli-Responsive Materials

From Molecules to Nature Mimicking Materials Design

Marek W. Urban

Clemson University, SC, USA
Email: mareku@clemson.edu

ROYAL SOCIETY
OF **CHEMISTRY**

THE QUEEN'S AWARDS
FOR ENTERPRISE:
INTERNATIONAL TRADE
2013

Print ISBN: 978-1-84973-656-5

A catalogue record for this book is available from the British Library

Published by The Royal Society of Chemistry,
Thomas Graham House, Science Park, Milton Road,
Cambridge CB4 0WF, UK

Registered Charity Number 207890

Visit our website at www.rsc.org/books

Printed in the United Kingdom by CPI Group (UK) Ltd, Croydon, CR0 4YY, UK

Preface

Combining logical gates and organizing them into specific executable sequences serves as the basis for programming electronic devices. For materials the programming is achieved by combining specific sequentially positioned chemical and/or physical motifs to achieve desirable functions. The most obvious programmable systems are the strands of DNA and RNA that serve as the huge and accurate information database with even more complex tools containing instructions for accurate regulation, replication and translation of encoded information.

In recent years there have been tremendous efforts put into developing synthetic programmable materials. Fewer efforts have been made to understand molecular processes leading to stimuli-responsive behavior. Until the turn of the 21st century only very few research articles considered the term 'stimuli responsiveness' in the context of materials science, chemistry, physics or engineering. This vocabulary was rarely part of symposia or scientific meetings. At that time, most would view polymers and other materials as functional. *Plastics, remember son, the future is in plastics*, the famous phrase from The Graduate movie with Dustin Hoffman, said it all. Indeed, polymers serve admirable functions, ranging from paints that protect substrates, or protective armor saving the lives of soldiers, or space industries, not to mention artificial organs, simple catheters or pacemakers, or even cosmetics. You name the field, polymers are there. Since 2002, when the National Science Foundation founded the first Materials Research Science Engineering Center (MRSEC) on Stimuli Responsive Polymeric Films and Coatings, many new research activities have emerged and continue, as does the amazing and

Stimuli-Responsive Materials: From Molecules to Nature Mimicking Materials Design
By Marek W. Urban
© Marek W. Urban, 2016
Published by the Royal Society of Chemistry, www.rsc.org

headline-making new wave of ideas and discoveries, as well as technological opportunities. Today, many researchers around the world are actively involved in searching for new materials with stimuli-responsive characteristics.

In the last decade at least 30 books have been edited and encompass a variety of topics related to stimuli responsiveness. So why write another one? While exciting findings detailing individual research activities can be found in many excellent review articles and edited monographs, the field is mature enough that new principles of design, synthesis and characterization have been established. Thus, for newcomers entering the field, as well as for those who are already immersed in it, this book will hopefully serve as a stimulating source of new knowledge and resourceful information for further advances. Senior undergraduate and graduate students, majoring in materials chemistry and materials science and engineering, as well as chemistry and physics, tested the content of the book. The content appeared to be attractive and this modern core course offered at Clemson University attracted the curiosity of new generations of scientists, yet offered modern views of the new ideas and technologies. This is reflected in the book content, which covers a broad range of topics, from controlled polymer synthesis to physicochemical aspects of stimuli responsiveness in nanomaterials, brushes, surfaces and interfaces, photonic and photochromic materials, field and bio-responsive materials as well as self-healing. As stimuli responsiveness typically requires an input, as a trigger to execute the instructions and perform desired functions, in many cases the output is required to end a given event.

Mother Nature is probably the best teacher and inspiration for stimuli responsiveness. Biological systems exhibit an extraordinary ability to heal wounds autonomously; for plants to heal mechanical damage, where different substances such as suberin, tannins and phenols are activated to prevent further lesions. Similarly, mechanical damage to human skin resulting in the outward flow of blood cells is arrested by the crosslinked network of fibrin, leading to self-repair. One common feature in these bio-events is the presence of heterogeneous, often multi-layered morphologies that interact with each other and respond to external or internal stimuli. These intriguing features stimulated the formation of this new and exciting field of science. Although the field is particularly close to polymer scientists because of inherent similarities of biological systems, other materials are not far behind. Almost any heterogeneous material may exhibit stimuli responsiveness, as long as interfacial regions can facilitate molecular responses and rearrangements. Polymers, due to their versatile features, exhibit a great degree of flexibility, and designing them from molecular blocks to form manufactured architectures at nano, micro, macro and larger scales is critical for creating stimuli-responsive materials capable of autonomous functions. This is reflected in the first introductory chapter, which uses the human eye as an example to delineate the complexity of challenges. The remaining parts of the introduction define general concepts that will be inter-woven throughout the book. The remaining chapters are divided into synthetic aspects of stimuli-responsive materials, whereas the re-maining chapters feature the development of novel stimuli-responsive materials with the ability of signaling, reorganization, sensing and self-repair at various length scales. In essence, the main objective of this book is the programmable design of materials that can retain a set of specifically coded instructions by virtue of a chemical structure in an effort to be able to perform desired functions on demand. Because these materials hold great promise in a variety of applications, understanding the fundamental concepts that govern stimuli-respon-sive behavior is critical; how chemical information can be translated into function is still to be understood. Considering the fact that so many excellent publications and ideas have been published in the last decade, the author apologizes that not all could be incorporated, and only selected examples could be used.

This book went through several tests, exams, and iterations. As in every effort of this nature, it could not be completed without the tremendous help of the end users, the students and postdocs. I am grateful to all Clemson University students who used the content of this book and provided useful insights in various capacities. All Urban Research Group members, past and present (www.clemson.edu/ces/urbanresearch/), who have taught me a lot over the years, and allowed me to be their scientific mentor and research advisor. I am particu-larly thankful to my current group members, Ying Yang, Chris Chornat, Dmitriy Davidovich, Chunling Lu, Tugba Demir, Laura Smith, Yanting Xing, Hiroyuki Mitsunaga and Zhanhua Wang, whose efforts are particularly appreciated. The author is also thankful to his family, Kasia, Ania and Mike, without whom this book could not have developed.

Marek W. Urban
Clemson University

Contents

Stimuli-Responsive Materials: From Molecules to Nature Mimicking Materials Design
By Marek W. Urban
© Marek W. Urban, 2016
Published by the Royal Society of Chemistry, www.rsc.org

8 Stimuli-responsive Materials in Medical Therapy 254

9 Photochromic Materials 287

10 Photorefractive Polymers 316

11 Self-healing Materials 348

1 What is Stimuli Responsiveness?

1.1 Introduction

It is well established that materials, and particularly polymeric materials, may serve many functions. These functions range from life-saving medical devices to paints that protect cars from rusting, or plastic utensils that rapidly fill up our landfill sites, to name just a few. Regardless of their function, there is a global need for functional materials, devices or complex systems to be either fully sustainable or entirely degradable. Making an object sustainable implies that all initial functions will be retained during its lifetime, but at some point it will lose its functions. While one term does not exclude the other, making materials degradable requires that at the end of its useful life, a given object will be turned into a new product of equal if not greater value. This concept, known as the *cradle-to-cradle* approach, seeks to create functional materials, devices or systems in such a way that they are not only efficient, but also waste-free. While it is environmentally advantageous to design all materials using the cradle-to-cradle approach, it is even more desirable to incorporate into materials design active functions that will be manifested by their property changes. For example, on a sunny hot day, it would be desirable to have a house with a white roof in order to minimize absorption of the sun's rays so the house would stay cool. But in winter we want the same white roof to change color to black in order to absorb sunlight to keep the house warmer. Thus, there is a stimulus, sun radiation, and there is a response, the change of color from white to black. If this process can be repeated many times, we have a roof that is stimuli-responsive to the external source of energy. Can internal sources lead to stimuli? If

Stimuli-Responsive Materials: From Molecules to Nature Mimicking Materials Design
By Marek W. Urban
© Marek W. Urban, 2016
Published by the Royal Society of Chemistry, www.rsc.org

there are internally built mechanical stresses or concentration gradients within a given material, molecular relaxations may lead to responses manifested by crack formation, which typically results from the non-equilibrium state of matter.

Although Mother Nature provides multiple inspirations for the design and development of new materials, creating synthetic systems capable of responding to stimuli in a controllable and predictable fashion presents significant but fascinating challenges. Particular challenges lie in mimicking biological systems where structural and compositional gradients at various length scales are necessary for orchestrated and orderly responsive behaviors. To tackle these challenges several stimuli-responsive systems have been developed, with the majority of studies dealing with polymeric solutions, gels, surfaces and interfaces, and to some extent, polymeric solids. These states of matter impose different degrees of restriction on the mobility of polymeric segments or chains, thus making dimensional responsiveness easily attainable for systems with a higher solvent content and minimal energy inputs. Significantly greater challenges exist when designing chemically or physically crosslinked gels and solid polymeric networks that require maintenance of their mechanical integrity. Restricted mobility within the network results from significant spatial limitations, thus imposing limits on obtaining stimuli responsiveness. The challenge in designing these stimuli-responsive polymeric systems is to create networks capable of inducing minute molecular yet orchestrated changes that lead to significant physicochemical responses upon external or internal stimuli.

To illustrate spatial restrictions on mobility in the x-, y- and z-directions in solutions, at surfaces and interfaces, in gels, and in solids, Figure 1.1 is a schematic diagram of the four states and relative dimensional restrictions within each state.[1] When going from a solution phase to surfaces and interfaces, or gels to solid state phases, segmental mobilities of polymer chains decrease due to significant spatial restrictions manifested by smaller displacement vectors in the x-, y- and z-directions. Consequently, the energetic requirements for responses to temperature, mechanical stimuli, electromagnetic irradiation, or electrochemical stimuli, pH, ionic strength, or bioactive species will be different for each physical state. Examples of responses are depicted in Figure 1.1 and are classified into chemical and physical categories, where multiple stimuli may result in one or more responses, or one stimulus may result in more than one response. Because spatial restrictions also dictate energetic requirements, the next section will discuss these relationships.

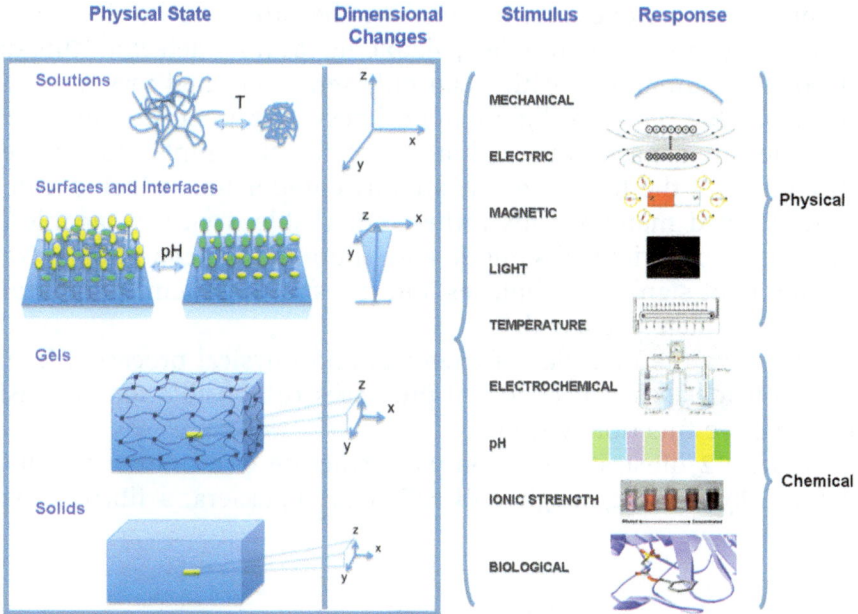

Figure 1.1 Physical state and dimensional changes in chemical and physical stimuli-responsive processes.
Reprinted from ref. 1, Copyright (2010), with permission from Elsevier.

1.2 An Interplay of Chemical and Physical Responsiveness

How do our eyes work, and why we are able to see? Why do flowers bloom every spring? Why does self-healing of wounds occur in mammals and plants? How do we age? Why do living organisms metabolize and materials do not? Why are grizzly bears able to hibernate in the winter and humans cannot? Some of these questions might be obvious, but for the most part they are not. Why? The majority of these processes are only partially understood and the complexity arises from the lack of correlations between chemical reactions and physical processes responsible for the overall outcome. Another way of looking at it is that using chemically the same material to make different objects will require different physical designs. Making a chair for a child and an adult from the same material will require different designs. On the contrary, using the same design for a child and an adult may require different materials. To take a closer look at stimuli responsiveness let us consider the processes involved in vision. To see we need an object, visible light and a detector. This detector is a human eye, which can only sense reflected rays when an object is illuminated by light – in darkness we cannot see. Some may consider the eye as a camera; but is it really that simple? As depicted in Figure 1.2 the human eye has many components, which act in a synchronized manner. This rather remarkable device is capable of adjusting to various distances and illumination conditions, converting light signals to impulses and transmitting them to the brain where an image is created.

To realize the interplay of chemical and physical processes let us now consider selected chemical processes responsible for the basic physical eye function, vision.

Figure 1.2 illustrates the complex structure of a human eye. It is covered by a white, tough wall called the episclera, a fibrous layer

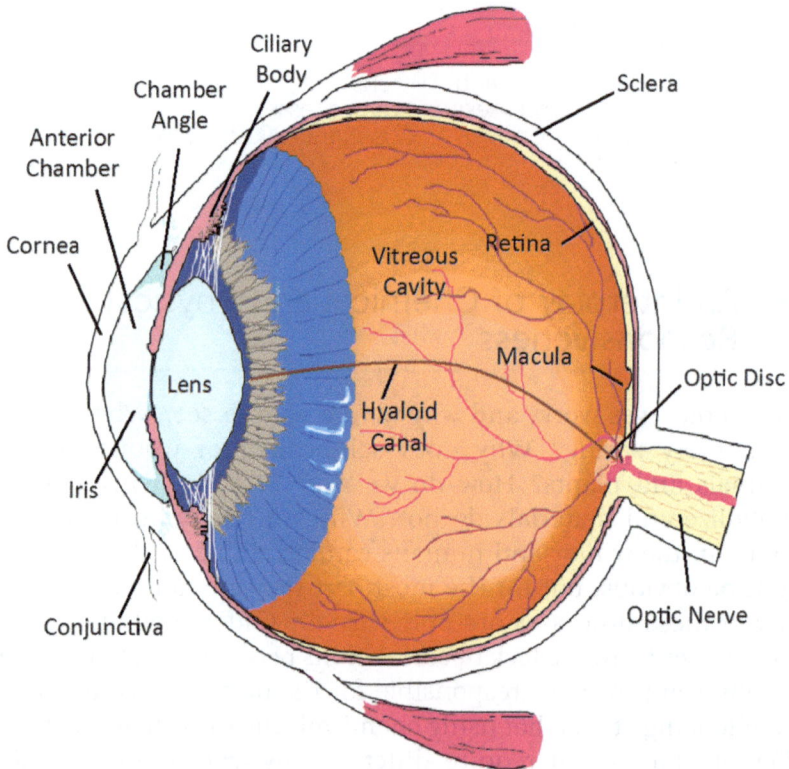

Figure 1.2   The complexity of a human eye structure.
(Eyeball cross-section image © 1989–2001 by Lippincott Williams & Wilkins, Baltimore, MD.)

between the conjunctiva and sclera. The eye muscles are connected to the conjunctiva. The cavity in the front part of the eye, between the lens and cornea, is the anterior chamber, which is filled with aqueous fluid that is recirculated every 100 minutes; this is produced by the ciliary body and drains back into the blood circulation through channels in the chamber angle. The structure behind the iris (mostly invisible), which produces the fluids filling the front part of the eye and maintains the eye pressure, is the ciliary body. Another important function of this part of the eye is to facilitate focusing. The iris acts like the diaphragm of a camera (and is responsible for the color of the eyes), allowing only a certain amount of light to enter the eye by dilating and constricting the pupil. The pupil is the dark opening in the center of the colored iris controlling how much light enters the eye. Immediately behind the iris is the lens, which is responsible for focusing light rays onto the retina. The white part of the eye, which is a thin lining over the sclera and inside the eyelids, is the conjunctiva. The cells of the conjunctiva produce mucous, which lubricates the eye. The primary focusing element of the eye is the cornea, also known as the epithelium. It is made of transparent cells capable of rapid regeneration. Its inner layer is transparent, allowing light to pass through. The narrow channel that runs from the optic disc to the back surface of the lens is called the hyaloid canal, and the body of the eye is filled with a jelly-like clear substance called vitreous humor. The retina is a layer of membrane lining the back of the eye. It contains photoreceptor cells that react to light intensity by sending impulses to the brain *via* the optic nerve. The light passes through the cornea and lens, creating an image of the visual world on the retina, which serves much the same function as the film in a camera. Light striking the retina initiates a cascade of chemical and electrical events that ultimately trigger nerve impulses. The macula is the most sensitive part of the retina and is responsible for the central (or reading) vision. Being near the optic nerve, directly at the back of the eye, it is also responsible for color vision. The optic disc is located in the back of the eye where the nerve, along with arteries and veins, enters the eye. The optic nerve consists of a bundle of a million nerves and is responsible for transferring information from the retina as electrical signals and delivering it to the brain, where this information is processed as a visual image.

There are two types of photoreceptor cell within the retina layer: rods and cones. As the light hits the photoreceptor cells, the first step is for chromophore 11-*cis*-retinal to isomerize to all-*trans*-retinal. The protein that is covalently bonded to 11-*cis*-retinal is

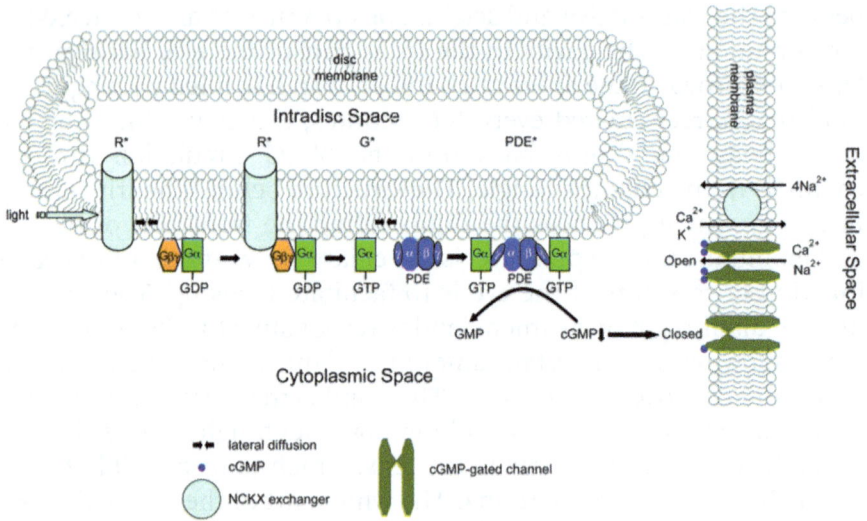

Figure 1.3 Schematic representation of the activation of rod phototransduction. Upon photon absorption, activated rhodopsin (R*) activates heterotrimeric G protein, and catalyzes the exchange of GDP for GTP, thus producing active Gα*-GTP. Two Gα*-GTP bind to the two inhibitory subunits of PDE, thereby releasing the inhibition on the catalytic α and β subunits, and forms PDE*. This, in turn, catalyzes the hydrolysis of cGMP. The consequent decrease in the cytoplasmic free cGMP concentration leads to the closure of the cGMP-gated channels on the plasma membrane and blockage of the influx of cations into the outer segments, which results in reduction of the circulating dark current.
Reproduced from ref. 2. © 2015 Webvision.

opsin, and the complex 11-*cis*-retinal–opsin is known as rhodopsin. Opsin has seven hydrophobic α-helical regions that pass through the lipid membrane of the pigment-containing discs (Figure 1.3), forming a hydrophobic pocket where 11-*cis*-retinal resides. In the dark, 11-*cis*-retinal binds the opsin as an inverse agonist and holds it in an inactive conformation. When light strikes the visual pigment, the isomerization of 11-*cis*-retinal to all-*trans*-retinal in the binding pocket pushes the opsin into an active conformation and initiates phototransduction, which ultimately leads to the generation of nerve impulses. The maximum absorbance radiation for rhodopsin is ~500 nm. Phototransduction is the most impressive signal transduction cascade leading to the generation of a nerve impulse to the brain. The changes in rhodopsin activate transducin, which, in turn, activates another enzyme, phosphodiesterase. The latter catalyzes the hydrolysis of cyclic GMP, as shown in Figure 1.4. Hydrolysis

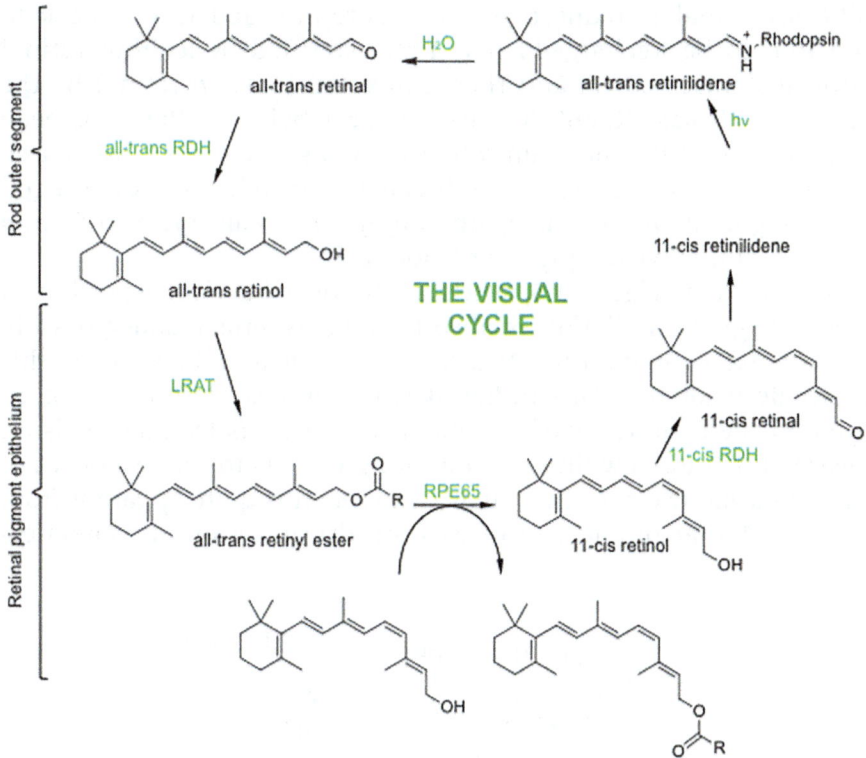

Figure 1.4 Anticipated chemical reactions during the visual cycle of a human eye.

of cyclic GMP will subsequently close the Na^+ channels. In the dark, the Na^+ channels are open. The influx of Na^+ is compensated for by an outflux of potassium ions (K^+) through the K^+ channel, resulting in depolarization of photoreceptors in the dark. The depolarized state of the membrane will trigger continual transmitter release from the synaptic terminals of the photoreceptor cells. Under light, Na^+ channels are closed so that a large charge difference across the rod's outer membrane can build up, caused by the outflow of K^+, a process known as membrane hyperpolarization, which decreases the release of neurotransmitters. Meanwhile, after all-*trans*-retinal is formed, it is released from opsin within nanoseconds $(10^{-9}$ s). Because opsin itself cannot absorb light, 11-*cis*-retinal needs to be regenerated and bind to opsin again. This is accomplished in the retinal pigment epithelium (RPE) *via* the visual cycle (Figure 1.4). RPE is a single layer of cells located between the retina and sclera. Following isomerization and release from the opsin protein, all-*trans*-retinal is reduced to all-*trans*-retinol and travels back to the RPE to be 'recharged'. It is first esterified by lecithin retinol acyltransferase (LRAT) and then converted to 11-*cis*-retinol by the isomerohydrolase RPE65. The isomerase activity of RPE65 has been shown; it is still uncertain whether it also acts as a hydrolase. Finally, it is oxidized to 11-*cis*-retinal before traveling back to the rod outer segment where it is again conjugated to an opsin to form a new, functional visual pigment (rhodopsin).

As shown in Figure 1.5, when a photon of a visible portion of electromagnetic radiation is absorbed, 11-*cis*-retinal undergoes all-*trans*-retinal configuration. As a result, the size and the shape of this molecule is altered. This rather sketchy physical description of an eye and the complexity of the chemical reactions leading to vision clearly illustrate how illumination of light leads to the responses of this remarkable device that allow us to see. It is quite apparent from Figure 1.1 that physical processes are driven by minute chemical

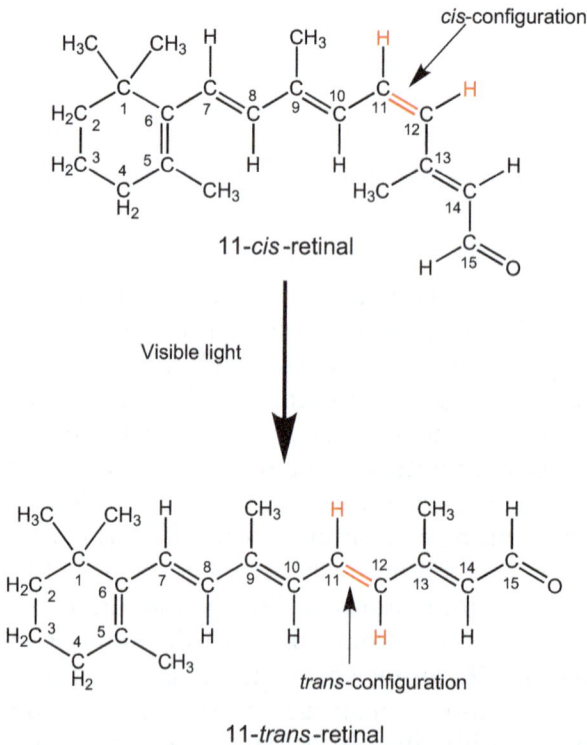

Figure 1.5 Upon absorption of electromagnetic radiation, 11-*cis*-retinal undergoes all-*trans*-retinal configuration.

changes that must occur in an orchestrated and sequential manner. One also needs to keep in mind that the event connectivity is time-dependent. Because 'cause' and 'effect' are typically connected or separated in time by intermediate processes, in designing stimuli-responsive materials one needs to identify what causes and links exist in a given time frame. If an event A is the cause and an event B is the effect, 'cause A' and the 'effect B' may be related by transient overlap or entirely separated, but connected in a hierarchical manner. There are a few fundamental classes of causes, which can be applicable to chemical processes or physical events in materials (Figures 1.6 and 1.7).

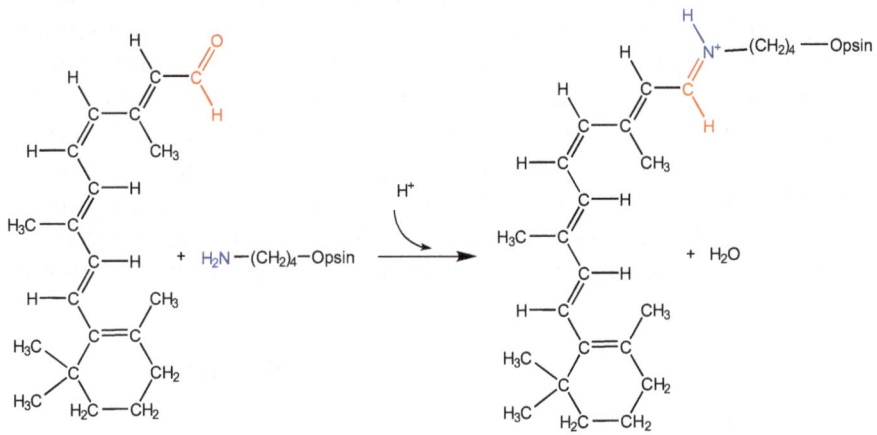

Figure 1.6   Reaction of 11-*cis*-retinal to opsin.

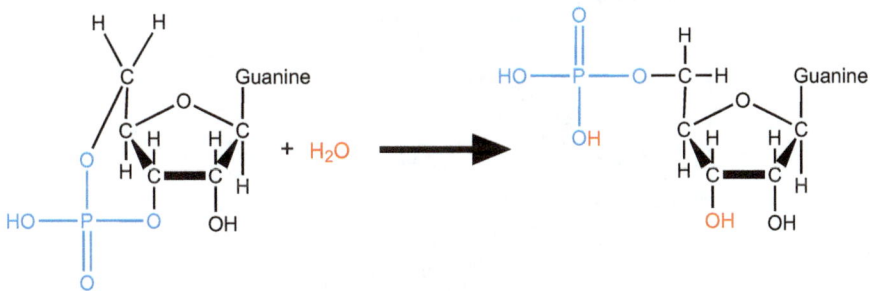

Figure 1.7   The addition of a water molecule (red) to GMP and breaking the bond between the phosphate group (blue) and carbon using a phosphor-diesterase catalyzed process leads to hydrolysis of cyclic GMP.

1.3   Energetic Requirements

To illustrate energy relationships in the context of stimuli responsiveness for a given state, Figure 1.8 depicts a series of equilibrium
energy diagrams as well as the relative stimuli energy input as a
function of response for each state. The main features of each diagram are the magnitude of the energy required to undergo transitions
from one state to another, while maintaining chemical/physical integrity and functionality of a given state represented by the equilibrium energy (ΔE_{eq}). For materials to exhibit stimuli responsiveness
it is essential to maintain the equilibrium state in order to preserve
the functionality and, at the same time, create a physical and/or
chemical environment that will require significantly lower amounts of
energy (ΔE_{SR}) for a system to undergo stimuli-responsive transitions.
The latter is represented by two usually smaller metastable energy
minima. Transitions between these minima will represent the energy
required for a system to go from one responsive state to another,
while maintaining the 'functional' equilibrium. Examples of such
stimuli-responsive transitions are molecular *cis–trans* rearrangements
or other conformational changes, hydrogen bonding-induced rearrangements,   aggregation–dissociation,   penetration–separation,

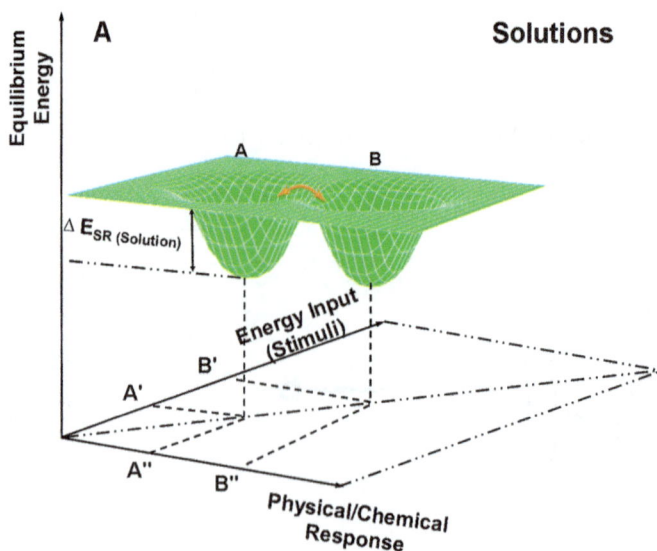

Figure 1.8   Equilibria energy for stimuli responsiveness in solutions.
Reprinted from ref. 1, Copyright (2010), with permission from
Elsevier.

order–disorder transitions, or protonation–deprotonation. These lower energy transitions may or may not be reversible, and their energy requirements will depend on initial physical and chemical states. The main challenge, however, is to obtain responsiveness at longer length scales while maintaining mechanical integrity of the network.

1.3.1 Stimuli-responsive Solutions

Stimuli-responsive behavior is easily obtained in polymeric solutions because the Brownian motion of solvent molecules requires relatively low energies for macromolecular segments to displace solvent molecules. For an ideal system, the kinetic energy required for the Brownian motion at room temperature is 1.5 k_BT (where k_B is the Boltzmann constant $(1.38 \times 10^{-23}$ J K$^{-1})$, and T is the temperature).[3,4] To illustrate stimuli-responsive transitions in solutions, Figure 1.8 depicts two energy minima, A and B, where in order for a given polymeric solution to go from energy state A to B, a certain amount of energy $\Delta E_{SR(solution)}$ is required. One example of these transitions in polymeric solutions is the lower critical solution temperature (LCST),[5] which is the lowest temperature of phase separation on the concentration–temperature phase diagrams. Below the LCST, polymer chains and solvent molecules are in one homogenous mixing phase, and exhibit favorable free energy $(\Delta G < 0)$, which is believed to be facilitated by hydrogen bonding interactions between the two phases. Above the LCST, a phase separation occurs as enthalpic (ΔH) energy overcomes the entropic (ΔS) contributions resulting in unfavorable free energy $(\Delta G > 0)$ of the entire system. For the majority of polymeric solutions,[6–11] temperature-induced LCST transitions result in a particle or aggregate size decrease above the LCST, and the reported size changes are substantial, typically in the range from 250–3000 nm below the LCST, to 100–1000 nm above. For polymeric solutions that exhibit LCST behavior, the relationship between the solution equilibrium energy (z-axis), the stimulus energy input (x-axis) and physical/chemical response (y-axis) is depicted in Figure 1.8. Initially, the system is in the equilibrium energy state A, and as a result of the stimulus energy input, such as temperature $A' \rightarrow B'$, the system undergoes transition to the equilibrium state B. As a consequence, the system will go from the physical state A'' below the LCST to a new state B'' above the LCST, manifested by a physical response in the form of collapse of the particles or aggregate size changes.

1.3.2   Stimuli-responsive Surfaces and Interfaces

As illustrated in Figure 1.1, the mobility of responsive chains on surfaces is restricted and depends on many factors. Due to the anchoring of one end of a polymeric segment to the surface, restricted freedom of movement is 'transmitted' along the chain. If we define the stimuli-responsive surface as a point on the anchored chain, the energy required for the segments further away from the anchoring point to respond is a function of the distance from the surface. Figure 1.9 illustrates the relationship between the surface layer equilibrium energy (z-axis), stimulus energy input (x-axis) and a physical/chemical response to a stimulus (y-axis). For segments that are close to the anchoring points of the surface, relatively higher energy ($\Delta E_{SR(surface)}$) input will be required to undergo A to B transitions, because more space and free volume are available further away from the anchored point, thus providing energetically and spatially favorable rearrangement conditions. This is illustrated in Figure 1.9, which depicts the magnitude changes between the energy minima A and B for segments close to the surfaces (anchored ends) and those further away (free ends). This will also be reflected in the T_g changes as a function of distance from the anchoring point at the surface. The first experimental evidence manifesting the importance of the T_g changes as a function of distance from the surface showed that the T_g values for 1000 nm thick poly(methyl methacrylate) (PMMA) films vary by as much as 20 K for the 25 nm thick layer at the free surface compared to the same layer in the bulk.[12] Generally, stimuli responsiveness at polymeric surfaces is an entropy (ΔS) driven process,[13] in which the disorder (mobility fluctuation) of anchored chains (ΔS) has greater contributions to the ΔG values than the conformational changes resulting from the enthalpic (ΔH) component of the free energy. Examples of this behavior are switching of surface wettability,[14,15] which involves transformation from one equilibrium state to another (A→B) in response to the external energy input (A' to B'), resulting in physical/chemical state changes from A'' to B''.

The energy diagram for polymer interfaces is illustrated in Figure 1.9, which in principle depicts similar energetic behavior, but the presence of two anchor points forming an interface will reduce chain mobility and consequently stimuli responsiveness. The energy input required for the stimuli-responsive A→B transitions will be significantly greater than for the surfaces, and will be a function of molecular distance between the two surface anchoring points. This is schematically depicted in Figure 1.9. One of the challenges will be to

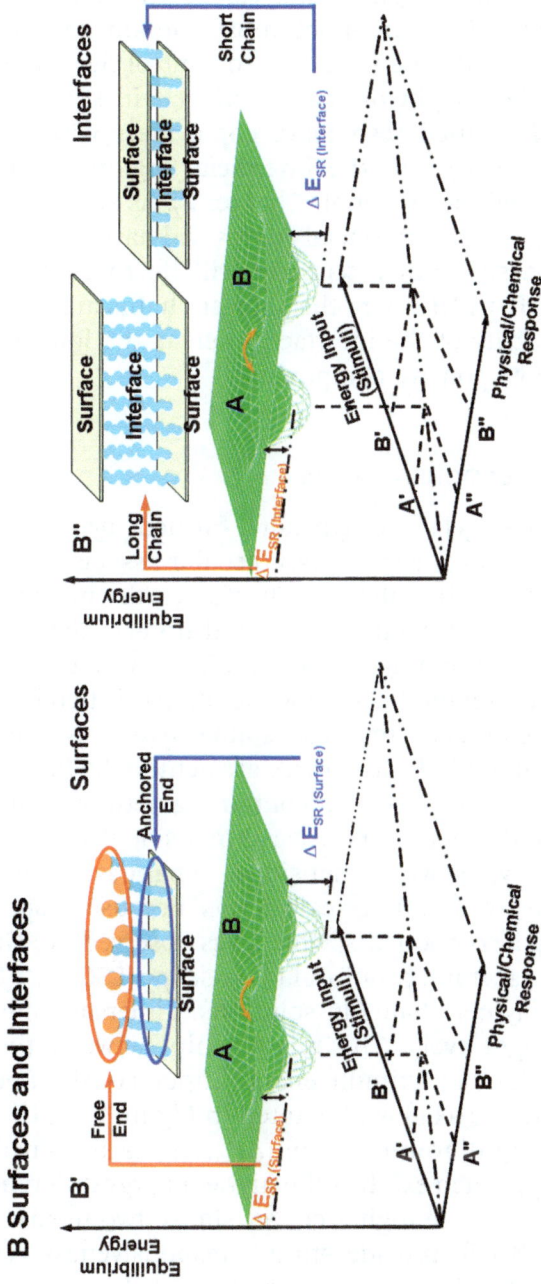

Figure 1.9 Equilibria energy for stimuli-responsiveness at surfaces and interfaces. Reprinted from ref. 1, Copyright (2010), with permission from Elsevier.

design stimuli-responsive interfaces with various molecular structures at the interfacial regions that exhibit stimuli responsiveness. While each chain of the interface must contain stimuli-responsive components, it will also be necessary to control the interfacial chain lengths, molecular weight of chains, and chain stiffness. Thus, incorporating well-defined responsive copolymers with desired polydispersity index (PDI) into the interfacial regions using precision polymerizations will be essential (Chapter 2). As shown in Figure 1.9, molecular weight of the interfacial chains will have a significant effect on responsiveness. Another scenario will be to control interfacial entities with higher PDI where the shorter chains are responsible for mechanical integrity of the interface whereas the longer chains will serve as stimuli-responsive components.

1.3.3   Stimuli-responsive Gels

Stimuli-responsive gels are typically formed by physical and/or chemical crosslinking, or by supramolecular associations of molecular chains dispersed in solvents. The forces driving the molecular makeup of gels are covalent bonds[16] and non-covalent interactions, such as hydrogen bonding, hydrophobic or van der Waals interactions, and π–π stacking.[17] As a polymer matrix is tied by crosslinked points or entanglements, relatively stable hydrogels and/or organogels with maintained bulk structures are achieved. However, because gels are usually porous or solvent-containing networks, their integrity may be destroyed when responsive dimensional changes (deformation) occur. Typically, dimensional changes within stimuli-responsive gels are at micro-scale levels because responses occur along the network components, which was observed in PNIPAAm gels by utilizing laser scanning confocal microscopy (LSCM).[18] In contrast, for dispersed polymer chains in solutions, dimensional changes are at the nano-range levels. The relationship between the gel equilibrium energy (z-axis), stimuli energy input (x-axis) and physical/chemical response (y-axis) is illustrated in Figure 1.10. If we consider again that the equilibrium energy ($\Delta E_{\text{eq(gel)}}$) is responsible for gel network stability, which exhibits the higher energy difference between the minimum E and the higher energy states, two metastable energy minima A and B will provide stimuli responsiveness. This is controlled by the entropic component (ΔS), as ΔH is smaller and contributes to a lesser degree to overall ΔG values. One example is the expansion and shrinkage of hydrogel networks which result in the

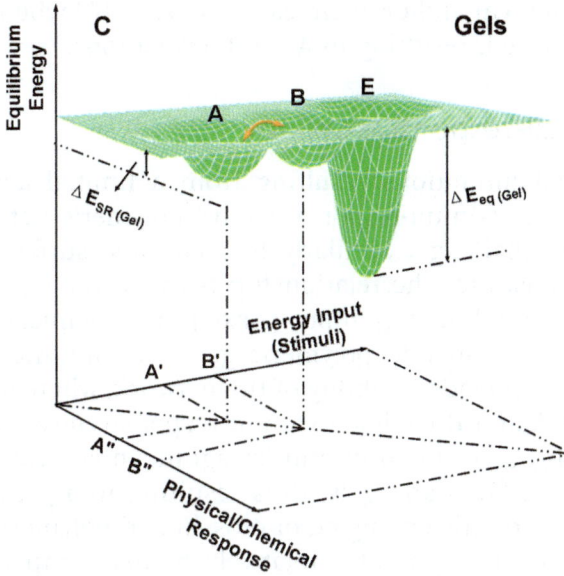

Figure 1.10 Equilibria energy for stimuli-responsiveness in gels. Reprinted from ref. 1, Copyright (2010), with permission from Elsevier.

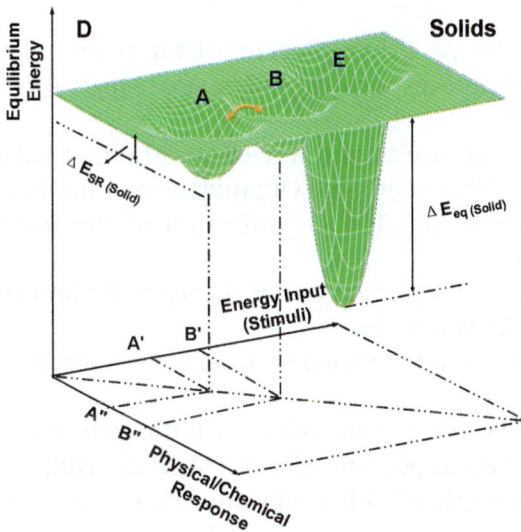

Figure 1.11 Equilibria energy for stimuli-responsiveness in solids. Reprinted from ref. 1, Copyright (2010), with permission from Elsevier.

changes of physical and/or chemical state $(A'' \rightarrow B'')$ when the stimuli energy is delivered, resulting in A' to B' transitions.

1.3.4 Stimuli-responsive Solids

Due to spatial limitations resulting from a limited access to free volume, creating stimuli-responsive solid polymeric networks represents a great challenge. Similarly to solutions, surfaces and gels, Figure 1.11 illustrates the relationship between the equilibrium energy (z-axis), stimuli energy input (x-axis) and chemical/physical response (y-axis). Similar to polymeric gels, the minimum energy at equilibrium (E) provides stability of the network, whereas metastable energy states A and B facilitate stimuli responsiveness. In polymeric solids, the $\Delta E_{eq(solid)}$ at equilibrium is significantly greater compared to solutions, surfaces and gels, thus resulting in a greater network integrity due to tightly entangled or crosslinked polymeric chains. As a consequence, the T_g is also relatively higher compared to other states. However, in solid polymeric networks, the entropic term (ΔS) does not contribute significantly to stimuli responsiveness because spatial mobility in these tighter networks is restricted.

References

1. F. Liu and M. W. Urban, Recent advances and challenges in designing stimuli-responsive polymers, *Prog. Polym. Sci.*, 2010, **35**(1), 3–23.
2. H. Kolb, E. Fernandez and R. Nelson, Phototransduction in Rods and Cones–Webvision: The Organization of the Retina and Visual System. Webvision: The Organization of the Retina and Visual System, 1995.
3. A. Einstein, *Investigations on the theory of the brownian movement*, New York, Dover, 1956.
4. J. Perrin, *Brownian motion and molecular reality*, New York, Dover, 2005.
5. S. Maeda, Lower critical solution temperature (L.C.S.T.) of the systems, glycerin-polypropylenglycol-ether (GP), or glycerin-polyethylenglycol-ether (GEP) and water, *Bull. Pharm. Res. Inst.*, 1966, **62**, 11–13.
6. S. Fujishige and K. K. I. Ando, Phase transition of aqueous solutions of poly(N-isopropylacrylamide) and poly(N-isopropylmethacrylamide), *J. Phys. Chem.*, 1989, **93**, 3311–3313.

7. R. H. Pelton and P. Chibante, Preparation of aqueous lattices with *N*-isopropylacrylamide, *Colloids Surf.*, 1986, **20**, 247–256.
8. T. Baltes, F. Garret-Flaudy and R. Freitag, Investigation of the LCST of polyacrylamides as a function of molecular parameters and the solvent composition, *J. Polym. Sci., Part A: Polym. Chem.*, 1999, **37**, 2977–2989.
9. A. C. W. Lau and C. Wu, Thermally sensitive and biocompatible poly(*N*-vinylcaprolactam): synthesis and characterization of high molar mass linear chains, *Macromolecules*, 1999, **32**, 581–584.
10. E. E. Makhaeva, H. Tenhu and A. R. Khokhlov, Behavior of poly(*N*-vinylcaprolactam-*co*-methacrylic acid) macromolecules in aqueous solution: interplay between coulombic and hydrophobic interaction, *Macromolecules*, 2002, **35**, 1870–1876.
11. Y. Maeda, T. Nakamura and I. Ikeda, Hydration and phase behavior of poly(*N*-vinylcaprolactam) and poly(*N*-vinylpyrrolidone) in water, *Macromolecules*, 2002, **35**, 217–222.
12. R. D. Priestley, J. C. Ellison, L. B. Broadbelt and J. M. Torkelson, Structural relaxation of polymer glasses at surfaces, interfaces, and in between, *Science*, 2005, **309**, 456–460.
13. F. Xia, Y. Zhu, L. Feng and L. Jiang, Smart responsive surfaces switching reversibly between super-hydrophobicity and super-hydrophilicity, *Soft Matter*, 2009, **5**, 275–281.
14. S. Wang, H. Liu, D. Liu, X. Ma, X. Fang and L. Jiang, Enthalpy-driven three-state switching of a superhydrophilic/super-hydrophobic surface, *Angew Chem., Int. Ed.*, 2007, **46**, 3915–3917.
15. S. Minko, M. Müller, M. Motornov, M. Nitschke, K. Grundke and M. Stamm, Two-level structured self-adaptive surfaces with reversibly tunable properties, *J. Am. Chem. Soc.*, 2003, **125**, 3896–3900.
16. F. Cellesi, N. Tirelli and J. A. Hubbell, Towards a fully-synthetic substitute of alginate: development of a new process using thermal gelation and chemical cross-linking, *Biomaterials*, 2004, **25**, 5115–5124.
17. K. P. Kamath and K. Park, Biodegradable hydrogels in drug delivery, *Adv. Drug Delivery Rev.*, 1993, **11**, 59–84.
18. Y. Hirokawa, H. Jinnai, Y. Nishikawa, T. Okamoto and T. Hashimoto, Direct observation of internal structures in poly(*N*-isopropylacrylamide) chemical gels, *Macromolecules*, 1999, **32**, 7093–7099.

2 Design of Stimuli-responsive Macromolecular Blocks

2.1 Basic Concepts

Chemical reactions occur when two or more molecules are in close proximity and exhibit an affinity to form chemical bonds. Initially controlled by diffusion, followed by a chemical reaction, the question is how can one precisely design the sequences of monomers linked to form a macromolecular chain with precisely defined architecture, controlled molecular weight and molecular weight distributions? This chapter will provide some basic knowledge on the synthetic aspects of polymerizations.

Polymers are macromolecules composed of a large number of repeating units. They are built up *via* polymerization of many 'mers' (small molecules called monomers). For example, polymerization of n ethylene monomers generates a polyethylene with n repeating units:

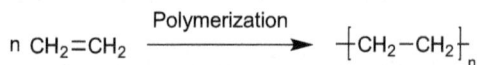

$$n \text{ CH}_2{=}\text{CH}_2 \xrightarrow{\text{Polymerization}} {\Big\{}\text{CH}_2{-}\text{CH}_2{\Big\}}_n$$

If a macromolecule only consists of a smaller number of repeating units, it is termed an oligomer. An oligomer becomes a polymer when the properties of the macromolecule do not change after the addition of one monomeric unit. There are both naturally occurring biopolymers, such as DNA, proteins and polysaccharides, and synthetic plastics and rubbers that are essential to our daily life. The unique properties of polymers are a result of their size, chemical

Stimuli-Responsive Materials: From Molecules to Nature Mimicking Materials Design
By Marek W. Urban
© Marek W. Urban, 2016
Published by the Royal Society of Chemistry, www.rsc.org

composition, arrangements of repeating units, their sizes and architecture.

2.1.1 Average Molecular Weight and Molecular Weight Distribution

The key factors determining polymer properties are the degree of polymerization and molecular weight distribution. In contrast to small molecules, polymers are made of chains that may have different molecular weights or different numbers of monomers. Thus, the terms average molecular weight and molecular weight distributions are used. There are different ways of calculating the average molecular weight. As shown in Figure 2.1, the number average molecular weight (M_n) is the total weight of the polymer (w) divided by the number of chains, where n_i is the number of chains with a molecular weight of M_i. The weight average molecular weight, on the other hand, is determined from the weight fraction (w_i) of chains with molecular weight of M_i. This is a second power average molecular weight. The Z-average molecular weight (M_z) is a third power average molecular weight. For any polymer, M_n, M_w and M_z will have different values. This is illustrated with an example of a tiger whose weight is 1000 lb, and nine cats which weigh 10 lb each. As you add up the numbers, you see that the number average (M_n) molecular weight is different from the weight average (M_w), and different again from the so-called Z-average. This is due to the polydispersity of the molecular weights of individual chains. M_n is more affected by the presence of small molecules, while large molecules have greater contributions to M_w and M_z. Typically, for a polydispersed polymer, $M_w > M_v > M_z$, as shown in Figure 2.2. Therefore, to characterize a polymer, the polydispersity index (PDI) described by the M_w/M_n ratio is used. So if the tiger weighs the same as the cat, PDI $= M_w/M_n = 1$. However, in a typical statistical polymerization, the PDI is usually above 1. The higher the PDI is, the more polydispersed the polymer mixture is, and the values may be as high as 10.

2.1.2 Architecture of Polymer Chains

Polymers may have different microstructures and large degrees of heterogeneities, giving a variety of unique properties. As shown in Figure 2.3, polymers are categorized into linear, branched and crosslinked. The simplest type is a linear homopolymer, whose chain is made up of identical units linked together in a linear sequence.

MOLECULAR WEIGHT - Number Average: M_n

$$\bar{M}_n = \frac{w}{\sum n_i} = \frac{\sum n_i M_i}{\sum n_i}$$

1000 lb   10 lb

$$\bar{M}_n = \frac{1 \times 1000 + 9 \times 10}{1 + 9}$$

$$\bar{M}_n = 109$$

MOLECULAR WEIGHT - Weight Average: M_w

$$\bar{M}_w = \sum w_i M_i = \frac{\sum n_i M_i^2}{\sum n_i M_i}$$

1000 lb   10 lb

$$\bar{M}_w = \frac{1 \times 10^6 + 9 \times 10^2}{1 \times 1000 + 9 \times 10}$$

$$\bar{M}_w = 918.25$$

MOLECULAR WEIGHT - Z Average: M_z

$$\bar{M}_z = \sum w_i M_i = \frac{\sum n_i M_i^3}{\sum n_i M_i^2}$$

1000 lb   10 lb

$$\bar{M}_z = \frac{1 \times 10^9 + 9 \times 10^3}{1 \times 10^6 + 9 \times 10^2}$$

$$\bar{M}_z = 999.11$$

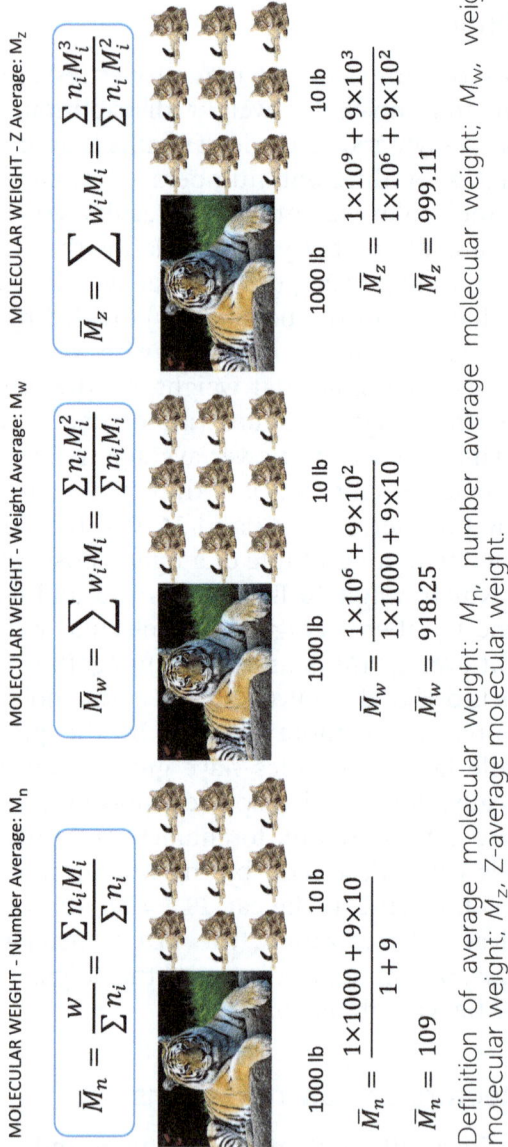

Figure 2.1 Definition of average molecular weight: M_n, number average molecular weight; M_w, weight average molecular weight; M_z, Z-average molecular weight.

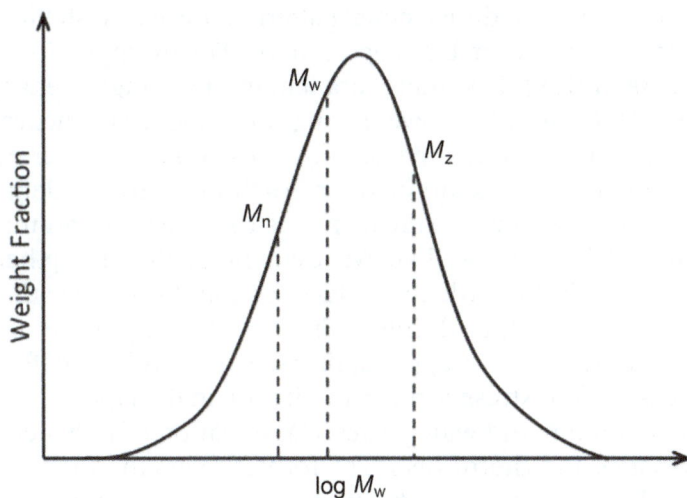

Figure 2.2 M_n, M_w and M_z molecular weights of a polydispersed polymer.

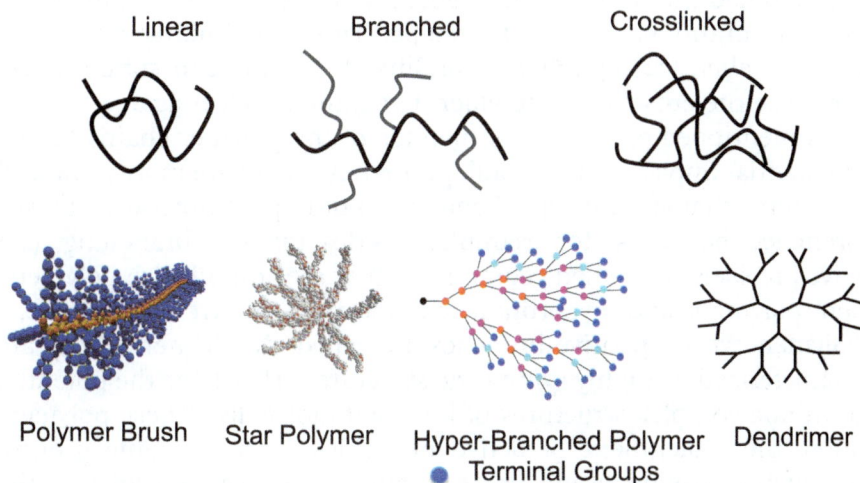

Figure 2.3 Examples of various polymer morphologies. The main challenge is to control each and every molecular component of these morphologies.

The linear chains can be imagined as spaghetti, in that they can coil up and become tangled with other chains. When one or more side chains are covalently bonded to the polymer chain (polymer backbone), it forms branches. Depending on the length and density of the side chains, branched polymers pack in different ways, which will impact on their bulk properties. For example, linear polymers can pack in a regular three-dimensional pattern to form a crystalline phase, but polymers with short branches cannot. For example, low density polyethylene (LDPE) has more branching than high density poly-ethylene (HDPE), so it has lower strength but higher resilience due to the looser packing and less crystallinity. Crosslinked polymer networks are formed when chains are linked to each other by covalent bonds. Formation of crosslinks usually increases the strength and toughness of polymers. The most well-known example is the strengthening of rubber by vulcanization, which produces crosslinks between the chains by sulfur. Once crosslinked, the chains cannot freely move relative to each other. When stress is applied, the vulcanized rubber deforms, but upon release of the stress it reverts to its original shape.

A more common and widely used classification of polymers is into thermoplastics and thermosets. The former refers to non-crosslinked polymers that can flow after heating to a certain temperature. They might be rigid under ambient temperatures, but upon heating can be melted and remolded. Thermosets are crosslinked polymers. They are first molded into a desirable shape, following by crosslinking under certain conditions to set the shape. This crosslinking process is usually called 'curing'. Once crosslinked, heating can soften or de-compose the crosslinked polymer, but cannot melt it again.

As mentioned earlier, the structures of the polymer chains have a substantial impact on their bulk properties. There are more advanced structures that show unique behaviors, but require precise synthesis. Branched polymers, for example, besides random branching, can have precisely controlled geometry such as the comb/brush polymers, star polymers and hyperbranched polymers shown in Figure 2.3. Polymers with repetitive branches are called 'dendrimers'. They are characterized by highly symmetric structures. They have the potential to mimic complex structures of biomacromolecules. These polymers allow the attachment of active molecules, such as ionic groups, conducting groups, or any other stimuli-responsive molecules, to the brushes or the arms of the star or the dendrimer, providing pro-grammable molecular interactions, large numbers of coupling or recognition sites, and maximum mobility. One can envisage various functionalities associated with a number of structural possibilities.

2.2 Homopolymers and Copolymers

A polymer synthesized from a single monomer is a homopolymer. However, combining two or more monomers is often desirable, to combine the properties of two homopolymers in one copolymer. For example, simply mixing rigid poly(methyl methacrylate) (PMMA) with soft poly(butyl acrylate) (nBA) results in a blend of homopolymers which exhibits very poor mechanical properties due to the immiscibility of the two. A better way of controlling properties and stability is to copolymerize methyl methacrylate (MMA) and nBA to obtain copolymers with different rigidities by varying the monomer ratios. Changing the structure of the copolymer can further vary the properties. Typical copolymer structures are shown in Figure 2.4. Block copolymers can be synthesized by sequential addition of different monomers. They are differentiated by the number of blocks and composition of each block. Shown in Figure 2.4, for example, are AB diblock, ABA triblock, and ABC triblock copolymers, where A, B and C represent different monomers. Statistical or random copolymers can be formed when two types of monomers with similar reactivity are polymerized. The resulting polymer has functional groups randomly distributed along the backbone. If the two monomers have higher reactivity toward each other than reacting with themselves, an alternating copolymer can be obtained. Properties of gradient copolymers lie between block copolymers and random or alternating copolymers. The monomer composition changes gradually along the chain, so that the two ends of the polymer have opposite compositions. Many of the polymers with unique structures, as shown in Figure 2.3, such as the brush polymers and star polymers, are copolymers composed of multi-components to give the desired chemical and physical

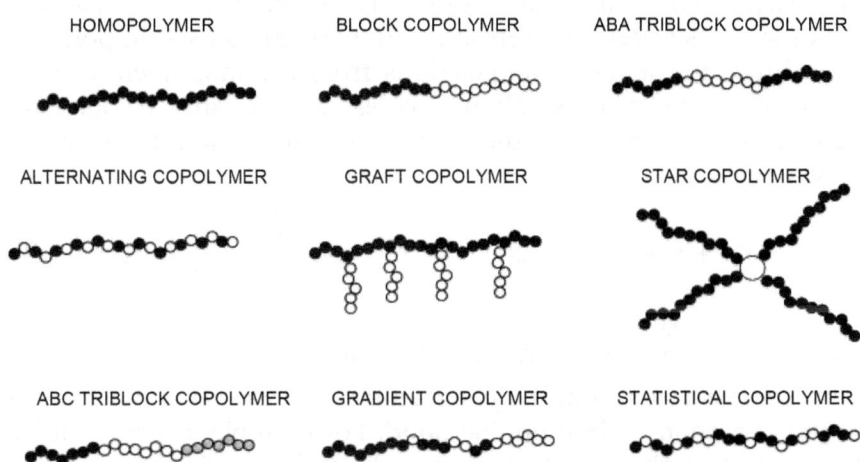

Figure 2.4 Typical structural features of copolymers: open, light filled, and filled circles represent different types of monomer.

Table 2.1 Nomenclatures for identifying copolymer morphologies.

Copolymer	Abbreviation	Example
unspecified	*co*	poly(styrene-*co*-isoprene)
random	*ran*	poly[(methyl methacrylate)-*ran*-(*n*-butyl acrylate)]
alternating	*alt*	poly[styrene-*alt*-(maleic anhydride)]
block	*block*	poly[styrene-*block*-(methyl methacrylate)]
graft	*graft*	polystyrene-*graft*-poly(ethylene oxide)

properties. Controlled living polymerization, which will be introduced in Section 2.7, was developed to build more complex structures (Table 2.1).

2.3 Synthesis of Polymers

The next question is how polymers with complex structural features can be synthesized; this will allow you to understand how to design a polymer, and polymer synthesis, and most importantly, how to introduce functional groups into polymers to obtain stimuli-responsive properties. These are basically two types of polymerization mechanisms. Originally, polymerizations were classified into condensation and addition reactions. Condensation polymerizations are reactions of monomers having at least two functional groups that can react with each other *via* a condensation reaction, eliminating small molecules such as water. The repeating units of condensation polymers have different chemical compositions from the monomers because of the elimination of small molecules. An example is polyester, formed from diacid and diol with the elimination of water, as shown below:

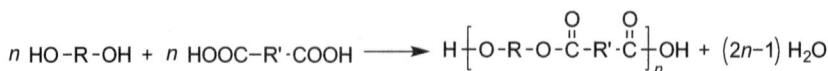

$$n \text{ HO-R-OH} + n \text{ HOOC-R'·COOH} \longrightarrow \text{H}\left[\text{O-R-O-}\overset{\overset{\text{O}}{\|}}{\text{C}}\text{-R'·}\overset{\overset{\text{O}}{\|}}{\text{C}}\right]_n\text{OH} + (2n-1)\text{ H}_2\text{O}$$

where $-\text{O-R-O-}\overset{\overset{\text{O}}{\|}}{\text{C}}\text{-R'·}\overset{\overset{\text{O}}{\|}}{\text{C}}-$ is the repeating unit.

In contrast, addition polymerization involves monomers containing a carbon–carbon double bond. For example, the free radical polymerization shown below proceeds by addition of a free radical to the double bond of the monomer, generating a free radical propagating center, which will add to another monomer. This process will go on until the free radical is terminated. The resulting polymers have the same molecular composition as their monomers.

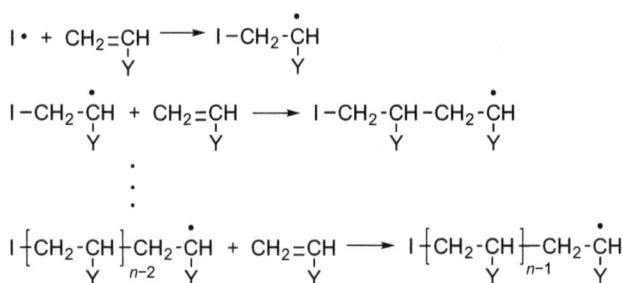

$$I\bullet + CH_2=\underset{Y}{CH} \longrightarrow I-CH_2-\underset{Y}{\overset{\bullet}{CH}}$$

$$I-CH_2-\underset{Y}{\overset{\bullet}{CH}} + CH_2=\underset{Y}{CH} \longrightarrow I-CH_2-\underset{Y}{CH}-CH_2-\underset{Y}{\overset{\bullet}{CH}}$$

$$I\left[CH_2-\underset{Y}{CH}\right]_{n-2}CH_2-\underset{Y}{\overset{\bullet}{CH}} + CH_2=\underset{Y}{CH} \longrightarrow I\left[CH_2-\underset{Y}{CH}\right]_{n-1}CH_2-\underset{Y}{\overset{\bullet}{CH}}$$

If the polymer backbone is only composed of C–C bonds, it can be concluded that this polymer is synthesized from addition polymerization. On the other hand, if the backbone contains groups such as ester and amides, it is a condensation polymer. However, there are polymerization reactions that belong to neither condensation nor addition reactions. These include polyurethane reactions, where the isocyanates react with –OH groups to form urethane bonds or react with –NH_2 to generate ureas. No small molecules are released from the reaction. It is more accurate to classify polymerization reactions as step-growth and chain-growth reactions based on their different reaction mechanisms.

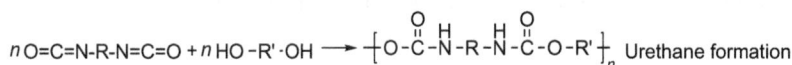

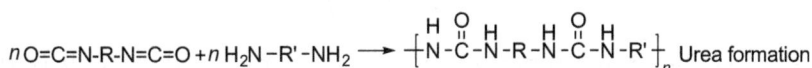

$$n\,O=C=N-R-N=C=O + n\,HO-R'\cdot OH \longrightarrow \left[O-\overset{O}{\overset{\|}{C}}-\overset{H}{N}-R-\overset{H}{N}-\overset{O}{\overset{\|}{C}}-O-R'\right]_n \text{ Urethane formation}$$

$$n\,O=C=N-R-N=C=O + n\,H_2N-R'-NH_2 \longrightarrow \left[\overset{H}{N}-\overset{O}{\overset{\|}{C}}-\overset{H}{N}-R-\overset{H}{N}-\overset{O}{\overset{\|}{C}}-\overset{H}{N}-R'\right]_n \text{ Urea formation}$$

2.3.1 Step-growth Polymerization

Many of the condensation reactions, and also reactions involving polyurethane and polyurea formations, are step-growth processes. They are classified into two types of reactions:

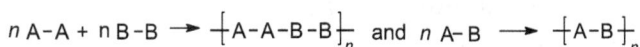

$$n\,A-A + n\,B-B \longrightarrow \left[A-A-B-B\right]_n \quad \text{and} \quad n\,A-B \longrightarrow \left[A-B\right]_n$$

where A and B are two functional groups on a monomer. Because the polymerization proceeds *via* reactions between functional groups A and B, dimers are formed at the beginning of the reaction, then trimers, followed by longer oligomers. Eventually, coupling of oligomers produces higher molecular weight polymers. Therefore, the molecular weight slowly increases with reaction time in a stepwise fashion as shown below:

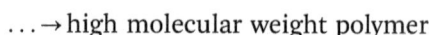

$$monomer + monomer \rightarrow dimer$$

$$dimer + monomer \rightarrow trimer$$

$$dimer + dimer \rightarrow tetramer$$

$$trimer + monomer \rightarrow tetramer$$

$$tetramer + monomer \rightarrow pentamer$$

$$trimer + trimer \rightarrow hexamer$$

$$\ldots \rightarrow high \ molecular \ weight \ polymer$$

The characteristics of step-growth polymerization are: (i) all the monomers are consumed at the early stage of polymerization; (ii) high molecular weight polymers are formed when the reaction reaches high conversion rates; (iii) molecular weight can be controlled by varying the ratio of A and B groups; (iv) high purity monomers are required. High molecular weight polymers can only be obtained when A and B are near stoichiometric ratios. Addition of monofunctional reactants, with one functional group A or B, can also decrease the molecular weight. The degree of polymerization can be calculated using the Carothers[1] equation:

$$\bar{X}_n = \frac{1 + r}{1 + r - 2rp}$$

where r is the stoichiometric ratio of reactants, and $r \leq 1$. p is the conversion, and can be calculated by:

$$p = \frac{N_0 - N}{N_0}$$

(N_0 = the number of molecules present initially as monomer, and N = the total number of molecules, including monomers, oligomers and polymers, present after time t.)

When p approaches 1:

$$X_n \rightarrow \frac{1 + r}{1 - r}$$

One of the best known polymers is Nylon™, which is the commercial name for polyamides. These materials can be synthesized from reactions of diacid with diamine, and the synthesis of Nylon 6,6 is shown below:

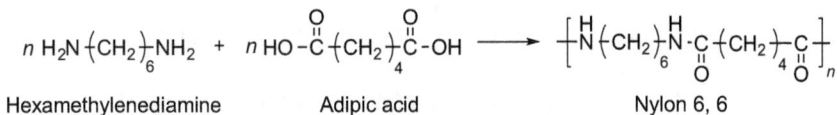

Hexamethylenediamine	Adipic acid	Nylon 6, 6

The numbers represent the number of carbons in the diamine monomer and diacid, respectively. Another approach to synthesize nylon 6,6 is by ring-opening polymerization, which will be discussed in later sections.

2.4 Chain-growth Polymerization

As step-growth polymerization follows a step-by-step, gradual process from monomers to oligomers and polymers, the mechanism for chain-growth polymerization is different. The polymerization is typically initiated by addition of the initiator to a monomer. As shown below, this step generates an active propagating center, and when monomers are added, leads to propagation, until the propagating centers are terminated by termination or chain transfer reactions.

$$I - I \rightarrow 2I$$

$$I^\bullet + M \rightarrow I^\bullet + M$$

During the propagation, monomers are added to the propagating center in a sequential order. Polymerizations that follow a chain-growth mechanism include free radical polymerization, ring-opening polymerization, ionic polymerization, and coordination polymerization. To illustrate the chain-growth process, free radical polymerization is used as an example:

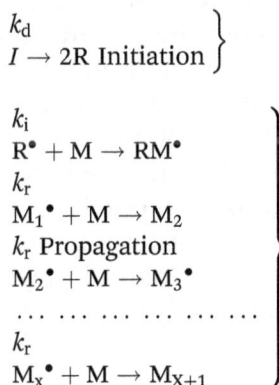

$$\left. \begin{array}{l} k_d \\ I \rightarrow 2R \text{ Initiation} \end{array} \right\}$$

$$\left. \begin{array}{l} k_i \\ R^\bullet + M \rightarrow RM^\bullet \\ k_r \\ M_1^\bullet + M \rightarrow M_2 \\ k_r \text{ Propagation} \\ M_2^\bullet + M \rightarrow M_3^\bullet \\ \cdots \cdots \cdots \cdots \cdots \cdots \\ k_r \\ M_x^\bullet + M \rightarrow M_{x+1} \end{array} \right\}$$

k_{1c} TERMINATION VIA

$M_x{}^{\bullet} + M_y{}^{\bullet} \rightarrow M_{x+y}$ COMBINATION

k_{1d}

$M_x{}^{\bullet} + M_y \rightarrow M_x + M_y$ DISPROPORTIONATION

k_{tm}

$M_x{}^{\bullet} + M \rightarrow M_x + M^{\bullet}$ TRANSFER OF MONOMER

k_{tp}

$M_x{}^{\bullet} + M_y \rightarrow M_x + M_y$ TRANSFER OF POLYMER

k_{ti}

$M_x{}^{\bullet} + I \rightarrow M_x + I$ TRANSFER OF INITIATOR

k_{ts}

$M_x{}^{\bullet} + S \rightarrow M_x + S^{\bullet}$ TRANSFER OF SOLVENT

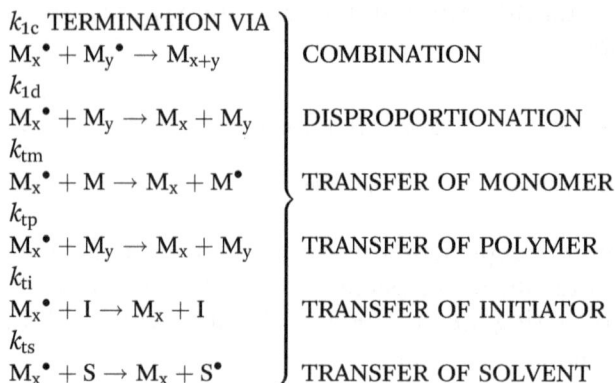

The most common free radical polymerization is a statistical process. As shown below, polymerization is initiated by the decomposition of initiator I, and its role is to lower the temperature of free radical monomer formation R. Each initiator molecule produces two initiator radicals R$^{\bullet}$ and each of these R$^{\bullet}$ radicals will react with a monomer. During the propagation stage, high molecular weight polymer chains will be produced. Termination can be accomplished by a variety of means – recombining two reactive polymer ends, disproportionation, monomer, solvent, polymer, or initiator transfer. As simple as it sounds, this process offers relatively little control over the polymer architecture and high molecular weight polymers are formed at the onset of polymerization reactions.

Figure 2.5 illustrates the characteristic molecular weight *vs.* monomer conversion curves for chain-growth and step-growth polymerizations. Notably, chain growth requires a very short induction period and high molecular weight is achieved almost instantly. In contrast, step growth requires very high conversion rates, often >99%, to achieve high molecular weight polymers. The high PDI values may sound like a bad thing, but high molecular weights can be achieved in a relatively short time, which ultimately leads to useful, potentially important properties. There are, however, other polymerization techniques that are critical in many applications, which are often referred to as 'living' radical polymerizations. The polymerization rate is controlled in 'living' radical polymerization systems so that termination and chain transfer reactions are minimized. Living radical polymerization is characterized by the linear increased molecular weight increase with conversion.

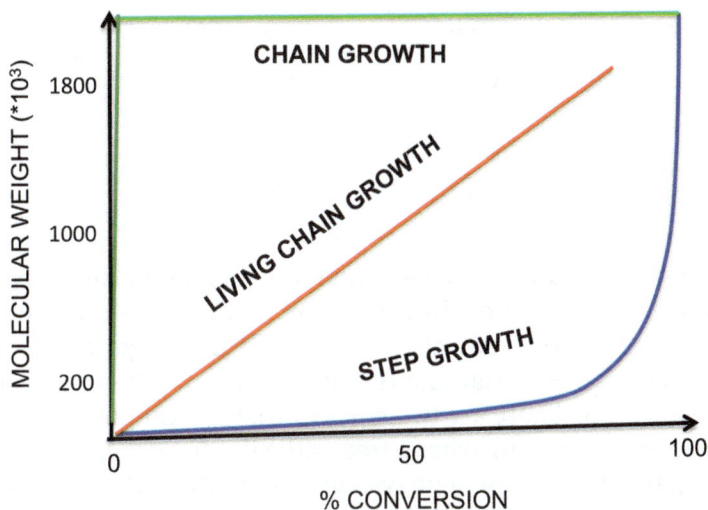

Figure 2.5 The relationship between molecular weight and % monomer conversion in chain growth (green), living chair growth (red), and step growth (navy blue) polymerizations. Note: initial stages of chain growth polymerization occur very rapidly and high molecular weight is formed instantly.

2.4.1 Free Radical Polymerization

2.4.1.1 Initiation

The breakdown of initiators should generate free radicals. The initiator itself needs to be stable enough for storage, but active enough to dissociate under relatively mild conditions. There are two major types of initiators – thermally activated initiators and photoinitiators. Shown below are the decomposition mechanisms for two of the most commonly used thermal initiators, 2,2′-azobisisobutyronitrile (AIBN) and benzoyl peroxide (BPO). AIBN dissociates by breaking the C–N bonds, releasing N_2, generating free radicals. BPO has a weak peroxide bond, which can easily dissociate upon heating, generating benzoate free radicals. While benzoate free radicals are ready to initiate polymerization, they can continue to dissociate, release a CO_2 molecule, turning into free phenyl radicals.

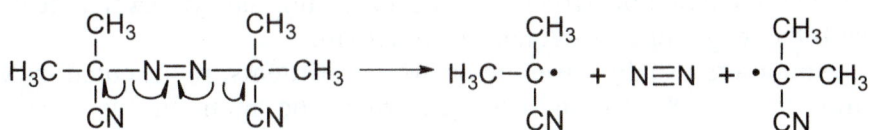

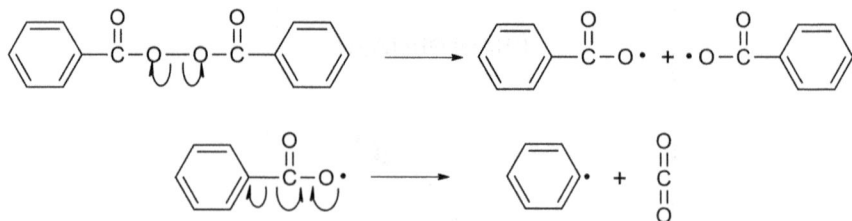

Photopolymerization occurs when the reactants are exposed to electromagnetic radiation in the UV or visible region, the initiators decompose into free radicals and activate the polymerization. The most commonly seen example is dental resin, which can be cured by visible blue light. There are two types of free radical photoinitiators: a one-component system where free radicals are generated by bond cleavage, and two-component systems where the radical is generated by electron transfer of a donor compound. For example, cleavage of benzyl ketal upon UV radiation generates two free radicals, both of which can initiate polymerizations.

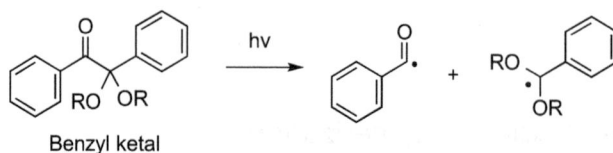

Benzyl ketal

Benzophenone prefers to accept electrons to generate free radicals. The most effective electron donor is a tertiary amine with an α-hydrogen. Free radicals are generated by electron transfer followed by proton transfer from the amine to the ketone. Such photoinitiators also include xanthones and quinones.

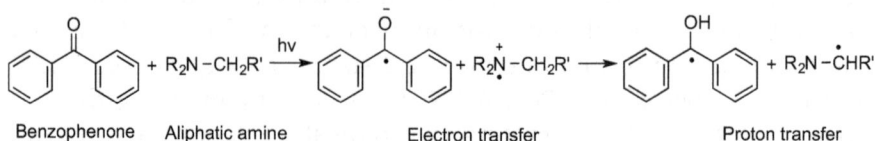

Benzophenone   Aliphatic amine   Electron transfer   Proton transfer

2.4.1.2 Propagation

The propagating step in free radical polymerization proceeds very rapidly by addition of the radicals to a carbon–carbon double bond. There is considerable structural variation in homopolymers that deals with tacticity and regiosequence isomerism.

Monomers usually have one or two substituents, represented by X and Y. X or Y can be hydrogen when the monomer is mono-substituted. Because X and Y are usually different, it gives rise to a chiral center and the polymer chains exhibit tacticity. It can affect flexibility, packing and crystallinity of the polymer.

$$\text{~~} \underset{\substack{| \\ Y}}{\left[CH_2 - \underset{}{C} \right]}_n \text{~~}$$

Chiral center

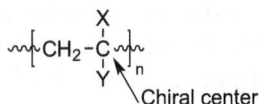

Ideally, there are four types of chain structure:

  (i) Isotactic chain where the relative chiral configurations of all the substituted carbons are the same.
 (ii) Syndiotactic chain where the relative configuration of the substituted carbon centers alternates along the chain.
(iii) Heterotactic chain where the dyad configuration alternates along the chain.
(iv) Atactic chain where the arrangement of carbon centers along the chain is random.

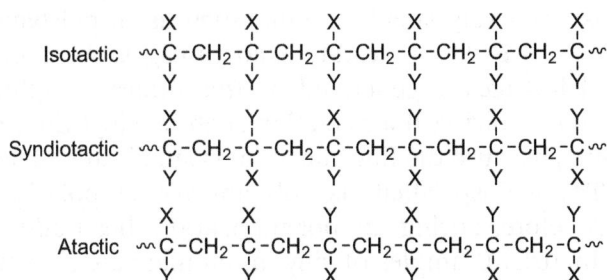

Isotactic

$$\text{~~} \underset{Y}{\overset{X}{C}} - CH_2 - \underset{Y}{\overset{X}{C}} - CH_2 - \underset{Y}{\overset{X}{C}} - CH_2 - \underset{Y}{\overset{X}{C}} - CH_2 - \underset{Y}{\overset{X}{C}} - CH_2 - \underset{Y}{\overset{X}{C}} \text{~~}$$

Syndiotactic

$$\text{~~} \underset{Y}{\overset{X}{C}} - CH_2 - \underset{X}{\overset{Y}{C}} - CH_2 - \underset{Y}{\overset{X}{C}} - CH_2 - \underset{X}{\overset{Y}{C}} - CH_2 - \underset{Y}{\overset{X}{C}} - CH_2 - \underset{X}{\overset{Y}{C}} \text{~~}$$

Atactic

$$\text{~~} \underset{Y}{\overset{X}{C}} - CH_2 - \underset{Y}{\overset{X}{C}} - CH_2 - \underset{X}{\overset{Y}{C}} - CH_2 - \underset{Y}{\overset{X}{C}} - CH_2 - \underset{X}{\overset{Y}{C}} - CH_2 - \underset{X}{\overset{Y}{C}} \text{~~}$$

In reality, idealized structures do not exist, but one of these structures predominates. The microstructures can be detected using NMR to look at the structural regularity. The chains are often characterized as diads, triads, tetrads, pentads, *etc.* For every dyad, there can be two types of configurations – meso (m) and racemic (r).

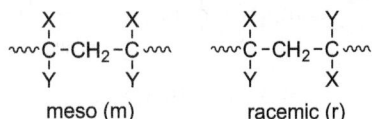

$$\text{~~} \underset{Y}{\overset{X}{C}} - CH_2 - \underset{Y}{\overset{X}{C}} \text{~~} \qquad \text{~~} \underset{Y}{\overset{X}{C}} - CH_2 - \underset{X}{\overset{Y}{C}} \text{~~}$$

meso (m) racemic (r)

Tacticity of a short segment of a polymer chain can then be identified as follows:

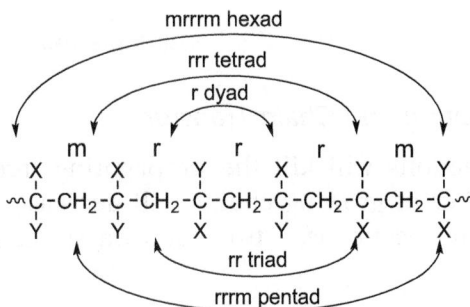

mrrrm hexad

rrr tetrad

r dyad

$$\overset{m}{} \quad \overset{r}{} \quad \overset{r}{} \quad \overset{r}{} \quad \overset{m}{}$$

$$\text{~~} \underset{Y}{\overset{X}{C}} - CH_2 - \underset{Y}{\overset{X}{C}} - CH_2 - \underset{X}{\overset{Y}{C}} - CH_2 - \underset{Y}{\overset{X}{C}} - CH_2 - \underset{X}{\overset{Y}{C}} - CH_2 - \underset{X}{\overset{Y}{C}} \text{~~}$$

rr triad

rrrm pentad

Tacticity is determined by the type of substituents and can be affected by solvent. Most polymers formed from free radical polymerization have more syndiotactic than isotactic dyads. Monomers with bulky substituents form polymers that are predominantly syndiotactic.

Also, because of the asymmetric substitution of monomers, the two ends of the double bond are distinctly different. The end with fewer substitutions is called the 'tail' and the more substituted end is the 'head'.

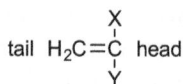

$$\text{tail } H_2C=\underset{Y}{\overset{X}{C}} \text{ head}$$

Because of the influence of steric, polar, resonance and bond strength factors, tail addition to the free radical is preferable. But head-to-head addition may also occur. For most of the polymer, head-to-head linkages exist by a very small amount, such as in poly(vinyl acetate), where there is $\sim$1–2% of head-to-head linkages. Another source of head-to-head linkages is generated by free radical coupling. But for polymerization involving fluoro-olefins such as vinyl fluoride and tri-fluoroethylene, because the fluorine atom is small, steric factors can be neglected. The regiospecificity is only affected by polarity and bond strength. Therefore, during the polymerization, head addition occurs by more than 10%. Examples of vinyl monomers are shown below.

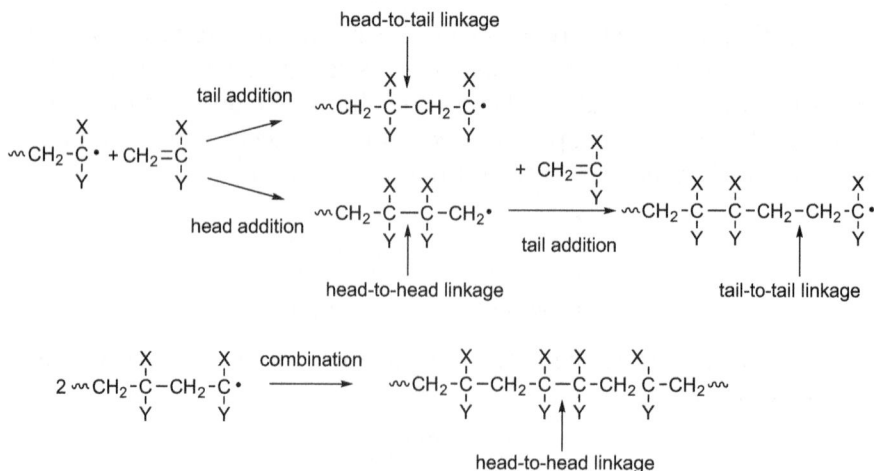

2.4.1.3 Termination and Chain Transfer

Termination reactions will kill the propagating free radicals. As a result, the number of growing chains will decrease. Radical–radical termination can occur *via* two pathways: combination and disproportionation. Combination is the coupling of two radicals. Disproportionation involves a proton transfer, generating a carbon–carbon double bond at one chain, and an aliphatic end on the other chain. Termination reactions can occur between two propagating chains, or between a propagating chain and any other free radicals existing in the reaction media, such as initiator-derived, oxygen-derived or impurity-derived free radicals.

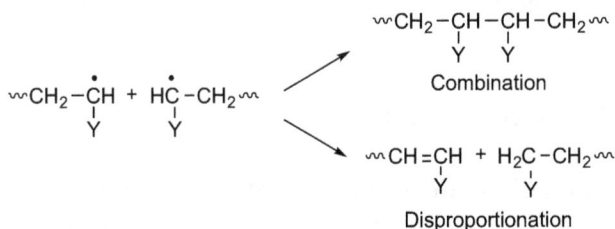

Chain transfer is a reaction between the propagating radical and a non-radical species to produce a dead polymer chain and a new radical capable of initiating polymerization. It causes the premature termination of propagating chains, resulting in lower average molecular weight. Chain transfer can occur with initiator, monomer, solvent, polymer, or chain transfer agent. Different initiators have very different tendencies for chain transfer. Peroxides generally have a higher chain transfer constant, while azonitriles such as AIBN are less prone to transfer reactions. Transfer to monomer, polymer, solvent, and many chain transfer agents takes place by transfer of a proton from them to the propagating radical. Monomer transfer constants are usually small due to the high energy needed to break the C–H bond, thus do not significantly affect molecular weight. Although chain transfer to polymer also has a small transfer constant, it generates branching, which may significantly affect the properties of the polymer. Knowing the probability of transfer of the free radical to the chain transfer agent is important. The chain transfer agent could be the solvent. Solvents such as benzene and cyclohexane, which have strong C–H bonds, show very low transfer constants. But toluene has a higher transfer constant because the free radical can be stabilized by the phenyl group. Acids, carbonyl compounds, ethers, amines and alcohols have high transfer constants due to a stabilization effect of the radicals by adjacent O, N or carbonyl group. This is important to know, in order to choose the correct solvent as the reaction media. The weak S–S bond has a very high transfer constant. When properly used, the S–S is an important component in controlling the polymerization process, as will be discussed later when we talk about living radical polymerization.

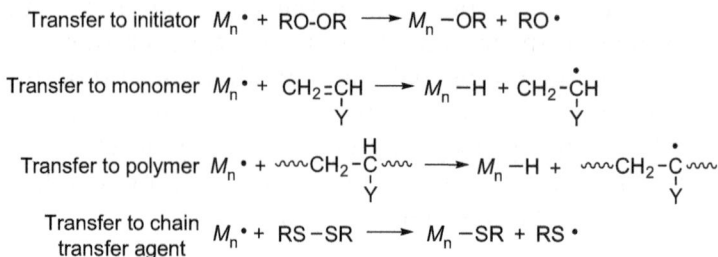

Transfer to initiator $M_n{}^\bullet$ + RO-OR ⟶ M_n -OR + RO$\bullet$

Transfer to monomer $M_n{}^\bullet$ + CH$_2$=CH ⟶ M_n -H + CH$_2$-ĊH
                                         | |
                                         Y   Y

Transfer to polymer $M_n{}^\bullet$ + ∿CH$_2$-C∿ ⟶ M_n -H + ∿CH$_2$-Ċ∿
                                          | |
                                          Y   Y

Transfer to chain $M_n{}^\bullet$ + RS-SR ⟶ M_n -SR + RS$\bullet$
transfer agent

2.5 Ionic Polymerization

Ionic polymerization proceeds in a chain reaction similar to free radical polymerization, but is initiated by cationic or anionic initiators instead of free radicals. Transfer of the charge from ionic initiator to a monomer generates the propagating center.

The reactions below show the initiation of cationic polymerization by addition of a positively charged ion A^+ to the monomer. Monomers that can be catatonically polymerized usually contain electron-donating substituents R. Below, X^- is the negatively charged counterion, which forms an ion pair or free ion with the propagating center. The propagation step proceeds as a chain reaction.

Initiation

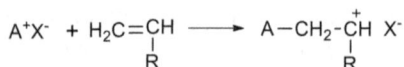

Propagation

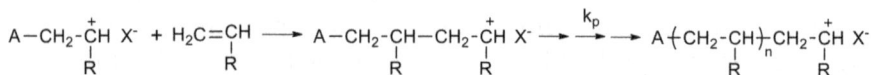

Anionic polymerization proceeds in a similar fashion, and is shown below. It is initiated with a negatively charged species B^-. Typical counterions X^+ are small metal ions, such as Li^+ and Na^+. Addition of B^- to a monomer containing an electron-withdrawing substituent R produces a negatively charged propagating center, where monomers are added through a chain reaction to produce a polymer. The main differences between cationic and anionic polymerizations are the faster rates for cationic, and being more reactive, cationic is harder to control and stabilize. Also, due to side reactions it is harder to control PDI.

$$B^-X^+ + H_2C=\underset{R}{\overset{|}{CH}} \longrightarrow B-CH_2-\underset{R}{\overset{|}{\overset{\cdot\cdot}{CH}}}\ X^+$$

$$B-CH_2-\underset{R}{\overset{|}{\overset{\cdot\cdot}{CH}}}\ X^+ + H_2C=\underset{R}{\overset{|}{CH}} \longrightarrow B-CH_2-\underset{R}{\overset{|}{\overset{\cdot\cdot}{C}}}\!-CH_2\text{-}CH_2X^+ \xrightarrow{\ \ k_p\ \ } B\big(CH_2\text{-}\underset{R}{\overset{|}{CH}}\big)_{\!n}CH_2\text{-}\underset{R}{\overset{|}{\overset{\cdot\cdot}{CH}}}\ X^+$$

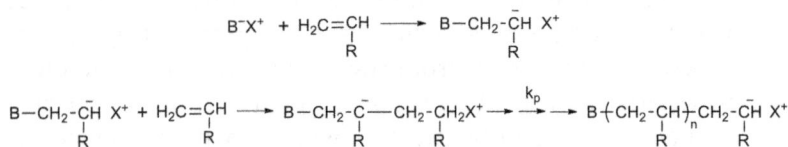

There are several characteristics that differentiate ionic polymerization from free radial polymerization:

(i) Ionic polymerization is highly selective toward monomers. While almost all monomers with C=C can be polymerized by free radicals, cationic polymerization is limited to olefins with electron-donating substituents, such as –OR, –NR$_2$, –O–C(=O)– R or –R. Anionic polymerization can only polymerize vinyl groups with electron-withdrawing substituents, such as –CN, –NO$_3$ and –CO$_2$R. Styrene is a unique monomer, which can be polymerized by both cationic and anionic polymerization. The very restricted selectivity toward monomers is due to stabilization of the cationic or anionic propagating species. If Y is an electron-donating group, the electron density on the C=C double bond is shown as below. Higher electron density on the head C will favor addition to a cationic propagating center. The electron-donating substituent at the propagating center delocalizes the positive charge, thus stabilizing the propagating center. Below, the same principle applies to anionic polymerization, where the anionic propagating center is stabilized by electron-withdrawing groups.

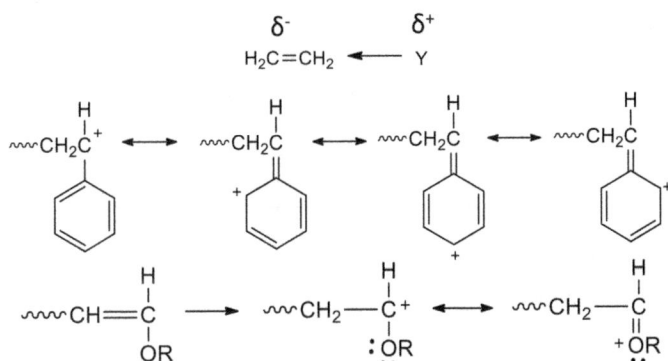

$$\overset{\delta^-}{H_2C}=\overset{\delta^+}{CH_2} \longleftarrow Y$$

(ii) The formation of ions with sufficiently long lifetimes for propagation to yield high molecular weight products requires stabilization of the propagating centers by solvation. Therefore, selection of the correct solvent is critical in reaction kinetics, molecular weight of the final polymer, and molecular weight distribution. Association of the propagating center with the counterion ranges from covalent to free ions. Usually, ionic polymerization involves the ion pair, free ions, or coexistence of the two, depending on the distance between the centers of charge. Increased solvent polarity results in a shift from ion pairs to free ions, and also from lower to higher reactivity. But highly polar solvents may inhibit initiation.

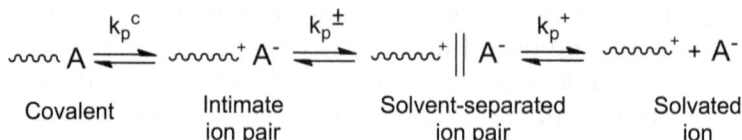

$$\sim\sim A \underset{k_p^c}{\rightleftharpoons} \sim\sim^+ A^- \underset{k_p^\pm}{\rightleftharpoons} \sim\sim^+ \| A^- \underset{k_p^+}{\rightleftharpoons} \sim\sim^+ + A^-$$

| Covalent | Intimate ion pair | Solvent-separated ion pair | Solvated ion |

(iii) Living polymerization can be achieved by using a properly selected reaction system, especially for anionic polymerization. Unlike radical polymerization, termination in ionic polymerization never involves the bimolecular reaction between two propagating polymer chains of like charge. Termination of a propagating chain occurs by its reaction with the counterion, solvent, or other species present in the reaction system. Relatively low or moderate temperatures are needed to suppress termination, transfer and other chain-breaking reactions, which destroy propagating centers.

2.5.1 Cationic Polymerization

The reaction sequence below shows the cationic polymerization of styrene.

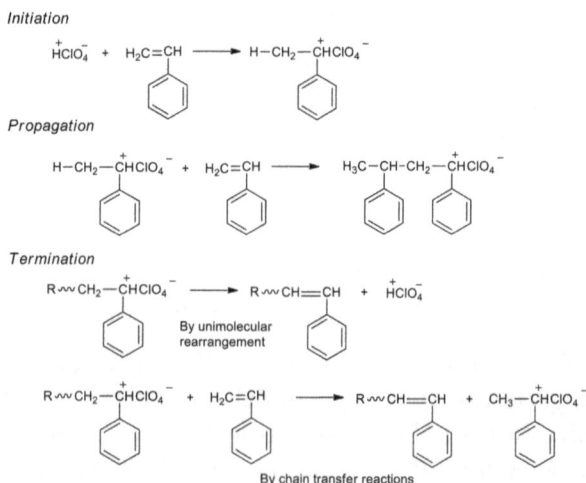

Protonic acid $HClO_4$ initiates the polymerization by protonation of the polystyrene. When a protonic acid is used as the initiator, the acid needs to be strong enough to produce a sufficient concentration of cations, while the anion of the acid should not be highly nucleophile, otherwise it will strongly bind to the carbocation and terminate the propagating center. Due to this requirement, use of protonic acid as initiator is only limited to a few species. Lewis acids, such as $SnCl_4$, $AlCl_3$, BF_3 and $TiCl_4$, are more frequently used initiators in cationic polymerization, producing high molecular weight polymers. The presence of a suitable cation source can facilitate the initiation process, where Lewis acid is referred to as a coinitiator while the cation source is the initiator. The cation source can be water, alcohols, or even a carbocation donor such as an ester or an anhydride. Upon reaction of the initiator with the coinitiator, an intermediate complex is formed, which then goes on to react with the monomer unit. The counterion produced by the initiator–coinitiator complex is less nucleophilic than that of the protonic acid A^- counterion. Other initiation methods such as photoinitiation or ionizing radiation have also been utilized.

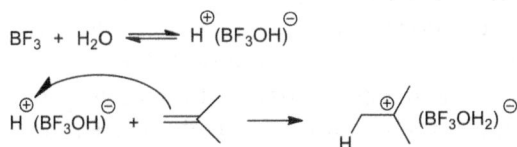

$$BF_3 + H_2O \rightleftharpoons \overset{\oplus}{H}(BF_3OH)^{\ominus}$$

Propagation proceeds by insertion of a monomer between the carbocation and its counterion. While monomers quickly add to the carbocation one by one, termination with counterions will kill the propagating cation *via* the following processes:

(i) Combination with counterion:

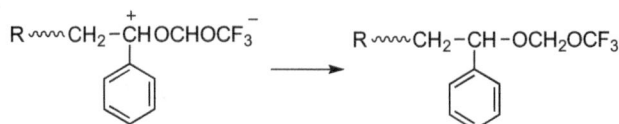

(ii) Anion splitting:

Termination with a counterion causes the concentration of propagating chains to decrease. Chain transfer is also likely to occur. The chain transfer reactions shown above are the most common ones, where the β proton adjacent to the carbocation can be transferred to the counterion or monomer. In such cases, the chain is terminated, but a new propagating center is generated, so that the concentration of carbocation does not change. However, chain transfer reactions will result in polymers of lower molecular weight.

2.5.2 Anionic Polymerization

Living polymerization was first discovered in the anionic polymerization of styrene in 1956. The term 'living' means there is no termination step, so the propagating centers are always present. Anionic polymerization is such a system – it does not have termination reactions. The anionic propagating center can stay stable in a wide range of conditions. This is difficult to achieve for cationic polymerizations. Although successful living cationic polymerizations have been demonstrated, the lifetime of the carbocation is significantly lower than the anionic system.

Initiation involves addition of a nucleophile to the monomer. Common initiators include ionic or covalent metal amides such as $NaNH_2$, alkoxides such as NaOR, hydroxides, cyanides, amines or organometallic compounds. Alkyl lithium compounds are the most commonly used initiator for anionic polymerization of styrene and 1,3-butadiene. Polymerization of styrene initiated by *n*-butyl lithium is shown below. The type of initiator used needs to match the reactivity of the monomers. Strong nucleophiles, such as amide and alkyl carboanions, are needed to polymerize monomers with relatively weak electron-withdrawing groups, such as styrene and 1,3-butadiene. Weak nucleophiles, such as alkoxides and hydroxides, can polymerize monomers with strong electron-withdrawing substituents, such as acrylonitrile and methyl methacrylate.

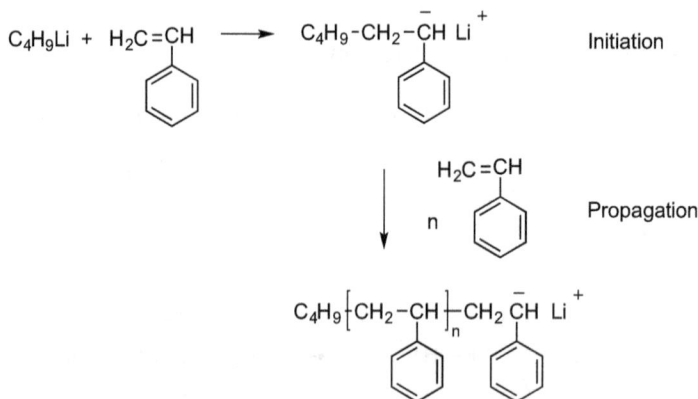

Termination and chain transfer reactions that occur in cationic polymerization do not exist in anionic systems. The anion does not combine with a metal ion. Impurities or added chain transfer agent can terminate anionic polymerization. H_2O, CO_2 and O_2 can terminate the reaction by adding to the propagating carbanions. H_2O, for example from moisture, can terminate the chain by the following reaction:

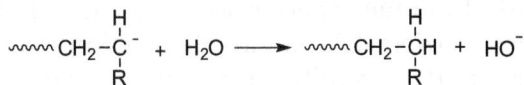

$$\text{wwwCH}_2-\underset{\underset{R}{|}}{\overset{\overset{H}{|}}{C}}{}^- \; + \; H_2O \; \longrightarrow \; \text{wwwCH}_2-\underset{\underset{R}{|}}{\overset{\overset{H}{|}}{C}}H \; + \; HO^-$$

The HO^- generated is not sufficiently nucleophilic to initiate polymerization. For this reason, most anionic polymerizations are carried out in an inert atmosphere with rigorous removal of any moisture. The same applies to cationic polymerization. When the reaction is complete, chain transfer agents such as water or alcohol are added to the system to terminate the carbanions.

The living nature of anionic polymerization allows the synthesis of block copolymers. After polymerizing the first monomer, the propagating centers remain alive, and restart the polymerization of the second monomer. Living radical polymerizations were invented later on with more versatility, but the discovery of living anionic polymerization is a big step in polymer history.

2.6 Coordination Polymerization

Before 1950, polyethylene (PE) was produced by a high-pressure process. In the mid-1950s, Ziegler discovered that the polyethylene prepared catalytically had much less branching, and showed a greater mechanical strength and a higher melting point. Furthermore, linear high molecular weight PE can be produced at low temperature (50–100 °C) and low pressure using a Ziegler catalyst. Following this work, Natta used Ziegler's catalyst and achieved stereoselective polymerization of propylene and other α-olefins. Before his discovery, α-olefins could not be polymerized to achieve high molecular weight by either free radical or ionic polymerizations. As discussed in the preceding section, stereochemistry or tacticity of polymers has a tremendous effect on a material's properties. This comes from the fact that chains with high structural regularity have a greater ability to pack and crystalize to gain greater physical strength. For example, isotactic polypropene is a crystalline polymer with high melting point and mechanical strength. However, atactic polypropene is a fully amorphous material. Before coordination polymerization, fully isotactic polymers could not be synthesized by either free radical or ionic polymerization. Coordination polymerizations have been industrialized for the commercial production of high density polyethylene (HDPE), polypropylene, and many other polyolefins.[2,3] For their contributions, Ziegler and Natta were jointly awarded the 1963 Nobel Prize in Chemistry.

The most important component during the initiation reaction is the transition metal compound, such as titanium tetrachloride, vanadium trichloride, and chromium triacetylacetonate. Group I to III metal compounds such as triethylaluminum, diethylaluminum chloride and diethyl zinc are used as coinitiator to activate the transition metal compound for initiation. The polymerization mechanism is shown below. The transition metal M forms a complex with the π electrons of the monomer. During this step, monomer can only be inserted in a specific orientation held by the coordination in the form of a four-membered ring. Coordination between the initiator fragment and the propagating center is broken simultaneously with the formation of new bonds both between the propagating center and the new monomer, and between the new monomer unit and the transition metal catalyst. Formation of tacticity is due to the fact that all monomers coordinate with the coordination center with the same orientation.

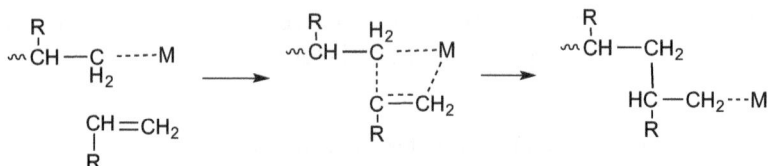

The diagram below illustrates the formation of isotactic polypropylene catalyzed by $TiCl_4$. Monomer coordinates at the vacant site of Ti form a four-membered ring transition state, followed by insertion of the monomer into the propagating center–transition metal bond. A critical step for generating the isotactic structure is migration of the newly formed propagating center back to the original coordination site. This is referred to as migratory insertion, and only the atoms of the first few repeat units move. If no migration occurs, syndiotactic placements will form.

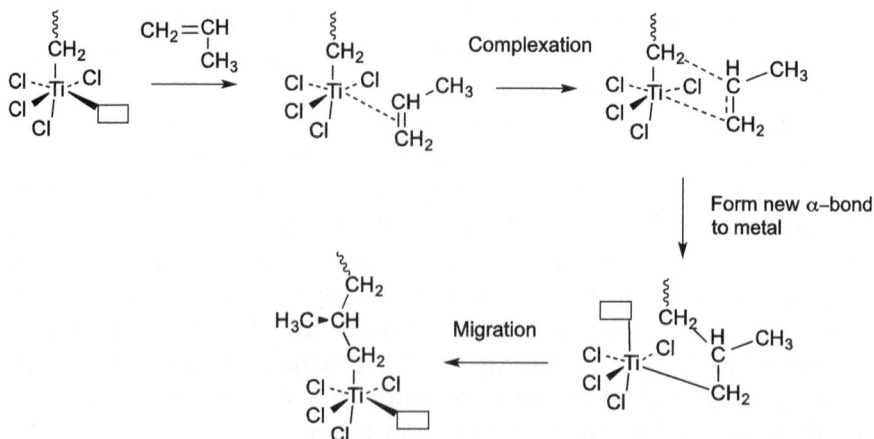

2.6.1 Ring-opening Polymerization (ROP)

Ring-opening polymerization (ROP) is classified as a third type of polymerization mechanism, which together with chain (free radical, ionic and coordination) polymerization and step polymerization, are the three polymerization methods governing the field of polymer synthesis. Industrial production of polydiols, Nylon 6 and poly-siloxane are by ROP. The propagating center can be radical, anionic or cationic. Some ROP can be considered as a chain polymerization but many reactions are more complicated. There is a wide range of monomers that can be polymerized using ROP, mainly heterocyclic compounds containing oxygen, sulfur, nitrogen, phosphorus or silicon in the ring. Polymerization is mainly driven by relief of the ring bond–angle strain. This section will use three typical monomers: epoxides, ε-caprolactam, and octamethylcyclotetrasiloxane as examples to illustrate the processes of ROP.[4]

Epoxides such as ethylene and propylene oxides are usually polymerized anionically or cationically. The anionic ROP can be initiated using metal hydroxides, alkoxides, oxides and amides. It proceeds by nucleophilic attack of the growing chain end on a heterocyclic monomer.

Initiation $\quad R^+A^- + H_2C\overset{O}{\overset{\diagup\diagdown}{-}}CH_2 \longrightarrow A\text{-}CH_2\text{-}CH_2\text{-}O^- R^+$

Propagation $\quad A\text{-}CH_2\text{-}CH_2\text{-}O^- R^+ + H_2C\overset{O}{\overset{\diagup\diagdown}{-}}CH_2 \longrightarrow A\text{-}CH_2\text{-}CH_2\text{-}O\text{-}CH_2\text{-}CH_2\text{-}O^- R^+$

$$A\Big[CH_2\text{-}CH_2\text{-}O\Big]_{n-1}CH_2\text{-}CH_2\text{-}O^- R^+ + H_2C\overset{O}{\overset{\diagup\diagdown}{-}}CH_2 \longrightarrow A\Big[CH_2\text{-}CH_2\text{-}O\Big]_n CH_2\text{-}CH_2\text{-}O^- R^+$$

The termination reaction is rare for anionic ROP, but chain transfer to monomers can be the limiting factor to obtain high molecular weight polymers for certain monomers, such as 1,2-propylene oxide. This leads to the formation of an alcohol end group and a new propagating chain.

$$\overset{CH_3}{\underset{|}{\sim\!\!\sim CH_2\text{-}CH\text{-}O^-}} + H_2C\overset{O}{\overset{\diagup\diagdown}{-}}CH\text{-}CH_3 \longrightarrow \overset{CH_3}{\underset{|}{\sim\!\!\sim CH_2\text{-}CH\text{-}OH}} + CH_2\text{=}CH\text{-}CH_2\text{-}O^-$$

$$\overset{CH_3}{\underset{|}{\sim\!\!\sim CH_2\text{-}CH\text{-}O^-}} + CH_2\text{=}CH\text{-}CH_2\text{-}OH$$

While anionic ROP is a form of chain-growth polymerization, cationic ROP is more complicated. Cationic ROP of epoxides is generally considered to proceed *via* a tertiary oxonium ion, but there are exceptions. A strong protonic acid such as trifluoroacetic acid is used as the initiator to form an oxonium ion. A^- is the counterion. Because the α-carbon of the oxonium ion is electron deficient, the oxygen of the monomer can react with the α-carbon *via* nucleophilic attack and open the ring. Therefore, the propagation proceeds by nucleophilic attack of each monomer to the activated propagating center. Termination is absent in cationic ROP systems at proper temperatures and monomer concentrations.

Another example is polymerization of lactams (cyclic amines). Nylon 6,6 can be produced from a condensation reaction. Polymers of the same structure as Nylon 6,6 can be produced by ROP of ε-caprolactam. The product formed by ROP is called Nylon 6. Along the same lines, Nylon 11 and Nylon 12 are also produced as specialty polymers. Initiation with water, referred to as hydrolytic polymerization, and anionic polymerization are both carried out commercially.

Anionic ROP of lactams can be initiated with strong bases such as alkali metals, metal hydrides or metal amides. Lactam reacts with an alkali metal, forms a lactam anion, which, in turn, attacks the C=O group of another monomer, leading to C–N bond cleavage. In the next step, a primary amine anion will abstract a proton from a monomer, and as a result, regenerate the lactam anion.

The lactam anion can be viewed as an activated monomer. The propagation proceeds by attack of the lactam propagating chain end by an activated monomer. The following proton transfer will activate another monomer for addition:

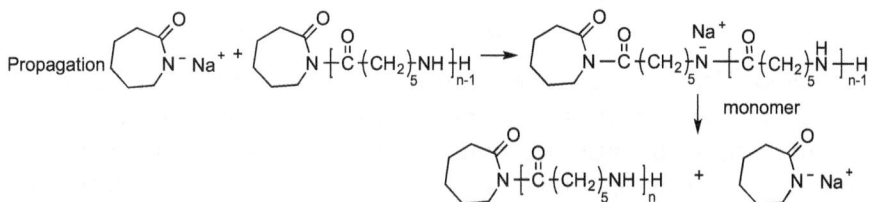

Hydrolytic polymerization starts with hydrolyzation of ε-caprolactam into ε-amino caproic acid. Then the COOH of the amino acid can protonate a monomer, followed by nucleophilic attack by amine groups at the propagating chain end. Condensations of amino acid itself also contribute to molecular weight increase. However, ROP is the dominant route.

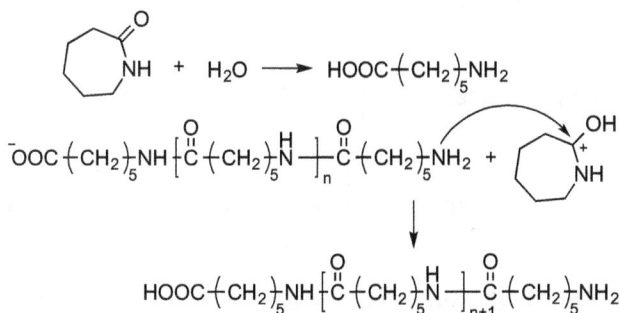

The last example is production of high molecular weight polysiloxanes from anionic or cationic polymerization of cyclic siloxanes. Shown below is the formation of poly(dimethylsiloxane) from octamethylcyclotetrasiloxane. Polymerization proceeds in a similar way, involving nucleophilic attack, so it will not be discussed in any detail.

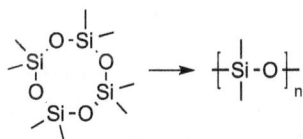

2.6.2 Ring-opening Metathesis Polymerization (ROMP)

Olefin metathesis is a well-developed synthetic route for production of various olefins. It allows olefins to exchange substituents catalyzed by Grubb's catalyst[5] or other transition metal–carbene complexes as shown below:

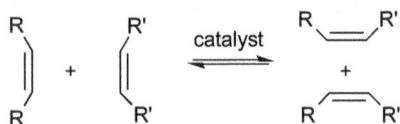

Cycloolefins can ring open and polymerize in the presence of the same catalyst *via* similar mechanisms, generating polymers containing highly unsaturated backbones. Such polymerization is referred to as ring-opening metathesis polymerization (ROMP). The unsaturated polymer can be partially or fully hydrogenated, or be modified with various functional groups. The best known industrial products synthesized *via* ROMP include polynorbornene (Norsorex®), polycyclooctene (Vestenamer®) and polycyclopentadiene (Metton®). Schrock, Grubbs and Chauvin were honored with a Nobel Prize in 2005 for their contributions in this area.[6–8]

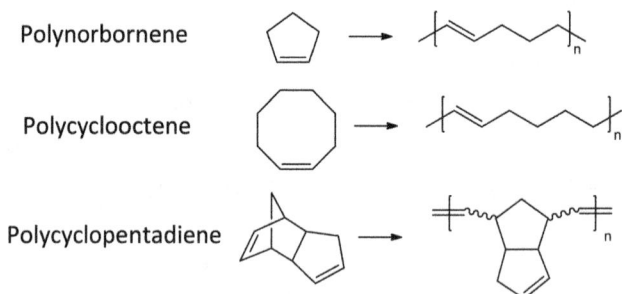

The reaction is driven by release of ring strain enthalpy. Grubbs initiators based on ruthenium (Ru) such as the one shown below are most commonly used for their adaptability to different monomers and reaction conditions.[9]

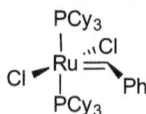

The ROMP proceeds *via* [2 + 2] cycloaddition. A critical step is formation of a metallocyclobutane intermediate from coordination of a cycloolefin with the catalytic metal center, followed by cleavage of a π bond of the metal–carbene. The metallocyclobutanes then rearrange, break the original metal–olefin bond, and generate a new metal–carbene propagating center. Propagation continues by insertion of the double bond of the monomer into the metal–carbene bond in the same way until the complete consumption of all monomer. Polymerizations can also be terminated by adding small olefin molecules.

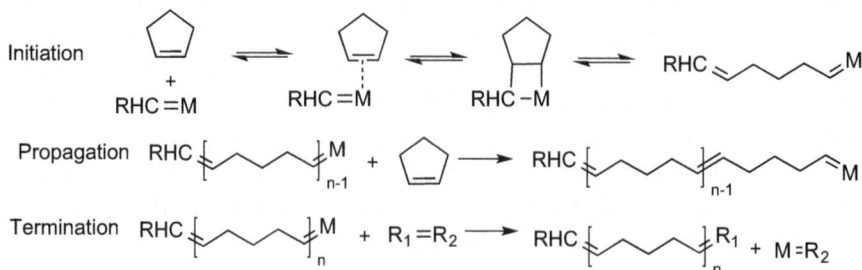

Polymerization rate and polymer molecular weight of ROMP can be affected by many factors, including electron-donating or withdrawing ligands attached to the metal initiator, using a different solvent, and the substituents on the monomer ring. ROMP is an equilibrium reaction and the system includes ring-closing metathesis (RCM) and cross-metathesis (CM) in the equilibrium. Therefore, cyclic molecules are always present in the product due to RCM reactions during the polymerization, where the growing chain end 'backbites' the intramolecular double bond along the polymer backbone. Meanwhile, CM may also occur, causing intermolecular exchange reactions between the active chain-end species and backbone double bond on other polymer chains, resulting in generation of shorter polymer chains. These chain transfer reactions are unavoidable in ROMP systems and are responsible for broad molecular weight distribution.

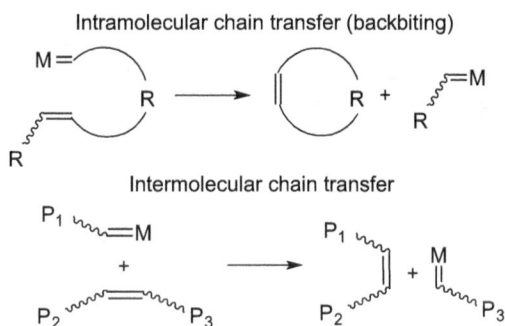

2.7 Controlled Radical Polymerization (CRP)

For many years so-called 'living' controlled polymerization methods were dominated by living cationic and anionic polymerization. Why? Because there were no other synthetic approaches available that would facilitate the control of PDI and copolymer architecture. Since the collection of monomers for both synthetic approaches is not very achievable, it was just a matter of time before new controlled radical polymerization (CRP) paths would be developed. Aside from the obvious limitations of anionic and cationic polymerizations, perhaps the main motivation was the ability to synthesize a wide range of polymer architectures that were otherwise unobtainable. The three most popular CRP are stable free radical polymerization (SFRP), atom transfer radical polymerization (ATRP), and reversible addition–fragmentation chain transfer (RAFT) polymerization.

2.7.1 Nitroxide-mediated Polymerization (NMP)

Nitroxides were well known for inhibiting polymerization reactions due to their ability to scavenge carbon-centered radicals to form alkoxyamines. Alkoxyamines as polymerization initiators and the use of NMP were explored in the mid-1980s.[10] One example of the NMP mechanism using 2,2,6,6-tetramethyl-1-piperidinyloxyl (TEMPO) is shown below, where I is initiator, and M is monomer.[11]

Initiation

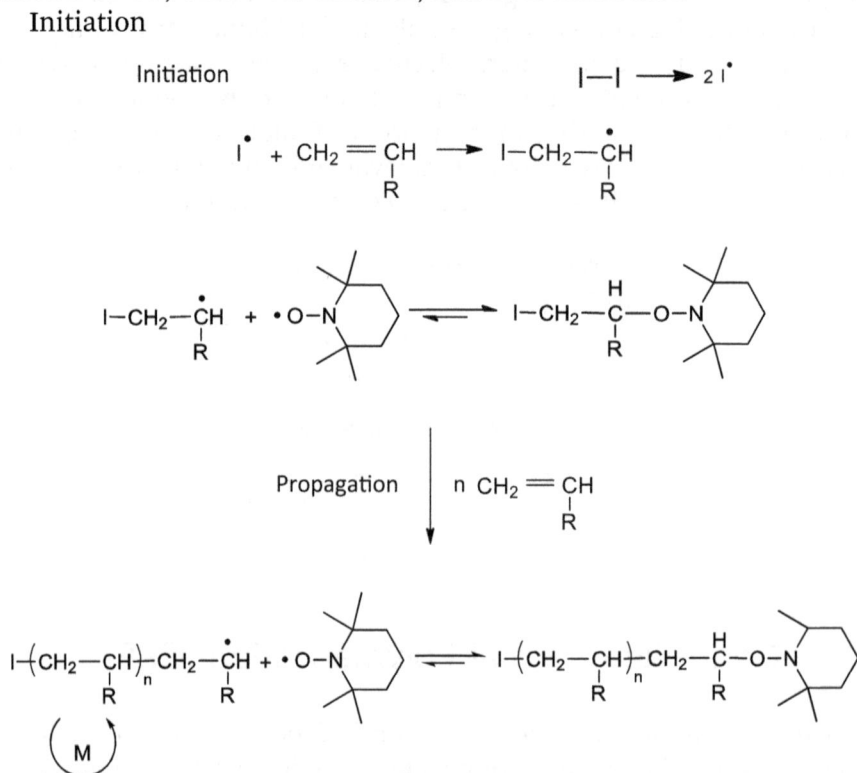

A wide range of nitroxides and derived alkoxyamines has been explored in NMP applications and the critical parameters are the activation–deactivation equilibrium constant and the combination of disproportionation ratio for the reaction of nitroxides with propagating radicals and the intrinsic stability of nitroxide and the alkoxyamine under given polymerization conditions.

2.7.2 Atom Transfer Radical Polymerization (ATRP)

Control of molecular architecture in a CRP was considered impossible at a level similar to other living ionic systems because two radicals always terminate at a very fast, diffusion-controlled rate.[12] When the concept of dynamic equilibrium was introduced to radical polymerization, the entire field changed, giving an opportunity for precisely controlled molecular weight, low PDI ($M_w/M_n < 1.1$), and more importantly, the control of molecular architecture. ATRP was developed by combining a proper catalyst (transition metal compound) and ligand, in the presence of an initiator with an appropriate structure, and proper polymerization conditions. As a consequence, molecular weight during polymerization increased linearly with conversion and the polydispersities were typical for a living process. The ATRP reaction mixture is hence a multi-component system consisting of an initiator (mostly alkyl halogenides or chlorosulfonic acids), transition metal catalyst, ligand and monomer, if necessary solvent and other compounds (activator or deactivator). The principles of ATRP equilibrium polymerization is schematically illustrated below:

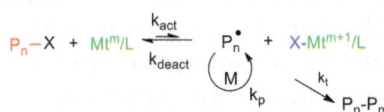

$$P_n-X \; + \; Mt^m/L \; \underset{k_{deact}}{\overset{k_{act}}{\rightleftarrows}} \; P_n^{\bullet} \; + \; X\text{-}Mt^{m+1}/L$$

The reaction is controlled by the equilibrium between propagating radicals and dormant species, typically in the form of initiating alkyl halides/macromolecular species (Pn–X). The dormant species periodically react at the rate constant of activation (k_{act}) with a transition metal complex Mt^m/L being at a lower oxidation state (where Mt^m is the transition metal at the oxidation state m and L is a ligand). Omitting the charges of ionic components and counterions, the complex functions as an activator forming growing radicals ($P_n^{\bullet}$), and deactivators, which are transition metal complexes in their higher oxidation states coordinated with halide ligands $X - Mt^{m+1}/L$. This catalytic process can be mediated by a number of redox-active transition metal complexes. The most popular are Cu^I/L and $X–Cu^{II}/L$ complexes, but other metal complexes may include Rh, Fe, Mo, Os, and others. An example of ATRP (shown below) illustrates styrene polymerization with the conversion of 100% styrene units polymerized with one unit of *N,N,N′,N,N* pentamethyldiethylenetriamine (PMDETA).

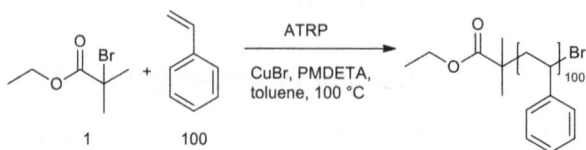

The rate of ATRP propagation, or polymerization, depends on the rate constant of propagation and the concentration ratio of monomers to free radicals. This is because the radical concentration depends on the ATRP equilibrium constant and the concentration of dormant species, activators, and deactivators, as shown by the equation below.

$$R_p = k_p\{M\}\left[P_n^*\right] = k_p K_{\mathrm{ATRP}} x \frac{[P_n X]\left[\dfrac{Cu^I}{L}\right][M]}{\left[X - \dfrac{Cu^{II}}{L}\right]}$$

It should be noted that there are many variables that may contribute to kinetics, molecular weight and PDI. Specifically, structural features of the ligand and monomer/dormant species as well as reaction conditions, such as solvent environment, temperature and pressure can strongly influence the values of the rate constants, k_{act} and k_{deact}, and their ratio. Notably, the rates increase with catalyst activity (k_{ATRP}), but due to radical termination resulting from low $[Cu^I/L]/[X–Cu^{II}/L]$ ratio, usually caused by a buildup in the concentration of deactivator, will decrease. Thus, the high molecular weights easily obtained by 'regular' free radical polymerization are not easily obtainable here.

Perhaps the best representation of ATRP is formation of selected copolymers with controllable architectures, such as illustrated in Figure 2.4. Typical monomers utilized in polymerization include, but are not limited to, styrenes, acrylates, methacrylates, acrylamides and acrylonitrile as well as vinyl acetate and vinyl chloride. Due to the controllable nature of this catalytic process a variety of compositions, topologies, functionalities and complex architectures can be developed.

Another intriguing feature that may lead to a number of unexplored commercial applications is replacing reducing agents required by ATRP by electrons. This can be achieved by the simple

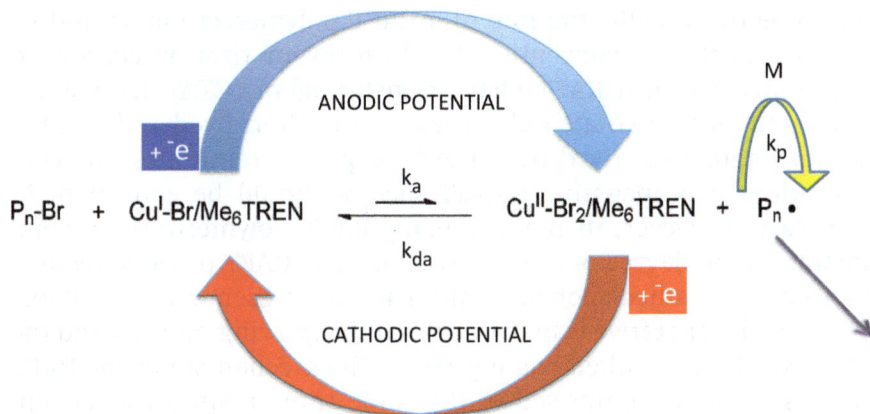

Figure 2.6 Schematic depiction of control of ATRP polymerization using anodic and cathodic potentials.
Adapted from ref. 13.

electrochemical process depicted in Figure 2.6. This concept is based on electrochemically mediated ATRP (eATRP) in which the ratio of the concentrations of activator to deactivator is controlled by electrochemical reactions.[13] Specific parameters are applied current, potential, and total charge passed. Controlling these parameters will allow the selection of the desired concentration of the redox-active catalytic species.

While there is no question that this approach opened up new synthetic opportunities, there are challenges. Among the major ones are metal catalysts that are often not environmentally friendly, their potential side reactions, the reduction of organic solvents and replacement by water, the use of reducing agents, and sensitivity to temperature/pressure conditions. But perhaps the most significant limits are molecular weight limitations and time-consuming reactions. After all, conventional radical polymerization produces ~100 million tons of polymers annually. In contrast, control radical polymerization offers a great forum for controlled architectures, but still too early to call for commercial products.

2.7.3 Reversible Addition–Fragmentation Chain Transfer (RAFT)

RAFT polymerization is particularly promising due to its applicability to a wide range of monomers and its tolerance to most functional groups and reaction solvents.[14,15] The proposed RAFT mechanism is shown below. Ideally, the rate of a RAFT polymerization should be identical to that of conventional radical polymerization conducted in the absence of a RAFT chain transfer agent (CTA) because assuming the intermediate radical species are short-lived and that the radical fragment derived from the CTA quickly reinitiates, the concentration of propagating radical species should be equivalent in each case. However, in practice many RAFT polymerizations demonstrate some degree of rate retardation. The RAFT process involves conventional radical polymerization in the presence of a suitable CTA. The degenerative transfer between the growing radicals and the CTA provides controlled chain growth. The mechanism of the RAFT process below is comprised of the same three main steps as conventional free radical polymerization: initiation, propagation and termination. Additionally the propagation step in RAFT consists of two stages – the RAFT pre-equilibrium and the main RAFT equilibrium. The first stage involves the activation of all added CTA along with some degree of propagation, while the second stage consists of chain equilibrium and propagation. Due to the presence of the CTA and the subsequent degenerative transfer, the termination step is largely suppressed. The key to the structural control in the RAFT process is careful selection of appropriate monomers, initiators and CTAs.

Initiation

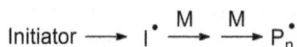

Reversible chain transfer

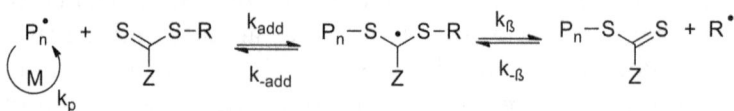

Reinitiation

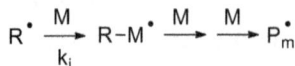

Chain equilibrium

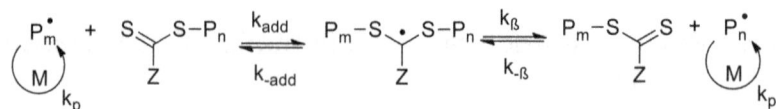

Termination

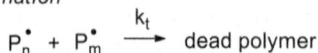

All these synthetic methods offer unique opportunities for controlling molecular architectures as well as the PDI, but one of the drawbacks is the lack of achieving high molecular weights in a reasonable time frame. Thus, realistically speaking, CRP takes roughly 3–4 days and the controlled molecular weight as well as the PDI are in the range of ~ 20–30k and close to unity, respectively. In contrast, so-called statistical free radical polymerization takes roughly 4 hours and molecular weights are in the range of millions, resulting in useful macromolecules. Thus, is seems practical to explore the high molecular weight synthesis and focus on controlling the PDI. One of these opportunities is the development of stimuli-responsive copolymers during a free radical colloidal process.

2.7.4 Heterogeneous Radical Polymerization (HRP)

One of the challenging aspects of CRP is obtaining ultrahigh molecular weight (>106 g mol^{-1}) block copolymers. Using CRP methods requires long synthesis times and is very often impractical. Recent studies have shown that amphiphilic monomers can be be copolymerized to form ultrahigh block copolymers. Specifically, poly(2-(N,N-dimethylamino) ethyl methacrylate-block-n-butyl acrylate (p(DMAEMA-b-nBA)) block copolymers can be synthesized using one-step surfactant-free emulsion polymerization. This synthesis is accomplished by formation of phase-separation conditions and proper choice of monomer/solvent conditions during polymerization. The unique feature of ultrahigh molecular weight block copolymers is their ability to self-assemble in non-polar solvents, but with the promoted solvation of p(DMAEMA) block by an aqueous phase. Due to the presence of high molecular weight segments, these copolymers form inverse polymeric micelles. Figure 2.7 illustrates basic principles of block copolymerization in which hydrophilic M1 and hydrophobic M2 monomers exist in two phases during emulsion polymerization, forming a heterogeneous mixture. Initially, M1 is dissolved in an aqueous phase forming concentrated homogeneous solution. When hydrophobic M2 is added, it will form droplets. When an initiator is continuously fed into the reaction system, polymerization is initiated. The radicals formed upon thermal decomposition in the continuous phase predominantly initiate M1 polymerization due to significantly higher concentrations compared to water insoluble M2. Thus, the propagation step predominantly occurs in aqueous phase upon addition of M1. As polymerization proceeds, the growing polymer chains

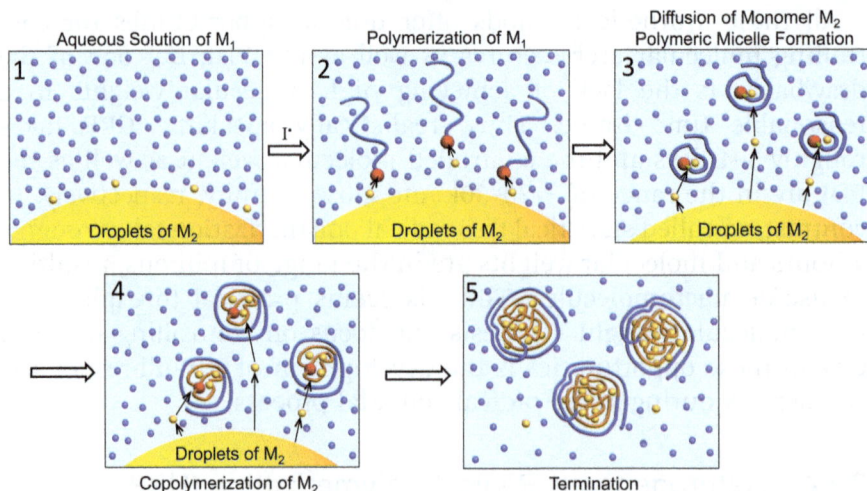

Figure 2.7 Schematic illustration of amphiphilic block copolymer synthesis. Hydrophilic radicals initiate polymerization of predominantly
water soluble Monomer 1 (M1) and propagation of polymer
chains of hydrophobic Monomer 2 (M2) continues in the segregated hydrophobic phase.
Reprinted with permission from C. Lu and M. W. Urban, One-
step Synthesis of Amphiphilic Ultrahigh Molecular Weight Block
Copolymers by Surfactant-Free Heterogeneous Radical Polymerization (HRP), *ACS Macro Lett.*, 2015, **4**(12), 1317–1320.
Copyright (2015) American Chemical Society.

become hydrophobic as soon as a few hydrophobic M2 monomers are
added (for most hydrophobic monomers, the critical number of entry
radicals are copolymerized), thus phase-separating into the hydrophobic environment for M2-ended radicals. The radicals are segregated and protected in the individually segregated phases, thus
limiting the possibility of bimolecular termination by other radicals.
Meanwhile, the kinetically driven hydrophobic M2 continuously diffuses from the monomer droplets into the segregated phases driven by
a high surface area. As a result, M2 predominantly resides in the
segregated phases, whereas M1 remains solvated in aqueous phase
and rarely diffuses into segregated phases. Therefore, the radicals that
enter the segregated phases only propagate with the addition of M2,
until another polymeric radical diffuses to the same region, resulting
in bimolecular termination. Chain termination by disproportionation
and combination will result in the p(M1-block-M2) diblock and p(M1-
block-M2-block-M1) triblock copolymer formation, respectively.[30] It is
critical that the polymerization is not allowed to reach the high

conversion rates so that the concentration of M1 monomer in aqueous phase is still predominant (at least 10 times higher) than M2.[16]

References

1. G. Odian, *Principles of polymerization*, John Wiley & Sons; 2004.
2. J. Boor, Ziegler-Natta catalysts and polymerizations, 1979.
3. P. Cossee, Ziegler-Natta catalysis I. Mechanism of polymerization of α-olefins with Ziegler-Natta catalysts, *J. Catal.*, 1964, **3**(1), 80–88.
4. W. J. Bailey, *Ring-opening polymerization, Comprehensive Polymer Science: the Synthesis, Characterization, Reactions & Applications of Polymers*, Pergamon Press plc, 1989, vol. 3, pp. 283–320.
5. T. M. Trnka and R. H. Grubbs, The development of L2×2Ru CHR olefin metathesis catalysts: an organometallic success story, *Acc. Chem. Res.*, 2001, **34**(1), 18–29.
6. R. R. Schrock, Living ring-opening metathesis polymerization catalyzed by well-characterized transition-metal alkylidene complexes, *Acc. Chem. Res.*, 1990, **23**(5), 158–165.
7. S. T. Nguyen, L. K. Johnson, R. H. Grubbs and J. W. Ziller, Ring-opening metathesis polymerization (ROMP) of norbornene by a group VIII carbene complex in protic media, *J. Am. Chem. Soc.*, 1992, **114**(10), 3974–3975.
8. C. W. Bielawski and R. H. Grubbs, Living ring-opening metathesis polymerization, *Prog. Polym. Sci.*, 2007, **32**(1), 1–29.
9. R. H. Grubbs, D. J. O'Leary, *Handbook of metathesis: Applications in organic synthesis*, John Wiley & Sons, 2015.
10. D. H. Solomon, E. Rizzardo and P. U. S. Cacioli, *US Pat.* 4,581,429, *Chem. Abstr.*, 1985, **102**, 221335q.
11. C. J. Hawker, A. W. Bosman and E. Harth, New polymer synthesis by nitroxide mediated living radical polymerizations, *Chem. Rev.*, 2001, **101**(12), 3661–3688.
12. K. Matyjaszewski, Controlled radical polymerization, *Curr. Opin. Solid State Mater. Sci.*, 1996, **1**(6), 769–776.
13. A. J. Magenau, N. C. Strandwitz, A. Gennaro and K. Matyjaszewski, Electrochemically mediated atom transfer radical polymerization, *Science*, 2011, **332**(6025), 81–84.
14. J. Chiefari, Y. Chong, F. Ercole, J. Krstina, J. Jeffery and T. P. Le *et al.*, Living free-radical polymerization by reversible addition-fragmentation chain transfer: the RAFT process, *Macromolecules.*, 1998, **31**(16), 5559–5562.

15. G. Moad, Y. Chong, A. Postma, E. Rizzardo and S. H. Thang, Advances in RAFT polymerization: the synthesis of polymers with defined end-groups, *Polymer*, 2005, **46**(19), 8458–8468.
16. C. Lu and M. W. Urban, One-step Synthesis of Amphiphilic Ultrahigh Molecular Weight Block Copolymers by Surfactant-Free Heterogeneous Radical Polymerization (HRP), *ACS Macro Lett.*, 2015, **4**(12), 1317–1320.

3 Thermally Responsive Materials

3.1 Polymeric Solutions

In polymeric solutions two transitions are highly critical as a function of temperature: the lower critical solution temperature (LCST) and the upper critical solution temperature (UCST). The LCST is the temperature below which macromolecular segments are miscible for a given composition. The UCST is the temperature above which macromolecular segments are miscible for a given composition. The phase behavior in polymers is an important property as it serves in the development and design of many polymer processes. When polymers exhibit LCST-like behavior in aqueous environments, their solubility will often change due to the presence of responsive components. The classical example is poly(isopropyl acrylamide) (PNIPAM), a polymer that has been studied since the 1960s[1] and still remains to be significant to the scientific community and of technological importance. In particular, when copolymerized with other monomers, LCST transitions can be tuned, thus enabling uses in drug release applications, membrane technologies, and others. As a result of heating or cooling, intra- and intermolecular interactions will be altered, which will be manifested in solubility changes.[2] These transitions are critical in numerous applications and the transition temperatures can be tuned by the monomer ratio (thermally responsive/non-responsive) during copolymerization. Perhaps the most popular ones are drug delivery systems around a physiological temperature. If a given copolymer exhibits thermo-responsive behavior at physiological temperatures (which can be easily adjusted by copolymerization of various monomers), disruption of inter- or intramolecular hydrogen

Stimuli-Responsive Materials: From Molecules to Nature Mimicking Materials Design
By Marek W. Urban
© Marek W. Urban, 2016
Published by the Royal Society of Chemistry, www.rsc.org

bonding will lead to a release of previously entrapped drug molecules. Both the LCST and UCST are dictated by a chemical make-up of polymer backbone and Table 3.1 provides a list of homo- and copolymers that exhibit broad ranges of LCSTs, thus facilitating an unlimited number of possibilities for tuning to a specific LCST.

Notably, the mechanisms of responsiveness will vary, depending on the type of homo- or copolymer, as well as the environment. For example, thermally induced formation and acid-triggered dissociation of micelles prepared from poly(ethylene glycol)-*b*-poly(trans-*N*-(2-ethoxy-1,3-dioxan-5-yl)-acrylamide) is shown in Figure 3.1A,[3] whereas thermally induced phase transition and acid-triggered hydrolysis of PNDMM is illustrated in Figure 3.1B.[4] For comparison, aqueous solution behavior of temperature- and light-responsive schizophrenic block copolymers and PNDMA is illustrated in Figure 3.1C.[5] Similarly, doubly responsive PMOVE-*b*-PAEVA, responding to changes in pH and/or temperature is depicted in Figure 3.1D.[6]

3.2 Polymeric Solids

Unlike in solutions, significantly greater challenges exist when designing chemically or physically crosslinked gels and solid polymeric networks that require maintenance of their mechanical integrity. Restricted mobility within the network results from significant spatial limitations, thus imposing limits on obtaining stimuli responsiveness. The challenge in designing these stimuli-responsive polymeric systems is to create networks capable of inducing minute molecular yet orchestrated changes that lead to significant physicochemical responses upon external or internal stimuli. As was discussed in Chapter 1, spatial restrictions on mobility in the *x*-, *y*- and *z*-directions in solutions, at surfaces and interfaces, in gels, and in solids are different for each state. Consequently, energetic requirements for responses to temperature, mechanical stimuli, electromagnetic irradiation, or electrochemical stimuli, pH, ionic strength, or bioactive species will be different for each physical state.[7] Because solids represent the most restrictive mobility, the glass transition temperature (T_g) and free volume will have a significant importance on responsiveness. Intuitively, above the T_g the access of free volume will allow greater responsiveness, as opposed to a solid state below the T_g. Therefore, if the presence of localized 'voids' in the solid networks provides space for polymer chain rearrangements, a combination of stimuli-responsive segments with low T_g components is one of the essential ingredients to achieve

Table 3.1 Structures and LCSTs of selected homo- and copolymers with thermal responsive properties (adapted from ref. 13).

	LCST (C°)	Ref.
Poly(N-isopropylacrylamide) (PNIPAM)	32	1
Poly(N-n-propylacrylamide) (PNNPAM)	10	14, 15
Poly(N-cyclopropylacrylamide) (PNCPAM)	53	16
Poly(N,N-diethylacrylamide) (PDEAM)	33	17
Poly(N-(N′-isobutylcarbamido)propyl methacrylamide) (PiBuCPMA)	13	18
Poly(N-(N′-ethylcarbamido)propyl methacrylamide) (PiEtCPMA)	49.5–56.5	18

Table 3.1 (*Continued*)

		LCST (C°)	Ref.
Poly[N-(1-hydroxymethyl)propyl methacrylamide] (PHMPMA)		30 (L-iso.) 34 (DL-mix.)	19–22
Poly[N-(2,2-dimethyl-1,3-dioxolane)methyl] acrylamide (PDMDOMA)		23	23
Poly[[N-(2,2-dimethyl-1,3-dioxolane methyl]acrylamide-co-[N-(2,3-dihydroxyl-n-propyl)]acrylamide]		23–49	23
Poly(N-(2-methoxy-1,3-dioxan-5-yl) methacrylamide) (PNMM)		22	24
Poly[N-(2-ethoxy-1,3-dioxan-5-yl) methacrylamide] (PNEM)		52	24

Poly(N-(2,2-dimethyl-1,3-dioxan-5-yl) methacrylamide) (PNDMM)	15.3	4
Poly(N-(2,2-dimethyl-1,3-dioxan-5-yl) acrylamide) (PNDMA)	17.7	4
N-isopropylmethacrylamide copolymer and a methacrylamide monomer containing labile hydrazone linkages	13–44	25
Poly(trans-N-(2-ethoxy-1,3-dioxan-yl) acrylamide) (PtNEA)	13–17	3

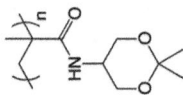

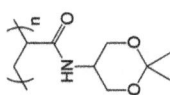

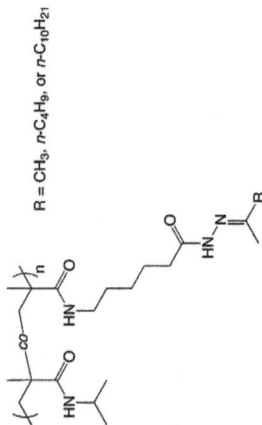

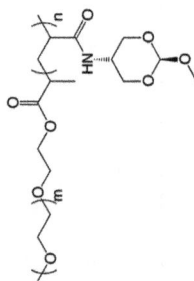

$R = CH_3$, $n\text{-}C_4H_9$, or $n\text{-}C_{10}H_{21}$

Table 3.1 (*Continued*)

		LCST (C°)	Ref.
Poly(*N*-vinylisobutyramide) (PNVIBA)		39	26, 27
Poly(*N*-vinylisobutyramide) (PNVBA)		32	28
Poly[*N*-acryloyl-*N*0-propylpiperazine) (PNANPP)		37	29
Poly(*N*-vinylcaprolactam) (PVCa)		32	30, 31
Poly(*N*-vinylpyrrolidone) (PVPy)		30	32
Poly[*N*-(2-methacryloyloxyethyl)pyrrolidone] (PNMP)		51.9	33

Polymer		
Poly(N-ethylpyrrolidine methacrylate) (PEPyM)	15	34
Poly(N-acryloylpyrrolidine) (PAPR)	51	35
Poly(dimethylaminoethyl methacrylate) (PDMAEMA)	14–50	36, 37
Poly(ethylene oxide) (PEO)		
Poly(ethylene glycol) (PEG)	85	38
Poly(propylene oxide) (PPO)		
Poly(propylene glycol) (PPG)	0–50	39
Poly(2-(2-methoxyethoxy)ethyl methacrylate) (PMEO$_2$MA)	26	40
Poly(2-[2-(2-methoxyethoxy)ethoxy]ethyl methacrylate) (PMEO$_3$MA)	52	40

Table 3.1 (*Continued*)

		LCST (C°)	Ref.
Poly(oligo(ethylene glycol)methacrylate (POEGMA)		60–90 (x = 4–9)	35, 41
Poly[[di(ethylene glycol) ethyl ether acrylate]-co-(oligoethylene glycol acrylate)] (P[DEGA-co-OEGA])		15–90	42
Poly[endo,exo-bicyclo[2.2.1] hept-5-ene-2,3-dicarboxylic acid, bis[2-[2-(2-ethoxyethoxy)ethoxy]ethyl] ester)		25	43
Poly(4-vinylbenzyl methoxytetrakis (oxyethylene)ether)		39	44
Oligo(ethylene oxide)-grafted polylactide		19 (x = 3) 27 (x = 4)	45

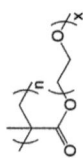

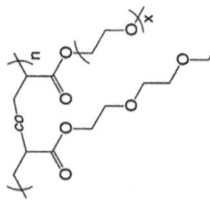

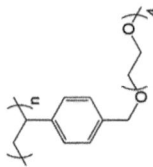

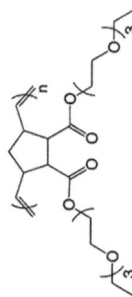

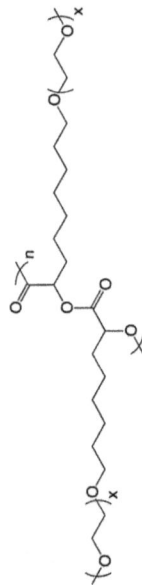

Polymer	Structure		
Poly[bis((ethoxyethoxy)ethoxy) phosphazene] (PBEEP)		38	46
Poly[bis(2,3-bis(2-methoxyethoxy) propanoxy)phosphazene] (PBBMEPP)		38	47
Poly(methyl vinyl ether) (PMVE)		35–36	48–50
Poly(2-(2-ethoxy)ethoxyethyl vinyl ether) (PEOEOVE)		41	51
Poly(2-methoxyethyl vinyl ether) (PMOVE)		70	52
Poly(ethoxyethyl glycidal ether)		29.6–40.4	
Poly(2-ethyl-2-oxazoline) (PEOx)		62–65	53, 54

Table 3.1 (*Continued*)

Name	Structure	LCST (C°)	Ref.
Poly(2-isopropyl-2-oxazoline) (PiPOx)		36	55, 56
Poly(2-n-propyl-2-oxazoline) (PnPOx)		36	13
Poly[[oligo(2-ethyl-2-oxazoline) methacrylate]-co-(methyl methacrylate)]	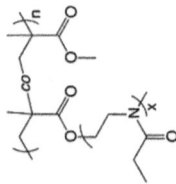	35–80	57
Poly(2-ethyl-2-oxazine) (PEtOZI)	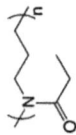	56 ($n > 100$)	58
Poly(2-n-propyl-2-oxazine) (PnPropOZI)		11–13 ($n = 15$–50)	59
P(Val-Pro-Gly-Val-Gly)	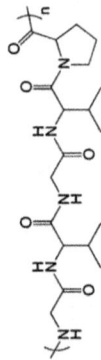	27	60

Val-Pro-Gly-Val-Gly and oligo(ethylene glycol) grafted polynorbornene — 16–30

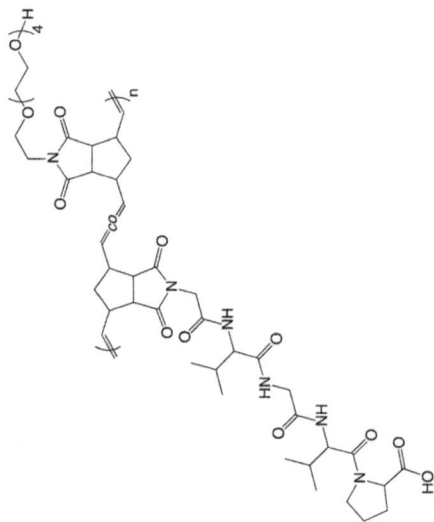

Val-Pro-Gly-Val-Gly derived polymethacrylate — 15–55 — 61

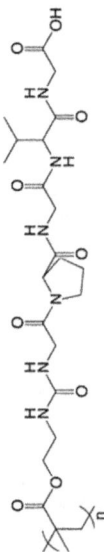

Poly(N-acryloyl-l-proline methyl ester) poly(A-Pro-OMe) — 15–20 — 62

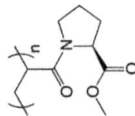

Derivatives of poly(N-substituted a/b-asparagine) — 5–100 — 63
R_1 = dodecylamine
R_2 = N,N-dimethyl-1,3-propanediamine

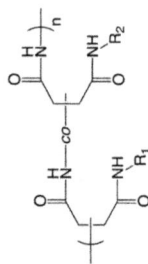

Table 3.1 (*Continued*)

	LCST (C°)	Ref.
Poly(*N*-acryloyl-L-valine-*N*0-methylamide) (PAVMA)	5.6–19.1	64
Ethyl and butyl modified polyglycine	20–60	65
PEG-ylated poly-L-glutamate	57 (*x* = 3) 30–50 (*x* = 2)	66
Poly(vinyl alcohol-*co*-vinyl acetal) (P(VOH-*co*-VAc))	17–41	67
Poly(glycidol-*co*-glycidol acetate)	4–100	68
Poly(2-hydroxypropylacrylate) (PHPA)	30–60	68

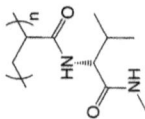

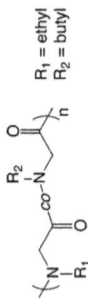

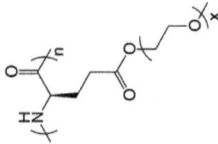

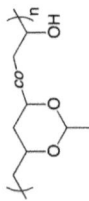

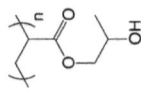

R₁ = ethyl
R₂ = butyl

Name	Structure		Ref
Butyl glycidyl ether modified Starch	R = H or -CH₂CH(OH)CH₂OCH₂CH₂CH₂CH₃	4.5–32.5	69
Poly[N-isopropylacrylamide)-b-poly[3-(N-(3-methacrylamidopropyl)-N,N-sulfonate] (PNIPAM-b-PSPP)dimethyl}ammoniopropane		8.6–19	70
Poly(N-acryloylglycinamide) (PNAGA)		22–23	71–73
Poly(N-acryloylasparaginamide) (PNAAAM)		4–28	74
Poly(acrylonitrile-co-acrylamide) (P(An-co-AM))		6–60	73
Poly(methacrylamide) (PMAAm)		57	73

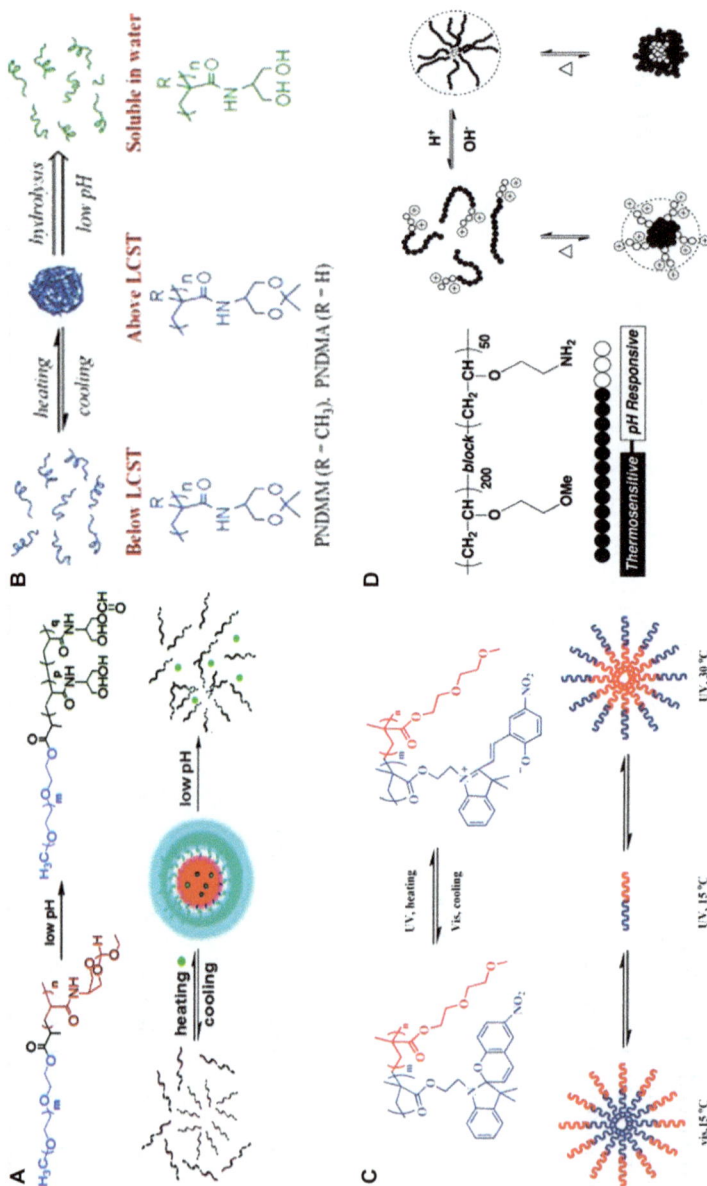

Figure 3.1 A. Formation and acid-triggered dissociation of micelles prepared from poly(ethylene glycol)-b-poly(trans-N-(2-ethoxy-1,3-dioxan-5-yl)acrylamide). B. Thermally induced phase transition and acid-triggered hydrolysis of PNDMM. C. Aqueous solution behavior of temperature- and light-responsive schizophrenic block copolymers and PNDMA. D. Doubly responsive PMOVE-b-PAEVA, responding to changes in pH and/or temperature changes.

A. Reprinted with permission from ref. 3. Copyright (2009) American Chemical Society. B. Reproduced from ref. 4 by permission of John Wiley and Sons. Copyright © 2008 Wiley Periodicals, Inc. C. and D. Reproduced from ref. 5 and 6 by permission of John Wiley and Sons. Copyright © 2010 Wiley Periodicals, Inc.

stable solid networks with stimuli-responsive characteristics. Following this concept, poly(N-(DL)-(1-hydroxymethyl)propylmethacrylamide-*co*-n-butyl acrylate) (p(DL-HMPMA-*co*-nBA))[8] and poly(2-(N,N-dimethylamino)-ethyl methacrylate-*co*-n-butyl acrylate) (p(DMAEMA-*co*-nBA))[9] colloidal particles were synthesized, which upon coalescence retain their stimuli-responsive properties. This is facilitated by the presence of low T_g nBA components. The temperature responsiveness is controlled in the solid state by DL-HMPMA or DMAEMA components, while the lower T_g nBA component provides sufficient free volume for copolymer chain rearrangements. One interesting macroscopic phenomenon resulting from these combinations of monomers is detectable 3D changes observed in p(DL-HMPMA-*co*-nBA) and p(DMAEMA-*co*-nBA) films. While p(DL-HMPMA-*co*-nBA) films shrink in the x–y plane and expand in the thickness (z) directions, p(DMAEMA-*co*-nBA) films shrink in all directions at elevated temperatures. Such dimensional differences in p(DL-HMPMA-*co*-nBA) and p(DMAEMA-*co*-nBA) copolymer films are attributed to preferential orientational changes of the side groups, as amide side groups in DL-HMPMA form preferential inter/intra-molecular interactions with itself or butyl ester pendant groups of nBA units, as compared to the ester side groups in DMAEMA. As a result, orientations of the side groups in p(DL-HMPMA-*co*-nBA) films change from preferentially parallel to perpendicular, which is responsible for the expansion in the z-direction. This orchestral macroscopic response to orientational changes of the copolymer side groups and structural features responsible for this behavior is illustrated in Figure 3.2. Computer modeling results shown in Figure 3.2A and B also confirmed the dimensional changes resulting from the buckling of the copolymer backbone and a collapse of the DL-HMPMA or DMAEMA components leading to macroscopic volume changes of the entire network.

Because the presence of free volume facilitates spatial conditions for polymer chain rearrangements, this is reflected in the T_g changes controlled by copolymer composition changes. If stimuli responsiveness results in conformational changes, these rearrangements should exhibit an endothermic character. This hypothesis has led to observations of compositional dependence of the new endothermic stimuli-responsive (T_{SR}) transition[10] for stimuli-responsive polymeric solids. As shown in Figure 3.3A, in addition to the lower T_g facilitating chain rearrangements, a series of DSC thermograms of p(DMAEMA/nBA) copolymer films recorded for different DMAEMA/nBA copolymer compositions show T_{SR}. As seen, similar to copolymer composition-dependent T_g transitions, the T_{SR} transitions also shift to higher temperatures as the amount of the stimuli-responsive DMAEMA

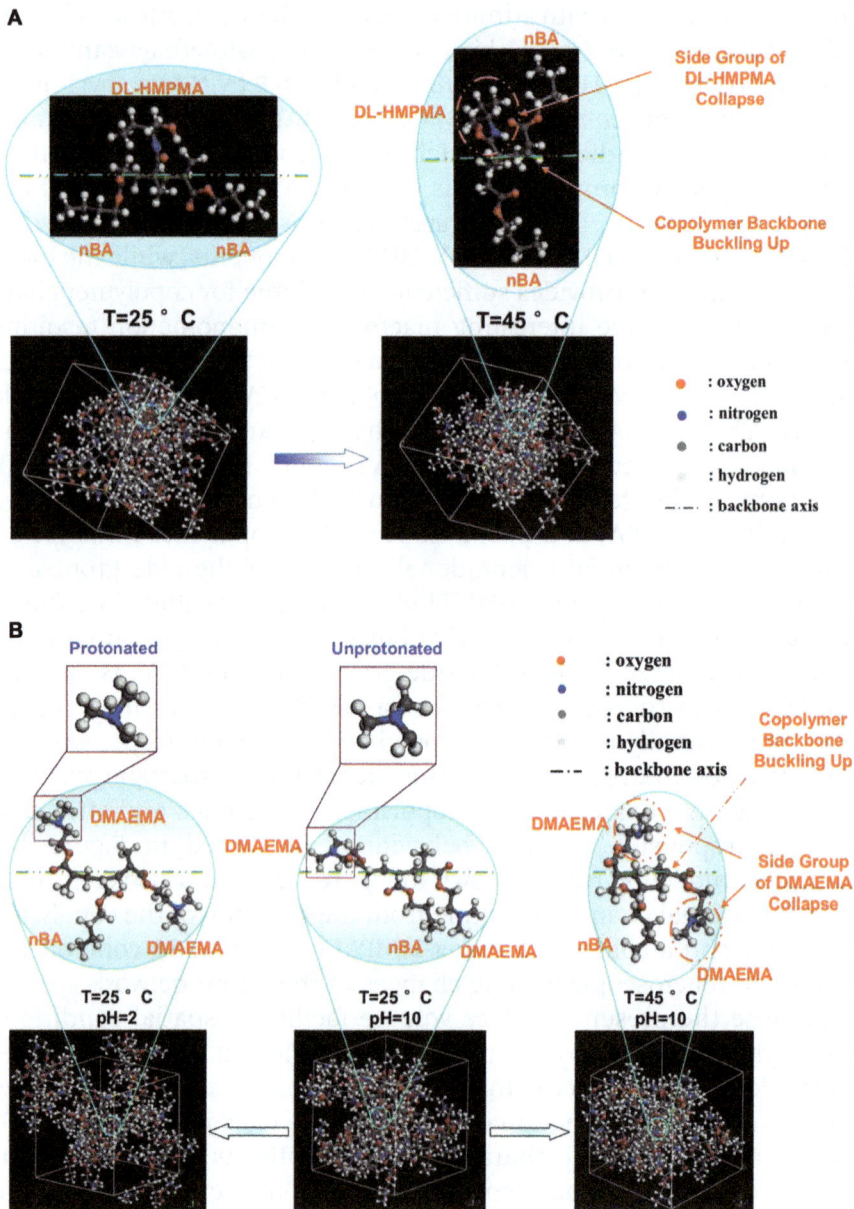

Figure 3.2   Computer simulation results of (A) p((DL)-HMPMA-*co-n*BA) and
(B) p(DMAEMA-*co-n*BA).
Reprinted with permission from ref. 8 and 9, Copyright (2008)
American Chemical Society.

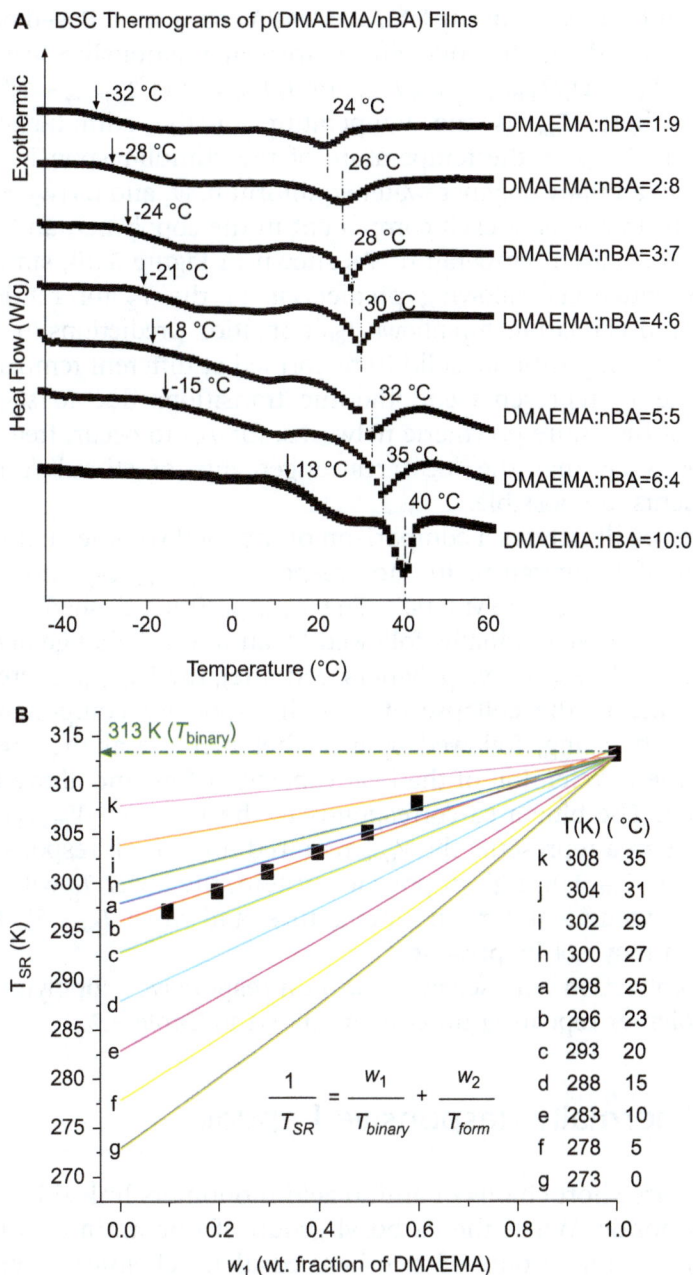

A DSC Thermograms of p(DMAEMA/nBA) Films

B

Figure 3.3   A. DSC thermograms of p(DMAEMA/nBA) copolymer films recorded for different DMAEMA/nBA weight ratios. B. Experimental T_{SR} values obtained from DSC measurements which allowed T_{SR} predictions for different T values plotted as a function of w_1: $1/T_{SR} = w_1/T_{binary} + w_2/T_{form}$.
Reprinted with permission from ref. 10. Copyright (2009) American Chemical Society.

component increases in the DMAEMA/nBA copolymer. Based on these experimental data, the following empirical relationship was established: $1/T_{SR} = w_1/T_{binary} + w_2/T_{form}$ or $1/T_{SR} = w_1(1/T_{binary} - 1/T_{form}) + 1/T_{form}$, where T_{SR} is the temperature of the stimuli-responsive transition, T_{binary} is the temperature of the stimuli-responsive homopolymer in a binary polymer–water equilibrium, w_1 and w_2 ($w_2 = 1 - w_1$) are weight fractions of each component in the copolymer, and T_{form} is the film-formation temperature. As shown in Figure 3.3B, similarly to the Fox equation[11] allowing predictions of the T_g for random copolymers, this relationship allows T_{SR} transition predictions in stimuli-responsive compositional solid films formed at different temperatures. While the T_g represents endothermic transitions due to segmental motion of the entire polymeric networks, for T_{SR} to occur, free volume must be such that the local rearrangements of stimuli-responsive components are possible.

Figure 3.4 illustrates a comparison of the total volume changes as a function of temperature for non-responsive ($V_{total\ NR}$) and stimuli-responsive ($V_{total\ SR}$) polymers.[12] The $V_{total\ NR}$ (blue dashed line) continues to increase gradually, followed by an upward change above the T_g. For stimuli-responsive polymers (red line), the $V_{total\ SR}$ decreases at the T_{SR} due to the collapse of stimuli-responsive components and backbone buckling, followed by a further increase at the same rate (slope) due to retention of the heat capacity before and above the T_{SR} transition. The blue area corresponds to the $V_{total\ NR} - V_{total\ SR}$, while the green area represents the V_{free} required for stimuli responsiveness to occur at T_{SR}. For a given polymer system, the lower T_g will provide more free volume at a given temperature, and when $T_{SR} < T_g$, the T_{SR} transition may not be present.

The collection of selected stimuli-responsive copolymer and homopolymer repeating units is tabulated in Table 3.2.

3.3   Thermally Responsive Peptides

Peptides are short chains of amino acid monomers linked by peptide (amide) bonds. When the carboxyl group of one amino acid reacts with the amino group of another, covalent chemical bonds are formed. When two amino acids are reacted, dipeptides are formed. When four peptide linkages are formed, a tetrapeptide as shown in Figure 3.5 can be synthesized.

The uniqueness of amino acid structures is that the individual amino acids define the properties of peptides. As shown in

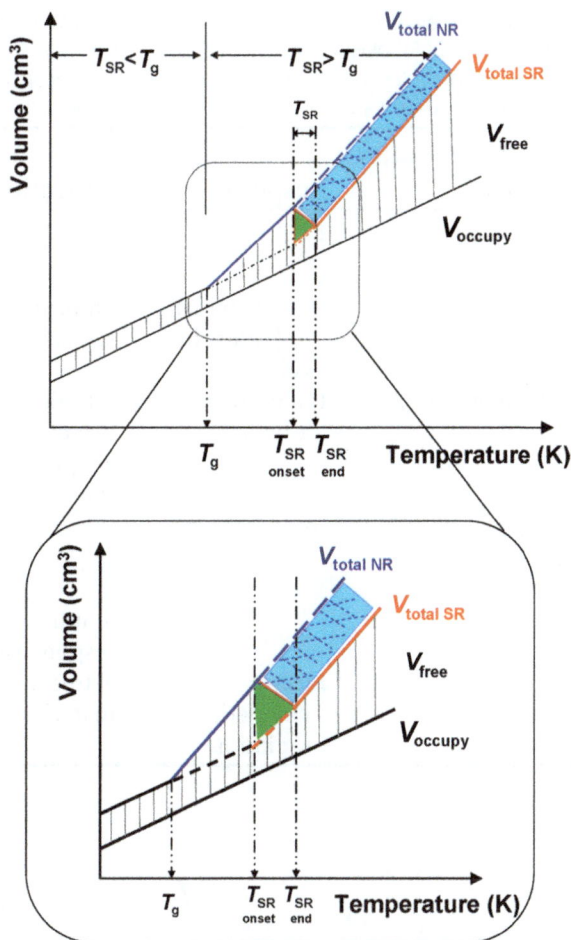

Figure 3.4   Schematic representation of the volume changes as a function of temperature for non-responsive and stimuli-responsive polymers. Reprinted with permission from ref. 2. Copyright (2010) American Chemical Society.

Figure 3.6, not only the sequence of amino acids, but also the R1–R4 substituents will determine their stimuli-responsive properties. Also, acidic and basic environments will be able to neutralize amino (green) of acid (purple) ends. As a consequence, the diversity of covalent and non-covalent bonds may facilitate hydrogen bonding, hydrophobic interactions, aromatic ring stocking, electrostatic interactions or disulfide bridges. Thus, a number of combinations of assemblies can be prepared that can be classified as electrostatic interactions, hydrophobic assemblies, hydrogen bonding, and

Table 3.2 Examples of temperature-sensitive peptide sequences based on helical and/or coiled-coil structures. The code used in the peptide chemistry is given in Appendix 3.1. Reproduced from ref. 75 with permission from The Royal Society of Chemistry.

Sequence	Temperature (°C)	Response	Application	Ref.
Ac-YGCVAALETK IAALETKKAALET IAALC–NH$_2$	60	Coiled-coil sol to β-hairpin gel	Conformational switches	80
IGKLKEEIDKLNR	0 at pH 5.75	Random coil to α-helix	Neurodegenera-tive cancer apoptosis research	81, 82
DLDDMEDENEQ LKQENKTLLKVV GKLTR		Ionic strength		
VKVKVKVKVD			Stimuli responsive	83, 84
PPTKVKVKVKV– NH$_2$	25	Electro-magnetic radiation	Sol-to-gel materials	

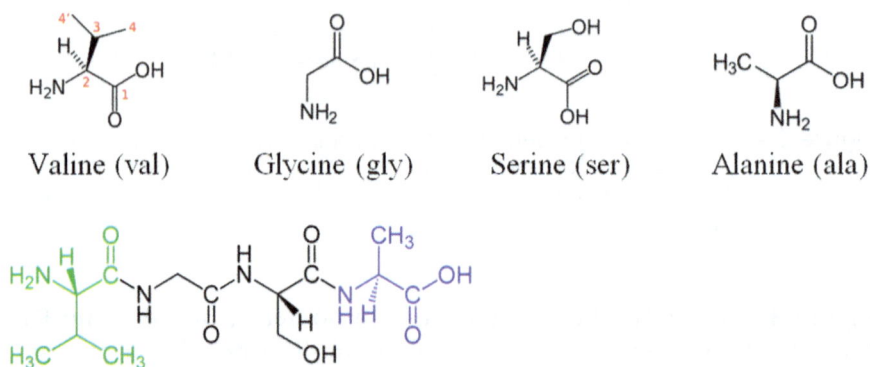

Valine (val)   Glycine (gly)   Serine (ser)   Alanine (ala)

Figure 3.5 Tetrapeptide prepared from val-gly-ser-ala amino acids. The green end is the amine end L-valine, whereas the blue end is the carboxyl end L-alanine.

higher structures. Figure 3.7 summarizes selected examples of these interactions along with the examples of amino acids that form these structures.

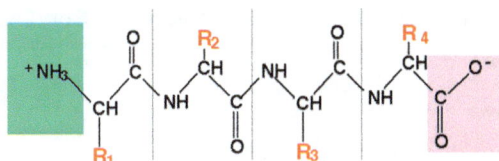

Figure 3.6 Terminal and 'internal' R1–R4 groups will determine assembly of peptide structures and their thermal responsiveness.

The following strategies are used in the design of a peptide macromonomer consisting of a primary sequence of either amphiphilic or secondary structural motifs, such as α-helix, β-sheet, β-turn (elastin-like sequence), in which responsive elements are rationally incorporated. One of the unique properties of peptides is that these entities are capable of forming secondary structures characterized by a single, spiral chain stabilized by hydrogen bonding. These are known as α-helices. Often, amino acids that exhibit three or four residues form these multidimensional structures shown in Figure 3.8. These entities, when rationally designed with stimuli-responsive amino acids, will form dynamic self-assemblies. Selected sequences are responsive to temperature and their sequences are listed in Table 3.2. The formation of α-helices is the key secondary structure stabilized by hydrogen bonding between a single string of spiral amino acids. These α-helix peptide chains appear to exhibit repeating segments every 3–4 residues and this spacing corresponds to on average 3.6 residues per each α-helical turn. To achieve stimuli responsiveness, amino acid stimuli-responsive components are incorporated within these structures.[38,76–79] For example, when cysteine residues were incorporated in hexadecapeptide to facilitate an azobenzene-based crosslinker peptide backbone, the *trans*-to-*cis* photoisomerization caused the decrease of α-helical content.[78]

Macroscopically detected responsiveness to external stimuli results from the presence of the switchable shapes of nano-fibers, spheres or nanotubes. Depicted in Figure 3.9 is an α-helix based quaternary structure of peptides called the coiled coil. These structures are characterized by two or more α-helices organized into a supercoil, each peptide a certain length motif repeating unit, but the interhelical interactions are captured by pair-wise interactions by key positions where hydrophobic residues form the hydrophobic core of a coiled coil. Various positions either side of the hydrophobic core can participate in electrostatic interhelical contacts, which may also alter core hydrophobicity. Changing the nature of these contacts by introducing responsive amino acids can alter the stability of the

Electrostatic Interactions (0.5-4 kJ/mol/bond)
pH and Ionic Strength Dependant
Acidic and Basic Amino Acids

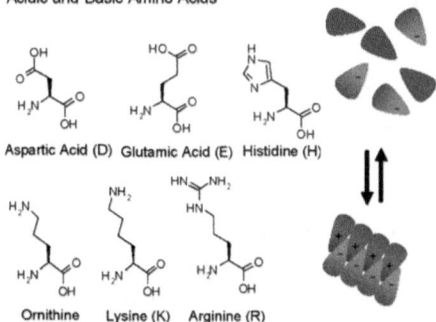

Aspartic Acid (D) Glutamic Acid (E) Histidine (H)

Ornithine   Lysine (K)   Arginine (R)

Also phosphorylation/dephosphorylation of serine or threonine residues.

Hydrophobic Effects (4-8 kJ/mol/residue)
Solvent Polarity, Ionic Strength and Temperature Dependant
Non-Polar Amino Acids

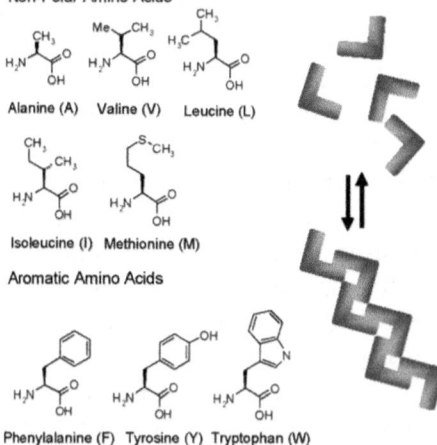

Alanine (A)   Valine (V)   Leucine (L)

Isoleucine (I) Methionine (M)

Aromatic Amino Acids

Phenylalanine (F) Tyrosine (Y) Tryptophan (W)

Hydrogen Bonding   (4-12 kJ/mol/bond)
Solvent/Temperature Dependant
Polar Amino Acids

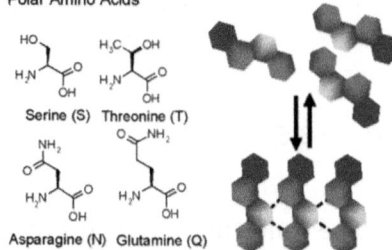

Serine (S) Threonine (T)

Asparagine (N) Glutamine (Q)

Strand Directing Residues
Can Induce Temperature Dependant Helicity
Conformationally Constrained/Flexible Amino Acids

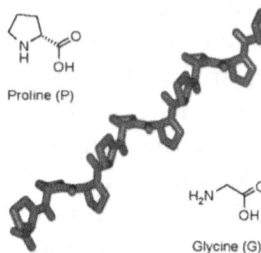

Proline (P)

Glycine (G)

Responsive to Electrochemical Stimuli
Disulfide Bridge Forming Amino Acids

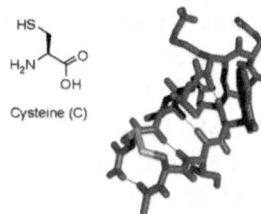

Cysteine (C)

Figure 3.7 Electrostatic, hydrogen bonding, hydrophobic interactions and strand directing residues formed by various amino acids. Reproduced from ref. 75 with permission from The Royal Society of Chemistry.

conformation and provide a mechanism for control of these dynamic materials. These interactions raise several fundamental questions regarding the role of competing polar and non-polar components, which are often attributed to ordering of oligoprolines.

The secondary structures are β-sheets which are formed by interactions of adjacent parallel and anti-parallel peptide strands of hydrogen bonded entities. One of the driving forces for understanding these interactions is their role in Alzheimer's and Parkinson's diseases, as well as during the formation of spider silk.[85–87] Their

A

B

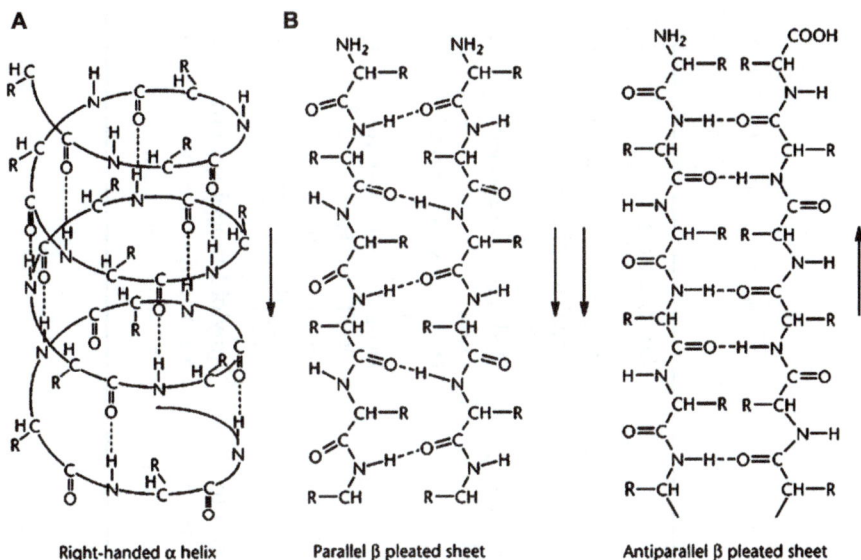

Right-handed α helix

Parallel β pleated sheet

Antiparallel β pleated sheet

Figure 3.8 Higher-order helix and sheets formed by peptides.

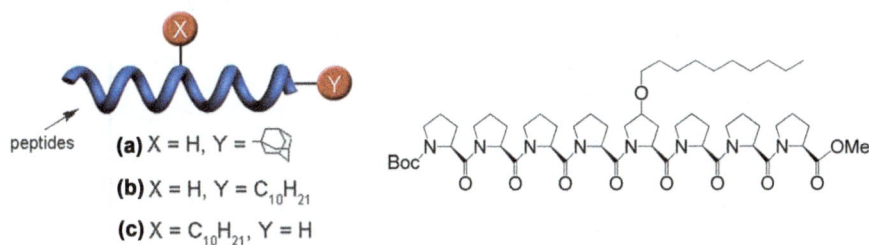

peptides

(a) X = H, Y = ⌬

(b) X = H, Y = $C_{10}H_{21}$

(c) X = $C_{10}H_{21}$, Y = H

Figure 3.9 Thermo-responsiveness in oligoprolines is achieved by covalent attachments of hydrophobic units.
Reproduced from ref. 97 with permission from The Royal Society of Chemistry.

formation occurs by H-bonding of adjacent parallel and anti-parallel peptide strands to form a weakly bonded sheet. Although the majority of these entities are static, only selected β-sheets are stimuli responsive (Table 3.3). As was shown,[88] the primary sequences may consist of alternating cationic and anionic amino acids, and hydrophobic entities, such as arginine-alanine-aspartic acid (RAD), phenyl-alanine-glutamic acid-lysine (FEK), or glutamic acid-alanine-lysine (EAK).

The other secondary peptide structures are β-hairpins, which develop when the amino acid sequence contains a pair of turns, such as proline, followed by glycine or threonine.[95] As a result of a small kink in the α-carbon backbone in the vicinity of flexible glycine or

Table 3.3 Examples of temperature-sensitive peptide sequences based on β-sheet structures. The code used in the peptide chemistry is included in Appendix 3.1. Reproduced from ref. 75 with permission from The Royal Society of Chemistry.

Sequence	Stimulus	Response	Application	Ref.
FEK16 FEFEFKFKFEFEFKFK	Ca^{2+} ions *via* temperature or light stimulated vesicle rupture	Soluble to aggregate	Drug delivery, wound healing, tissue engineering	89
P$_{11}$–4 Ac-QQRFEWEFEQQ-NH$_2$	pH 7.0	Nematic to isotropic sol to gel	Hydrogels, organogels, liquid	90
P$_{11}$–5 Ac-QQXFXWXFQQQ-NH$_2$ (where X denotes ornithine)	pH 7.5		Drug delivery, wound healing	
KFE12 FKFEFKFEFKFE	Ionic strength (1 mM NaCl) pH 5–10	Sol to gel	Drug delivery, wound healing, tissue engineering	91
KFQE12 FKFQFKFQFKFQ	Ionic strength (50 mM NaCl)	Sol to gel	Drug delivery, wound healing	92
$L_4K_8L_4$ (all L or all D)	pH > 9	Random coil to	Amyloid model system	93
DDDAAAVVV-NH$(CH_2)_{14}CH_3$ KKKVVVV$_e$D-NH$(CH_2)_{14}CH_3$ DDDAAAVVV$_e$D-NH$(CH_2)_{14}CH_3$	pH or Ca^{2+} ions	Random coil to β-sheet	Advanced medicine, cell culture	94

threonine, a reversible directionality of the α-carbon backbone occurs. Table 3.4 show selected examples of responsive sequences based on β-hairpin formation.

This intriguing class of peptides that is capable of adopting two distinctly different helical conformations is the oligoprolines. Depending on the polarity of the solvent environment, in aliphatic less polar solvents right-handed polyproline I (PPI) will be the dominant morphology, whereas in polar solvents like water, left-handed polyproline II (PPII) is present.[96] PPII has been found to be a common secondary structure in natural proteins, and plays important roles in many biological processes. However, oligo- or polyprolines themselves are not stimuli-responsive. Their thermo-responsiveness can be achieved by covalent attachments of hydrophobic units, X and Y, as shown in Figure 3.9.[97]

Another group of peptides is based on amphiphiles and consist of polar and non-polar aliphatic tails. These structural differences and diversified interactions make these peptides particularly suitable in many stimuli-responsive interactions. The responsiveness is primarily achieved by a hydrophobic effect, thus making these materials suitable for biorecognition and enzyme responsiveness. Again, Table 3.5 illustrates selective examples of responsive peptide sequences based on amphiphilic structures.

When aromatic motifs are incorporated into peptide structures, the presence of PI–PI interactions of aromatic rings in the presence of polar H-bonding and ionic interactions facilitate self-assembly in aqueous environments. The beauty of these self-assemblies is the possibility of the formation of nanotubes,[104] hollow spherical structures,[105] or amyloid-like fibers.[106] Table 3.6 summarizes selected examples of stimuli-responsive peptides containing aromatic groups.

3.4 Molecular Design, Entropy and Stimuli Responsiveness

An often articulated rule in the molecular design of stimuli-responsive polymeric systems is that in order to incorporate two or more molecular entities that respond to different stimuli, regardless of physical, chemical or biological stimuli, incorporated entities should be mutually compatible during synthesis and response processes. What does this mean? Nobody really knows. One thing is certain. In order for responses to occur, there must be a sufficient amount of local space to move-to and to move-from molecular entities. Thus, the entire issue of compatibility carries limited merit. As

Table 3.4 Examples of responsive peptide sequences based on β-hairpin structures. The code used in the peptide chemistry is included in Appendix 3.1. Reproduced from ref. 75 with permission from The Royal Society of Chemistry.

Sequence	Stimulus	Response	Application	Ref.
MAX7 VKVKVKVKVDPPTKVKXKVKV-NH$_2$ (X = cys or cys(a-carboxy-2-nitrobenzyl)	Light	Sol to gel	Tissue engineering regeneration drug delivery	84
MAX1 VKVKVKVKVDPPTKVKVKVKV-NH$_2$ MAX2 VKVKVKVKVDPPTKVKTKVKV-NH$_2$ MAX3 VKVKVKTKVDPPTKVKTKVKV-NH$_2$	Heat ($\sim$25 °C) Heat ($\sim$40 °C) Heat ($\sim$60 °C)	Sol to gel	Stimuli-responsive materials	83

Table 3.5 Examples of responsive peptide sequences based on amphiphilic structures. The code used in the peptide chemistry is included in Appendix 3.1. Reproduced from ref. 75 with permission from The Royal Society of Chemistry.

Sequence	Stimulus	Response	Application	Ref.
PA-1: $CH_3(CH_2)_{14}CO$-AAAGGGS-(PO_4)KGE	pH 9	Sol to gel	Regenerative medicine	98
12 peptides, *e.g.* PA-4: $CH_3(CH_2)_{14}$ COCCCCGGGS (PO_4)RGD	pH (various acidic); Di- and trivalent metal biomineralisation ions (20 mM)	Sol to nanofibres	Cell culture, regenerative medicine	99–102
KK(DOTA-e-K-e-K) LLCCCK-$(CO(CH_2)_{14}CH_3)$	pH >7	Sol to gel	Magnetic resonance imaging, metabolic studies	103
KK(DOTA-e-KGRGDS) LLLAAA-$(CO(CH_2)_{14}CH_3)$ PA-1: $CH_3(CH_2)_{14}CO$-AAAGGGS-(PO_4)KGE	pH <9; Ionic strength (.30 mM $M_{2\pm3}$)	Sol to gel	Regenerative medicine	98

Table 3.6 Examples of responsive peptide sequences containing aromatic structures. The code used in the peptide chemistry is included in Appendix 3.1. Reproduced from ref. 75 with permission from The Royal Society of Chemistry.

Sequence	Stimulus	Response	Application	Ref.
FF	Proteinase K	Disassembly of nanotubes	45 Nanowire templating, micro/nano-electronics	106
Fmoc-GG,AA,FG,GF,FF,FF/K$_a$, FF/GG$_a$ No gel formed for Fmoc-GF	pH 4–8	Sol to gel	46 3D cell culture	107
Fmoc-AA (L,L and D,D),GG,GA,GS No gel formation for Fmoc-GT	pH 3–5	Sol to gel	47 Sensing	108
Naphthalene-FFGEY	i. Phosphatase ii. Kinase	i. Sol to gel ii. Gel to sol	51 *in vivo* gelation, regenerative medicine	109

a matter of fact, on the contrary, the best self-healing polymeric systems consist of hydrophobic and hydrophilic components (see Chapter 11) combined into one network. These are entropic contributions that are essential in rearrangements of macromolecular segments and play a critical role in externally or internally stimulated responses.

In view of the thermally responsive monomers and oligomers let us consider the role of entropy in molecular rearrangements. After all, understanding entropic contributions to responsiveness is essential for designing these systems. Thus, starting with the definition, the second law of thermodynamics resulted from the simple empirical observation that heat does not spontaneously flow from a cooler to a hotter material. Clausius was the first to show that there is a quantity called entropy that always increases during irreversible changes in isolated systems. It was Boltzmann who established the relationship between entropy and the atomistic description of nature, captured by stating that entropy, S, is equal to the logarithm of the number of states accessible to the system, W, multiplied by a constant, k_B ($S = k_B \ln W$). Therefore, the second law of thermodynamics implies that irreversible changes in closed systems are only possible if the number of states upon completion of the process is much greater than the initial state. In a classical sense, nothing changes and the number of states implies the excessive volume in phase space. In other words, spontaneous assemblies do not happen from a large volume resulting in smaller volumes. As we recall, the energy input will be necessary for stimuli-responsive transitions to take place, and the volume changes will be quite drastic. Interestingly, it was estimated that if the entropy of a system increases by only 1 J K^{-1} (which corresponds to >0.1% for a liter of water), the number of accessible states increases by the astonishing number of $10^{10\ 22.5}$.[110] Needless to say, minute entropy changes have a staggering effect on changes within the system. Although experimentally it is quite challenging to show, simulations have shown that entropic effects alone can account for a wide variety of lyotropic liquid crystalline phases.[111] An interesting relationship can be conceptually described using translational and orientational entropy of a rod with length L and diameter D. If one envisages a macromolecular block as a rod, the volume will be minimal when macromolecular blocks are parallel, and maximal when they are perpendicular to each other. Thus, for parallel macromolecular chains translational entropy will be maximal, but orientational entropy will be minimal, and above a certain threshold of density, translational entropy will overcome orientational entropy.

Thermally responsive polymers represent an excellent example of the potential for programmed assemblies of structures containing molecular instructions that guide reproducible formation of complex dimensional features. There are two essential components of information that are critical: location of connection and the means of connectivity. Regardless whether this is top-down or bottom-up assembly, location will identify structural components that exhibit affinity of conformational changes or rebonding, whereas connectivity will be determined by the type of attachment. For example, supramolecular interactions, or their combination, may serve this purpose. The connectivity of individual components into a specific sequence which use non-specific interactions may include, but are not limited to, shape matching resulting in mechanical interlocking and other long-range interactions.

Appendix 3.1   Codes Utilized in Peptide Chemistry

Peptide	3-Letter Code	1-Letter Code
Alanine	Ala	A
Arginine	Arg	R
Asparagine	Asn	N
Aspartic acid	Asp	D
Cysteine	Cys	C
Glutamine	Gln	Q
Glutamic acid	Glu	E
Glycine	Gly	G
Histidine	His	H
Isoleucine	Ile	I
Leucine	Leu	L
Lysine	Lys	K
Methionine	Met	M
Phenylalanine	Phe	F
Proline	Pro	P
Serine	Ser	S
Threonine	Thr	T
Tryptophan	Trp	W
Tyrosine	Tyr	Y
Valine	Val	V

References

1. J. S. Scarpa, D. D. Mueller and I. M. Klotz, Slow hydrogen-deuterium exchange in a non-. alpha.-helical polyamide, *J. Am. Chem. Soc.*, 1967, **89**(24), 6024–6030.
2. S. Fujishige, K. Kubota and I. Ando, Phase transition of aqueous solutions of poly (N-isopropylacrylamide) and poly (N-isopropylmethacrylamide), *J. Phys. Chem.*, 1989, **93**(8), 3311–3313.
3. X. Huang, F. Du, J. Cheng, Y. Dong, D. Liang and S. Ji *et al.*, Acid-sensitive polymeric micelles based on thermoresponsive block copolymers with pendent cyclic orthoester groups, *Macromolecules*, 2009, **42**(3), 783–790.
4. X. N. Huang, F. S. Du, B. Zhang, J. Y. Zhao and Z. C. Li, Acid-labile, thermoresponsive (meth) acrylamide polymers with pendant cyclic acetal moieties, *J. Polym. Sci., Part A: Polym. Chem.*, 2008, **46**(13), 4332–4343.
5. Q. Jin, G. Liu and J. Ji, Micelles and reverse micelles with a photo and thermo double-responsive block copolymer, *J. Polym. Sci., Part A: Polym. Chem.*, 2010, **48**(13), 2855–2861.
6. Y. Oda, S. Kanaoka and S. Aoshima, Synthesis of dual pH/temperature-responsive polymers with amino groups by living cationic polymerization, *J. Polym. Sci., Part A: Polym. Chem.*, 2010, **48**(5), 1207–1213.
7. M. Urban, Stratification, stimuli-responsiveness, self-healing, and signaling in polymer networks, *Prog. Polym. Sci.*, 2009, **34**, 679–687.
8. F. Liu and M. Urban, 3D directional temperature responsive (N-(DL)-(1-hydroxymethyl) propylmethacrylamide-co-n-butyl acrylate) colloids and their coalescence, *Macromolecules*, 2008, **41**, 352–360.
9. F. Liu and M. Urban, Dual temperature and pH responsiveness of poly(2-(N,N-dimethylamino)ethyl methacrylate-co-n-butyl acrylate) colloidal dispersions and their films, *Macromolecules*, 2008, **41**, 6531–6539.
10. F. Liu and M. Urban, New thermal transitions in stimuli-responsive polymers, *Macromolecules*, 2009, **42**, 2161–2167.
11. T. G. FoX, Influence of diluent and of copolymer composition on the glass temperature of a polymer system, *Bull. Am. Phys. Soc.*, 1956, **2**, 123.
12. F. Liu, W. L. Jarrett and M. W. Urban, Glass (T g) and Stimuli-Responsive (T SR) Transitions in Random Copolymers, *Macromolecules*, 2010, **43**(12), 5330–5337.

13. D. Roy, W. L. Brooks and B. S. Sumerlin, New directions in thermoresponsive polymers, *Chem. Soc. Rev.*, 2013, **42**(17), 7214–7243.

14. D. Ito and K. Kubota, Solution properties and thermal behavior of poly (N-n-propylacrylamide) in water, *Macromolecules*, 1997, **30**(25), 7828–7834.

15. D. Ito and K. Kubota, Thermal response of poly (Nn-propylacrylamide), *Polym. J.*, 1999, **31**(3), 254–257.

16. Y. Maeda, T. Nakamura and I. Ikeda, Changes in the hydration states of poly (N-alkylacrylamide) s during their phase transitions in water observed by FTIR spectroscopy, *Macromolecules*, 2001, **34**(5), 1391–1399.

17. I. Idziak, D. Avoce, D. Lessard, D. Gravel and X. Zhu, Thermosensitivity of aqueous solutions of poly (N, N-diethylacrylamide), *Macromolecules*, 1999, **32**(4), 1260–1263.

18. Y. Akiyama, Y. Shinohara, Y. Hasegawa, A. Kikuchi and T. Okano, Preparation of novel acrylamide-based thermoresponsive polymer analogues and their application as thermoresponsive chromatographic matrices, *J. Polym. Sci., Part A: Polym. Chem.*, 2008, **46**(16), 5471–5482.

19. T. Aoki, M. Muramatsu, T. Torii, K. Sanui and N. Ogata, Thermosensitive phase transition of an optically active polymer in aqueous milieu, *Macromolecules*, 2001, **34**(10), 3118–3119.

20. T. Aoki, M. Muramatsu, A. Nishina, K. Sanui and N. Ogata, Thermosensitivity of Optically Active Hydrogels Constructed with N-(L)-(1-hydroxymethyl) propylmethacrylamide, *Macromol. Biosci.*, 2004, **4**(10), 943–949.

21. Y. Seto, T. Aoki and S. Kunugi, Temperature-and pressure-responsive properties of l-and dl-forms of poly (N-(1-hydroxymethyl) propylmethacrylamide) in aqueous solutions, *Colloid Polym. Sci.*, 2005, **283**(10), 1137–1142.

22. F. Liu and M. W. Urban, Recent advances and challenges in designing stimuli-responsive polymers, *Prog. Polym. Sci.*, 2010, **35**(1), 3–23.

23. Y. Zou, D. E. Brooks and J. N. Kizhakkedathu, A novel functional polymer with tunable LCST, *Macromolecules*, 2008, **41**(14), 5393–5405.

24. X. Huang, F. Du, R. Ju and Z. Li, Novel Acid-Labile, Thermoresponsive Poly(methacrylamide)s with Pendent Ortho Ester Moieties, *Macromol. Rapid Commun.*, 2007, **28**(5), 597–603.

25. M. Hruby, J. Kucka, O. Lebeda, H. Mackova, M. Babic and C. Konak *et al.*, New bioerodable thermoresponsive polymers for

possible radiotherapeutic applications, *J. Controlled Release*, 2007, **119**(1), 25–33.

26. K. Suwa, Y. Wada, Y. Kikunaga, K. Morishita, A. Kishida and M. Akashi, Synthesis and functionalities of poly (N-vinylalkylamide). IV. Synthesis and free radical polymerization of N-vinylisobutyramide and thermosensitive properties of the polymer, *J. Polym. Sci., Part A: Polym. Chem.*, 1997, **35**(9), 1763–1768.

27. K. Suwa, K. Morishita, A. Kishida and M. Akashi, Synthesis and functionalities of poly (N-vinylalkylamide). V. Control of a lower critical solution temperature of poly (N-vinylalkylamide), *J. Polym. Sci., Part A: Polym. Chem.*, 1997, **35**(15), 3087–3094.

28. S. Kunugi, T. Tada, Y. Yamazaki, K. Yamamoto and M. Akashi, Thermodynamic studies on coil-globule transitions of poly (N-vinylisobutyramide-co-vinylamine) in aqueous solutions, *Langmuir*, 2000, **16**(4), 2042–2044.

29. L. Gan, Y. Gan and G. R. Deen, Poly (N-acryloyl-N'-propylpiperazine): A new stimuli-responsive polymer, *Macromolecules*, 2000, **33**(21), 7893–7897.

30. A. C. Lau and C. Wu, Thermally sensitive and biocompatible poly (N-vinylcaprolactam): synthesis and characterization of high molar mass linear chains, *Macromolecules*, 1999, **32**(3), 581–584.

31. A. Tager, A. Safronov, E. Berezyuk and I. Y. Galaev, Lower critical solution temperature and hydrophobic hydration in aqueous polymer solutions, *Colloid Polym. Sci.*, 1994, **272**(10), 1234–1239.

32. Y. Maeda, T. Nakamura and I. Ikeda, Hydration and phase behavior of poly (N-vinylcaprolactam) and poly (N-vinylpyrrolidone) in water, *Macromolecules*, 2002, **35**(1), 217–222.

33. J. Deng, Y. Shi, W. Jiang, Y. Peng, L. Lu and Y. Cai, Facile synthesis and thermoresponsive behaviors of a well-defined pyrrolidone based hydrophilic polymer, *Macromolecules*, 2008, **41**(9), 3007–3014.

34. N. González, C. Elvira and J. S. Román, Novel dual-stimuli-responsive polymers derived from ethylpyrrolidine, *Macromolecules*, 2005, **38**(22), 9298–9303.

35. M. Mertoglu, S. Garnier, A. Laschewsky, K. Skrabania and J. Storsberg, Stimuli responsive amphiphilic block copolymers for aqueous media synthesised via reversible addition fragmentation chain transfer polymerisation (RAFT), *Polymer*, 2005, **46**(18), 7726–7740.

36. M. Okubo, H. Ahmad and T. Suzuki, Synthesis of temperature-sensitive micron-sized monodispersed composite polymer

particles and its application as a carrier for biomolecules, *Colloid Polym. Sci.*, 1998, **276**(6), 470–475.

37. F. A. Plamper, M. Ruppel, A. Schmalz, O. Borisov, M. Ballauff and A. H. Müller, Tuning the thermoresponsive properties of weak polyelectrolytes: aqueous solutions of star-shaped and linear poly (N, N-dimethylaminoethyl methacrylate), *Macromolecules*, 2007, **40**(23), 8361–8366.

38. C. de las Heras Alarcón, S. Pennadam and C. Alexander, Stimuli responsive polymers for biomedical applications, *Chem. Soc. Rev.*, 2005, **34**(3), 276–285.

39. A. C. Colin, S. M. Cancho, R. G. Rubio and A. Compostizo, Equation of state of aqueous polymer systems: poly (propylene glycol)+ water, *Phys. Chem. Chem. Phys.*, 1999, **1**(2), 319–322.

40. S. Han, M. Hagiwara and T. Ishizone, Synthesis of thermally sensitive water-soluble polymethacrylates by living anionic polymerizations of oligo (ethylene glycol) methyl ether methacrylates, *Macromolecules*, 2003, **36**(22), 8312–8319.

41. J. F. Lutz, Polymerization of oligo (ethylene glycol)(meth) acrylates: toward new generations of smart biocompatible materials, *J. Polym. Sci., Part A: Polym. Chem.*, 2008, **46**(11), 3459–3470.

42. C. Boyer, M. R. Whittaker, M. Luzon and T. P. Davis, Design and synthesis of dual thermoresponsive and antifouling hybrid polymer/gold nanoparticles, *Macromolecules*, 2009, **42**(18), 6917–6926.

43. B. Zhao, D. Li, F. Hua and D. R. Green, Synthesis of thermosensitive water-soluble polystyrenics with pendant methoxyoligo (ethylene glycol) groups by nitroxide-mediated radical polymerization, *Macromolecules*, 2005, **38**(23), 9509–9517.

44. T. Bauer and C. Slugovc, The thermo responsive behavior of glycol functionalized ring opening metathesis polymers, *J. Polym. Sci., Part A: Polym. Chem.*, 2010, **48**(10), 2098–2108.

45. X. Jiang, E. B. Vogel, M. R. Smith and G. L. Baker, "Clickable" polyglycolides: tunable synthons for thermoresponsive, degradable polymers, *Macromolecules*, 2008, **41**(6), 1937–1944.

46. H. R. Allcock, S. R. Pucher, M. L. Turner and R. J. Fitzpatrick, Poly (organophosphazenes) with poly (alkyl ether) side groups: a study of their water solubility and the swelling characteristics of their hydrogels, *Macromolecules*, 1992, **25**(21), 5573–5577.

47. H. R. Allcock and G. K. Dudley, Lower critical solubility temperature study of alkyl ether based polyphosphazenes, *Macromolecules*, 1996, **29**(4), 1313–1319.

48. B. Verdonck, E. J. Goethals and F. E. Du Prez, Block Copolymers of Methyl Vinyl Ether and Isobutyl Vinyl Ether With Thermo-Adjustable Amphiphilic Properties, *Macromol. Chem. Phys.*, 2003, **204**(17), 2090–2098.

49. W. Z. Zhang, X. D. Chen, W.-a. Luo, J. Yang, M. Q. Zhang and F. M. Zhu, . Study of phase separation of poly (vinyl methyl ether) aqueous solutions with Rayleigh scattering technique, *Macromolecules*, 2009, **42**(5), 1720–1725.

50. Y. Maeda, IR spectroscopic study on the hydration and the phase transition of poly (vinyl methyl ether) in water, *Langmuir*, 2001, **17**(5), 1737–1742.

51. S. Aoshima, H. Oda and E. Kobayashi, Synthesis of thermally-induced phase separating polymer with well-defined polymer structure by living cationic polymerization. I. Synthesis of poly (vinyl ether) s with oxyethylene units in the pendant and its phase separation behavior in aqueous solution, *J. Polym. Sci., Part A: Polym. Chem.*, 1992, **30**(11), 2407–2413.

52. S. Inoue, H. Kakikawa, N. Nakadan, S.-i. Imabayashi and M. Watanabe, Thermal response of poly (ethoxyethyl glycidyl ether) grafted on gold surfaces probed on the basis of temperature-dependent water wettability, *Langmuir*, 2009, **25**(5), 2837–2841.

53. D. Christova, R. Velichkova, W. Loos, E. J. Goethals and F. D. Prez, New thermo-responsive polymer materials based on poly (2-ethyl-2-oxazoline) segments, *Polymer*, 2003, **44**(8), 2255–2261.

54. T. T. Chiu and B. P. Thill, Water Soluble Polymers, in *Advances in Chemistry Series*, ed. J. E. Glass, ACS, Washington, DC, 1986, p. 213.

55. H. Uyama and S. Kobayashi, A novel thermo-sensitive polymer. Poly (2-iso-propyl-2-oxazoline), *Chem. Lett.*, 1992, **9**, 1643–1646.

56. C. Diab, Y. Akiyama, K. Kataoka and F. M. Winnik, Microcalorimetric study of the temperature-induced phase separation in aqueous solutions of poly (2-isopropyl-2-oxazolines), *Macromolecules*, 2004, **37**(7), 2556–2562.

57. C. Weber, C. R. Becer, R. Hoogenboom and U. S. Schubert, Lower critical solution temperature behavior of comb and graft shaped poly [oligo (2-ethyl-2-oxazoline) methacrylate] s, *Macromolecules*, 2009, **42**(8), 2965–2971.

58. M. M. Bloksma, R. M. Paulus, H. P. van Kuringen, F. van der Woerdt, H. M. Lambermont-Thijs and U. S. Schubert *et al.*, Thermoresponsive Poly (2-oxazine) s, *Macromol. Rapid Commun.*, 2012, **33**(1), 92–96.

59. D. E. Meyer, B. Shin, G. Kong, M. Dewhirst and A. Chilkoti, Drug targeting using thermally responsive polymers and local hyperthermia, *J. Controlled Release*, 2001, **74**(1), 213–224.

60. K. Bebis, M. W. Jones, D. M. Haddleton and M. I. Gibson, Thermoresponsive behaviour of poly [(oligo (ethyleneglycol methacrylate)] s and their protein conjugates: importance of concentration and solvent system, *Polym. Chem.*, 2011, **2**(4), 975–982.

61. F. Fernandez-Trillo, A. Dureault, J. P. Bayley, J. C. van Hest, J. C. Thies and T. Michon *et al.*, Elastin-based side-chain polymers: improved synthesis via RAFT and stimulus responsive behavior, *Macromolecules*, 2007, **40**(17), 6094–6099.

62. H. Mori, H. Iwaya, A. Nagai and T. Endo, Controlled synthesis of thermoresponsive polymers derived from L-proline via RAFT polymerization, *Chem. Commun.*, 2005, **38**, 4872–4874.

63. ed. E. Watanabe, N. Tomoshige and H. Uyama, New biodegradable and thermoresponsive polymers based on amphiphilic poly (asparagine) derivatives, *Macromolecular Symposia*, 2007, Wiley Online Library.

64. Z. Liu, J. Hu, J. Sun, G. He, Y. Li and G. Zhang, Preparation of thermoresponsive polymers bearing amino acid diamide derivatives via RAFT polymerization, *J. Polym. Sci., Part A: Polym. Chem.*, 2010, **48**(16), 3573–3586.

65. S. H. Lahasky, X. Hu and D. Zhang, Thermoresponsive poly (α-peptoid) s: tuning the cloud point temperatures by composition and architecture, *ACS Macro Lett.*, 2012, **1**(5), 580–584.

66. C. Chen, Z. Wang and Z. Li, Thermoresponsive polypeptides from pegylated poly-l-glutamates, *Biomacromolecules*, 2011, **12**(8), 2859–2863.

67. D. Christova, S. Ivanova and G. Ivanova, Water-soluble temperature-responsive poly (vinyl alcohol-co-vinyl acetal) s, *Polym. Bull.*, 2003, **50**(5-6), 367–372.

68. A. Dworak, B. Trzebicka, A. Utrata and W. Walach, Hydrophobically modified polyglycidol–the control of lower critical solution temperature, *Polym. Bull.*, 2003, **50**(1-2), 47–54.

69. B. Ju, D. Yan and S. Zhang, Micelles self-assembled from thermoresponsive 2-hydroxy-3-butoxypropyl starches for drug delivery, *Carbohydr. Polym.*, 2012, **87**(2), 1404–1409.

70. M. Arotçaréna, B. Heise, S. Ishaya and A. Laschewsky, Switching the inside and the outside of aggregates of water-soluble block copolymers with double thermoresponsivity, *J. Am. Chem. Soc.*, 2002, **124**(14), 3787–3793.

71. S. Glatzel, N. Badi, M. Päch, A. Laschewsky and J.-F. Lutz, Well-defined synthetic polymers with a protein-like gelation behavior in water, *Chem. Commun.*, 2010, **46**(25), 4517–4519.
72. J. Seuring, F. M. Bayer, K. Huber and S. Agarwal, Upper critical solution temperature of poly (N-acryloyl glycinamide) in water: a concealed property, *Macromolecules*, 2011, **45**(1), 374–384.
73. J. Seuring and S. Agarwal, First example of a universal and cost-effective approach: polymers with tunable upper critical solution temperature in water and electrolyte solution, *Macromolecules*, 2012, **45**(9), 3910–3918.
74. S. Glatzel, A. Laschewsky and J.-F. Lutz, Well-defined uncharged polymers with a sharp UCST in water and in physiological milieu, *Macromolecules*, 2010, **44**(2), 413–415.
75. R. J. Mart, R. D. Osborne, M. M. Stevens and R. V. Ulijn, Peptide-based stimuli-responsive biomaterials, *Soft Matter*, 2006, **2**(10), 822–835.
76. B. Jeong and A. Gutowska, Lessons from nature: stimuli-responsive polymers and their biomedical applications, *Trends Biotechnol.*, 2002, **20**(7), 305–311.
77. B. D. Ratner and S. J. Bryant, Biomaterials: where we have been and where we are going, *Annu. Rev. Biomed. Eng.*, 2004, **6**, 41–75.
78. C. Boulègue, M. Löweneck, C. Renner and L. Moroder, Redox potential of azobenzene as an amino acid residue in peptides, *ChemBioChem*, 2007, **8**(6), 591–594.
79. A. Khan, C. Kaiser and S. Hecht, Prototype of a photoswitchable foldamer, *Angew. Chem., Int. Ed.*, 2006, **45**(12), 1878–1881.
80. B. Ciani, E. G. Hutchinson, R. B. Sessions and D. N. Woolfson, . A designed system for assessing how sequence affects α to β conformational transitions in proteins, *J. Biol. Chem.*, 2002, **277**(12), 10150–10155.
81. K. Dutta, A. Alexandrov, H. Huang and S. M. Pascal, pH-induced folding of an apoptotic coiled coil, *Protein Sci.*, 2001, **10**(12), 2531–2540.
82. S. C. West, Molecular views of recombination proteins and their control, *Nat. Rev. Mol. Cell Biol.*, 2003, **4**(6), 435–445.
83. D. J. Pochan, J. P. Schneider, J. Kretsinger, B. Ozbas, K. Rajagopal and L. Haines, Thermally reversible hydrogels via intramolecular folding and consequent self-assembly of a de novo designed peptide, *J. Am. Chem. Soc.*, 2003, **125**(39), 11802–11803.
84. L. A. Haines, K. Rajagopal, B. Ozbas, D. A. Salick, D. J. Pochan and J. P. Schneider, Light-activated hydrogel formation via the

triggered folding and self-assembly of a designed peptide, *J. Am. Chem. Soc.*, 2005, **127**(48), 17025–17029.

85. J. R. Silveira, G. J. Raymond, A. G. Hughson, R. E. Race, V. L. Sim and S. F. Hayes *et al.*, The most infectious prion protein particles, *Nature*, 2005, **437**(7056), 257–261.

86. J. P. Taylor, J. Hardy and K. H. Fischbeck, Toxic proteins in neurodegenerative disease, *Science*, 2002, **296**(5575), 1991–1995.

87. J. M. Kenney, D. Knight, M. J. Wise and F. Vollrath, Amyloidogenic nature of spider silk, *Eur. J. Biochem.*, 2002, **269**(16), 4159–4163.

88. S. Zhang, Emerging biological materials through molecular self-assembly, *Biotechnol. Adv.*, 2002, **20**(5), 321–339.

89. J. H. Collier, B.-H. Hu, J. W. Ruberti, J. Zhang, P. Shum and D. H. Thompson *et al.*, Thermally and photochemically triggered self-assembly of peptide hydrogels, *J. Am. Chem. Soc.*, 2001, **123**(38), 9463–9464.

90. A. Aggeli, M. Bell, L. M. Carrick, C. W. Fishwick, R. Harding and P. J. Mawer *et al.*, pH as a trigger of peptide β-sheet self-assembly and reversible switching between nematic and isotropic phases, *J. Am. Chem. Soc.*, 2003, **125**(32), 9619–9628.

91. M. R. Caplan, P. N. Moore, S. Zhang, R. D. Kamm and D. A. Lauffenburger, Self-assembly of a β-sheet protein governed by relief of electrostatic repulsion relative to van der Waals attraction, *Biomacromolecules*, 2000, **1**(4), 627–631.

92. M. R. Caplan, E. M. Schwartzfarb, S. Zhang, R. D. Kamm and D. A. Lauffenburger, Control of self-assembling oligopeptide matrix formation through systematic variation of amino acid sequence, *Biomaterials*, 2002, **23**(1), 219–227.

93. T. Koga, M. Matsuoka and N. Higashi, Structural control of self-assembled nanofibers by artificial β-sheet peptides composed of D-or L-isomer, *J. Am. Chem. Soc.*, 2005, **127**(50), 17596–17597.

94. H. A. Behanna, J. J. Donners, A. C. Gordon and S. I. Stupp, Coassembly of amphiphiles with opposite peptide polarities into nanofibers, *J. Am. Chem. Soc.*, 2005, **127**(4), 1193–1200.

95. C. Wilmot and J. Thornton, Analysis and prediction of the different types of β-turn in proteins, *J. Mol. Biol.*, 1988, **203**(1), 221–232.

96. P. M. Cowan and S. McGavin, Structure of poly-L-proline, *Nature*, 1955, **176**, 501–503.

97. F. Chen, X. Zhang, W. Li, K. Liu, Y. Guo and J. Yan *et al.*, Thermoresponsive oligoprolines, *Soft Matter*, 2012, **8**(18), 4869–4872.

98. J. C. Stendahl, M. S. Rao, M. O. Guler and S. I. Stupp, Intermolecular forces in the self-assembly of peptide amphiphile nanofibers, *Adv. Funct. Mater.*, 2006, **16**(4), 499–508.

99. J. D. Hartgerink, E. Beniash and S. I. Stupp, Peptide-amphiphile nanofibers: a versatile scaffold for the preparation of self-assembling materials, *Proc. Natl. Acad. Sci.*, 2002, **99**(8), 5133–5138.

100. E. Beniash, J. D. Hartgerink, H. Storrie, J. C. Stendahl and S. I. Stupp, Self-assembling peptide amphiphile nanofiber matrices for cell entrapment, *Acta Biomater.*, 2005, **1**(4), 387–397.

101. J. D. Hartgerink, E. Beniash and S. I. Stupp, Self-assembly and mineralization of peptide-amphiphile nanofibers, *Science*, 2001, **294**(5547), 1684–1688.

102. E. D. Sone and S. I. Stupp, Semiconductor-encapsulated peptide-amphiphile nanofibers, *J. Am. Chem. Soc.*, 2004, **126**(40), 12756–12757.

103. S. R. Bull, M. O. Guler, R. E. Bras, T. J. Meade and S. I. Stupp, Self-assembled peptide amphiphile nanofibers conjugated to MRI contrast agents, *Nano Lett.*, 2005, **5**(1), 1–4.

104. M. Reches and E. Gazit, Casting metal nanowires within discrete self-assembled peptide nanotubes, *Science*, 2003, **300**(5619), 625–627.

105. M. Reches and E. Gazit, Formation of closed-cage nanostructures by self-assembly of aromatic dipeptides, *Nano Lett.*, 2004, **4**(4), 581–585.

106. M. Reches and E. Gazit, Self-assembly of peptide nanotubes and amyloid-like structures by charged-termini-capped diphenylalanine peptide analogues, *Isr. J. Chem.*, 2005, **45**(3), 363–371.

107. V. Jayawarna, M. Ali, T. A. Jowitt, A. F. Miller, A. Saiani and J. E. Gough *et al.*, Nanostructured Hydrogels for Three-Dimensional Cell Culture Through Self-Assembly of Fluorenylmethoxycarbonyl–Dipeptides, *Adv. Mater.*, 2006, **18**(5), 611–614.

108. Z. Yang, H. Gu, D. Fu, P. Gao, J. K. Lam and B. Xu, Enzymatic formation of supramolecular hydrogels, *Adv. Mater.*, 2004, **16**(16), 1440–1444.

109. Z. Yang, G. Liang, L. Wang and B. Xu, Using a kinase/phosphatase switch to regulate a supramolecular hydrogel and forming the supramolecular hydrogel in vivo, *J. Am. Chem. Soc.*, 2006, **128**(9), 3038–3043.

110. D. Frenkel, Order through entropy, *Nat. Mater.*, 2015, **14**(1), 9–12.

111. D. Frenkel and B. Smit, *Understanding Molecular Simulation: From Algorithms to Applications*, Academic press, 2001.

4 Stimuli-responsive Surfaces and Interfaces

4.1 Surface/Interface *vs.* Bulk

A surface is typically defined as a physical zone within which intrinsic materials properties change in the direction normal to the surface. Most agree that molecular conformations at the surface are considerably different from those in the bulk. If one defines an interface (or a zone) as the boundary between two phases, the question is what is the size and composition of this zone. Several scenarios are depicted in Figure 4.1 for liquid–vapor (A), solid–vapor (B), solid–liquid (C), liquid–liquid (D) and solid–solid (E) interfaces, and polymer surfaces (F).

This raises a couple of practical and scientific questions. For example, why is ice slippery so that we can skate on it, as shown in Figure 4.2; or is the glass transition temperature of a surface the same as that of a bulk? In 1859 Faraday postulated that a thin layer of water covers the ice surface, even though the temperature is well below freezing. For almost a century it was believed a skater exerts pressure of the order of several hundred atmospheres, which was sufficient to reduce the melting temperature by a few degrees. Right? Not necessarily. Neither pressure melting nor frictional heating can explain why a person standing on a surface of ice still can slip. Surprisingly, experiments conducted in the 1950s revealed that when two spheres of ice were placed in contact with each other, adhesion occurred at $-4\,^\circ$C, leading to the conclusion that the surface roughness was

Stimuli-Responsive Materials: From Molecules to Nature Mimicking Materials Design
By Marek W. Urban
© Marek W. Urban, 2016
Published by the Royal Society of Chemistry, www.rsc.org

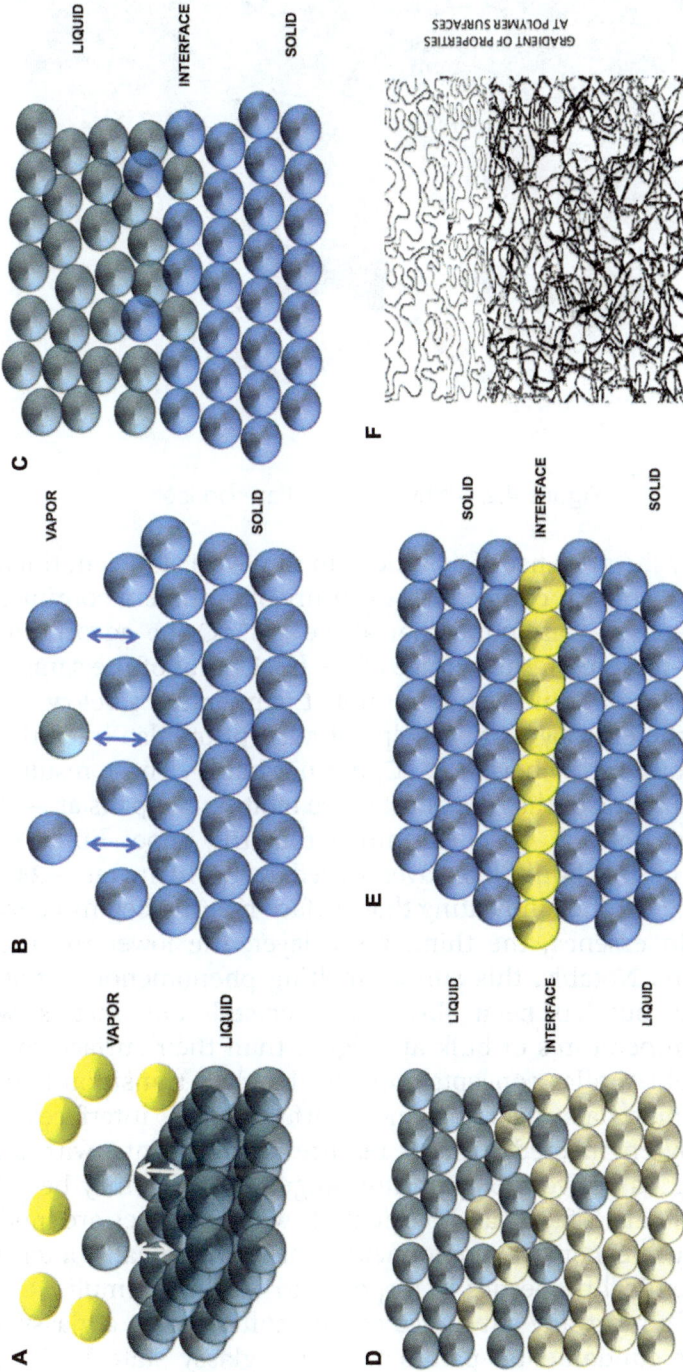

Figure 4.1 Examples of (A) liquid–vapor, (B) solid–vapor, (C) solid–liquid, (D) liquid–liquid and (E) solid/solid interfaces and (F) polymer surfaces.

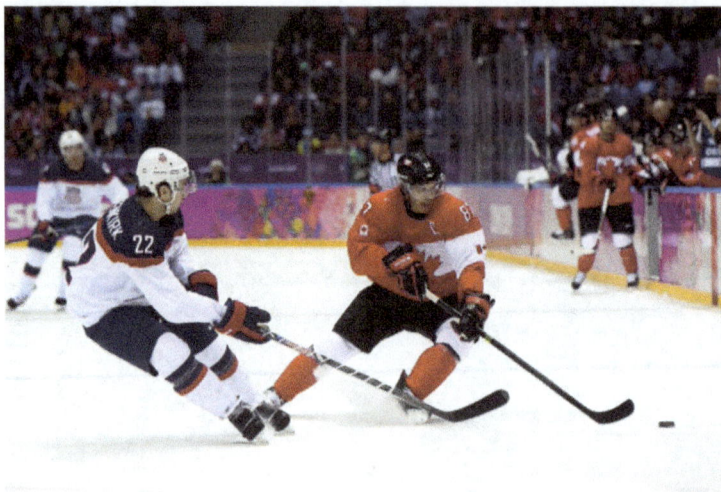

Figure 4.2   Why can we skate on ice?

removed by the presence of a liquid thin film with sufficient thickness to provide a smooth contact.[1,2] As century-old problem continued, in the late 1960s it was found that above $-35\,°C$ the adsorption isotherm of *n*-hexane on the surface of ice tracks that of the same vapor on the surface of liquid water, but not at temperatures below $-35\,°C$. However, the entropy and enthalpy of adsorption also track the pattern of liquid water above $-35\,°C$, but not below. These results were interpreted to mean that the onset of ice melting happens at $-35\,°C$,[3] and in the 1990s it was determined that the upper limits of the thickness of the liquid-like layer varied from $12\,nm$ at $-24\,°C$ to $70\,nm$ at $0.7\,°C$, thus indicating that surface melting begins at around $-33\,°C$.[4] In essence, the thinner the layer, the lower the melting temperature. Notably, this surface-melting phenomenon is not limited to ice, but has been observed in crystals and metals, where melting temperatures of bulk are higher than their surface counterpart.[5] Would similar concepts apply to the glass transition temperatures of materials near polymer surfaces and interfaces? Will molecularly engineered surfaces require no lubricants, with a minimum wear in a given temperature range? If so, we may be able to skate on metal surfaces, and particularly when surfaces are modified with stimuli-responsive components that dynamically alter their physicochemical properties in response to external stimuli.

Because almost any substance can be converted to a glassy state, one of the fundamental parameters of a glassy state is the glass transition temperature (T_g). For temperatures above the T_g, the material is a viscous liquid. As the liquid is cooled down below the T_g, an amorphous solid phase is being formed. While the role of the T_g of bulk materials is somewhat understood, the role of the T_g near the surface is not, although one can argue that if the T_g is a thermo-dynamic[6] or kinetic[7] phenomenon, intuitively the T_g near the surface will be lower than that of the bulk. This is because polymeric chain ends have more freedom to move and with the free volume being high, a lower energy input is required to achieve the same free volume changes as in the bulk. Just like ice surfaces, the thickness of the surface layers that exhibit lower T_g values will be affected by thermodynamics and kinetics. To unify the role of both processes the concept of cooperative rearrangements was introduced,[8] which demonstrated that kinetics coupled with thermodynamics leads to the system dynamics described by the WLF[†] equation[9] near the T_g. Interestingly enough, it has been estimated that the size of such co-operatively rearranging regions is in the order of 10 Å for a number of glass-forming materials,[10] thus suggesting the length scales required for experiments.

In essence, not only the localization of chain ends near the surface, which typically arises from conformational entropy, but the extent of chain end separation with respect to a surface is critical. It was suggested that the difference between the surface tension of an infinitely long polymer chain and that of the end groups is a measure of surface segregation, but at these length scales experimental evidence was not sufficient.[11] In an effort to appreciate the role of surface T_g on surface and interfacial stimuli responsiveness, let us estimate what factors will impact the reduction of T_g at the boundaries of the property change in the direction normal to the surface/interface.

[†]The Williams–Landel–Ferry (WLF) equation is an empirical formula typically shown in this format:

$$\log(a_T) = \frac{-C_1(T - T_r)}{C_2 + (T - T_r)}$$

where T is the temperature, T_r is a reference temperature chosen to construct the compliance master curve and C_1, C_2 are empirical constants adjusted to fit the values of the superposition parameter a_T. This relationship allows the estimation of polymer properties (for example, viscosity) for temperatures other than those for which the material was tested. Using this approach the master curve can be constructed and applied to other temperatures. It should be kept in mind that when the constants are obtained with data at temperatures above the glass transition temperature (T_g), the WLF equation is applicable to temperatures at or above T_g. The constants are positive and represent Arrhenius behavior. However, at temperatures below the T_g negative values of C_1, C_2 are obtained, which are not applicable above T_g and do not represent Arrhenius behavior.

Assuming the δa^2 values with respect to thermal energy (kT) (where a is the length of the polymer segment N), the surface end concentration should be equal to that of the bulk, $\Phi_b = 2/N$ (subscripts b and s refer to the bulk and surface, respectively). When $\delta a^2/kT = 1$, all chain ends with a distance R from the surface will be localized on the surface. Under these assumptions, the concentration of the chain ends (Φ_s) in the near surface layer of thickness d will be:

$$\Phi_s = \Phi_b(R/d) = 2/N^{-1/2}(a/d) \tag{4.1}$$

This relationship is valued to accommodate stretched chains near the surface, where surface attractive forces are moderate. When considering a near surface zone of thickness d, one can define a new *effective degree* of polymerization, N_e, which is the equivalent molecular weight of polymer chains in the bulk. This relationship for the end concentration at the surface is expressed by:

$$N_s = 2/N_s \quad \text{or} \quad N_s = N^{1/2}(d/a) \tag{4.2}$$

This relationship has significant implications on the relationship on the T_g near polymer surfaces. For amorphous polymers, the T_g in the bulk can be empirically defined by:

$$T_g = T_{g,\infty} - C/N \tag{4.3}$$

where $T_{g,\infty}$ is the T_g of polymer chains with infinitely high molecular weight and C is a material constant. Following the free volume concept, the $1/N$ dependence arises from the contribution of chain ends to the total free volume of a polymer. One can use the analogy of the molecular weight dependence on the free volume: the free volume increases with decreasing molecular weight. If the T_g *vs.* C/N dependence can be applied at the materials boundary conditions, the chain end segregation will reduce the T_g near the surface zone. Thus, the reduction of the T_g near the surface can be estimated by substituting N in eqn (4.3) by N_s in eqn (4.2), giving:

$$\Delta T_{gs} = T_{g,\infty} - T_{gs} = C/N^{-1/2}(a/d) \tag{4.4}$$

For example, for polystyrene, the reduction of T_g in polymers can be estimated using the following assumptions: molecular weight is equal to 1×10^5, so $N = 960$. Assuming a segregated surface layer thickness $d \sim a$, the surface equivalent molecular weight will be ~ 6500, or $N_s = 31$. Because the T_g of bulk polystyrene is ~ 373 K and C is 1.1×10^3, the T_{gs} is estimated to be 337 K. Several theoretical[12] and experimental[13] studies considered conformational changes of polymer chains in the melt state near impenetrable surface boundaries and indeed showed an enhancement of the chain end densities within two polymer segment lengths. Furthermore, thickness-dependent T_{gs} have been measured by ellipsometry,[14,15] X-ray reflectivity,[16] dielectric spectroscopy,[17] Brillouin scattering,[18] fluorescence,[19] and other methods,[20–23] but there are still many unanswered questions. This interest is primarily driven by the lack of understanding the T_g gradients within structural nano-confinements of materials. Understanding of the distribution of T_{gs} will have a tremendous impact of the development of new stimuli-responsive composites or other materials with dynamic properties. Going back to the ice analogy, a 25 nm thick layer at a free surface of a 1000 nm thick film will show reduction of T_g by 5–6 K compared to the PMMA bulk T_g (393 K). However, a 25 nm thick layer at the substrate of a 1000 nm thick film will exhibit an increase of the T_g by 12 K.[24] These data indicate that surface and interfacial processes may be designed to control relaxation processes in polymer nanocomposites and other materials.

4.2 Surface/Interfacial Switchability

It should be made clear that the previous discussion on surfaces and interfaces has relatively little to do with stimuli responsiveness. However, it is important to know what interfacial and surface properties are critical in designing stimuli-responsive attributes. These attributes that differentiate surfaces and interfaces from the bulk may be constructive (or sometimes destructive) in designing stimuli responsiveness. Unless 'things happen' at surfaces or interfaces as a result of chemistry or physics in the surroundings, surfaces and interfaces exhibit their own static properties and are not considered to be stimuli responsive. For example, surface modifications may alter surface wettability. However, if the same surface is able to dynamically change wettability in response to, for example, humidity or temperature changes, it becomes stimuli responsive. Because switchability of surface and interfacial regions plays a critical role in many applications, let us consider some examples.

Switchability of surfaces offers many advantages, which are used in numerous applications. One example is a surface that can repel water when it rains, and absorb moisture under dry conditions. The stimuli that can effectively influence surface and interfacial properties are electrical and chemical potentials, electromagnetic radiation, pH, temperature, or mechanical forces. As we have learned earlier, the key component of the stimuli responsiveness is the ability of macro-molecular components to change their conformations. Again, the strategy is to design surface macromolecular entities with a dual character. For example, if one end is hydrophilic and the other hydrophobic, and both have the ability of conformational changes as a result of stimuli, such surface properties as wettability, optical and electric properties, or opacity may be altered. These are the most studied phenomena, but there are others – for example, one can en-visage the color of a house roof changing from dark in winter to ab-sorb sun radiation to much brighter during the summer months to reflect the sunlight. Most of us experience deicing of airplanes during snowfall. If an aircraft's fuselage was covered with a coating that can cause instantaneous snow melt and water evaporation, it would make flying faster and safer. In essence, the ability of surface stimuli re-sponsiveness offers numerous advantages in many areas, and more applications are being explored in tissue engineering, microfluidics, biosensors, molecular electronics, colorimetric displays, and many others. The specific chemical entities responsible for responses to electrical potential, light, pH and temperature are listed in Table 4.1.

A self-assembled monolayer (SAM) is a single layer of amphiphilic molecules that spontaneously organize themselves on a substrate due to affinity between the amphiphile and the substrate. One of the ways of creating responsive surfaces is the use of low- and high-density

Table 4.1 Selected categories of molecular entities responsive to external stimuli.

	Electrical potential	Light	pH	Temperature
Self-assembled monolayers (SAMs)	X		X	
Azobenzene		X		
Spiropyran		X		
Polyelectrolyte brushes			X	X
Rotaxane	X	X	X	
Catenane	X			
DNA monolayers	X		X	
Peptide monolayers	X		X	X

SAMs. A precursor monolayer is often hydrolyzed to form a hydrophilic head monolayer.[25] Upon application of an electrical potential to the gold substrate, hydrophobic alkyl chains are exposed to the environment, thus making the switchable hydrophobic–hydrophilic surface.[26] A typical monolayer–substrate combination is a gold–alkanethiolate system in which gold serves as an electrically conducting electrode. However, such high-density self-assembled monolayers (HDSAMs) may have limited spatial freedom due to tight packing. To achieve greater stimuli responsiveness, low-density self-assembled mono-layers (LDSAMs) can be generated, allowing conformational mobility. SAMs are typically created *via* physisorption onto a substrate. Most common approaches involve immersing in a SAM solution for a sufficient period of time to form a homogeneous layer. As a result hydrophilic and hydrophobic responses can be developed.[27–29] To generate LDSAMs modulated by an electric potential, bulky end group SAMs can be attached, but densely packed end groups are usually eliminated in order to form loosely packed chains capable of reversible conformational changes upon application of an electric current.[30–35] It should also be realized that aside from chain surface packing, the molecular weight of oligomeric chains sticking out of the surface, as well as their 'stiffness', will have a significant impact on the dynamics of switchability. Thus, the choice of a backbone will be critical, especially so that these properties may and will lead to impedance or resistance to current flow changes.[34] This may be advantageous in many technological advances ranging from sensing devices for diagnostics, to cell adhesion/motility studies and tissue engineering.

There are other methods of creating LDSAMs, including the assembly of complexes comprised of alkanothiols and cyclodextrin. These are so-called inclusion complexes (ICs) which rely on the insertion of alkanothiols into cyclodextrin moieties. This approach can be used to generate reversible changes in surface wettability and protein adsorption stimulated by electric currents[32] and extended to low-density acid- and amino-terminated SAMs used to coat micro-fluidic devices to achieve reversible and selective protein adsorption. When a negative potential is applied, acid-terminated LDSAMs are able to adsorb the positively charged protein and release it again upon the application of a positive potential. One can envisage protein separation applications in proteomics and sensing technologies.

The incorporation of electroactive hydroquinone moieties into SAMs can also be used to generate electroactive monolayers. Because hydroquinone moieties can be easily reduced by electrochemical oxidation to generate benzoquinone, one can envisage using this approach for releasing ligands attached to a substrate surface.[36] This may include the detachment of peptides that function as a cell adhesive from the substrate surface, triggering the release of cells adhered to the peptides.[37] When benzoquinone units resulting from this reduction are then utilized by use of a Diels–Alder reaction, selective immobilization of diene-functionalized peptides can be achieved.

4.3   Photoresponsive Surfaces

A highly attractive feature of on-and-off switching polymeric surfaces is the use of light. As a result, many applications such as hydrophilicity/hydrophobicity, structural arrangements and shape changes have been developed. The most common photoresponsive molecules are azobenzene and spiropyran molecules.

4.3.1   Azobenzene Sensors

Upon the application of a specific wavelength of light, azobenzene units switch their structure from the *trans* (straight) to the *cis* (bent) isomerization. This conformational change corresponds to different dipole moments, with the *cis*-isomer having a higher dipole moment. In an effort to mimic motion similar to living organisms, azobenzene as shown in Figure 4.3A was copolymerized into poly(2-(*N*,*N*-dimethyl-amino)ethylmethacrylate-*co*-*n*-butyl acrylate-*co*-*N*,*N*-(dimethylamino)-azobenzene acrylamide) (p(DMAEMA/*n*BA/DMAAZOAm)) copolymer.[38] Figure 4.3B illustrates shape changes of the p(DMAEMA/*n*BA/DMAAZOAm) copolymer as a function of UV exposure, where unexposed specimen (a) is exposed to left (b) and right (c) 302 nm UV radiation, followed by left (d) and right (e) exposure to 403 nm. As seen, an individual filament changes its shape, depending upon the wavelength of light and direction of the illumination. When films are exposed to the same radiation, which is illustrated by SEM inserts b′–e′, the surface will wrinkle due to shrinkage of the exposed side.

Alterations in molecular spatial arrangements can be transformed into macroscopic changes in surface wettability.[39,40] For example, azobenzene-treated surfaces have demonstrated the ability to control liquid droplet and liquid crystal (LC) alignment.[41,42] Azobenzene monolayers on quartz substrates form SAMs that upon exposure to ultraviolet light resulted in a reversible conformational change that

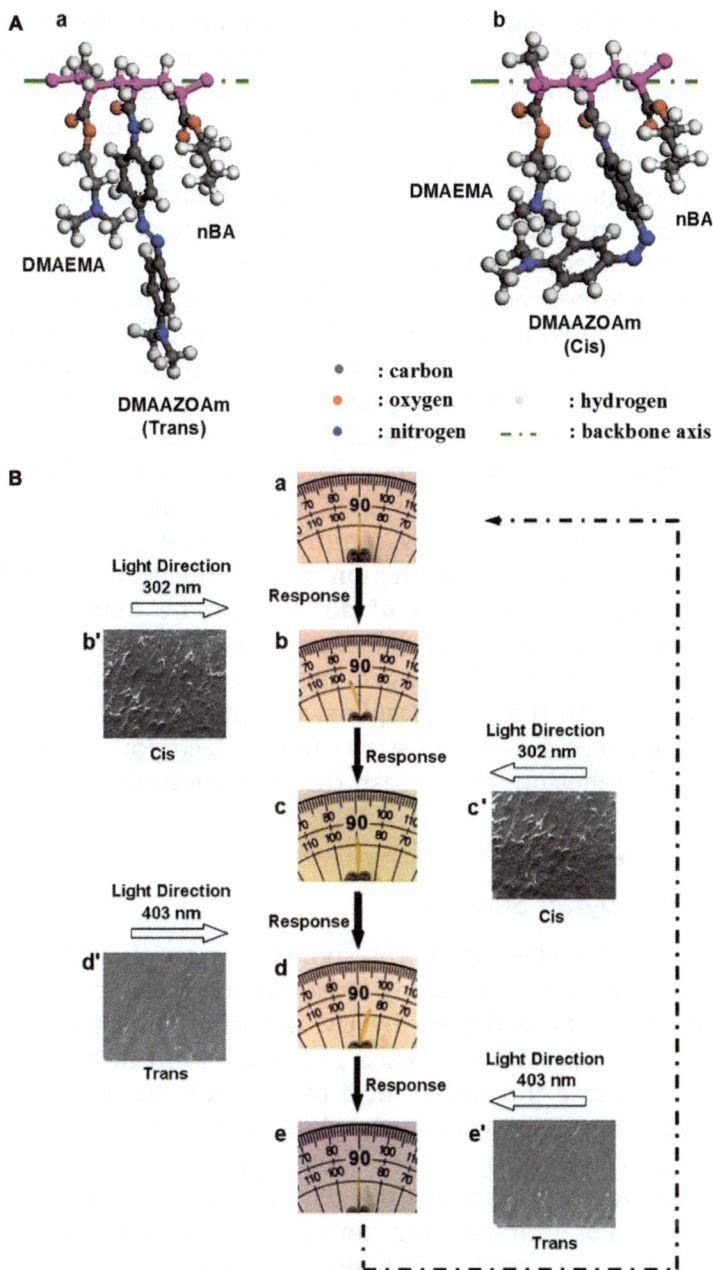

Figure 4.3 (A) *Trans* (a) and *cis* (b) conformations of p(DMAEMA/nBA/ DMAAZOAm). (B) Shape changes of p(DMAEMA/nBA/ DMAAZOAm) copolymer as a function of UV exposure: (a) unexposed specimen; (b, c) left and right side exposures to 302 nm; (d, e) left and right side exposure to 403 nm. Inserts b'–e' illustrate SEM images of exposed sides.

Ref. 38 reproduced by permission of John Wiley and Sons, Copyright © 2010 WILEY-VCH Verlag GmbH & Co. KGaA, Weinheim.

led to the parallel alignment of LCs in contact with the monolayer. Changes in the molecular shape of azobenzene molecules also result in mechanical actuation. Ji *et al.* demonstrated the efficacy of azobenzenes in actuation by coating a micro-cantilever with a thiol-terminated azobenzene derivative.[43] Application of UV light caused the downward deflection of the coated cantilever because alterations in azobenzene arrangement from the *trans-* to *cis*-conformation resulted in the repulsion of molecules within the monolayer.

The photoresponsiveness of azobenzenes has been exploited in biology *via* the incorporation of azobenzene units into peptides. For example, synthesized peptides with an azobenzene backbone were utilized with the corresponding RNA binding aptamer as an *in vitro* selection tool for RNA–ligand pairing.[44] The concept here was to use peptides by reversible photoresponsive binding to target RNA that was turned 'on' in the presence of visible light and turned 'off' upon application of UV light. Similarly, arginine-glycine-aspartic acid (RGD) peptides were used in conjunction with azobenzene to influence cellular adhesion.[45] Addition of an azobenzene derivative to the peptide sequence enables spatial control because the *trans*-conformation of azobenzene is 3 Å longer than the *cis*-conformation. Thus, upon coating with the photoresponsive peptide, a reduction in spacing between peptides and substrates in the *cis*-conformation resulted in lower cell adhesion. In contrast, the *trans*-conformation enhanced adhesion.

4.3.2 Spiropyran Sensors

Spiropyrans are a class of photoresponsive materials with reversible switching capabilities. Switching is modulated by the photochemical cleavage of a C–O facilitated by UV radiation, which changes the molecular structure from a closed non-polar form to an open polar form.[46–51] Because the closed form of the molecule is hydrophobic and the merocyanine form is hydrophilic, exposure to UV light alters wettability. Consequently, incorporating spiropyrans into a substrate or using them as a surface coating allows for the control of substrate wettability.[52] For example, the ability of spiropyrans to modulate wettability in conjunction with nanopatterning to control volumetric changes has been proposed,[53,54] where nanopatterned poly(ethyl methacrylate)-*co*-poly(methyl acrylate), P(EMA)-*co*-P(MA) was doped with spiropyran and then exposed to UV illumination.

When incorporated into copolymers, spiropyrenes will serve as color-changing sensors. Figure 4.4 illustrates an example of the

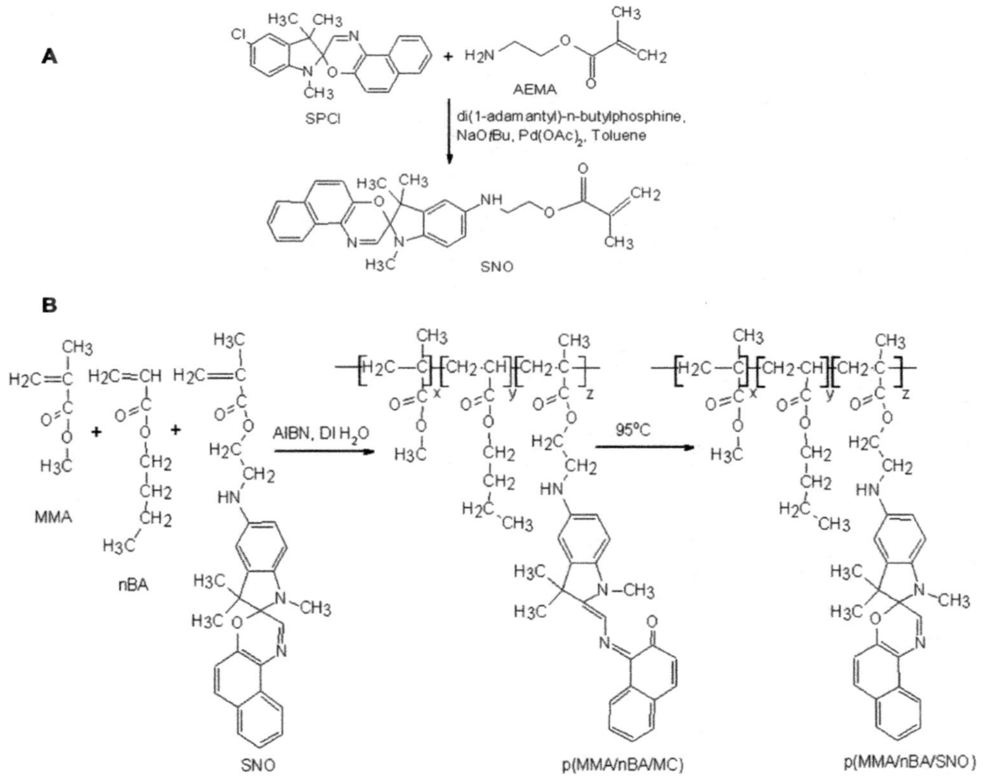

Figure 4.4 (A) Synthesis of 2-[(1,3,3-trimethyl-1,3-dihydrospiro[indole-2,39-naphtho[2,1-*b*][1,4]oxazin]-5-yl)aminoethyl 2-methyl-acrylate (SNO) monomer. (B) Copolymerization of methylmethacrylate (MMA), *n*-butyl acrylate (*n*BA), and 2-[(1,3,3-trimethyl-1,3-dihydrospiro[indole-2,39-naphtho[2,1-*b*][1,4]oxazin]-5-yl)aminoethyl-2-methylacrylate (SNO) monomers in 0.07/0.05/0.0013 molar ratios resulted in pink color open ring poly(methyl methacrylate/*n*-butylacrylate/2-{[(1,3,3-trimethyl-2-{[(1-Z)-2-oxonaphthalen-1(2-*H*)-ylidene]amino}methylene)-2.3-dihydro-1-*H*-indol-5-ylamino]ethyl-2-methylacrylate) p(MMA/*n*BA/MC) copolymer, which was then heated at 95 °C for 5 min to obtain colorless films. Reproduced from ref. 55 with permission from The Royal Society of Chemistry.

synthesis of 2-[(1,3,3-trimethyl-1,3-dihydrospiro[indole-2,39-naph-tho[2,1-*b*][1,4]oxazin]-5-yl)amino]ethyl 2-methylacrylate (SNO) mono-mer (A) as well as copolymerization of methyl methacrylate (MMA), *n*-butyl acrylate (*n*BA) and 2-[(1,3,3-trimethyl-1,3-dihydrospiro[indole-2,39-naphtho[2,1-*b*][1,4]oxazin]-5-yl)amino]ethyl-2-methylacrylate (SNO) monomers to form a light-sensitive copolymer. One of the unique ap-plications of spiropyrenes is in color change upon mechanical damage.[55] The optical images shown in Figure 4.5 illustrate color changes and disappearance upon exposure to visible light as well as temperature. This is perhaps one of the most versatile tools in the analysis of surface and interfacial molecular events in IR and Raman imaging.[56,57] The spatial resolution in mid-IR (under 1 µm) with molecular signatures of events that are not easily accessible is quite powerful.[58]

Control of wettability is also critical in controlling cell attachment. A copolymer of nitrobenzospiropyran and methyl methacrylate (poly(NSP-*co*-MMA)) was synthesized to control platelet and mesenchymal stem cell adhesion.[59] UV exposure of cells or platelets previously attached to the copolymer-coated substrate were detached. Fibrinogen adsorption is known to play a significant role in platelet adhesion, and so the impact of UV exposure on fibrinogen adhesion was measured. Because no substantial differences in fibrinogen adhesion were observed in the presence of UV light, cell and platelet detachment resulted from a change in surface energy and/or an alteration in the surface conformation derived from the switch of spiropyran from the closed non-polar state to the polar merocyanine

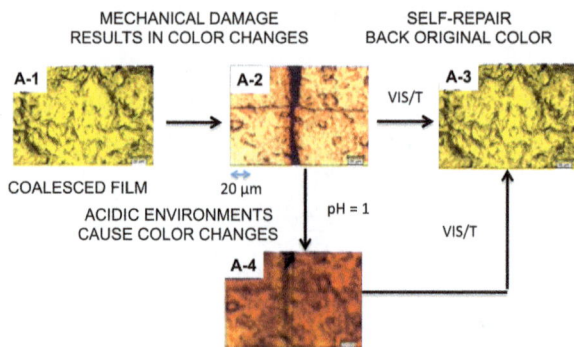

Figure 4.5   Optical images of p(MMA/*n*BA/SNO) films: undamaged (A-1), mechanically damaged (A-2), after exposure to VIS radiation or temperature (A-3) and after exposure to acidic vapors (A-4). Repair is achieved by either one of these conditions: visible light, temperature, or acidic pH vapors.
Reproduced from ref. 55 with permission from The Royal Society of Chemistry.

state upon UV light exposure. Wettability control by spiropyrans was achieved to create a photoresponsive cell culture surface[60] by incorporating spiropyrans as side chains to a poly(N-isopropylacrylamide) polymer. A unique application of spiropyrans is in molecular gating. Because spiropyrans are able to control nanoarchitecture through the creation of a gating system consisting of mesoporous MCM-41 incorporated with spiropyran moieties (which function as a 3D support) and 1.5 poly(amidoamine) dendrimers, these systems can serve as nanomechanical stoppers.[61] Specifically, UV light exposure will lead to the attraction of the negatively-charged dendrimers to the positively-charged spiropyrans in the polar conformation, causing the entrapment of $Ru(bipy)_3Cl_2$ dye molecules. This is schematically depicted in Figure 4.6. During this process dendrimers as well as dye molecules can be released to the bulk solution upon the application of visible light upon spiropyran conversion to its neutrally charged non-polar conformation.

Chemically specific sensing systems based on receptor–guest complexation are one of the methods of non-destructive analysis of biological and environmental species. Synthetic receptors integrated with chromophores and fluorophores are particularly useful for colorimetric and fluorometric sensing.[62–65] To achieve selective and sensitive colorimetric sensors for metal ions, oligoether-linked bis(spiropyran) podands were developed to colorimetrically sense alkaline earth metal ions.[66] Also, colorimetric chemosensors for metal ions, macrocyclic,[67,68] as well as acyclic[69] oligoethers have employed the same concept of an analyte-binding site.

4.4 pH-responsive Surfaces

Unlike other stimuli considered in this chapter, pH requires alteration of the chemical environment. This is particularly relevant for biological applications for a narrow pH range. The reversibility is another drawback that arises with the use of pH-responsive materials because it typically requires solvent replacement, which may also alter the physical environment. In spite of these limitations, pH-responsive materials have shown promise in fields such as drug delivery and microprocessing.[70]

4.4.1 pH-sensitive SAMs

Surface and interfacial forces may be constructive or destructive. One example is release or attachment of two essential components, like a

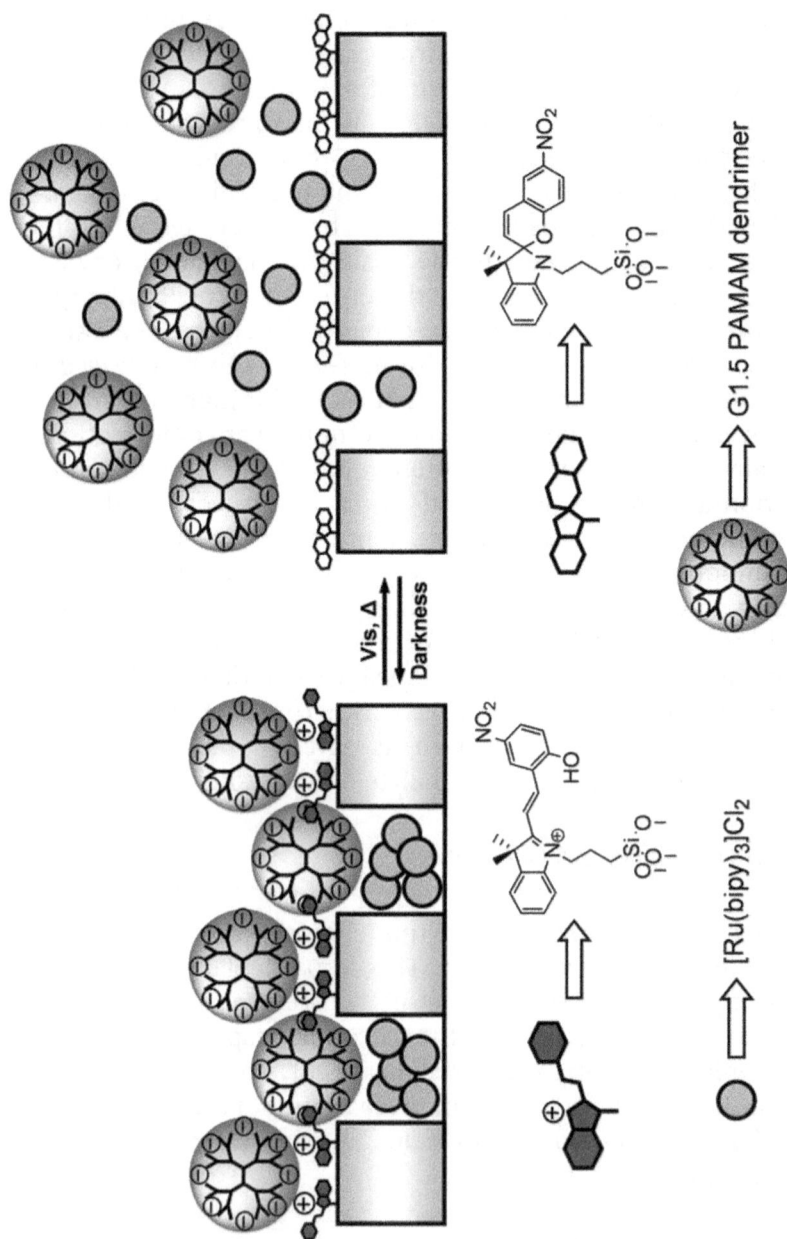

Figure 4.6 Schematic diagram of a gating system that consists of a mesoporous framework containing a photo-switchable anchored spiropyran, a dye entrapped in the inner pores, and carboxylate-terminated dendrimers as molecular caps.

From ref. 58, reproduced with permission from John Wiley and Sons, Copyright © 2007 WILEY-VCH Verlag GmbH & Co. KGaA, Weinheim.

tool and an object being manufactured during assembly. Ideally, a tool grabs an object to move to the next step of production, at which point the object should be released. Thus, one of the primary objectives is to control the interaction of the assembly tools with the surface forces of the object (*e.g.* van der Waals forces) that result in adhesion of the object to the assembly tool(s), thereby preventing object release.[71] Recently this problem has been addressed through the use of pH-modulated SAMs[72] in which a glass microsphere and the silica gripper in the form of a microcantilever were chemically modified with SAMs consisting of aminosilane-grafted 3-(ethoxydimethylsilyl) propylamine (APDMES) and (3-aminopropyl)triethoxysilane (APTES). Because both SAMs components are amine-functionalized, their protonation under acidic pH can be easily accomplished. By changing the pH environment the attractive and repulsive forces between the tool and the micro-object were also altered. In particular, as the liquid pH was increased from acidic to basic conditions, the repulsive forces increased, enabling object release. Although the concept is not new, precise control of the chemistry and physics of surface and interfacial regions will be required to create submergible tools that could grasp and release objects on demand by pH environmental changes.

4.4.2 pH-switchable Surfaces Containing Polymer Brushes

Although polymer brushes are extensively described in Chapter 5,[73–76] here we just focus on polyelectrolyte brushes. Polyelectrolyte brushes are pH-responsive materials that undergo structural changes at interfaces when their chains are charged and/or discharged because of the protonation/dissociation of acid/base groups. As a result, upon an alteration in pH, polyelectrolyte brushes transform from the swollen state to a shrunken state in which the polymer chains are collapsed.[77,78]

Using this concept, polyelectrolyte brushes to create a responsive interface with tunable/switchable redox properties were developed,[79] in which indium tin oxide (ITO) served as the substrate and a modified poly(4-vinylpyridine) modified with an Os-complex redox unit was grafted to the surface. As shown in Figure 4.7, at an acidic pH (pH = 4), the polymer brush was swollen and thus the redox units were in direct contact with the conducting ITO surface resulting in an active electrode. At a neutral pH (pH > 6), the brush shrank, which restricted the mobility of the polymer chains and therefore access of the redox units to the conducting surface, yielding a non-active electrode state. One can envisage numerous applications in self-repairing

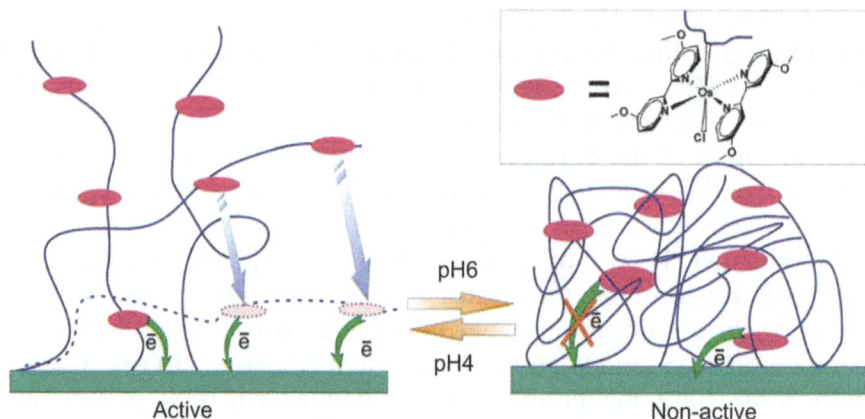

Figure 4.7 Reversible pH-controlled transformation of the redox-polymer brush on the electrode surface between electrochemically active and inactive states.
Reproduced with permission from ref. 68. Copyright (2008) American Chemical Society.

polymers, which may have brushes as side chains, which may expand to grab another end and initiate self-healing.

The pH responsiveness of polymer brushes has also been exploited as a molecular gate. In this case, a mixed polyelectrolyte brush consisting of poly(2-vinylpyridine) (P2VP) and poly(acrylic acid) (PAA) that had switchable permeability for anions and cations was formed.[80] Using ITO as the electrode substrate, mixed brushes were assessed *via* two soluble redox probes $[Fe(CN)_6]^{4-}$ and $[Ru(NH_3)_6]^{3+}$. As illustrated in Figure 4.8, when the environment pH was acidic (pH < 3), positively-charged P2VP chains became permeable to the anionic probe. However, the redox process for the cationic probe was inhibited due to the lack of transport of positively-charged ions. When the solvent environment was neutral to basic (pH > 6), the process was reversed. This switchable and selective gating approach may find numerous applications in drug delivery, membrane and biosensor technologies.

Charged polyelectrolyte mixed brushes are particularly interesting because they consist of oppositely charged polymers. When mixed brushes are used, they typically constitute two incompatible polymers grafted to the same substrate. This is schematically depicted in Figure 4.9. Theory predicts that,[81] in contrast to uniformly charged homopolymer brushes in which the intra- and inter-Coulombian repulsions cause the chain stretching, the oppositely changed mixed brushes depicted in Figure 4.9 may exhibit other modes of reducing the electrostatic repulsion. Depending on the charge ratio, defined as

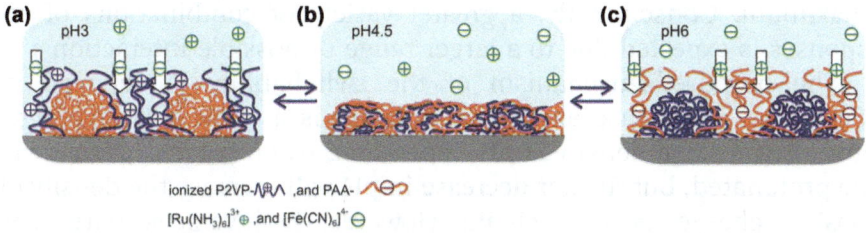

ionized P2VP-/\/\ ,and PAA- /\⊖/\

[Ru(NH$_3$)$_6$]$^{3+}$⊕ ,and [Fe(CN)$_6$]$^{4-}$ ⊖

Figure 4.8   Proposed mechanism for switching of permselective properties of the mixed P2VP–PAA brush: the brush with swollen
positively-charged P2VP domains (dark, in blue) is permeable
for anions (a), the brush with swollen negatively-charged PAA
domains (gray, in red) is permeable for cations (c), and the
brush in the state of P2VP–PAA uncharged PE complex is not
permeable at $4 < pH < 6$ (b).
From ref. 72. © IOP Publishing. Reproduced with permission.
All rights reserved.

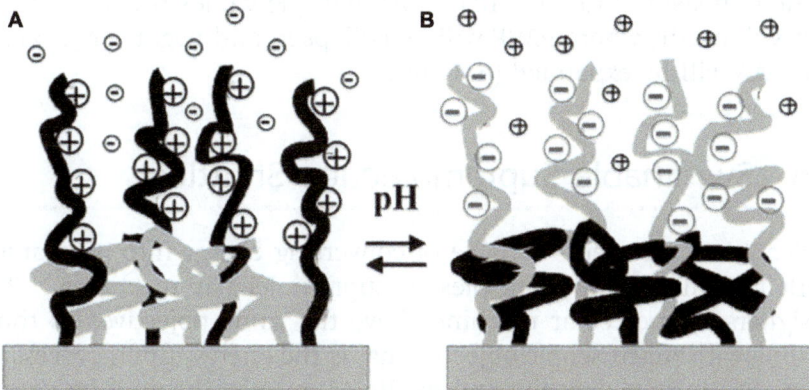

Figure 4.9   Schematic representation of switching behavior of mixed poly-
electrolyte brushes upon change of pH: below isoelectric point
(A) and above the isoelectric point (B).
Reproduced with permission from ref. 71. Copyright (2003)
American Chemical Society.

the degree of compensation of the total charge of the homopolymer A
consisting of N_A segments by the total opposite charge of the homopolymer B consisting of N_B segments, their structural features may be
altered. For $N_A > N_B$, polymer A chains are coiled due to the electrostatic attraction between A and B. When $N_A = N_B$, both chains form a
compact brush, but addition of salt leads to brush expansions. For
brushes with a low charge ratio, further increase in salt concentration
will result in a decrease of brush thickness due to screening of the
repulsion between equally charged segments, thus exhibiting a

maximum. Consequently, a greater variety of combinations of responses is expected due to a larger range of possible interactions.

The proposed mechanism of the switchability is depicted in
Figure 4.9.[82] Because each homopolymer is a weak polyelectrolyte,
and their charge density is pH dependent, when pH < 6.7, P2VP will
be protonated, but further decrease in pH will increase the density of
positive charges on P2VP chains. However, the inverse scenario may
occur on PAA chains. When pH > 3.2, PAA will be negatively charged.
Between pH 3.2 and 6.7, the charged P2VP an d PAA interact so that at
pH = 4.9 the charges will cancel out and be completely compensated.
As a result, the surface will be neutral. The measured thickness of the
brush at the isoelectric point is 7 nm, thus closely corresponding to
the thickness of the dry film. Outside this pH region, at low pH values,
PAA chains will adopt a collapsed, compact conformation near the
surface of the film, whereas P2VP chains will be protonated and extended, measured to be ~18 nm. At high pH values the inverse process will occur, where P2VP will be collapsed and negatively charged
and PAA will be expanded (~15 nm).

4.5   Switchable Supramolecular Shuttles

Molecular components capable of converting energy into mechanical
work formulated the principles of supramolecular machines.[83] The
designers of molecular machines have the same objectives as those
dealing with macroscopic objects: what is the energy input, processing
time, and output quantity and quality. Typically, inspired by nature,
supramolecular machines often utilize adenosine triphosphate (ATP)
synthase, myosin and kinesin.[84] Myosin and kinesin are linear motor
proteins capable of traversing microtubules and actin filaments, respectively, thus converting energy from ATP hydrolysis to molecular
motion.[85] Molecular architects and designers have been fascinated by
the construction of stimuli-responsive molecular devices and molecular machines. In particular, molecular shuttles in which a ring moves
back and forth like a shuttle between two or more 'stations' in response
to external stimuli has been a target of rotaxane synthesis; these are
compounds in which a ring is threaded by a linear chain bearing bulky
end units. On the other hand, cyclodextrins (CyDs) are known to form
ICs with a variety of aromatic compounds in water.

In this context, synthetic supramolecular assemblies for nanotechnology applications include rotaxanes and catenanes, shown below.

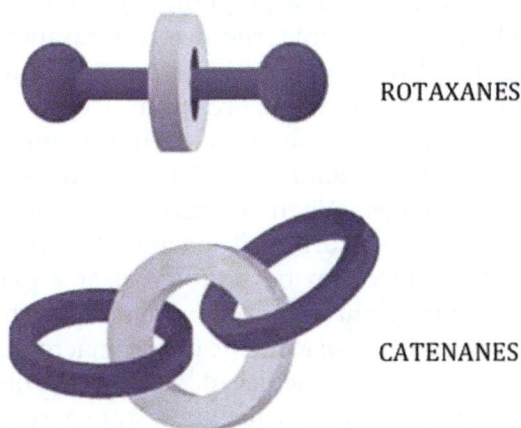

ROTAXANES

CATENANES

4.5.1 Rotaxanes

Rotaxanes can be used to control the hydrophilicity and hydrophobicity of surfaces. These supramolecular structures are comprised of an axis dumbbell-shape molecule and a ring molecule called a macrocycle.[86] Molecular switching is achieved by rotaxane and pseudorotaxane monolayers formed by the presence of interlocking cavities within the monolayers.[87–91] The sequence of reactions below illustrates the first design and synthesis of a rotaxane that functions as a light-driven molecular shuttle.[92] In this design α-CyD moves back and forth from—to the azobenzene moiety to—from the methylene spacer by using alternating UV and VIS electromagnetic radiation.

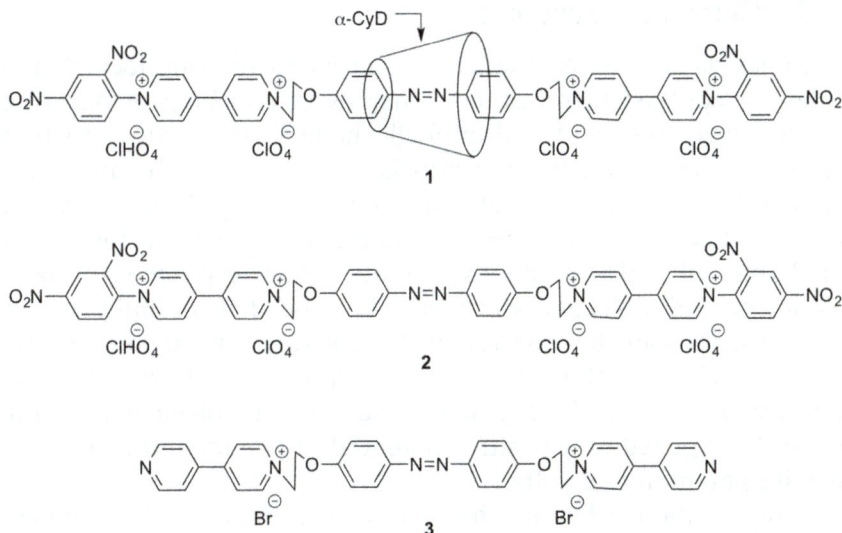

These structural features allow programmable surfaces that can be reversibly switched. For example, the use of rotaxanes in control of wettability is achieved by incorporating rotaxanes into SAMs[93] comprised of rotaxane combined with mercaptoundecanoic acid on a gold surface. These surfaces can be used to move liquid droplets in response to light-induced changes. pH may also modulate the responsiveness of rotaxanes. When rotaxane is comprised of a fluorenyltriamine axle and a cucurbituril macrocycle, nitrogen atoms in the fluorenyltriamine axle can be protonated at a pH as low as 1. When exposed to fluorescent light intensity changes, this molecule is pH responsive. As a result, placement of the macrocycle at the protonated diaminohexane site is achieved.[94] Upon deprotonation at pH as high as 8, the macrocycle can facilitate transport to the diprotonated diaminobutane site, resulting in color changes. Redox-active rotaxanes may find a number of applications in bioelectronics. There are numerous applications, for example enzyme electrodes utilize electron transfer between the redox enzyme and the surface of the electrode, which requires rapid communication. Aligning the enzyme with the electrode and utilizing the redox relay units has been exploited by associating an apoprotein, apo-glucose oxidase (apo-GOx), onto relay-functionalized materials. These may include flavin adenine dinucleotide (FAD) monolayers, nanoparticles, and carbon nanotubes with electrodes,[95–98] as well as reversible redox-active rotaxane shuttles in the bioelectrocatalyzed oxidation of glucose.[89]

4.5.2 Catenanes Switches

Rotaxanes as well as catenanes are mechanically interlocked molecules. In contrast to rotaxanes, whereby interlocking of one ring-shaped macrocycle and a dumbbell shape is the main structural feature, catenanes consist of interlocked macrocycles. The number of macrocycles contained in a catenane is indicated by the numeral that precedes it. Catenanes exhibit bistable and multistable forms and switchable, bistable catenanes are commonly exploited in nanotechnology and molecular electronics because their mobility and rearrangements can be controlled by electrochemical processes.[99] Collier *et al.* were the first to demonstrate the electroactivity of interlocked catenanes.[100] This work resulted in a molecular switching device that opened at a positive potential of 2 volts and closed at a negative potential of 2 volts.

A single-station catenane that exhibited fundamental differences from the bi- and multi-station catenane switches[101] has been prepared from an electrochemically switchable unit, tetra-thiafulvalene (TTF). Very often, electrically sensitive molecules are incorporated into catenanes to yield architectures that can be controlled electrically. Normally, electrochemically switchable catenanes are comprised of a primary binding station consisting of tetracationic cyclophane cyclobis(paraquat-p-phenylene) ($CBQT^{4+}$), which releases oxidized TTF^{2+} *via* electrostatic repulsion, and a secondary binding station that provides reversibility following the release of TTF2. However, in this work a single-station catenane that functions as a reversible switch between two translational states was created, eliminating the need for the secondary station and thus reducing system complexity. Applications for this technology may include solid-state electronic devices.

4.6 DNA Switchable Surfaces

The specificity of DNA base pairs is advantageous in genetic screening because it results in highly specific binding between targets.[102–104] DNA can be analyzed with a myriad of surface analysis techniques including those that are label-free, or methods that do not require fluorescent tagging of the DNA. They may include atomic force microscopy,[105] electrochemical methods[106,107] or surface plasmon resonance spectroscopy.[108,109] Mass, conductivity and electric field measurements may also be utilized to characterize DNA biosensing.[110] In spite of the fact that DNA on surfaces are attractive in genetic sensing, peptide monolayers are starting to be extensively studied for many biotechnology applications.[111]

DNA molecules can be directed by electrical potentials when tethered to electrode surfaces and this interaction is commonly exploited with gold as the substrate.[112–114] One way to alter surfaces is to modulate the conformation of a thiol-modified oligonucleotide on gold surfaces by altering electric potential exposure.[115] This approach allows tilting of monolayers and control of their thickness by changing electrical potential. Electrical modulation has also been combined with fluorescence to facilitate real-time optical monitoring of switchable DNA monolayers.[116] In a typical experiment single-stranded oligonucleotides are labeled with a fluorophore and DNA hybridization is initiated by the addition of unlabeled complementary strands. As a result of the phosphates in the sugar–phosphate backbone of the negatively charged oligonucleotides, the DNA monolayers can be switched from a tilted or flat to vertical conformations. This orientation difference was indicated by changes in fluorescence intensity because the tilted conformation partially quenches the fluorescence signal.

Electric field changes have been used in conjunction with pH changes to alter the conformation of DNA monolayers.[117-120] For example, in DNA comprised of two distinct domains, one domain consisted of a single strand linked to the gold substrate and the other domain contained four strands called an i-motif.[117] The i-motif can be modulated by modifications of the pH. As DNA molecules form a SAM, the i-motif functions as a bulky head group, resulting in a monolayer that is densely packed with respect to the i-motif, but loose with respect to the single-stranded DNA. As a result, small molecules cannot penetrate the monolayer, but when pH is increased, the i-motif can transition to a single-stranded conformation, thus increasing the permeability to small molecules.

4.7   Aptamer Switches

An aptamer is a short, single-stranded DNA and RNA sequence. These molecules typically demonstrate a high affinity towards a multitude of molecules, from proteins to cells.[121] The sensitivity and selectivity with which aptamers bind make them promising options in diagnostic and therapeutic applications.[121,122] Because aptamers experience reversible denaturation and regeneration, their use is particularly attractive in reusable sensing technologies.[123] Many of the potential uses for aptamers require that these molecules are immobilized on the surface. This requirement complicates their use because the retaining method employed must sustain the high binding affinity aptamers demonstrate in solution. Another challenge concerns the need to acquire a flexible, ordered receptor molecule while maintaining the appropriate aptamer folding that induces binding.

Numerous techniques are employed to study the reversible binding of aptamers and various biomolecules including cantilever-based sensors, surface plasmon resonance (SPR) spectroscopy, and electrochemical impedance.[124] An aptamer that binds thrombin was assembled on a gold nanowire by Huang and Chen.[125] In this work, the aptamer probe was controlled by changes in electric potential and was utilized in a fluorescent protein test. The probe was affixed to the nanowire and specifically interacted with a biotinylated, fluorescently labeled thrombin. Upon the application of a potential, this complex was either attracted or repelled, depending on the sign of the applied potential. Because the complex was negatively charged, a positive potential resulted in a reduction in the fluorescence intensity due to increased distance between the fluorescent label and the nanowire.

4.8 Molecular Motors

In contrast to internal heat combustion engines, which convert heat into mechanical work, biomolecular motors achieve the cold conversion of chemical energy into mechanical energy. The genesis of biomolecular motors goes back many years, but the key to their success is the great expertise of biochemists and biophysicists in observing and purifying motor proteins *in vitro*. Also, understanding muscle physiology as well as the role of proteins and interactions with cations such as calcium, has resulted in a bank of knowledge that even without conclusive answers, continues to fascinate. Furthermore, the large number of well-understood building blocks including DNA, RNA, peptides, lipids and proteins is a treasure chest for old and new scientists and engineers. Mother Nature has mastered these processes and, with just a few building blocks, is capable of signaling and recognizing subtle changes through manipulations of non-covalent, typically weak hydrogen bonding interactions. One biopolymer that is capable of encoding genetic data is DNA, and among other functions its sequence determines the type of organism or disease susceptibility, but its critical feature is the ability to precisely encode a specific protein sequence. The key component in these processes is molecular recognition and the ability of selective binding to specific molecular segments. Before we consider molecular motors, let us consider one example of molecular matching between synthetic well-known polymer and lipid molecules. The first example to mimic potential lipid–protein interactions was the synthesis of poly-(methyl methacrylate/*n*-butyl acrylate) (p-MMA/*n*BA) colloidal particles stabilized by phospholipids (PLs), such as 1,2-dilauroyl-*sn-glycero*-3-phosphocholine (DLPC).[126] Upon particle coalescence, PL stratification resulted in the formation of surface localized ionic clusters (SLICs). These entities are capable of recognizing MMA/*n*BA monomer interfaces along the p-MMA/*n*BA copolymer backbone and form crystalline SLICs at the monomer interface. As seen in Figure 4.10, almost perfect alignment of PL in a diester well is facilitated by surface area selective non-covalent bonds of anionic phosphate and cationic quaternary ammonium segments of DLPC that interact with two neighboring

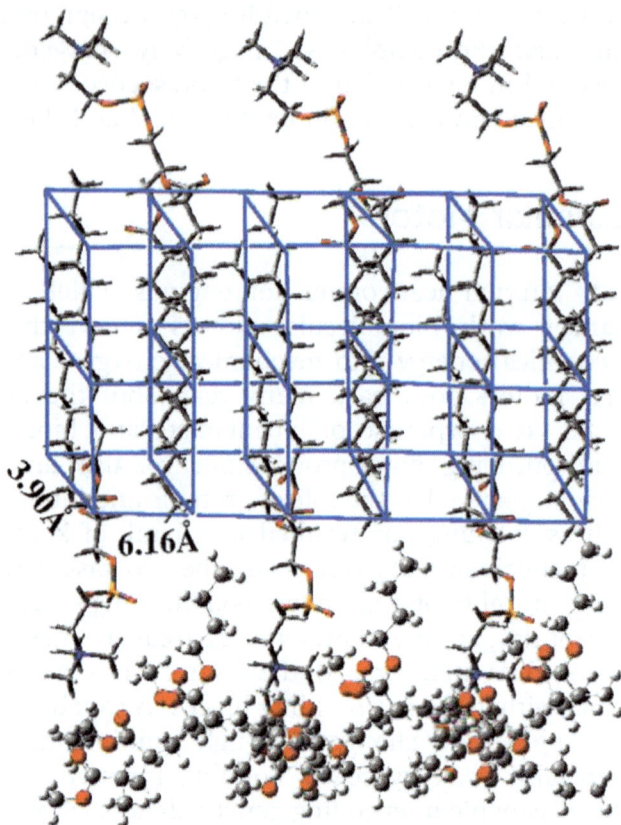

Figure 4.10   Recognition of DLPC on p-MMA/*n*BA polymer backbone. Reprinted with permission from ref. 115. Copyright (2008) American Chemical Society.

carbonyl groups of *n*BA and MMA monomers of the p-MMA/*n*BA polymer backbone. The dominating non-covalent bonds responsible for the molecular recognition are a combination of H-bonding and electrostatic interactions.

As one can envisage, molecular recognition will be a key feature of molecular motors. Figure 4.11 illustrates a typical molecular shuttle design which relies on surface adhered kinesin motors to transport functionalized microtubes that may carry different types of cargo.[127] Kinesin is a dimer, where each amino acid chain folds into a head (the ATP-hydrolyzing and microtubule-binding motor domain) and a tail domain. The two tails form a coiled coil, which holds the two subunits together. The coordinated movement of the two heads enables a 'walking' motion. Microtubules are hollow cylinders

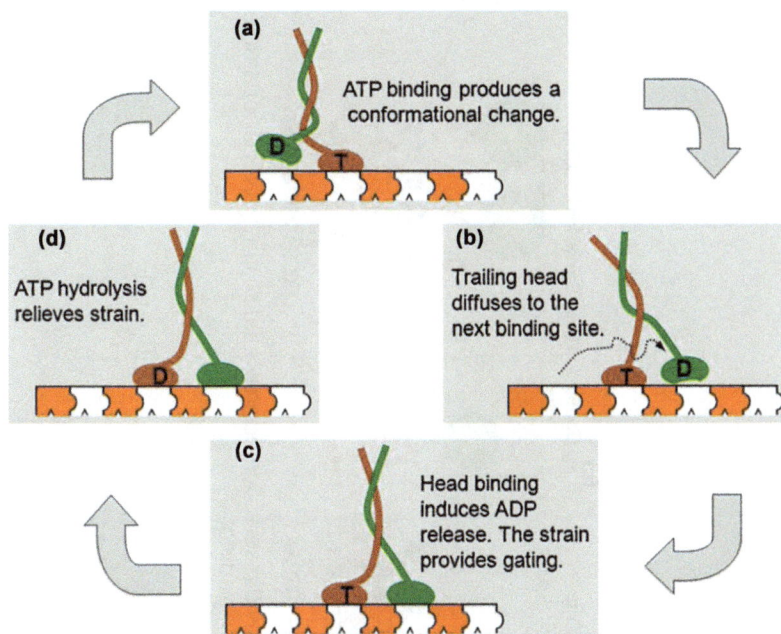

Figure 4.11 Asymmetric hand-over-hand mechanochemical cycle of kinesin walk.
Reproduced from ref. 116 with permission from Elsevier.

composed of tubulin heterodimers,[128] and tubulin dimers bind head-to-tail to form linear protofilaments with an 8 nm repeat distance and a variable number (10–18, often 13) of protofilaments to assemble into a cylindrical structure with an outer diameter of about 25 nm and a length of many micrometers. Cargo loading onto molecular shuttles was first achieved by attaching streptavidin-coated polystyrene microspheres to kinesin-propelled microtubules[129] which is attached to filaments using covalent as well as non-covalent chemistries. Examples of loading strategies are illustrated in Figure 4.12.[130] Non-covalent interactions, such as antibody/antigen bonds, are preferred due to their selectivity and reversibility. In particular the strong non-covalent bond between biotin and streptavidin is readily integrated into molecular shuttle systems due to the commercial availability of biotinylated tubulin and actin. Applications are endless, from delivery of protein and drugs to trapping and loading viruses and anti-bodies.[131] While these fascinating biological motors sparked significant interest, there are tremendous challenges in the precise design of these molecular motors which can be classified into the following categories: how to control motor activity, loading and unloading

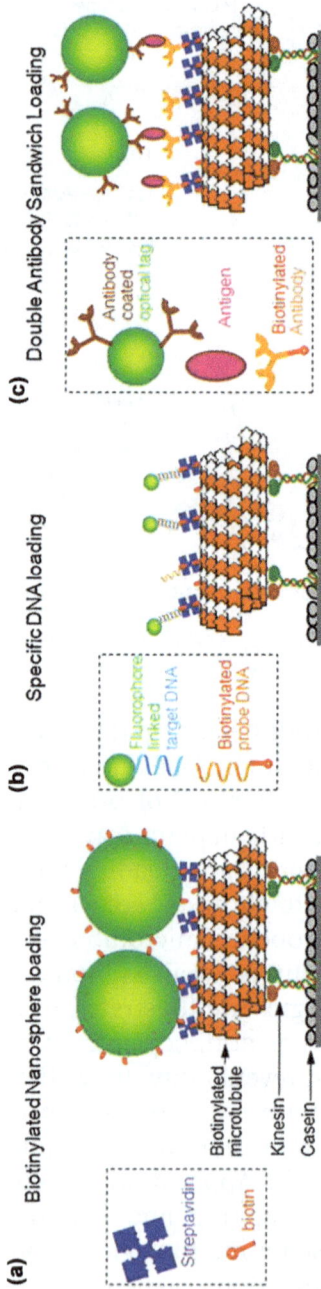

Figure 4.12 Examples of loading strategies utilizing the biotin-streptavidin selective bonding: (a) biotinylated nanoparticle loading; (b) DNA loading; (c) double antibody sandwich loading. Reproduced from ref. 119 with permission from Elsevier.

cargo, and controlling the shuttle movement. The control of these processes will lead to the development of novel 'living' materials capable of their own existence. The concept of mimicking many forms of cytoplasmic motility, such as mitotic chromosome segregation, the transport of vesicular organelles as well as the beating of cilia or flagella, are all generated by interactions between microtubules and mechanochemical enzymes termed 'motors'. Many new materials will rely on these interactions, especially when the opposite polarity motors, kinesin and dynein, will impose forces in the same proximity.

As a final section to this chapter, it should be realized that although many excellent cartoons visualizing various surface-interfacial phenomena were created, it is undeniably critical to realize that characterization of these dynamic processes is not trivial. Historically, Schrödinger stated in 1952 that "we will never see an experiment with one electron, atom, or molecule", but 8 years later Feynman said that "there are no limits in arranging atoms", and in the early 1980s scanning tunneling microscopy (STM) changed views on single atoms and molecules. Although one could argue that such probes are available, and experiments on individual molecules have been successfully conducted, thus permitting deeper insights into the quantum electronics of molecular arrays of phthalocyanines and lipid layers or benzene, the problem arises when stimuli-responsive molecules or macromolecules are placed in complex and often similar, yet molecularly diverse environments. Polymers and other materials with stimuli-responsive attributes are complex heterogeneous entities, and selectivity as well as sensitivity of the existing and future analytical probes to distinguish intimate structural features will play a crucial role in their future developments. Among numerous available methods of surface and interfacial analysis, several methods are used more often than others. The techniques that are often used include mass spectrometry (MS) and its various derivations (GC-MS, MALDI),[132] atomic force microscopy (AFM),[133] nuclear magnetic resonance (NMR),[134] and vibrational spectroscopic methods (infrared and Raman).[56,57]

References

1. U. Nakaya and A. Matsumoto, Simple experiment showing the existence of "liquid water" film on the ice surface, *J. Colloid Sci.*, 1954, **9**(1), 41–49.
2. C. L. Hosler, D. Jensen and L. Goldshlak, On the aggregation of ice crystals to form snow, *J. Meteorol.*, 1957, **14**(5), 415–420.

3. M. W. Orem and A. W. Adamson, Physical adsorption of vapor on ice: II. n-alkanes, *J. Colloid Interface Sci.*, 1969, **31**(2), 278–286.

4. A. Döppenschmidt and H.-J. Butt, Measuring the thickness of the liquid-like layer on ice surfaces with atomic force microscopy, *Langmuir*, 2000, **16**(16), 6709–6714.

5. R. Rosenberg, Why is ice slippery?, *Phys. Today*, 2005, **58**(12), 50.

6. J. H. Gibbs and E. A. DiMarzio, Nature of the glass transition and the glassy state, *J. Chem. Phys.*, 1958, **28**(3), 373–383.

7. W. Gotze and L. Sjogren, Relaxation processes in supercooled liquids, *Rep. Prog. Phys.*, 1992, **55**(3), 241.

8. G. Adam and J. H. Gibbs, On the temperature dependence of cooperative relaxation properties in glass-forming liquids, *J. Chem. Phys.*, 1965, **43**(1), 139–146.

9. M. L. Williams, R. F. Landel and J. D. Ferry, The temperature dependence of relaxation mechanisms in amorphous polymers and other glass-forming liquids, *J. Am. Chem. Soc.*, 1955, 77(14), 3701–3707.

10. E. Donth, The size of cooperatively rearranging regions at the glass transition, *J. Non-Cryst. Solids*, 1982, **53**(3), 325–330.

11. P. De Gennes, Weak adhesive junctions, *J. Phys.*, 1989, **50**(18), 2551–2562.

12. A. M. Mayes, Glass transition of amorphous polymer surfaces, *Macromolecules*, 1994, **27**(11), 3114–3115.

13. W. Zhao, X. Zhao, M. Rafailovich, J. Sokolov, R. Composto and S. Smith *et al.*, Segregation of chain ends to polymer melt surfaces and interfaces, *Macromolecules*, 1993, **26**(3), 561–562.

14. J. L. Keddie, R. A. Jones and R. A. Cory, Size-dependent depression of the glass transition temperature in polymer films, *EPL (Europhys. Lett.)*, 1994, **27**(1), 59.

15. S. Kawana and R. A. Jones, Character of the glass transition in thin supported polymer films, *Phys. Rev. E: Stat., Nonlinear, Soft Matter Phys.*, 2001, **63**(2), 021501.

16. J. H. van Zanten, W. E. Wallace and W.-L. Wu, Effect of strongly favorable substrate interactions on the thermal properties of ultrathin polymer films, *Phys. Rev. E: Stat. Phys., Plasma, Fluids, Relat. Interdiscip. Top.*, 1996, **53**(3), R2053.

17. K. Fukao and Y. Miyamoto, Glass transitions and dynamics in thin polymer films: dielectric relaxation of thin films of polystyrene, *Phys. Rev. E: Stat. Phys., Plasma, Fluids, Relat. Interdiscip. Top.*, 2000, **61**(2), 1743.

18. J. Forrest, K. Dalnoki-Veress, J. Stevens and J. Dutcher, Effect of free surfaces on the glass transition temperature of thin polymer films, *Phys. Rev. Lett.*, 1996, **77**(10), 2002.
19. C. J. Ellison and J. M. Torkelson, The distribution of glass-transition temperatures in nanoscopically confined glass formers, *Nat. Mater.*, 2003, **2**(10), 695–700.
20. G. DeMaggio, W. Frieze, D. Gidley, M. Zhu, H. Hristov and A. Yee, Interface and surface effects on the glass transition in thin polystyrene films, *Phys. Rev. Lett.*, 1997, **78**(8), 1524–1527.
21. C. Frank, V. Rao, M. Despotopoulou, R. Pease, W. Hinsberg and R. Miller *et al.*, Structure in thin and ultrathin spin-cast polymer films, *Science*, 1996, **273**(5277), 912–915.
22. D. B. Hall, J. C. Hooker and J. M. Torkelson, Ultrathin polymer films near the glass transition: effect on the distribution of α-relaxation times as measured by second harmonic generation, *Macromolecules*, 1997, **30**(3), 667–669.
23. C. Ellison, S. Kim, D. Hall and J. Torkelson, Confinement and processing effects on glass transition temperature and physical aging in ultrathin polymer films: Novel fluorescence measurements, *Eur. Phys. J. E*, 2002, **8**(2), 155–166.
24. R. D. Priestley, C. J. Ellison, L. J. Broadbelt and J. M. Torkelson, Structural relaxation of polymer glasses at surfaces, interfaces, and in between, *Science*, 2005, **309**(5733), 456–459.
25. J. Lahann, S. Mitragotri, T. N. Tran, H. Kaido, J. Sundaram and I. S. Choi *et al.*, A reversibly switching surface, *Science*, 2003, **299**(5605), 371–374.
26. R. K. Smith, P. A. Lewis and P. S. Weiss, Patterning self-assembled monolayers, *Prog. Surf. Sci.*, 2004, **75**(1-2), 1–68.
27. N. L. Abbott, C. B. Gorman and G. M. Whitesides, Active Control of Wetting using Applied Electrical Potentials and Self-assembled Monolayers, *Langmuir*, 1995, **11**(1), 16–18.
28. C. B. Gorman, H. A. Biebuyck and G. M. Whitesides, Control of the Shape of Liquid Lenses on a Modified Gold Surface Using an Applied Electrical Potential Across a Self-assembled Monolayer, *Langmuir*, 1995, **11**(6), 2242–2246.
29. N. L. Abbott and G. M. Whitesides, Potential-Dependent Wetting of Aqueous-Solutions on Self-assembled Monolayers formed from 15-(ferrocenylcarbonyl)pentadecanethiol on Gold, *Langmuir*, 1994, **10**(5), 1493–1497.
30. B. Berron and G. K. Jennings, Loosely packed hydroxyl-terminated SAMs on gold, *Langmuir*, 2006, **22**(17), 7235–7240.

31. X. M. Wang, A. B. Kharitonov, E. Katz and I. Willner, Potential-controlled molecular machinery of bipyridinium monolayer-functionalized surfaces: an electrochemical and contact angle analysis, *Chem. Commun.*, 2003, **13**, 1542–1543.

32. Y. Liu, L. Mu, B. H. Liu, S. Zhang, P. Y. Yang and J. L. Kong, Controlled protein assembly on a switchable surface, *Chem. Commun.*, 2004, **10**, 1194–1195.

33. D. K. Peng and J. Lahann, Chemical, electrochemical, and structural stability of low-density self-assembled monolayers, *Langmuir*, 2007, **23**(20), 10184–10189.

34. D. K. Peng, S. T. Yu, D. J. Alberts and J. Lahann, Switching the Electrochemical Impedance of Low-Density Self-Assembled Monolayers, *Langmuir*, 2007, **23**(1), 297–304.

35. D. K. Peng, A. A. Ahmadi and J. Lahann, A Synthetic Surface that Undergoes Spatiotemporal Remodeling, *Nano Lett.*, 2008, **8**(10), 3336–3340.

36. W. S. Yeo and M. Mrksich, Electroactive self-assembled monolayers that permit orthogonal control over the adhesion of cells to patterned substrates, *Langmuir*, 2006, **22**(25), 10816–10820.

37. W. S. Yeo, M. N. Yousaf and M. Mrksich, Dynamic interfaces between cells and surfaces: Electroactive substrates that sequentially release and attach cells, *J. Am. Chem. Soc.*, 2003, **125**(49), 14994–14995.

38. F. Liu, D. Ramachandran and M. W. Urban, Colloidal films that mimic cilia, *Adv. Funct. Mater.*, 2010, **20**(18), 3163–3167.

39. C. J. Barrett, J. I. Mamiya, K. G. Yager and T. Ikeda, Photomechanical effects in azobenzene-containing soft materials, *Soft Matter*, 2007, **3**(10), 1249–1261.

40. P. B. Wan, Y. G. Jiang, Y. P. Wang, Z. Q. Wang and X. Zhang, Tuning surface wettability through photocontrolled reversible molecular shuttle, *Chem. Commun.*, 2008, **44**, 5710–5712.

41. K. Ichimura, Y. Suzuki, T. Seki, A. Hosoki and K. Aoki, Reversible change in alignment mode of nematic liquid crystals regulated photochemically by command surfaces modified with an azobenzene monolayer, *Langmuir*, 1988, **4**(5), 1214–1216.

42. K. Ichimura, S. K. Oh and M. Nakagawa, Light-driven motion of liquids on a photoresponsive surface, *Science*, 2000, **288**(5471), 1624–1626.

43. H. F. Ji, Y. Feng, X. H. Xu, V. Purushotham, T. Thundat and G. M. Brown, Photon-driven nanomechanical cyclic motion, *Chem. Commun.*, 2004, **22**, 2532–2533.

44. G. Hayashi, M. Hagihara, C. Dohno and K. Nakatani, Photo-regulation of a Peptide-RNA Interaction on a Gold Surface, *J. Am. Chem. Soc.*, 2007, **129**(28), 8678–8679.
45. J. Auernheimer, C. Dahmen, U. Hersel, A. Bausch and H. Kessler, Photoswitched Cell Adhesion on Surfaces with RGD Peptides, *J. Am. Chem. Soc.*, 2005, **127**(46), 16107–16110.
46. M. Yoshida and J. Lahann, Smart nanomaterials, *ACS Nano*, 2008, **2**(6), 1101–1107.
47. W. H. Jiang, G. J. Wang, Y. N. He, X. G. Wang, Y. L. An and Y. L. Song *et al.*, Photo-switched wettability on an electrostatic self-assembly azobenzene monolayer, *Chem. Commun.*, 2005, **28**, 3550–3552.
48. L. M. Siewierski, W. J. Brittain, S. Petrash and M. D. Foster, Photoresponsive monolayers containing in-chain azobenzene, *Langmuir*, 1996, **12**(24), 5838–5844.
49. S. T. Wang, Y. L. Song and L. Jiang, Photoresponsive surfaces with controllable wettability, *J. Photochem. Photobiol., C*, 2007, **8**(1), 18–29.
50. R. Rosario, D. Gust, A. A. Garcia, M. Hayes, J. L. Taraci and T. Clement *et al.*, Lotus effect amplifies light-induced contact angle switching, *J. Phys. Chem. B*, 2004, **108**(34), 12640–12642.
51. B. C. Bunker, B. I. Kim, J. E. Houston, R. Rosario, A. A. Garcia and M. Hayes *et al.*, Direct observation of photo switching in tethered spiropyrans using the interfacial force microscope, *Nano Lett.*, 2003, **3**(12), 1723–1727.
52. F. Xia, Y. Zhu, L. Feng and L. Jiang, Smart responsive surfaces switching reversibly between super-hydrophobicity and super-hydrophilicity, *Soft Matter*, 2009, **5**(2), 275–281.
53. A. Athanassiou, M. I. Lygeraki, D. Pisignano, K. Lakiotaki, M. Varda and E. Mele *et al.*, Photocontrolled variations in the wetting capability of photochromic polymers enhanced by surface nanostructuring, *Langmuir*, 2006, **22**(5), 2329–2333.
54. A. Athanassiou, M. Kalyva, K. Lakiotaki, S. Georgiou and C. Fotakis, All-optical reversible actuation of photochromic-polymer microsystems, *Adv. Mater.*, 2005, **17**(8), 988–992.
55. D. Ramachandran, F. Liu and M. W. Urban, Self-repairable copolymers that change color, *RSC Adv.*, 2012, **2**(1), 135–143.
56. M. W. Urban, *Vibrational Spectroscopy of Molecules and Macromolecules on Surfaces*, 1993.
57. M. W. Urban, *Attenuated Total Reflectance Spectroscopy of Polymers: Theory and Practice*, American Chemical Society, 1996.

58. D. B. Otts, P. Zhang and M. W. Urban, High fidelity surface chemical imaging at 1000 nm levels: internal reflection IR imaging (IRIRI) approach, *Langmuir*, 2002, **18**(17), 6473–6477.
59. A. Higuchi, A. Hamamura, Y. Shindo, H. Kitamura, B. O. Yoon and T. Mori *et al.*, Photon-modulated changes of cell attachments on poly(spiropyran-co-methyl methacrylate) membranes, *Biomacromolecules*, 2004, **5**(5), 1770–1774.
60. J. Edahiro, K. Sumaru, Y. Tada, K. Ohi, T. Takagi and M. Kameda *et al.*, In situ control of cell adhesion using photoresponsive culture surface, *Biomacromolecules*, 2005, **6**(2), 970–974.
61. E. Aznar, R. Casasus, B. Garcia-Acosta, M. D. Marcos and R. Martinez-Manez, Photochemical and chemical two-channel control of functional nanogated hybrid architectures, *Adv. Mater.*, 2007, **19**(17), 2228–2231.
62. A. P. Silva, H. NimaláGunaratne, P. MarkáLynch, E. Glenn and K. SamankumaraáSandanayake, Molecular fluorescent signalling with 'fluor–spacer–receptor' systems: approaches to sensing and switching devices *via* supramolecular photophysics, *Chem. Soc. Rev.*, 1992, **21**(3), 187–195.
63. T. D. James, K. Sandanayake and S. Shinkai, Saccharide sensing with molecular receptors based on boronic acid, *Angew. Chem., Int. Ed. Engl.*, 1996, **35**(17), 1910–1922.
64. A. P. De Silva, H. N. Gunaratne, T. Gunnlaugsson, A. J. Huxley, C. P. McCoy and J. T. Rademacher *et al.*, Signaling recognition events with fluorescent sensors and switches, *Chem. Rev.*, 1997, **97**(5), 1515–1566.
65. Y. Kubo, Binaphthyl-appended chromogenic receptors: Synthesis and application to their colorimetric recognition of amines, *Synlett*, 1999, **2**, 161–174.
66. S. Yagi, S. Nakamura, D. Watanabe and H. Nakazumi, Colorimetric sensing of metal ions by bis (spiropyran) podands: Towards naked-eye detection of alkaline earth metal ions, *Dyes Pigm.*, 2009, **80**(1), 98–105.
67. V. Král and J. L. Sessler, Molecular recognition *via* base-pairing and phosphate chelation. Ditopic and tritopic sapphyrin-based receptors for the recognition and transport of nucleotide monophosphates, *Tetrahedron*, 1995, **51**(2), 539–554.
68. M. Laspéras, H. Cambon, D. Brunel, I. Rodriguez and P. Geneste, Cesium oxide encapsulation in faujasite zeolites effect of framework composition on the nature and basicity of intrazeolitic species, *Microporous Mater.*, 1996, 7(2), 61–72.

69. A. Ajayaghosh, E. Arunkumar and J. Daub, A Highly Specific Ca2 + -Ion Sensor: Signaling by Exciton Interaction in a Rigid–Flexible–Rigid Bichromophoric "H" Foldamer, *Angew. Chem., Int. Ed.*, 2002, **41**(10), 1766–1769.
70. N. Murthy, J. Campbell, N. Fausto, A. S. Hoffman and P. S. Stayton, Bioinspired pH-responsive polymers for the intracellular delivery of biomolecular drugs, *Bioconjugate Chem.*, 2003, **14**(2), 412–419.
71. M. Gauthier, S. Régnier, P. Rougeot and N. Chaillet, Analysis of forces for micromanipulations in dry and liquid media, *J. Micromechatronics*, 2006, **3**, 389–413.
72. J. Dejeu, M. Gauthier, P. Rougeot and W. Boireau, Adhesion Forces Controlled by Chemical Self-Assembly and pH: Application to Robotic Microhandling, *ACS Appl. Mater. Interfaces*, 2009, **1**(9), 1966–1973.
73. J. N. Kizhakkedathu, R. Norris-Jones and D. E. Brooks, Synthesis of Well-Defined Environmentally Responsive Polymer Brushes by Aqueous ATRP, *Macromolecules*, 2004, **37**(3), 734–743.
74. J. Pyun, T. Kowalewski and K. Matyjaszewski, Synthesis of polymer brushes using atom transfer radical polymerization, *Macromol. Rapid Commun.*, 2003, **24**(18), 1043–1059.
75. O. Prucker and J. Ruhe, Mechanism of radical chain polymerizations initiated by azo compounds covalently bound to the surface of spherical particles, *Macromolecules*, 1998, **31**(3), 602–613.
76. R. Ducker, A. Garcia, J. M. Zhang, T. Chen and S. Zauscher, Polymeric and biomacromolecular brush nanostructures: progress in synthesis, patterning and characterization, *Soft Matter*, 2008, **4**(9), 1774–1786.
77. S. Minko, Responsive polymer brushes, *Polym. Rev.*, 2006, **46**(4), 397–420.
78. R. R. Netz and D. Andelman, Neutral and charged polymers at interfaces, *Phys. Rep.*, 2003, **380**(1–2), 1–95.
79. T. K. Tam, M. Ornatska, M. Pita, S. Minko and E. Katz, Polymer brush-modified electrode with switchable and tunable redox activity for bioelectronic applications, *J. Phys. Chem. C*, 2008, **112**(22), 8438–8445.
80. M. Motornov, T. K. Tam, M. Pita, I. Tokarev, E. Katz and S. Minko, Switchable selectivity for gating ion transport with mixed polyelectrolyte brushes: approaching 'smart' drug delivery systems, *Nanotechnology*, 2009, **20**(434006), 1–10.

81. N. Shusharina and P. Linse, Oppositely charged polyelectrolytes grafted onto planar surface: Mean-field lattice theory, *Eur. Phys. J. E: Soft Matter Biol. Phys.*, 2001, **6**(2), 147–155.

82. N. Houbenov, S. Minko and M. Stamm, Mixed polyelectrolyte brush from oppositely charged polymers for switching of surface charge and composition in aqueous environment, *Macromolecules*, 2003, **36**(16), 5897–5901.

83. V. Balzani, A. Credi, B. Ferrer, S. Silvi and M. Venturi, Artificial molecular motors and machines: Design principles and prototype systems, in *Molecular Machines*, Topics in Current Chemistry, Springer-Verlag Berlin, Berlin, 2005, vol. 262, pp. 1–27.

84. M. Schliwa and G. Woehlke, Molecular motors, *Nature*, 2003, **422**(6933), 759–765.

85. R. D. Vale and R. A. Milligan, The way things move: Looking under the hood of molecular motor proteins, *Science*, 2000, **288**(5463), 88–95.

86. E. R. Kay, D. A. Leigh and F. Zerbetto, Synthetic molecular motors and mechanical machines, *Angew. Chem., Int. Ed.*, 2007, **46**(1-2), 72–191.

87. V. Balzani, A. Credi and M. Venturi, Molecular machines working on surfaces and at interfaces, *ChemPhysChem*, 2008, **9**(2), 202–220.

88. K. Hiratani, M. Kaneyama, Y. Nagawa, E. Koyama and M. Kanesato, Synthesis of [1]rotaxane *via* covalent bond formation and Its unique fluorescent response by energy transfer in the presence of lithium ion, *J. Am. Chem. Soc.*, 2004, **126**(42), 13568–13569.

89. E. Katz, O. Lioubashevsky and I. Willner, Electromechanics of a redox-active rotaxane in a monolayer assembly on an electrode, *J. Am. Chem. Soc.*, 2004, **126**(47), 15520–15532.

90. G. Fioravanti, N. Haraszkiewicz, E. R. Kay, S. M. Mendoza, C. Bruno and M. Marcaccio *et al.*, Three state redox-active molecular shuttle that switches in solution and on a surface, *J. Am. Chem. Soc.*, 2008, **130**(8), 2593–2601.

91. B. C. Bunker, D. L. Huber, J. G. Kushmerick, T. Dunbar, M. Kelly and C. Matzke *et al.*, Switching surface chemistry with supramolecular machines, *Langmuir*, 2007, **23**(1), 31–34.

92. H. Murakami, A. Kawabuchi, K. Kotoo, M. Kunitake and N. Nakashima, A light-driven molecular shuttle based on a rotaxane, *J. Am. Chem. Soc.*, 1997, **119**(32), 7605–7606.

93. J. Berna, D. A. Leigh, M. Lubomska, S. M. Mendoza, E. M. Perez and P. Rudolf *et al.*, Macroscopic transport by synthetic molecular machines, *Nat. Mater.*, 2005, **4**(9), 704–710.

94. S. I. Jun, J. W. Lee, S. Sakamoto, K. Yamaguchi and K. Kim, Rotaxane-based molecular switch with fluorescence signaling, *Tetrahedron Lett.*, 2000, **41**(4), 471–475.

95. I. Willner, V. Heleg-Shabtai, R. Blonder, E. Katz, G. Tao and A. F. Buckmann *et al.*, Electrical wiring of glucose oxidase by reconstitution of FAD-modified monolayers assembled onto Au-electrodes, *J. Am. Chem. Soc.*, 1996, **118**(42), 10321–10322.

96. E. Katz, A. Riklin, V. Heleg-Shabtai, I. Willner and A. F. Bückmann, Glucose oxidase electrodes *via* reconstitution of the apo-enzyme: tailoring of novel glucose biosensors, *Anal. Chim. Acta*, 1999, **385**(1-3), 45–58.

97. Y. Xiao, F. Patolsky, E. Katz, J. F. Hainfield and I. Willner, Plugging into enzymes: nanowiring of redox enzymes by a gold nanoparticle, *Science*, 2003, **299**, 1877–1881.

98. F. Patolsky, Y. Weizmann and I. Willner, Long-range electrical contacting of redox enzymes by SWCNT connectors, *Angew. Chem., Int. Ed.*, 2004, **43**(16), 2113–2117.

99. A. H. Flood, R. J. A. Ramirez, W. Q. Deng, R. P. Muller, W. A. Goddard and J. F. Stoddart, Meccano on the nanoscale - A blueprint for making some of the world's tiniest machines, *Aust. J. Chem.*, 2004, **57**(4), 301–322.

100. C. P. Collier, G. Mattersteig, E. W. Wong, Y. Luo, K. Beverly and J. Sampaio *et al.*, A [2]Catenane-Based Solid State Electronically Reconfigurable Switch, *Science*, 2000, **289**(5482), 1172–1175.

101. J. M. Spruell, W. F. Paxton, J. C. Olsen, D. Benitez, E. Tkatchouk and C. L. Stern *et al.*, A Push-Button Molecular Switch, *J. Am. Chem. Soc.*, 2009, **131**(32), 11571–11580.

102. S. Cosnier and P. Mailley, Recent advances in DNA sensors, *Analyst*, 2008, **133**(8), 984–991.

103. J. S. Daniels and N. Pourmand, Label-free impedance bio-sensors: Opportunities and challenges, *Electroanalysis*, 2007, **19**(12), 1239–1257.

104. K. J. Odenthal and J. J. Gooding, An introduction to electro-chemical DNA biosensors, *Analyst*, 2007, **132**(7), 603–610.

105. K. Wang, C. Goyer, A. Anne and C. Demaille, Exploring the motional dynamics of end-grafted DNA oligonucleotides by *in situ* electrochemical atomic force microscopy, *J. Phys. Chem. B*, 2007, **111**(21), 6051–6058.

106. T. G. Drummond, M. G. Hill and J. K. Barton, Electrochemical DNA sensors, *Nat. Biotechnol.*, 2003, **21**(10), 1192–1199.

107. F. B. Meng, Y. X. Liu, L. Liu and G. X. Li, Conformational Transitions of Immobilized DNA Chains Driven by pH with

Electrochemical Output, *J. Phys. Chem. B*, 2009, **113**(4), 894–896.

108. R. Georgiadis, K. P. Peterlinz and A. W. Peterson, Quantitative measurements and modeling of kinetics in nucleic acid monolayer films using SPR spectroscopy, *J. Am. Chem. Soc.*, 2000, **122**(13), 3166–3173.

109. X. H. Yang, Q. Wang, K. M. Wang, W. H. Tan, J. Yao and H. M. Li, Electrical switching of DNA monolayers investigated by surface plasmon resonance, *Langmuir*, 2006, **22**(13), 5654–5659.

110. C. E. Immoos, S. J. Lee and M. W. Grinstaff, Conformationally gated electrochemical gene detection, *ChemBioChem*, 2004, **5**(8), 1100–1103.

111. D. Chow, M. L. Nunalee, D. W. Lim, A. J. Simnick and A. Chilkoti, Peptide-based biopolymers in biomedicine and biotechnology, *Mater. Sci. Eng., R*, 2008, **62**(4), 125–155.

112. U. Rant, K. Arinaga, M. Tornow, Y. W. Kim, R. R. Netz and S. Fujita *et al.*, Dissimilar kinetic behavior of electrically manipulated single- and double-stranded DNA tethered to a gold surface, *Biophys. J.*, 2006, **90**(10), 3666–3671.

113. U. Rant, K. Arinaga, S. Fujita, N. Yokoyama, G. Abstreiter and M. Tornow, Electrical manipulation of oligonucleotides grafted to charged surfaces, *Org. Biomol. Chem.*, 2006, **4**(18), 3448–3455.

114. U. Rant, K. Arinaga, S. Fujita, N. Yokoyama, G. Abstreiter and M. Tornow, Dynamic electrical switching of DNA layers on a metal surface, *Nano Lett.*, 2004, **4**(12), 2441–2445.

115. S. O. Kelley, J. K. Barton, N. M. Jackson, L. D. McPherson, A. B. Potter and E. M. Spain *et al.*, Orienting DNA helices on gold using applied electric fields, *Langmuir*, 1998, **14**(24), 6781–6784.

116. U. Rant, K. Arinaga, S. Scherer, E. Pringsheim, S. Fujita and N. Yokoyama *et al.*, Switchable DNA interfaces for the highly sensitive detection of label-free DNA targets, *Proc. Natl. Acad. Sci. U. S. A.*, 2007, **104**(44), 17364–17369.

117. Y. D. Mao, D. S. Liu, S. T. Wang, S. N. Luo, W. X. Wang and Y. L. Yang *et al.*, Alternating-electric-field-enhanced reversible switching of DNA nanocontainers with pH, *Nucleic Acids Res.*, 2007, **35**(5), 8.

118. S. T. Wang, H. J. Liu, D. S. Liu, X. Y. Ma, X. H. Fang and L. Jiang, Enthalpy-driven three-state switching of a superhydrophilic/superhydrophobic surface, *Angew. Chem., Int. Ed.*, 2007, **46**(21), 3915–3917.

119. T. Liedl, M. Olapinski and F. C. Simmel, A surface-bound DNA switch driven by a chemical oscillator, *Angew. Chem., Int. Ed.*, 2006, **45**(30), 5007–5010.
120. Y. D. Mao, S. Chang, S. X. Yang, Q. Ouyang and L. Jiang, Tunable non-equilibrium gating of flexible DNA nanochannels in response to transport flux, *Nat. Nanotechnol.*, 2007, **2**(6), 366–371.
121. S. Balamurugan, A. Obubuafo, S. A. Soper and D. A. Spivak, Surface immobilization methods for aptamer diagnostic applications, *Anal. Bioanal. Chem.*, 2008, **390**(4), 1009–1021.
122. W. Mok and Y. F. Li, Recent Progress in Nucleic Acid Aptamer-Based Biosensors and Bioassays, *Sensors*, 2008, **8**(11), 7050–7084.
123. S. P. Song, L. H. Wang, J. Li, J. L. Zhao and C. H. Fan, Aptamer-based biosensors, *TrAC, Trends Anal. Chem.*, 2008, **27**(2), 108–117.
124. M. C. Rodriguez, A. N. Kawde and J. Wang, Aptamer biosensor for label-free impedance spectroscopy detection of proteins based on recognition-induced switching of the surface charge, *Chem. Commun.*, 2005, **34**, 4267–4269.
125. S. X. Huang and Y. Chen, Ultrasensitive fluorescence detection of single protein molecules manipulated electrically on Au nanowire, *Nano Lett.*, 2008, **8**(9), 2829–2833.
126. M. Yu, M. W. Urban, Y. Sheng and J. Leszczynski, Molecular recognition at methyl methacrylate/n-butyl acrylate (MMA/nBA) monomer unit boundaries of phospholipids at p-MMA/nBA copolymer surfaces, *Langmuir*, 2008, **24**(18), 10382–10389.
127. S. M. Block, Kinesin motor mechanics: binding, stepping, tracking, gating, and limping, *Biophys. J.*, 2007, **92**(9), 2986–2995.
128. H. Li, D. J. DeRosier, W. V. Nicholson, E. Nogales and K. H. Downing, Microtubule structure at 8 Å resolution, *Structure*, 2002, **10**(10), 1317–1328.
129. H. Hess, J. Clemmens, D. Qin, J. Howard and V. Vogel, Light-controlled molecular shuttles made from motor proteins carrying cargo on engineered surfaces, *Nano Lett.*, 2001, **1**(5), 235–239.
130. A. Agarwal and H. Hess, Biomolecular motors at the intersection of nanotechnology and polymer science, *Prog. Polym. Sci.*, 2010, **35**(1), 252–277.
131. H. Hess, Engineering applications of biomolecular motors, *Annu. Rev. Biomed. Eng.*, 2011, **13**, 429–450.
132. S. Hanton, Mass spectrometry of polymers and polymer surfaces, *Chem. Rev.*, 2001, **101**(2), 527–570.
133. E. Meyer, Atomic force microscopy, *Prog. Surf. Sci.*, 1992, **41**(1), 3–49.
134. F. D. Blum, Magnetic resonance of polymers at surfaces, *Colloids Surf.*, 1990, **45**, 361–376.

5 Stimuli-responsive Polymer Brushes

5.1 Introduction

Initially, polymer brushes were defined as macromolecules attached by one end to an interface, and the density at the attachment points was sufficiently high so the chains could not collapse, but stretched away from the interface.[1,2] In other words, if tethering is sufficiently dense, polymer chains will be forced to change their conformations to stretch away from the interface. While this definition is still valid, a diversity of experimental approaches resulted in the creations of 'fuzzy' brushes, thus redefining polymer brush morphologies as any macromolecular entity that looks like 'a broom' or 'a rope' with whiskers sticking out. As long as these macromolecular entities are thicker than a single macromolecular chain, *i.e.* containing side chains or pendant groups, they can be considered as a brush. Another common feature is chain deformability, which results in stretched morphologies. Considering linear polymer chains as a benchmark for further developments, Figure 5.1 illustrates several examples of linear polymer brushes that are composed of homopolymers and random or block copolymers as well as liquid crystalline materials.

While the real exploration of synthetically controlled polymer brushes began in the early 1990s, back in the 1950s the studies of polymer and surfactant adsorption on various colloidal surfaces initiated practical interest in controlling flocculation.[3,4] The concept was to utilize surface brushes in colloidal particles to prevent flocculation due to repulsive forces arising from the high osmotic pressures inside

Stimuli-Responsive Materials: From Molecules to Nature Mimicking Materials Design
By Marek W. Urban
© Marek W. Urban, 2016
Published by the Royal Society of Chemistry, www.rsc.org

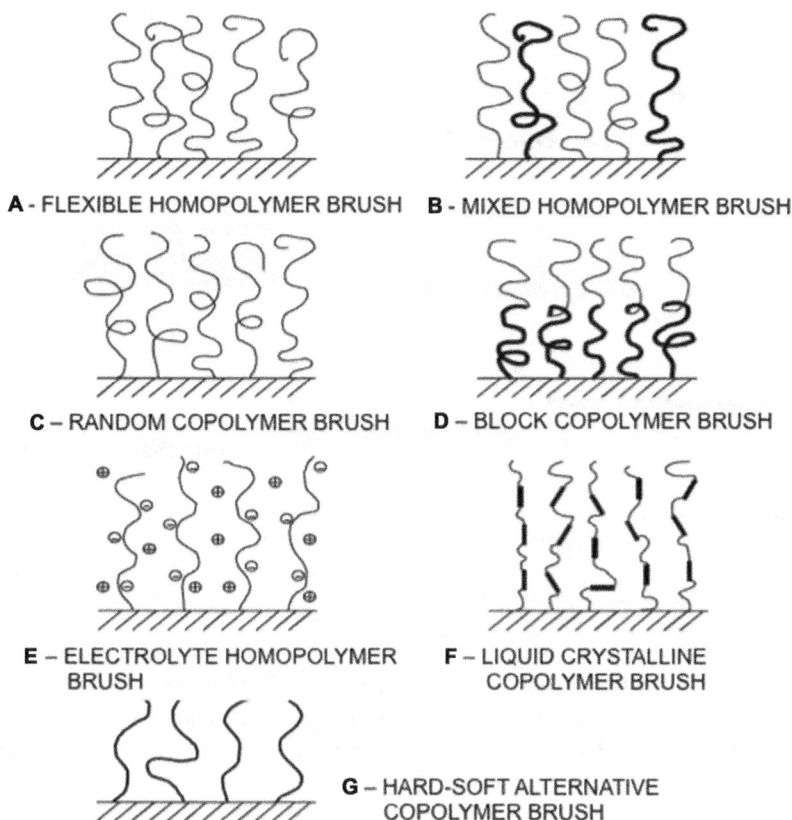

A - FLEXIBLE HOMOPOLYMER BRUSH   **B** - MIXED HOMOPOLYMER BRUSH

C – RANDOM COPOLYMER BRUSH   **D** – BLOCK COPOLYMER BRUSH

E – ELECTROLYTE HOMOPOLYMER BRUSH   **F** – LIQUID CRYSTALLINE COPOLYMER BRUSH

G – HARD-SOFT ALTERNATIVE COPOLYMER BRUSH

Figure 5.1 Classification of linear polymer brushes: (A) flexible homopolymer; (B) mixed homopolymer; (C) random homopolymer; (D) block copolymer; (E) electrolyte homopolymer; (F) liquid crystalline copolymer; (G) hard–soft alternating copolymer.

the brushes. This phenomenon resulted in many useful applications ranging from antifouling surfaces and new adhesives to lubricants, surfactants and chemical gates controlling mobility, to name just a few. Thus, the central model of polymer brushes depicted in Figure 5.2A can be expanded to many diversified geometries and this is why controllable synthetic approaches played a key role in the development of stimuli-responsive brushes. Before we explore stimuli-responsive brushes it is useful to realize how distinctly different polymer brushes are from 'regular' polymers. The presence of close proximity tethered segments plays a key role in the conformational changes of deformed coils, as well as interactions of repeating units of the neighboring chains. However, as polymer segments stretch, lowering the interaction energy per chain, the elastic energy density

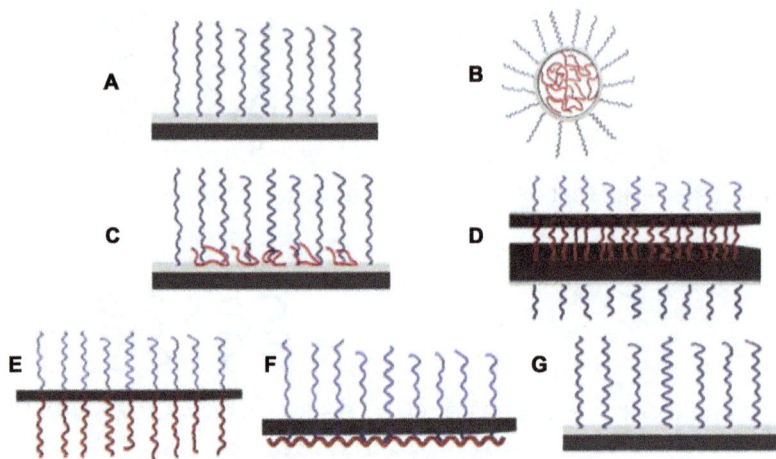

Figure 5.2 Polymer systems that use polymer brushes: (A) original configuration of a polymer brush; (B) polymer micelles; (C) adsorbed diblock copolymer; (D) diblock copolymer melt; (E) fluid–fluid block copolymer interface; (F) fluid–fluid graft copolymer; (G) end-grafted copolymer.

increases, thus determining the equilibrium thickness of the layer. This concept was used to theoretically predict the relationship between the dimensions of polymer chains under various conditions.

5.2 Theoretical Considerations

If one considers a non-adsorbing flat surface to which monodisperse polymer chains are tethered, such as depicted in Figure 5.1, and assuming that each chain consists of N statistical segments of diameter a, the average distance between the tethering point is d. This distance is much smaller than the radius of gyration of a free, undeformed chain. Thus, the free energy per chain will consist of two terms:

$$F = F_{int} + F_{el}$$

where F_{int} is the interaction energy between two statistical segments and F_{el} is the elastic free energy. While detailed discussion of this model can be found in the literature,[5] two assumptions are critical in these considerations: (i) the depth profile of statistical segments is step-like and the concentration of statistical segments remains the same and is constant within brushes, $\varphi = Na^3/d^2L$ and (ii) free ends of tethered polymer chains are located in the single plane at a distance L from the tethering surface. Using the Flory theory approximation[6] to obtain the free energy, one can estimate the reduction of configurational entropy resulting from the random walk chain constrained by one end to travel distance L to the outer edge of the brush. This is expressed by the following free energy per chain relationship:

$$F/kT = v\varphi^2 d^2 L/a^3 + L^2/R_0^2$$

where v is a dimensionless excluded volume parameter and R_0 is the radius of an unperturbed, ideal coil.

The interaction energy between statistical segments is described by the first terms, whereas the second represents the elasticity of Gaussian chains. Thus, the equilibrium brush thickness can be predicted by minimizing F with respect to L, which leads to the flowing relationship:

$$L/a = N(a/d)^{2/3}$$

This relationship shows that the equilibrium thickness will vary linearly with the degree of polymerization. The behavior under different environmental conditions, such as theta solvents where the interactions between statistical segments vanish, results in the following relationship:

$$F/kT = w\varphi^3 d^2 L/a^3 + L^2/Na^2$$

where w is the dimensionless third virtual coefficient. The relationship between the equilibrium thickness and N can be obtained by minimization of free energy with respect to L:

$$L/a = N(a/d)$$

The linearity is maintained in theta and poor solvents, in contrast to the behavior of free polymer chains in a good solvent $(R \sim N^{3/5})$ and in a theta solvent $(R_0 \sim N^{1/2})$ is non-linear.

For brushes without solvent, like melt brush, the relationship between the thickness and the degree of polymerization can also be obtained and was found to be as follows:

$$L = N^{2/3}$$

Although this simple model can be used to describe the hydrodynamic properties of polymer brushes as well as other properties, they will depend on perturbing the balance between chain stretching and chain–chain repulsion and include hydrodynamic thickness, permeability and the force per area required to compress a brush. In spite of these advantages, there are questions that these theoretical treatments did not address the location and makeup of chain ends, chain segregation, different chemical composition and interpenetration.

5.3 Stimuli-responsive Brushes

Stimuli-responsive polymers may respond to a variety of external stimuli, but the degree of response will depend upon the amount of input and the ability of polymer components to respond. For example, the presence of the secondary transitions in polymers will typically occur at low temperatures, and often are attributed to a crank-shaft motion of a polymer backbone,[7] just like those depicted in Figure 5.3A, or simply a rotation of an acid group along the C–C axis. These transitions typically require small amounts of energy and occur at -100 to $-50\,°C$. Synthetic efforts also facilitated the formation of expandable networks depicted in Figure 5.3B.[8] As was discussed earlier, the composition and architecture of homopolymers as well as

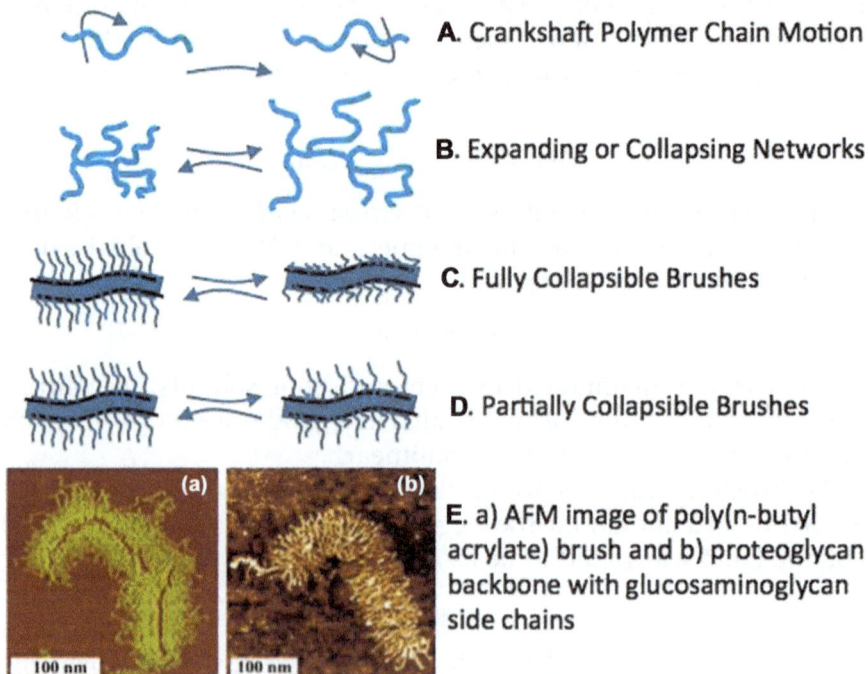

A. Crankshaft Polymer Chain Motion

B. Expanding or Collapsing Networks

C. Fully Collapsible Brushes

D. Partially Collapsible Brushes

E. a) AFM image of poly(n-butyl acrylate) brush and b) proteoglycan backbone with glucosaminoglycan side chains

(a)   (b)

100 nm   100 nm

Figure 5.3 (A) Crankshaft motion of a polymer chain. (B) Expanding and collapsing polymer networks. (C) Fully collapsible brushes. (D) Partially collapsible brushes. (E) AFM images of (a) poly(n-butyl acrylate) (Reproduced with permission from ref. 14. Copyright (2001) American Chemical Society) and (b) proteoglycan backbone with glucosaminoglycan side chains (Reproduced with permission from ref. 18. Copyright (2002) American Chemical Society).

statistical block or graft copolymers may lead to different responses. However, one unique morphology, known as molecular brushes, is characteristic for its molecular segregation stabilized by steric repulsion of densely grafted sidel chains.[9-11] For example, if the reversible conformational changes in response to external stimuli are limited to the single molecule observed by atomic force microscopy (AFM),[12,13] the response of the side chains of molecular brushes bound to a single strand of polymer backbone with an extremely high grafting density will reach four chains per 1 nm of the backbone or possibly higher. These congested morphologies may not only show low energy transitions, but fascinating properties, typically not achievable with other polymers. As shown in Figure 5.3E for poly(*n*-butyl acrylate),[14] the density of side groups causes steric repulsions between the side chains, thus generating significant mechanical tension along the backbone and resulting in chain folding. If solvent quality or the side chain length can tune grafting density, many properties can be achieved by amplification of tension from pico- to nano-Newtons upon adsorption. In addition, molecular brushes with amphiphilic block copolymer side chains can form a stable unimolecular micelle of cylindrical shape, which cannot be dissociated due to covalent attachment of the side chains to the backbone.[15-17] Thus, the magnitude of these interactions is significant. Since 1998, molecular brushes have attracted attention and can possibly be used in designing membranes for biomedical applications as they resemble a proteoglycan protein backbone with glucosaminoglycan side chains, as shown in Figure 5.3E.[14,18] In essence, Figure 5.3E illustrates that polymer brushes are graft copolymers with multiple side polymer chains grafted to a linear polymer.[19-22] The main chain is commonly referred to as the backbone and the pendant groups are the side chains. Notably, if the length of the backbone is significantly longer than that of the side chains, the polymer adopts a cylindrical shape with the backbone polymer in the core from which the side chains emanate radially. If the backbone is short, of the order of the length of the side chains, these polymers adopt compact, spherical dimensions resembling stars.

A typical classification of polymer brushes is shown in Figure 5.4 and depends upon the chemical composition of the side chains. There are brushes with homopolymer side chains that may be linear or gradients in grafting density along the copolymer backbone,[23,24] brush-block-coil copolymers[25,26] and brushes with copolymer side chains, with random/block copolymer side chains,[27-30] heterografted brushes,[31] and AB-type brush-block-brush copolymers.[32,33]

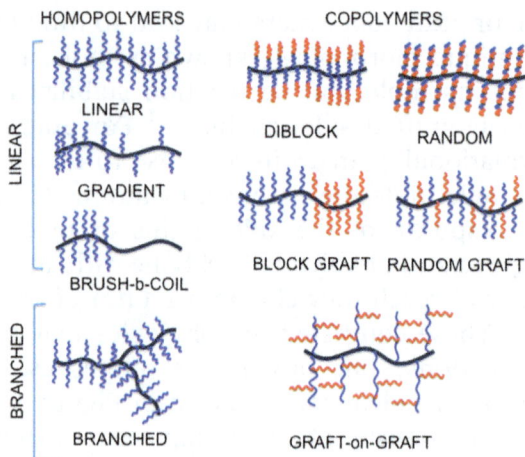

Figure 5.4 Classification of molecular brushes by topologies and chemical composition of side chains.

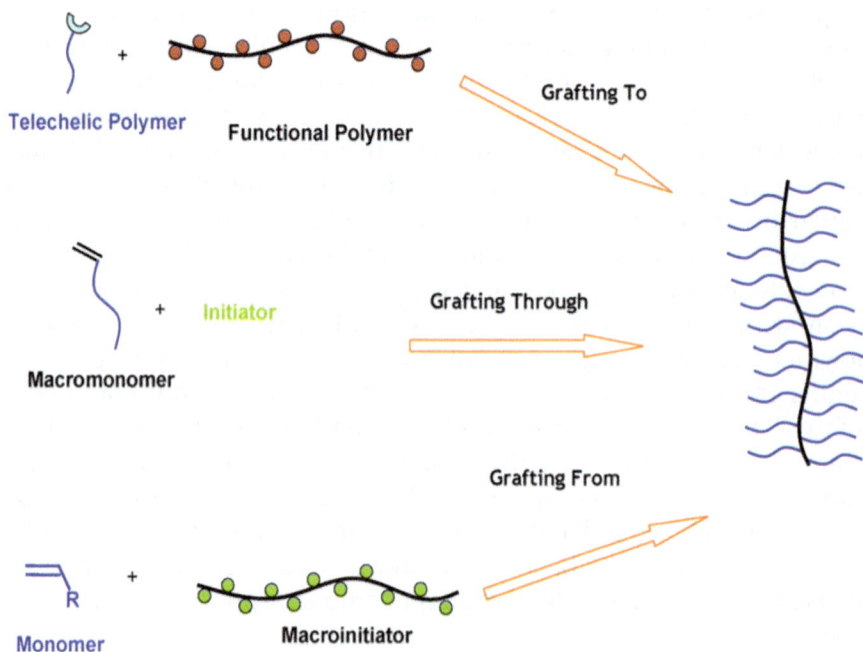

Figure 5.5 Synthetic methods used in preparing molecular brushes: 'grafting to', 'grafting through' and 'grafting from'.
Reproduced from ref. 47 with permission from Elsevier.

Figure 5.5 depicts synthetic methods for preparing molecular brushes by 'grafting to', 'grafting through' and 'grafting from'. The 'grafting to'[34,35] requires the coupling of individual side chains to a common backbone polymer, whereas 'grafting through'[21,36–40] consists of the polymerization of macromonomers. In contrast, 'grafting from' involves growth of the side chains from a well-defined backbone macroinitiator.[41–46] Because ATRP and RAFT facilitate low radical concentrations, thus causing significant suppression of inter- and intramolecular events (Chapter 2), they are highly suitable for the synthesis of molecular brushes.[47]

The synthesis of diblock copolymer brushes, for example poly(styrene) (PSty)-*b*-PSty, PSty-*b*-poly(acrylic acid), PSty-*b*-poly(*N*-isopropylacrylamide), and poly(methyl acrylate)-*b*-poly(*N*,*N*-(dimethy-lamino)ethyl acrylate) can be achieved by using a combination of surface-mediated atom transfer radical polymerization (ATRP) and reversible addition-fragmentation chain transfer (RAFT) polymer-ization techniques.[48] This is schematically depicted in Figure 5.6. For example, conversion of bromine end groups of homopolymer brushes formed by ATRP *via* a modified atom transfer addition reaction to a RAFT agent and diblock extension *via* RAFT polymerization allows the direct formation of well-defined stimuli-responsive diblock copolymer brushes. The growth of side chains can be achieved *via* ATRP-initiated pendant bromoester groups on a polymethacrylate backbone.[41]

Another approach was to use the concept of click chemistry for the synthesis of double hydrophilic coil-rod diblock copolymers with dually responsive asymmetric centipede-shaped polymer brushes as the rod segment. This is shown in Figure 5.7, which illustrates the synthesis of multifunctional initiator PEO-*block*-[PGMA-(N3)(Br)] bearing one azide and one bromine moiety, consisting of PMEO3MA

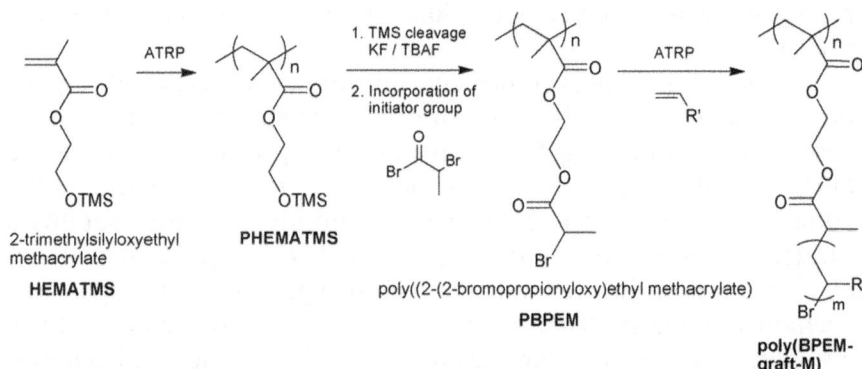

Figure 5.6 Synthetic methods used in preparing block brushes: blue block (ATRP), red block (RAFT).
Reproduced from ref. 47 with permission from Elsevier.

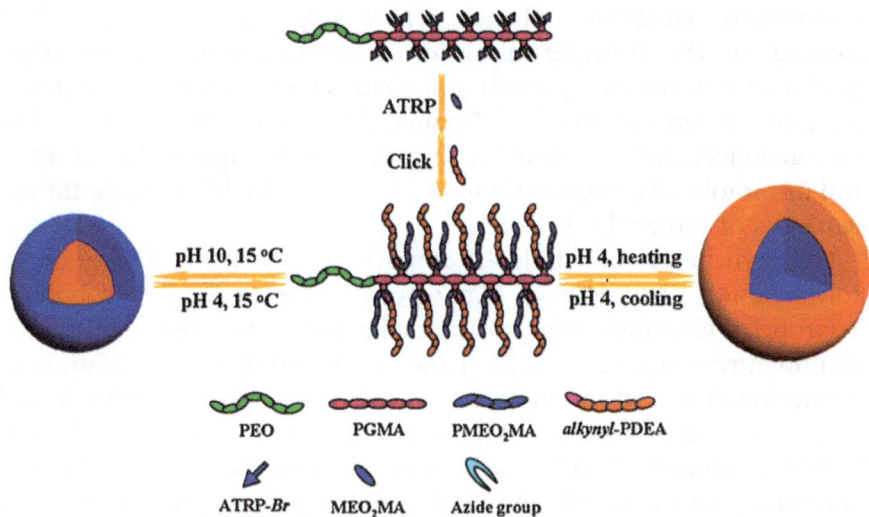

Figure 5.7 Schematic diagram of the synthesis of multi-responsive supramolecular self-assembly of coil-rod double hydrophilic diblock copolymer, PEO-*block*-[PGMA-*graft*-(PDEA)(PMEO2MA)]. Reproduced from ref. 47 with permission from Elsevier.

and PDEA homopolymers.[49] The lower critical solution temperature (LCST) of PEO113-*block*-[PGMA-*graft*-(N3)(PMEO2MA)16]75 in aqueous solution (1.0 wt.%) was detected at 19 °C, but the protonation of PDEA in PEO113-*block*-[PGMA-*graft*-(PDEA)23(PMEO2MA)16]75 at pH 4 rendered the copolymer soluble at room temperature. Heating the aqueous solution induces the thermal phase transition of PMEO3MA grafts at ∼37 °C. Upon adjusting the pH to above 8, the aggregates with hydrophobic PDEA as the core and well-solvated PMEO3MA as the corona were observed with phase transition temperature ∼16 °C (pH 10).

Conjugated polymers represent an important class of organic semiconductor materials. There are numerous useful applications of these materials ranging from sensors and optical/electrical devices (LED), to solar cells and transistors. As one can imagine, a prerequisite for these applications is the solubility and processability of conjugated polymers. With highly rigid conjugated backbones, unsubstituted conjugated polymers often aggregate, resulting in poor mechanical integrity. To overcome these obstacles one approach is to incorporate flexible pendant groups along the conjugated backbone. The use of ionic and non-ionic substituents attached to the conjugated backbone to achieve water solubility was also explored. The substituents include oligoethylene oxides[50] and crown ethers.[51]

Notably, the use of pendant ionic groups in conjugated polymers (called conjugated polyelectrolytes) resulted in highly sensitive fluorescence-based sensors for biological targets like ATP and DNA as well as for temperature, solvent polarity and pH.[52] For pH-responsive anionic poly(phenylene ethylene) with pendant phosphonate groups a significant red shift of both the absorption and fluorescence (FL) spectra with a decrease of pH from 12 to 7.5 was observed.[53] These observations were attributed to the transition of the polymer from a relatively less aggregated or monomeric state at high pH to an aggregated state at low pH. Along the same lines, pH-responsive anionic conjugated polyelectrolyte-containing substituted fluorine and phenylene units were also prepared.[54] The pendant groups of the phenylene units with carboxylic functionalities allow the red shift tuning as a function of pH.

The conjugated polymers of particular interest contain molecular brushes. The main advantage of the use of polymer brushes in conjugated polymers is to flexibilize a typically rigid backbone. Several flexible polymers have been grafted along the conjugated backbone, including poly(methyl acrylate),[55] polystyrene,[56] poly(ε-caprolactone) (PCL),[57] poly(N-isopropylacrylamide) (PNIPAAm)[58] and polyquinoline.[59] Of particular interest is a water-soluble, pH-sensitive molecular brush of poly(N,N-dimethylaminoethyl methacrylate) (PDMA) grafted polythiophene (PT). The good solubility of this polythiophene-based brush in a wide range of solvents is an attractive feature for fabrication of composites.[60] As shown in Figure 5.8, this polymer brush forms a more extended conformation with a decrease in pH from 8 to 2 due to the protonation of the Me$_2$N-groups and

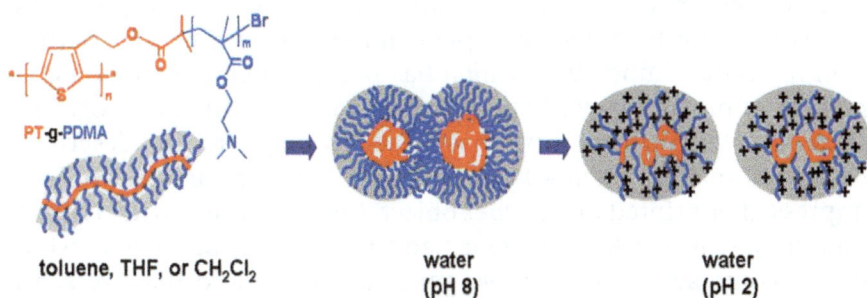

PT-g-PDMA

toluene, THF, or CH$_2$Cl$_2$ water (pH 8)   water (pH 2)

Figure 5.8 Proposed mechanism for molecular conformational transitions resulting from solvent polarity or pH changes in water. Reprinted with permission from ref. 60. Copyright (2008) American Chemical Society.

increased repulsive interactions among the PDMA side chains. As a consequence the red shift of the absorption spectra of the PT backbone are observed.

In designing stimuli-responsive polymer brushes the following considerations are typically taken into the account: transient responsiveness, concentration-dependent responsiveness, temperature, pH, solubility and reversibility. For example, reversible solubilization of thermo-responsive polymer brushes upon temperature changes will result in collapse by dehydration or expansion by dehydration. One can design two or more responsive elements into one polymer backbone. The most common are dual thermo-responsive polymers that exhibit LCST, upper critical solution temperature (UCST), or both. These properties are typically combined with pH responsiveness and may mimic natural responsive macromolecules. Although thermal response can be combined with magnetic or electrical responsiveness, usually applied to hydrogels or nanoparticles, it will take some time for these materials to be utilized in commercial applications. Although light remains an attractive source of stimuli, combining thermal and pH responses is quite appealing, particularly in generating dual responsive systems, when combined with thermal[61–64] and pH sources.[65]

Placement of monomers during polymerization into designated macromolecular chain locations formulates the best opportunity for creating specificity. Nature mastered this process by creating immunoresponses between antibodies and antigens, ligand–receptor interactions, and enzyme catalysis. These processes are governed by molecular recognition. A significant challenge for contemporary materials chemists is to develop synthetic materials that could serve as receptors with an affinity and specificities similar to nature. If existing synthetic methods are capable of placing specific monomeric units that, in combination with other units, will be able to recognize specific sites, many technological opportunities are feasible. One of them is in molecular imprinting, which has proven to be a highly versatile method of preparing synthetic polymers with tailor-made recognition sites. The term 'imprinting' has been initially used to describe a process of making a mark by applying pressure or a mark of a figure impressed or printed on another object. Genomic imprinting refers to inheritance of one kind of genes and silencing another one, which may become active later. If two genes are active and inactive at the same time, this may lead to severe developmental abnormalities. In psychobiology, on the other hand, imprinting is a time-sensitive process of learning that occurs at a particular age or life stage. In physical sciences, molecular imprinting of polymers (MIPs) refers to the ability to selectively bind to a given substrate. Because precisely designed and synthesized polymer brushes facilitate an opportunity for selective binding, there are numerous technological opportunities in separation methods, chemical sensors, drug delivery and separation, enzyme-mimicking catalysis, and others. For example the use of monomers, such as *N*-isopropylacrylamide (NIPAAm), acrylic acid and azobenzenes in molecular imprinting, along with precisely orchestrated crosslinkers, may offer stimuli responsiveness and water solubility. For example, Figure 5.9 illustrates the preparation of water-compatible and stimuli-responsive MIP microspheres containing surface grafted functional polymer brushes.[66] In this particular case stimuli-responsive MIPs were synthesized by the facile grafting of poly(NIPAAm) brushes onto the seed MIP particles *via* surface-initiated RAFT) polymerization. Taking advantage of the controllability of RAFT by the use of reversible chain transfer agents (dithioester and others) and the dynamic equilibrium between active radical species and dormant species thiocarbonylthio-terminated chains, good control over the polymer structures for a wide range of monomers and the mild reaction conditions can be achieved.

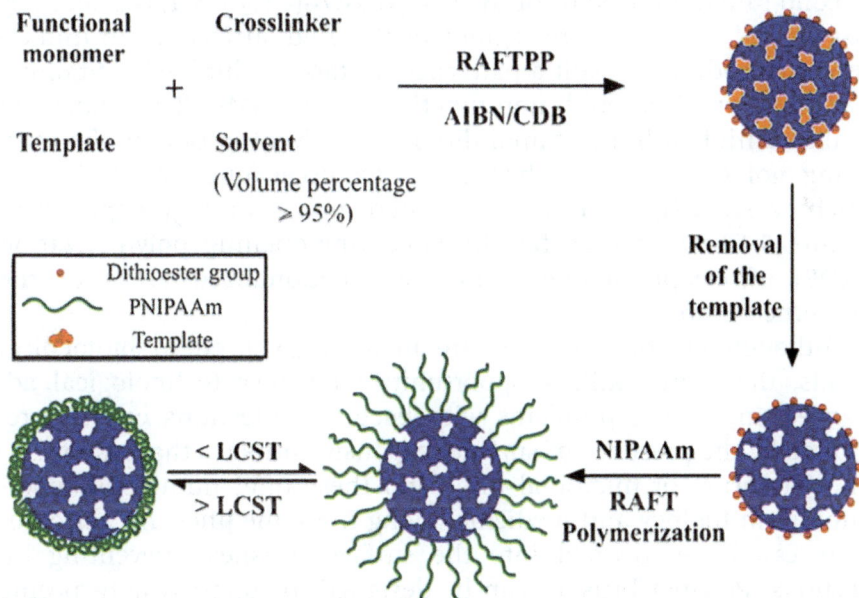

Figure 5.9 Schematic diagram illustrating the synthesis of water-compatible and thermally responsive MIP microspheres with grafted polymer brushes.
Reproduced from ref. 66 with permission from Elsevier.

5.4   Examples of Applications

Stimuli-responsive polymers respond to minute external changes with drastic responses, thus offering many technological advantages. Polymer brushes are no exception.[67] With the presence of highly congested and constrained structural features of molecular brushes one can envisage many potential applications. Remarkable scientific progress has been made in the development of various stimuli-responsive polymer brushes. The studies of responsive biointerfaces that may offer similarities to natural surfaces, controlled drug delivery and release systems, anti-fog surfaces and sensor devices, coatings that are capable of interacting and responding to their environment, composite materials that actuate and mimic the action of muscles, and thin films and particles that are capable of sensing minute concentrations of analytes, are just a few examples of technological advances that have emerged from this research. Patterned polymer brushes can be obtained using lithographic methods. As shown in Figure 5.10,[68] initially a non-reactive self-assembly monomer (SAM) formed from $CH_3-(CH_2)_{15}SH$ is placed in microcontact printed onto a gold surface and the second functionalized thiol $(HO(CH_2CH_2O)_2(CH_2)_{11}SH)$ is selectively assembled onto the bare regions of the gold surface by an immersion method. As a result a patterned surface is obtained containing OH-functionalized and non-functionalized SAMs. The choice of ε-alkanethiol with a terminal di(ethylene glycol) group as the initiating moiety was dictated by the use of simpler functionalized thiols such as $HO(CH_2)_{11}SH$. The final step in the strategy depicted in Figure 5.10 was the surface-initiated ring-opening polymerization (ROP) of ε-caprolactone from the functionalized areas of the patterned SAM.

  Although polymer brushes are increasingly used in biomedical fields, there are endless opportunities for new technological advances. Increasing problems with bacterial infections can be prevented by the presence of surface-responsive brushes that respond to the formation of microbial films and these offer numerous opportunities in biology and medicine. Along the same lines, Ag^+ or other ions can be embedded into the surface brushes, preventing infections. Polymer brushes can be designed to suppress non-specific adhesion of biomolecules to surfaces, or promote specific attachment of cell types to surfaces. Their use as sensing devices or sensors with well-defined surface morphologies can be envisaged in many applications, or simply to enhance the performance of existing sensors.

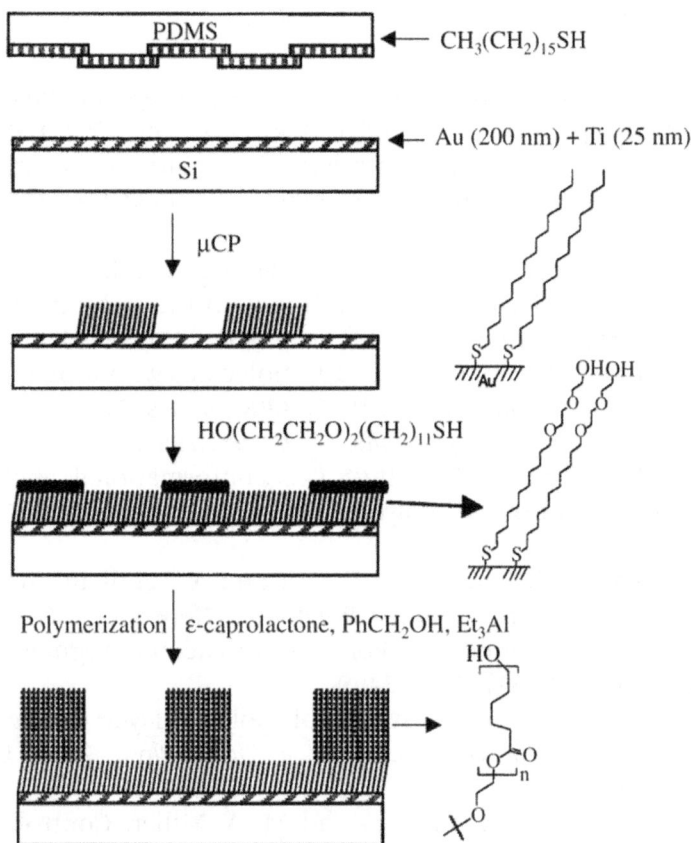

Figure 5.10 Process of forming SAMs using a lithographic approach. Reproduced from ref. 68 with permission from Elsevier.

Another huge area of interest is membrane technologies, where stimuli-responsive polymer brushes may regulate flow, diffusivity and selectivity. While these technologies have endless opportunities, the most challenging aspect in polymer brush developments is their characterization on a molecular level. Should these methods become routinely available, tremendous advances would be made. Current approaches have explored an array of well-established surface characterization methods, including attenuated total reflectance and chemical FT-IR imaging, X-ray photoelectron spectroscopy (XPS) (often referred to as electron spectroscopy for chemical analysis; ESCA), light scattering, near edge X-ray absorption fine structure (NEXAFS), scattering experiments, and secondary ion mass spectroscopy (SIMS), X-ray reflectometry, and AFM.

References

1. S. Milner, Polymer brushes, *Science*, 1991, **251**(4996), 905–914.
2. A. T. M. Halperin and T. P. Lodge, *Adv. Polym. Sci.*, 1992, **100**, 31–71.
3. M. Van der Waarden, Adsorption of aromatic hydrocarbons in nonaromatic media on carbon black, *J. Colloid Sci.*, 1951, **6**(5), 443–449.
4. E. Mackor and J. Van der Waals, The statistics of the adsorption of rod-shaped molecules in connection with the stability of certain colloidal dispersions, *J. Colloid Sci.*, 1952, **7**(5), 535–550.
5. S. Alexander, Adsorption of chain molecules with a polar head a scaling description, *J. Phys.*, 1977, **38**(8), 983–987.
6. P. J. Flory, *Principles of polymer chemistry*, 1953.
7. R. H. Boyd and S. Breitling, The conformational analysis of crankshaft motions in polyethylene, *Macromolecules*, 1974, **7**(6), 855–862.
8. J. Zotzmann, M. Behl, D. Hofmann and A. Lendlein, Reversible Triple-Shape Effect of Polymer Networks Containing Polypentadecalactone-and Poly (ε-caprolactone)-Segments, *Adv. Mater.*, 2010, **22**(31), 3424–3429.
9. D. Heine and D. T. Wu, A switchable polymer layer: Chain folding in end-charged polymer brushes, *J. Chem. Phys.*, 2001, **114**(12), 5313–5321.
10. B. M. Rubenstein, I. Coluzza and M. A. Miller, Controlling the folding and substrate-binding of proteins using polymer brushes, *Phys. Rev. Lett.*, 2012, **108**(20), 208104.
11. S. Edmondson, K. Frieda, J. E. Comrie, P. R. Onck and W. T. Huck, Buckling in Quasi-2D Polymers, *Adv. Mater.*, 2006, **18**(6), 724–728.
12. S. S. Sheiko and M. Möller, Visualization of Macromolecules a First Step to Manipulation and Controlled Response, *Chem. Rev.*, 2001, **101**(12), 4099–4124.
13. S. S. Sheiko, M. da Silva, D. Shirvaniants, I. LaRue, S. Prokhorova and M. Moeller *et al.*, Measuring molecular weight by atomic force microscopy, *J. Am. Chem. Soc.*, 2003, **125**(22), 6725–6728.
14. S. S. Sheiko, S. A. Prokhorova, K. L. Beers, K. Matyjaszewski, Potemkin II and A. R. Khokhlov *et al.*, Single molecule rod-globule phase transition for brush molecules at a flat interface, *Macromolecules*, 2001, **34**(23), 8354–8360.
15. R. Djalali, S.-Y. Li and M. Schmidt, Amphipolar core-shell cylindrical brushes as templates for the formation of gold clusters and nanowires, *Macromolecules*, 2002, **35**(11), 4282–4288.

16. M. Zhang, T. Breiner, H. Mori and A. H. Müller, Amphiphilic cylindrical brushes with poly (acrylic acid) core and poly (n-butyl acrylate) shell and narrow length distribution, *Polymer*, 2003, **44**(5), 1449–1458.
17. G. Cheng, A. Böker, M. Zhang, G. Krausch and A. H. Müller, Amphiphilic cylindrical core-shell brushes via a "grafting from" process using ATRP, *Macromolecules*, 2001, **34**(20), 6883–6888.
18. J. Seog, D. Dean, A. Plaas, S. Wong-Palms, A. Grodzinsky and C. Ortiz, Direct measurement of glycosaminoglycan inter-molecular interactions via high-resolution force spectroscopy, *Macromolecules*, 2002, **35**(14), 5601–5615.
19. A. Bhattacharya and B. Misra, Grafting: a versatile means to modify polymers: techniques, factors and applications, *Prog. Polym. Sci.*, 2004, **29**(8), 767–814.
20. S. S. Sheiko, B. S. Sumerlin and K. Matyjaszewski, Cylindrical molecular brushes: Synthesis, characterization, and properties, *Prog. Polym. Sci.*, 2008, **33**(7), 759–785.
21. M. Wintermantel, M. Gerle, K. Fischer, M. Schmidt, I. Wataoka and H. Urakawa *et al.*, Molecular bottlebrushes, *Macromolecules*, 1996, **29**(3), 978–983.
22. S. Kawaguchi, K. Akaike, Z.-M. Zhang, H. Matsumoto and K. Ito, Water soluble bottlebrushes, *Polym. J.*, 1998, **30**(12), 1004–1007.
23. H. G. Börner, D. Duran, K. Matyjaszewski, M. da Silva and S. S. Sheiko, Synthesis of molecular brushes with gradient in grafting density by atom transfer polymerization, *Macromolecules*, 2002, **35**(9), 3387–3394.
24. H.-i. Lee, K. Matyjaszewski, S. Yu and S. S. Sheiko, Molecular brushes with spontaneous gradient by atom transfer radical polymerization, *Macromolecules*, 2005, **38**(20), 8264–8271.
25. N. Khelfallah, N. Gunari, K. Fischer, G. Gkogkas, N. Hadjichristidis and M. Schmidt, Micelles Formed by Cylindrical Brush-Coil Block Copolymers, *Macromol. Rapid Commun.*, 2005, **26**(21), 1693–1697.
26. M. W. Neiser, S. Muth, U. Kolb, J. R. Harris, J. Okuda and M. Schmidt, Micelle formation from amphiphilic "cylindrical brush"—coil block copolymers prepared by metallocene catalysis, *Angew. Chem., Int. Ed.*, 2004, **43**(24), 3192–3195.
27. M. Zhang, A. Müller, P. Teissier, V. Cabuil, M. Krekhova, Poly-chelates of amphiphilic cylindrical core–shell polymer brushes with iron cations, *Trends in Colloid and Interface Science XVII*, Springer, 2004. pp. 35–39.

28. M. Zhang, C. Estournes, W. Bietsch and A. Müller, Super-paramagnetic hybrid nanocylinders, *Adv. Funct. Mater.*, 2004, **14**(9), 871–882.

29. K. Ishizu and H. Kakinuma, Synthesis of nanocylinders consisting of graft block copolymers by the photo-induced ATRP technique, *J. Polym. Sci., Part A: Polym. Chem.*, 2005, **43**(1), 63–70.

30. H.-i. Lee, W. Jakubowski, K. Matyjaszewski, S. Yu and S. S. Sheiko, Cylindrical core-shell brushes prepared by a combination of ROP and ATRP, *Macromolecules*, 2006, **39**(15), 4983–4989.

31. D. Neugebauer, Y. Zhang, T. Pakula and K. Matyjaszewski, Heterografted PEO–PnBA brush copolymers, *Polymer*, 2003, **44**(22), 6863–6871.

32. K. Ishizu, Architecture of multi-component copolymer brushes: synthesis, solution properties and application for nanodevices, *Polym. J.*, 2004, **36**(10), 775–792.

33. K. Ishizu, J. Satoh and A. Sogabe, Architecture and solution properties of AB-type brush–block–brush amphiphilic copolymers via ATRP techniques, *J. Colloid Interface Sci.*, 2004, **274**(2), 472–479.

34. A. Subbotin, M. Saariaho, O. Ikkala and G. Ten Brinke, Elasticity of comb copolymer cylindrical brushes, *Macromolecules*, 2000, **33**(9), 3447–3452.

35. S. W. Ryu and A. Hirao, Anionic synthesis of well-defined poly (m-halomethylstyrene) s and branched polymers via graft-onto methodology, *Macromolecules*, 2000, **33**(13), 4765–4771.

36. Y. Tsukahara, S. Kohjiya, K. Tsutsumi and Y. Okamoto, On the intrinsic viscosity of poly (macromonomer) s, *Macromolecules*, 1994, **27**(6), 1662–1664.

37. A. Subbotin, M. Saariaho, R. Stepanyan, O. Ikkala and G. Ten Brinke, Cylindrical brushes of comb copolymer molecules containing rigid side chains, *Macromolecules*, 2000, **33**(16), 6168–6173.

38. M. Gerle, K. Fischer, S. Roos, A. H. Müller, M. Schmidt and S. S. Sheiko *et al.*, Main chain conformation and anomalous elution behavior of cylindrical brushes as revealed by GPC/MALLS, light scattering, and SFM, *Macromolecules*, 1999, **32**(8), 2629–2637.

39. M. Wintermantel, M. Schmidt, Y. Tsukahara, K. Kajiwara and S. Kohjiya, Rodlike combs, *Macromol. Rapid Commun.*, 1994, **15**(3), 279–284.

40. G. C. Berry, S. Kahle, S. Ohno, K. Matyjaszewski and T. Pakula, Viscoelastic and dielectric studies on comb- and brush-shaped poly (n-butyl acrylate), *Polymer*, 2008, **49**(16), 3533–3540.

41. K. L. Beers, S. G. Gaynor, K. Matyjaszewski, S. S. Sheiko and M. Möller, The synthesis of densely grafted copolymers by atom transfer radical polymerization, *Macromolecules*, 1998, **31**(26), 9413–9415.
42. B. S. Sumerlin, D. Neugebauer and K. Matyjaszewski, Initiation efficiency in the synthesis of molecular brushes by grafting from via atom transfer radical polymerization, *Macromolecules*, 2005, **38**(3), 702–708.
43. A. Zhang, J. Barner, I. Göessl, J. P. Rabe and A. D. Schlüter, A Covalent-Chemistry Approach to Giant Macromolecules and Their Wetting Behavior on Solid Substrates, *Angew. Chem.*, 2004, **116**(39), 5297–5300.
44. R. Venkatesh, L. Yajjou, C. E. Koning and B. Klumperman, Novel brush copolymers via controlled radical polymerization, *Macromol. Chem. Phys.*, 2004, **205**(16), 2161–2168.
45. D. Neugebauer, B. S. Sumerlin, K. Matyjaszewski, B. Goodhart and S. S. Sheiko, How dense are cylindrical brushes grafted from a multifunctional macroinitiator?, *Polymer*, 2004, **45**(24), 8173–8179.
46. A. B. Lowe and C. L. McCormick, Reversible addition–fragmentation chain transfer (RAFT) radical polymerization and the synthesis of water-soluble (co) polymers under homogeneous conditions in organic and aqueous media, *Prog. Polym. Sci.*, 2007, **32**(3), 283–351.
47. H.-i. Lee, J. Pietrasik, S. S. Sheiko and K. Matyjaszewski, Stimuli-responsive molecular brushes, *Prog. Polym. Sci.*, 2010, **35**(1), 24–44.
48. M. D. Rowe, B. A. Hammer and S. G. Boyes, Synthesis of surface-initiated stimuli-responsive diblock copolymer brushes utilizing a combination of ATRP and RAFT polymerization techniques, *Macromolecules*, 2008, **41**(12), 4147–4157.
49. Z. Ge and S. Liu, Supramolecular Self-Assembly of Nonlinear Amphiphilic and Double Hydrophilic Block Copolymers in Aqueous Solutions, *Macromol. Rapid Commun.*, 2009, **30**(18), 1523–1532.
50. Y. Wang, W. B. Euler and B. L. Lucht, Unusual chromic and doping behavior of ether substituted polythiophenes, *Chem. Commun.*, 2004, **6**, 686–687.
51. D. T. McQuade, A. E. Pullen and T. M. Swager, Conjugated polymer-based chemical sensors, *Chem. Rev.*, 2000, **100**(7), 2537–2574.
52. R. D. McCullough, P. C. Ewbank and R. S. Loewe, Self-assembly and disassembly of regioregular, water soluble polythiophenes:

chemoselective ionchromatic sensing in water, *J. Am. Chem. Soc.*, 1997, **119**(3), 633–634.

53. M. R. Pinto, B. M. Kristal and K. S. Schanze, A water-soluble poly (phenylene ethynylene) with pendant phosphonate groups. Synthesis, photophysics, and layer-by-layer self-assembled films, *Langmuir*, 2003, **19**(16), 6523–6533.
54. F. Wang and G. C. Bazan, Aggregation-mediated optical properties of pH-responsive anionic conjugated polyelectrolytes, *J. Am. Chem. Soc.*, 2006, **128**(49), 15786–15792.
55. P. J. Costanzo and K. K. Stokes, Synthesis and characterization of poly (methyl acrylate) grafted from poly (thiophene) to form solid-state fluorescent materials, *Macromolecules*, 2002, **35**(18), 6804–6810.
56. C. A. Breen, T. Deng, T. Breiner, E. L. Thomas and T. M. Swager, Polarized Photoluminescence from Poly (p-phenylene-ethynylene) via a Block Copolymer Nanotemplate, *J. Am. Chem. Soc.*, 2003, **125**(33), 9942–9943.
57. I. B. Kim, B. Erdogan, J. N. Wilson and U. H. Bunz, Sugar–Poly (para-phenylene ethynylene) Conjugates as Sensory Materials: Efficient Quenching by Hg2+ and Pb2+ Ions, *Chem. – Eur. J.*, 2004, **10**(24), 6247–6254.
58. S. S. Balamurugan, G. B. Bantchev, Y. Yang and R. L. McCarley, Highly Water-Soluble Thermally Responsive Poly (thiophene)-Based Brushes, *Angew. Chem.*, 2005, **117**(31), 4950–4954.
59. S. P. Economopoulos, C. L. Chochos, V. G. Gregoriou, J. K. Kallitsis, S. Barrau and G. Hadziioannou, Novel brush-type copolymers bearing thiophene backbone and side chain quinoline blocks. Synthesis and their use as a compatibilizer in thiophene-quinoline polymer blends, *Macromolecules*, 2007, **40**(4), 921–927.
60. M. Wang, S. Zou, G. Guerin, L. Shen, K. Deng and M. Jones *et al.*, A water-soluble pH-responsive molecular brush of poly (N, N-dimethylaminoethyl methacrylate) grafted polythiophene, *Macromolecules*, 2008, **41**(19), 6993–7002.
61. M. Irie and D. Kunwatchakun, Photoresponsive polymers. 8. Reversible photostimulated dilation of polyacrylamide gels having triphenylmethane leuco derivatives, *Macromolecules*, 1986, **19**(10), 2476–2480.
62. K. Sugiyama and K. Sono, Characterization of photo- and thermoresponsible amphiphilic copolymers having azobenzene moieties as side groups, *J. Appl. Polym. Sci.*, 2001, **81**(12), 3056–3063.

63. R. Kröger, H. Menzel and M. L. Hallensleben, Light controlled solubility change of polymers: Copolymers of N, N-dimethylacrylamide and 4-phenylazophenyl acrylate, *Macromol. Chem. Phys.*, 1994, **195**(7), 2291–2298.
64. H. Menzel, B. Weichart, A. Schmidt, S. Paul, W. Knoll and J. Stumpe *et al.*, Small-angle X-ray scattering and ultraviolet-visible spectroscopy studies on the structure and structural changes in Langmuir-Blodgett films of polyglutamates with azobenzene moieties tethered by alkyl spacers of different length, *Langmuir*, 1994, **10**(6), 1926–1933.
65. L. Wu, X. Tuo, H. Cheng, Z. Chen and X. Wang, Synthesis, photoresponsive behavior, and self-assembly of poly (acrylic acid)-based azo polyelectrolytes, *Macromolecules*, 2001, **34**(23), 8005–8013.
66. G. Pan, Y. Zhang, X. Guo, C. Li and H. Zhang, An efficient approach to obtaining water-compatible and stimuli-responsive molecularly imprinted polymers by the facile surface-grafting of functional polymer brushes via RAFT polymerization, *Biosens. Bioelectron.*, 2010, **26**(3), 976–982.
67. S. Minko, Responsive polymer brushes, *J. Macromol. Sci., Polym. Rev.*, 2006, **46**(4), 397–420.
68. M. Husemann, D. Mecerreyes, C. J. Hawker, J. L. Hedrick, R. Shah and N. L. Abbott, Surface-initiated polymerization for amplification of self-assembled monolayers patterned by microcontact printing, *Angew. Chem., Int. Ed.*, 1999, **38**(5), 647–649.

6 Stimuli Responsiveness in Nano and Micro Materials

6.1 Introduction

It is interesting to look back and realize that it took over half a century to comprehend that arranging atoms by atoms will formulate today's science that feeds technological advances. Chemists and molecular physicists did this all along. Since Feynman's 1959 lecture on technology and engineering, entitled *There's Plenty of Room at the Bottom*, his vision of atomic fabrication was cited in the 2000 US presidential address, which turned into action by establishing the National Nanotechnology Initiative. Interestingly, the first patents on colloidal nanoparticle synthesis were filed in the 1940s to 1950s.[1,2] In essence, nanotechnology is not the ability to link atoms and molecules at the atomic level, which chemists do on a daily basis, but to create higher-level structural organizations that may offer unique properties in a controllable manner. If one looks at the complexity of the second-by-second events governing the process of vision facilitated by our eyes, the sequences of bio-driven reactions facilitate this process during which orchestrated events are responsible for sequential events. While the objectives that launched the field of nanotechnology are very powerful, the challenges are enormous and often frightening, due to an imbalance between synthetic and analytical methods. The lack of methods capable of obtaining highly selective chemical and morphological spatial information is one of the limiting factors governing further developments, particularly at nano- or sub-nanoscale levels of detection. As a result, a significant amount of the literature is

Stimuli-Responsive Materials: From Molecules to Nature Mimicking Materials Design
By Marek W. Urban
© Marek W. Urban, 2016
Published by the Royal Society of Chemistry, www.rsc.org

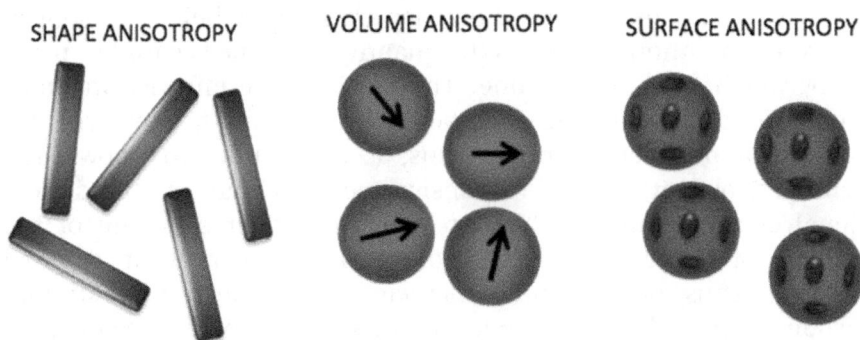

Figure 6.1 Shape, volume and surface anisotropy in nano-objects.

loaded with too many guesses and often irreproducible results. One could argue that Schrödinger stated we will never see an experiment with one electron, atom or molecule, while 8 years later Feynman advocated that there are no limits in arranging atoms. Neither of them knew that it took two decades to discover scanning tunneling microscopy (STM), which changed views on the visualization of selected single atoms and molecules. Is this enough, however? Synthetic polymer chemists realize that molecular weight determination is based on the 1950s gel permeation chromatography (GPC) concept. As the quest for more accurate and more sensitive probes continues, this chapter focuses only on selected synthetic aspects of stimuli-responsive developments on nano-objects and limited characterization approaches, with a particular focus on shape, volume and surface anisotropies schematically depicted in Figure 6.1.

6.2 Nano Stimuli-responsive Systems

Not too long ago, inorganic nanoparticles were considered to be simply 'dirt'. In fact, in many applications it was a nightmare to figure out why 'dirt' particles led to film defects. The answer is very simple: if 'dirt' particles landed on a wet surface, or if nanoparticles were not compatible with a liquid phase, due to surface tension differences between the low surface energy 'dirt' particles and a wet, high surface tension mixture of chemicals, the defect known as cratering would develop. Cratering results from the flow of liquid caused by the gradients of surface tension during film formation with mismatched miscibility. In an attempt to obtain a minimum energy, a liquid will flow from an area of low surface tension to compensate for the area with the higher surface energy. Thus, in the design of nanoparticles one has to take into account how they are going to behave in a given environment. Another issue is the quantity. The rule is simple – never go by weight, but use volume. The particles are effective in many applications if they replace a certain volume of another material, for example a polymeric binder. Thus, it is essential to know their density. However, stimuli-responsive particles that exhibit dimensional changes will also have to reorganize their structure or the degrees of aggregation in response to an internal or external stimulus. Thus volume content may vary. Furthermore, multi-stimuli-responsive particles that respond to a combination of three or more responses are particularly critical to many new developments in biology and medicine. Considering significant advances in stimuli-responsive polymeric micelles, nanoparticles, vesicles and hybrids, the next section will focus on these materials.

6.2.1 Polymeric Micelles

Stimuli-responsive polymeric micelles are formed by self-assembly of amphiphilic copolymers and are among the most investigated nanoparticles. These micelles are created by a spontaneous self-assembly of individual polymeric molecules, also known as unimers, which are synthetic amphiphilic diblock or triblock copolymers comprised of hydrophilic and hydrophobic blocks. Figure 6.2 illustrates a classical example of the assembly of micelle-like particles from polyelectrolyte–ionic surfactant complex and their disassembly upon protein binding.[3] This simple strategy for pattern recognition of proteins through micellar disassembly allows non-covalently assembled receptors sense binding of proteins. Their shape can be spherical core–shell type, typically with a hydrophobic core and hydrophilic shell.

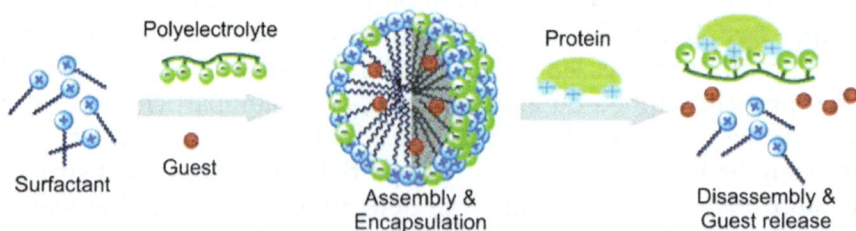

Figure 6.2   Schematic diagram illustrating micelle-like particles assembled from a polyelectrolyte–ionic surfactant complex, followed by disassembly due to protein binding.
Reprinted with permission from ref. 3. Copyright (2008) American Chemical Society.

The most commonly used hydrophilic blocks are poly(ethylene oxide) (PEO) (or poly(ethylene glycol) (PEG), and the choice of hydrophobic block is mostly dictated by thermodynamic and kinetic stability of the micelle. The following hydrophobic groups have often been used in polymeric micelles anticipated as drug carriers:[4–7] poly(L-amino acids);[8] poly(esters), such as poly(glycolic acid), poly(L-lactic acid), poly(D,L-lactic acid), lactide/glycolide copolymers, and poly(e-capro-lactone); phospholipids/long-chain fatty acids.[9] Examples of triblocks include the center block of poly(propylene oxide) (PPO).[6] The thermodynamic stability is facilitated by molecular interactions in a given micellar system while the kinetic stability is dependent upon the rate of disassembly of hydrophobic blocks of the core. The latter will depend on a block length and the block glass transition temperature (T_g).

Primarily driven by applications in drug delivery and gene transport systems,[10–12] it is worth pointing out that wormlike micelles can be triggered by redox reactions, UV/vis light, temperature, pH, CO_2, and a combination of stimuli. An illustrative example of multi-stimuli-responsive polymeric micelles is shown in Figure 6.3.[13] These

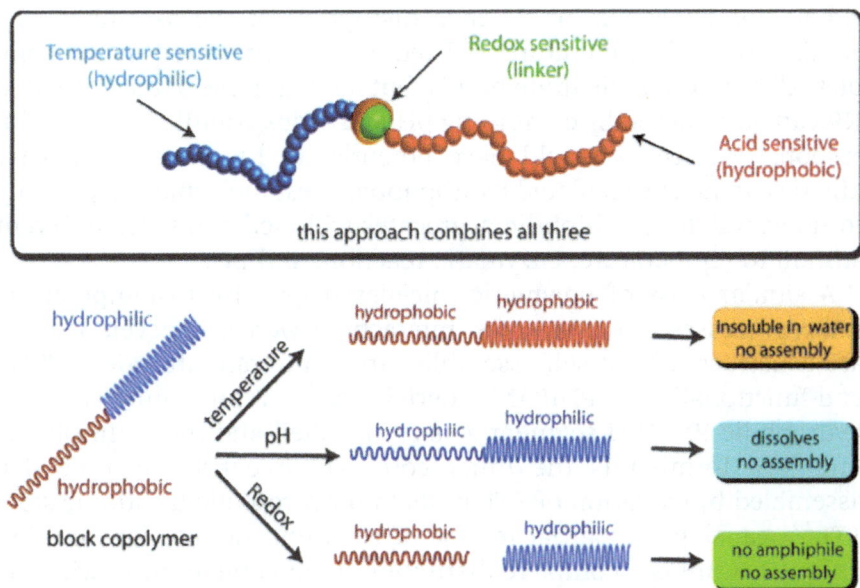

Figure 6.3 Schematic depiction of amphiphilic block copolymer capable of responding to pH, temperature and redox.
Reprinted with permission from ref. 13. Copyright (2009) American Chemical Society.

multi-stimuli-responsive supramolecular aggregates result from the complexation of a surfactant and an oppositely charged homopolymer consisting of a disulfide bond. Nile red (NR) is a model hydrophobic dye encapsulated in supramolecular structures. It turns out that the stimulus-induced disassembly of these micelles could be achieved by high salt concentrations and lowering the pH by diminishing the electrostatic interactions between the polymer and the surfactant. Interestingly, the addition of glutathione disrupts these aggregates through a reductive disulfide bond cleavage reaction. The triple stimuli sensitive block copolymer containing thermo-responsive poly(*N*-isopropylacrylamide) (PNIPAM) as hydrophilic part is attached to acid-responsive tetrahydropyran (THP)-protected 2-hydroxyethyl (HEMA) as hydrophobic part with a redox-responsive disulfide linker.

One example of multi-stimuli responsiveness of the amphiphilic block copolymer is schematically shown in Figure 6.3. By lowering the pH the acid sensitive hydrophobic blocks are converted to hydrophilic entities, resulting in the dissolution. However, above the lower critical solution temperature (LCST), the thermo-responsive PNIPAM block is transformed from hydrophilic to hydrophobic, making the polymer insoluble in water, thus disrupting the assembly. a reducing environment afforded the scission of the block copolymer into individual homopolymers and hence disruption of the assembly. The combination of temperature and redox stimuli resulted in the complete disruption of disulfide bonds, but faster release of encapsulated NR can be achieved by combining pH and redox stimuli. These multiresponsive micelles would have potential to fine-tune the release kinetics of the encapsulated hydrophobic guest molecules. Expending to non-covalent amphiphilic aggregates will facilitate three different stimuli to temperature, enzymatic reaction, and pH.

A similar class of polymeric micelles responsive to temperature, redox processes, and host–guest interactions were also developed and are constructed from self-assemblies from a tetrathiafulvalene (TTF) end-functionalized PNIPAM derivative.[14] Temperature-sensitive hydrophilic PNIPAM segment acts as the shell and the hydrophobic unit TTF forming is the inner core. These entities can be disassembled by oxidation of TTF units into hydrophilic dicationic state TTF^{2+}, causing the partial release of NR from the interior. Another unique feature is a complete disruption upon complexation with the tetracationic macrocycle cyclobis (paraquat-*p*-phenylene) ($CBPQT^{4+}$). $CBPQT^{4+}$ end-functionalized poly (*n*-butyl acrylate) *via* RAFT polymerization can be also assembled by pseudorotaxane-like linkages. These polymeric micelles will disassemble in response to electrochemical oxidation/reduction, temperature, and the addition of competitive molecular guests.[15]

Boronic acids easily react with dialcohols to give boronate esters.[16] Despite the fact that the covalent B–O bonds are stronger than typical C–O bonds, an interesting feature of this condensation reaction is that it is fast and reversible.[17] Polymer boronic acids are well known for their pH- and diol-responsive solubility, and the hydrophilicity can be increased under alkaline conditions or upon esterification with diols (*e.g.*, glucose). However, well-defined organoboran copolymers were prepared by employing atom transfer radical polymerization (ATRP)[18–20] and reversible addition-fragmentation chain transfer (RAFT) polymerization.[21–24] There are numerous follow-up studies that used a variety of responsive units that self-assemble to form polymeric micelles.

Polyether amines (PEAs) is another class of multi-stimuli-responsive polymers which exhibit strong temperature, pH, and ionic strength responses. Their uniqueness is a tunable cloud point (CP).[25] The grafted PEAs consist of hydrophilic poly(ethylene oxide) chains and hydrophobic poly (propylene oxide) chains. Amphiphilic PEAs comprised of hydrophilic PEO and hydrophobic octadecyl alkyl chains as side-chain poly(dimethylsiloxane)-containing poly(ether amine) as well as fluorinated (ether amine)s can be also prepared.[26] These materials can self-assemble into micelles and exhibit responses to temperature, pH, and ionic strength. Nanoparticles produced from amphiphilic azobenzene-contained hyperbranched PEA exhibit sensitivity not only to temperature, pH, and ionic strength, but also to UV light.[27] As anticipated, this is achieved by incorporating the light sensitive azobenzene units into the hyperbranched polymer. Star-graft quarterpolymer $PS_n[P2VP-b-(PAA-g-PNIPAM)]_n$ (PS: polystyrene, P2VP: poly (2-vinylpyridine)), is responsive to pH, temperature, and ionic strength.[28] The presence of ampholytic P2VP-*b*-PAA arms and grafted PNIPAM blocks make this copolymer responsive and the intramolecular interactions between different blocks and spatial distribution of the arms can be controlled by the external stimuli.

Poly(2-*N*,*N*-dimethylamino) ethyl methacrylate) (PDMAEMA) exhibits intriguing structural and property changes in response to CO_2/pH and temperature. The pH responsiveness of PDMAEMA is well established and attributed to protonation–deprotonation of the tertiary amine side groups. The PDMAEMA homopolymer exhibits a LCST in the range ~ 32–$46\,^\circ\mathrm{C}$. Responsive polymer micelle self-assembled from a pyrene-functionalized PDMAEMA can be synthesized by quaternization of the polymer PDMAEMA with (bromo-methyl) pyrene, which results in micelles that can be dissociated upon irradiation with UV light.[29] This is attributed to photo induced cleavage reaction of PDMAEMA containing pyrene and the disruption of hydrophilic–hydrophobic balance. Above LCST the micelles will shrink because the hydrophilicity of PDMAEMA segments is decreased, but acidic conditions will make them swollen due to the presence of protonated tertiary amines.

Hierarchical self-assembly of the supramolecular entities can be synthesized under different conditions with an objective to release drugs upon temperature, pH, or other guest molecules. Supramolecular micelles can readily self-assembled from ternary complexes consisting of end-functionalized PDMAEMA and PNIPAM with CB[8] acting as a dynamic 'handcuff'. Upon titration with a competitive guest for the CB[8], such as adamantaneamine, polymeric micelles shrink.[30] This is an example of how encapsulation and controlled drug release can be utilized with the chemotherapeutic drug doxorubicin (DOX). This triple stimuli-responsive double hydrophilic block copolymer micelle system represents step forward over conventional double stimuli-responsive covalent diblock copolymer systems with a significant reduction in the viability of HeLa cells upon triggered release of DOX from the supramolecular micellar nanocontainers. Noteworthy, there are many multifunctional stimuli-responsive polymeric micelle studies that integrate targeted therapeutic and diagnostic functions within a single drug delivery system.

Multi-stimuli responsive micellar systems can be prepared from photocleavable block copolymers. As shown in Figure 6.4, light irradiation at $\lambda = 300$ nm will cause the transformation of poly (*para*-methoxyphenacyl methacrylate) (PMPMA) into poly (methacrylic acid) (PMAA) and to the disruption of the initial micelles.[31] The double hydrophilic PMAA-*b*-POEGMA copolymer obtained after irradiation exhibits stimuli-responsive behavior in solution manifested by the formation of micelles with a POEGMA core and a PMAA corona can be induced by a thermal stimulus and by the addition of PO_4^{3-} anions. The formation of the inverse micelles with a PMAA core and a POEGMA corona can be promoted by a variation of pH and by the addition of Ca^{2+} cations into the solution.

Many emerging applications require the development of multifunctional stimuli-responsive anisotropic polymeric micelles that integrate targeted therapeutic and diagnostic functions. Figure 6.5 illustrates multi-stimuli-responsive fluorescence anisotropic micelles obtained through the direct aqueous self-assembly of the amphiphilic diblock copolymers Py-PPDO$_7$-*b*-PEG$_{16}$.[32] During

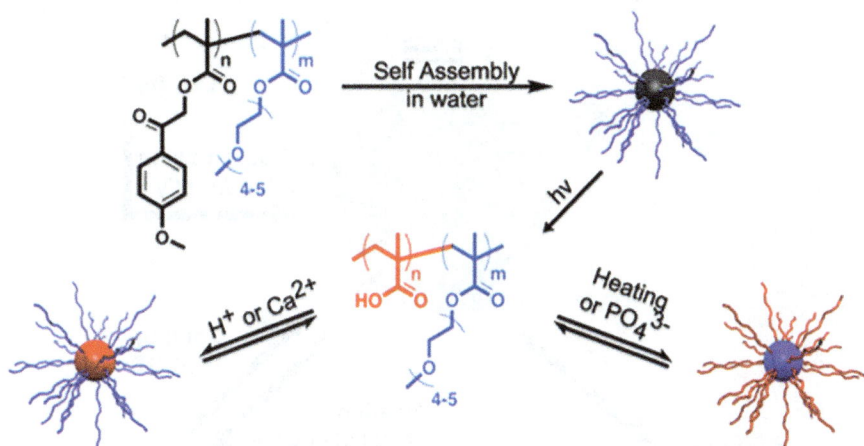

Figure 6.4 Response of poly(*para*-methoxyphenacyl methacrylate)-*block*-poly[(oligo ethylene glycol) methacrylate] (PMPMA-*b*-POEGMA) copolymer to pH, temperature, Ca^{2+}, and PO_4^{3-} ions in water. Reprinted with permission from ref. 31. Copyright (2012) American Chemical Society.

addition of Fe^{3+}, the resulting fluorescent anisotropic micelles were transformed to semicircular nanosheets, leading to 'switch off' of fluorescence due to the specific complexation of Fe^{3+} and pyrene. The addition of DL-dithiothreitol (DDT) will turn on the fluorescence and lead to the formation of new nanoparticles. Notably, the on–off switchable fluorescence and morphological transition can be reversibly repeated multiple times *via* sequentially adding DTT and Fe^{3+}. Furthermore, the micelles exhibit reversible dual-thermal responsiveness of fluorescence and morphology corresponding to the upper critical solution temperature and a LCST.

Although there are numerous excellent studies in this area, volume limitations forced us to include only selected processes and applications. The reader is directed to several excellent review articles on this topic which primarily deal with nanocarriers in drug delivery systems.[33–37] This chapter will focus will be on the development of novel stimuli-responsive nanoparticle morphologies.

6.2.2 Stimuli-responsive Nanoparticles

There are two aspects of stimuli-responsive nanoparticles: surface stimuli-responsive nanoparticles, which are achieved by surface modifications, and solid stimuli-responsive nano particles. In both categories, there are organic and inorganic particles. This section

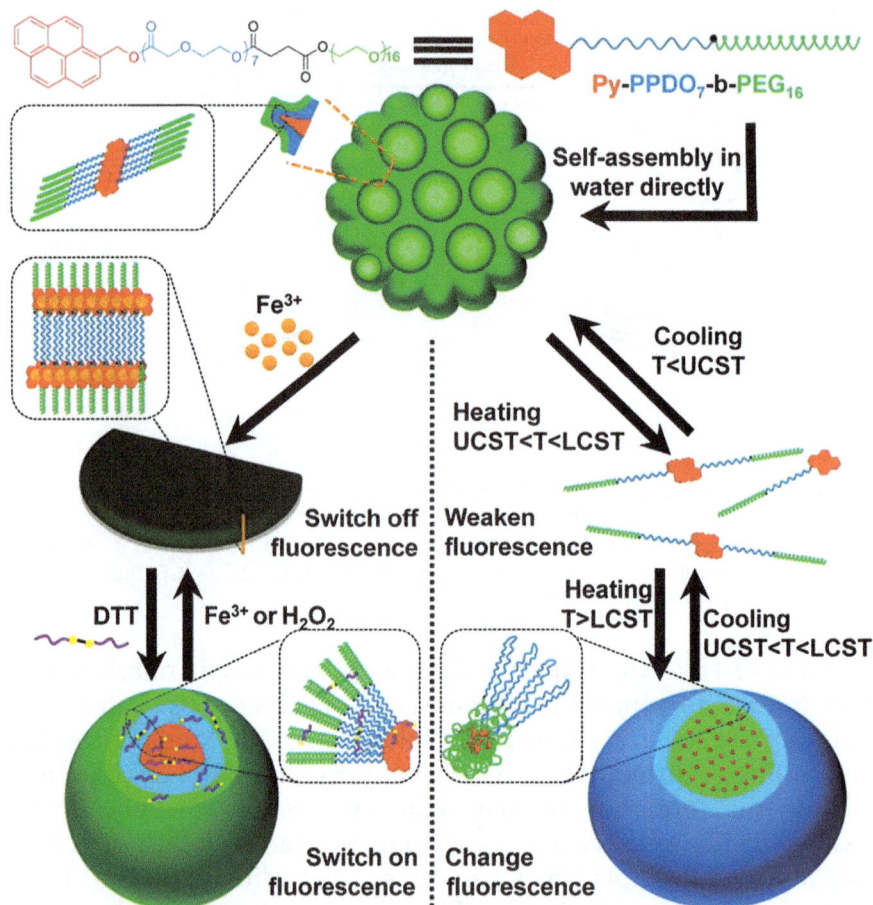

Figure 6.5 Self-assembly of pyrene-functionalized amphiphilic copolymers (Py-PPDO$_7$-b-PEG$_{16}$) and the multi-stimuli (Fe^{3+}, DTT, H$_2$O$_2$ and thermal) responsive fluorescence and morphology of the fluorescent anisotropic micelles.
Reprinted with permission from ref. 32. Copyright (2015) American Chemical Society.

focuses on solid nanoparticles with built-in stimuli responsiveness because they have many potential applications as they may serve as individual entities, or form organized 2–3D structures. They may be organic, inorganic or hybrids.

6.2.3 Nanoparticle Morphologies

The most common particle morphologies resulting from synthetic efforts exhibit spherical shapes,[38,39] which may or may not be

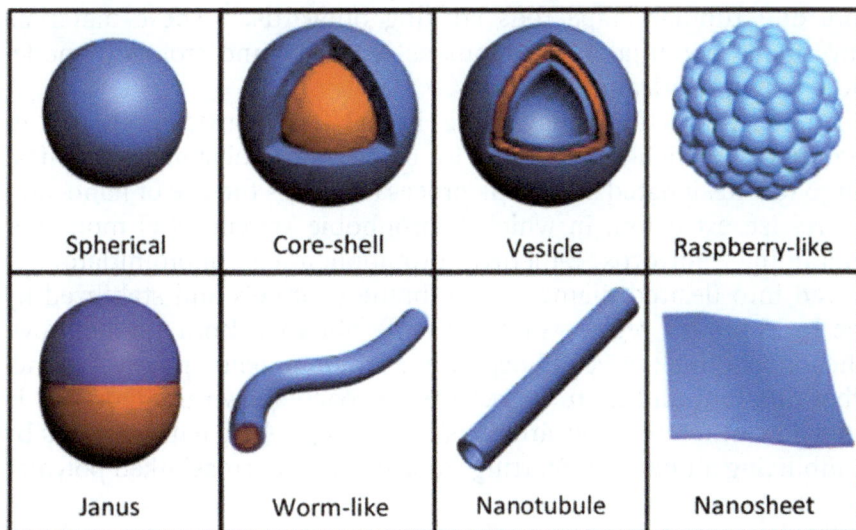

Figure 6.6 Example of various shapes of nanomaterials.

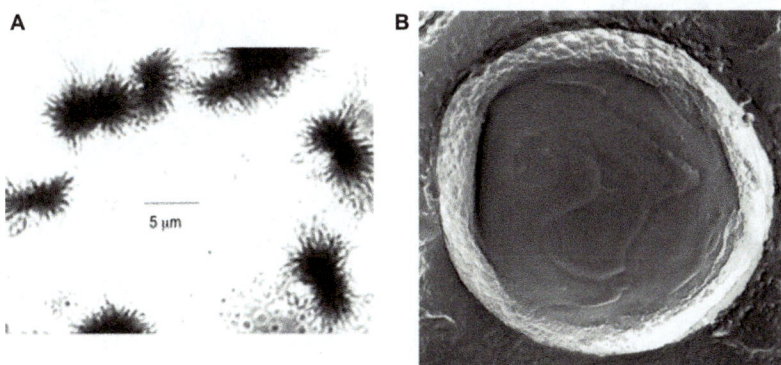

Figure 6.7 (A) TEM micrograph at 3000× magnification of large MMA/nBA copolymer particles stabilized by SDOSS/DCPC mixture. Reprinted with permission from ref. 43. © 2005 American Chemical Society. (B) Hollow colloidal nanoparticles can store a cargo in the interior. Reprinted with permission from ref. 44. Copyright (2005) American Chemical Society.

homogeneous. Figure 6.6 illustrates several examples of various nano particles and other shapes that are of particular interest. As shown in Figure 6.6, core–shell type particles, or similar morphologies can be easily prepared by semi-continuous polymerization, but obtaining anisotropic shapes requires more advanced approaches.[39–42]

Figure 6.7A shows scanning electron microscopy (SEM) images of particles that exhibit a cocklebur-shape and consist of the spherical

core and tubular-shape rods sticking outwards.[43] These materials
have many potential applications as receptors and storage nanodevices for biomedical applications.

Different functions may serve hollow nanoparticles should in
Figure 6.7B, which may be used for delivering a substantial chemical
cargo to a designated area. This process involves the use of nano- and
micro-size extrusion, in which hydrophobic styrene (Sty) monomer
containing 2-hydroxy-2-methyl propiophenone photoinitiator is
forced into desired diameter membrane channels and stabilized by
the hydrophobic regions of a liposome obtained from 1,2-dilauroyl-
phosphocholine (DLPC) phospholipid in an aqueous phase.[44] Somewhat different and more aggressive approach shown in Figure 6.8 is
driven by different applications of obtaining hollow nanoparticles by
solubilizing an interior. Starting from silica core, crosslinked polymer

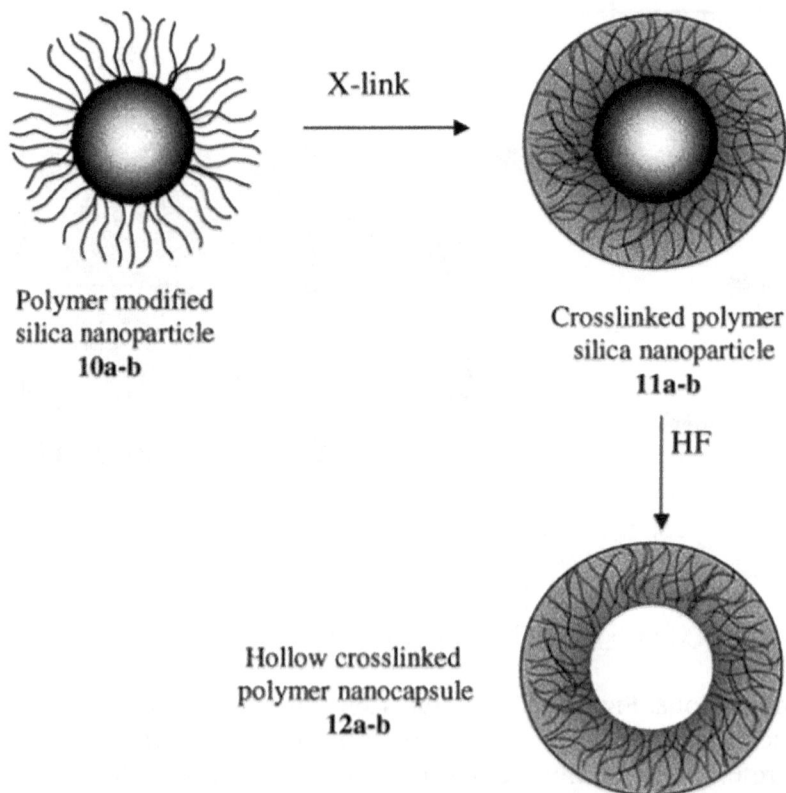

Polymer modified
silica nanoparticle
10a-b

X-link

Crosslinked polymer
silica nanoparticle
11a-b

HF

Hollow crosslinked
polymer nanocapsule
12a-b

Figure 6.8 The formation of crosslinked hollow nanoparticles by dissolv-
ing an inorganic silica core.
Reproduced with permission from ref. 45. Copyright © 2002
Wiley Periodicals, Inc.

formed a nanocapsule by dissolving silica core using hydrogen fluoride solution.[45] It should be noted that the strategies of creating hollow or voided particles is not new, as reflected in the patent literature from the 50s[46] and later,[47] as well as the review articles.[48,49]

6.2.4 Janus, Tunable Shape and Color-tunable Nanoparticles

While imparting distinctly different physical and/or chemical properties into single colloid, Janus particles (JPs) have attracted widespread attention because of their numerous technological advantages.[50–53] Janus, the Roman God of Gates and Doors has a distinctive artistic appearance commonly depicted with two faces, one regarding what is behind and the other one looking towards what lies ahead. As shown in Figure 6.9, there are two beautiful faces with the distinctly mirror image features pointing in opposite directions. So the one face remembers the past, while another is engaged in another activity. If nanoparticles can do that, stimuli responsiveness can be achieved at nanoscale levels. A few methods of preparation of Janus nanoparticles are being used. They include microfluidics,[50,54,55] block

Figure 6.9 Janus the Roman God is the god of beginnings and transitions, gates, doors, doorways, passages, and endings.

copolymer assembly,[56–59] masking technique,[60–63] heterogeneous nucleation,[64,65] flame synthesis,[66] and emulsion polymerization.[67–69] Of particular scientific interest, and perhaps most challenging is the synthesis of large quantities of Janus nanoparticles (JNPs) with precisely defined morphologies capable of responding to a variety of external or internal stimuli.[70,71] The presence of stimuli responsiveness built into JNPs may be beneficial in a variety of applications, particularly, if the particles are able to self-assemble in complex hierarchical morphologies.[57,72,73] Building upon controllable synthesis of 'acorn shape' JNPs using seeded emulsion polymerization,[67] shape evolution control of JNPs was achieved by adjusting the glass transition temperature (T_g) *via* compositional gradients during copolymerization.[68] For example, incorporating photochromic entities into shape-adjustable JNPs also facilitate tunable color changes.[69]

One of the interesting properties of JPs are their enhanced interfacial activities[74] and Janus balance (JB). The JB is defined as the ratio of hydrophilic and hydrophobic components,[75] which quantify the geometry of the JPs as well as their interfacial activities. However, only a few studies have demonstrated procedures capable of controlling JB values of JPs during the synthesis process, such as controlling the flow rate of monomers in microfluidic synthesis of JPs,[54] controlling the exposed area of particles to be chemically modified,[75–78] controlling of the block lengths of block terpolymers that self-assemble into JNPs.[79]

Although the anisotropic nature of JNPs stimulated significant scientific and technological interest,[72,80,81] synthesis of large reproducible quantities of JNPs with precisely defined morphologies still remains a challenge.[82,83] Stimuli responsiveness of individual JNP can be generated by introducing controllable heterogeneities capable of responding to a variety of external or internal stimuli. One of the approaches for controllable synthesis of 'acorn shape' JNPs involves conventional emulsion polymerization,[68] where the shape evolution control can be achieved by adjusting the glass transition temperature (T_g) *via* compositional gradients occurring during copolymerization. In essence, when T_g of the particle cores decreases, the shape of nanoparticles may be controlled, ranging from acorn to ellipsoidal and inverse core–shell morphologies. Taking advantage of this concept, a two step-wise emulsion copolymerization of methyl methacrylate (MMA), pentafluorostyrene (PFS) and 2-(dimethylamino)ethyl methacrylate (DMAEMA) with *n*-butyl acrylate (*n*BA) was developed, resulting in the dual phase colloidal JNPs.[84]

Due to unique morphology, self-assembly capability, and potential applications in materials science, biotechnology and surface

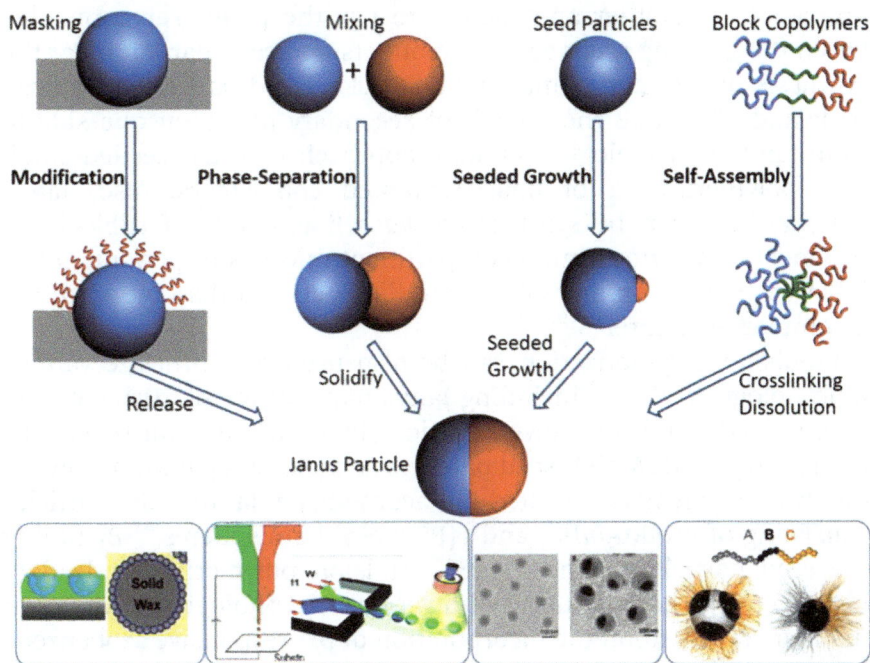

Figure 6.10 Common synthetic strategies of producing Janus particles: masking, mixing, seed particle and block copolymer.

chemistry, JPs with two sides of different chemistry or polarity have been of interest. Synthesis of JPs can be achieved using several techniques, which can be categorized into four strategies: masking, phase separation, seeded growth, and self-assembly. As shown in Figure 6.10, the most versatile strategy is selective chemical modification of exposed surface on temporarily immobilized spherical particles on 2D planar substrates or at the interface of Pickering emulsion droplets. Top surfaces of particles immobilized on 2D substrates can be chemically modified in various methods, such as metal deposition,[85–87] plasma treatments,[88] ligand exchanges,[89–91] chemical reactions,[92] electrostatic binding,[93] electrochemical growth,[94] and others.[95] The particles can also be immobilized on the surface of electrospun fibers[96] or Pickering emulsion droplets,[60,75] followed by chemical modification and release of the resulting asymmetric particles.

Another widely used synthetic procedure is phase separation of a two-component mixture in one single particle, which can be realized through electro-hydrodynamic co-jetting,[80] microfluidic co-flow,[54,97] and solvent assisted phase separation in polymer solution droplets.[98,99] Another approach is to use the phase separation between the growing secondary components and seed particles. For the synthesis of inorganic Janus nanoparticles, seed particles stabilized by ligands facilitate the growth of secondary phase on one side to form Janus morphologies. Another approach is to use seeded emulsion polymerization of phase-separated copolymers. Also, Janus nanoparticles can be synthesized *via* self-assembly of triblock copolymers into various multi-compartment micelles or periodic films, followed by crosslinking of the middle block and dissolution of the assembled structure.[56,79,100]

Emulsion polymerization can be also used to synthesize various morphologies,[43,101,102] including Janus nanoparticles.[83] Using a stepwise seeded emulsion polymerization, PFS and *n*BA can be copolymerized on p(MMA/*n*BA) seed particles and phase separation between the two copolymers resulted in acorn-shape Janus nanoparticles consisting of p(MMA/*n*BA) and p(PFS/*n*BA) hemispheres.[83] By tuning the seed particle T_g, the seeded emulsion polymerization lead to heterogeneous nanoparticles with various morphologies.[68] As shown in Figure 6.11A, upon the incorporation of pH-responsive azobenzene compound during synthesis, Janus nanoparticles exhibiting different colors at neutral and acidic conditions were developed.[103] Furthermore, when the p(MMA/*n*BA)-p(PFS/*n*BA) Janus nanoparticles are used as seed particles, DMAEMA and *n*BA were copolymerized semicontinuously, triphasic Janus nanoparticles with a stimuli-responsive hemispherical shell can be synthesized.[84] As shown in Figure 6.11B, such triphasic Janus nanoparticles are capable of changing their shape at different temperature and/or pH conditions.[84] A potential application of these nanoparticles is smart solid surfactants, because they are capable of stabilizing oil droplets at basic conditions while releasing oil at acidic conditions. Each JNP consists of three phase-separated copolymers: p(MMA/*n*BA) core, temperature and pH responsive (p(DMAEMA/*n*BA)) phase capable of reversible size and shape changes, and shape-adoptable (p(PFS/*n*BA)) phase. Due to built-in second-order lower critical solution temperature (II-LCST) transition of p(DMAEMA/*n*BA) copolymer, macromolecular segments collapse when temperature increases from 30 to 45 °C, resulting in the size and shape changes. The p(DMAEMA/*n*BA) and p(MMA/*n*BA) phases within each JNP assume concave, flat, or convex shapes, forcing p(PFS/*n*BA) phase to adopt convex, planar, or concave interfacial curvatures, respectively. As a result, the JB can be tuned from 3.78 to 0.72.

Due to the collapse of p(DMAEMA/*n*BA) phase during the II-LCST transition which can be tuned between 30 and 45 °C, the size of the

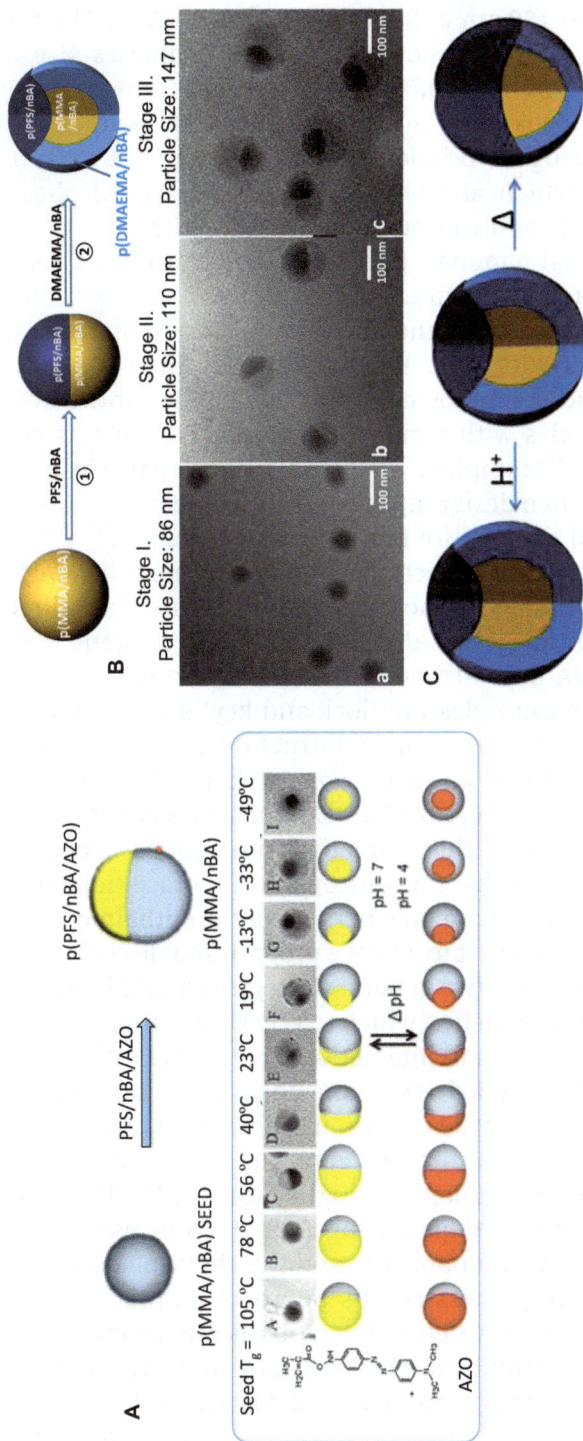

Figure 6.11 (A) Color changing Janus nanoparticles. Reproduced with permission from ref. 103. (B) Schematic diagram of copolymerization of stimuli-responsive JNPs. (a, b, c) TEM images of p(MMA-*n*BA), p(MMA-*n*BA/PFS-*n*BA) and p(MMA-*n*BA/PFS-*n*BA/DMAEMA-*n*BA) particles, respectively. The inset picture is the stimuli-responsive JNP stained by KMnO₄ aqueous solution for 5 min. (C) Dimensional changes of JNPs as a function of pH and temperature. Reprinted with permission from ref. 84. Copyright (2014) American Chemical Society.

JNPs changes from 150 nm at 30 °C to 130 nm at 45 °C. The presence of pH-responsive DMAEMA component also facilitates size and JB changes due to protonation of the tertiary amine groups of p(DMAEMA/nBA) backbone. Synthesized in this manner JNPs are capable of stabilizing oil droplets in water at high pH to form Pickering emulsions, which at lower pH values release oil phase. This process is reversible and can be repeated many times.

Although colloidal nanoparticles have been synthesized for several decades, due to many potential applications in biology, medicine, materials chemistry, and engineering, precisely designed shapes and sizes of organic and inorganic nanoparticles continue to be of the scientific interest and technological importance. The synthesis of well-defined nanoparticles with various morphologies, such as cubes,[104] clustered spheres,[105] dimpled,[106,107] and gibbous particles,[108] is particularly relevant when designing 2D/3D hierarchical structures, which require directional guidance of reproducible formation.[57,76,109] Due to the absence of anisotropic directional bonds, the most challenging and intriguing aspect of nanoparticle assemblies is encoding directional information. Typically, van der Waals,[110] electrostatic,[111,112] depletion[113] and DNA hybridization,[114] and other forces,[115] were used. The representative examples are 'lock-and-key' shape matching colloids resulting in the higher organized structures,[113,116] and assemblies of biotin/DNA decorated particles *via* 'patch–patch' interactions.[114] While the formation of organized 2–3D nanoarrays is not trivial, theoretical studies suggested that attractive interactions between nanoparticles are balanced by the entropy for surface non-adsorbing polymers, whereas adsorbing polymers are of the enthalpic origin.[117] It has also been theorized that to achieve programmable self-assembly of colloidal nanostructures, two general strategies should be considered: puzzle and folding approaches, which rely on short-range (electrostatic) and long range (geometrical) interactions, respectively.[118]

Using a step-wise seeded emulsion polymerization, gibbous and inverse-gibbous nanoparticles shown in Figure 6.12 (a–a''' and b–b''', respectively) can be synthesized by controlling core–shell phase separation and surface tension during synthesis.[108] These particle morphologies can be developed using the same process in which the core is a hydrophobic homopolymer and the shell consists of random copolymers. The uniqueness of this approach is that the choice of core–shell monomers may determine gibbous and/or inverse gibbous particle surface topographies. Due to shape-matching topographies reinforced by short-range surface electrostatic interactions these nanoparticles can be aligned to form 2D 'gear-like' directional

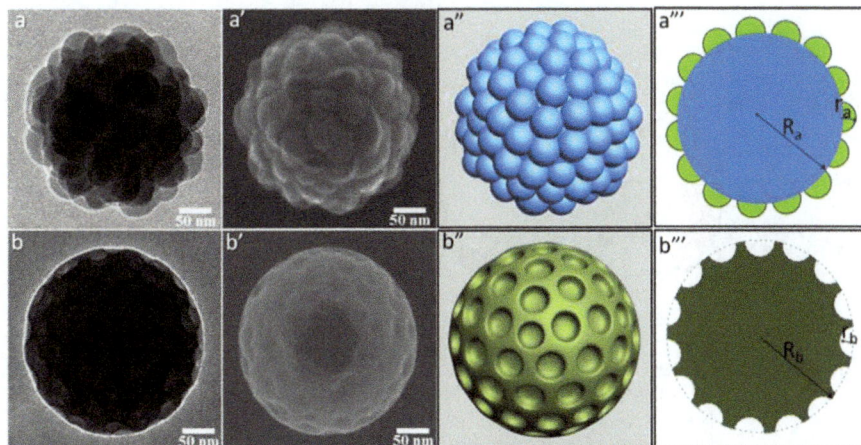

Figure 6.12 TEM (a/b) and SEM (a'/b') images and graphical representations of gibbous (a"–a''') and inverse gibbous particles (b"–b'''). Reproduced with permission from ref. 108. Copyright (2007) American Chemical Society.

assemblies. When gibbous and inverse-gibbous nanoparticles are deposited on a surface to form 3D colloidal films, crack-free colloidal films are formed due to combined short-range electrostatic repulsions and long-range 'gear-like' mechanical interlocking.

6.3 Inorganic Nanoparticles

The primary focus in the developments of inorganic stimuli-responsive nanoparticles deals with is surface modifications, such as grafting-to, grafting-from, or grafting-onto. The approached were discussed in the earlier chapter. The inorganic cores of interest are silica, gold, iron oxide, and quantum dots. Driven by many applications, the usage of these nanoparticles has attracted significant interest in biomedicine and biomedical engineering, including fluorescence imaging, magnetic carriers for drug delivery systems and contrast enhancement agents in magnetic resonance imaging (MRI). The most acceptable approaches to functionalize nanoparticles with polymer shells to achieve a core–shell structure are discussed earlier grafting-from and grafting-to particle surfaces. The most significant quantity is the grafting density which for superparamagnetic iron oxide nanoparticles is less than 0.5 chains/nm^2 of PEG in the 3000–10 000 range.[119]

Iron oxide nanoparticles are often sterically stabilized using dispersants, such as dextran. Unfortunately, this approaches lack of a

ALTERNATIONG
MAGNETIC FIELD

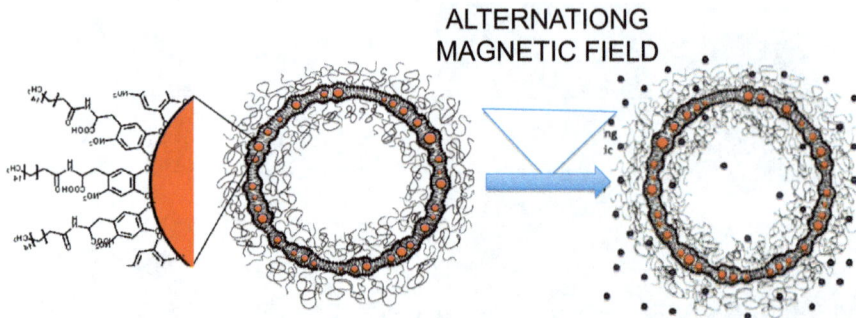

Figure 6.13 Schematic representation of liposomes containing iron oxide NPs inside the bilayer. NitroDOPA-palmityl stabilized iron oxide NPs are embedded in liposome membranes consisting of PEGylated and unmodified lipids.
Reprinted with permission from ref. 123. Copyright (2011) American Chemical Society.

well-defined anchor groups that results in a broad range of particle distributions exceeding many times the size of the iron oxide core diameter. In contrast, dopamine, a derivative of the amino acid DOPA abundantly present in the mussel adhesive protein Mytilus edulis,[120] exhibits high affinity binding group to stabilize iron oxide nanoparticles in water[121] and physiologic buffers, although there are arguments related to the suitability of dopamine.[122] Figure 6.13 illustrates the formation of liposomes comprising of self-assembled super paramagnetic iron oxide nanoparticles stabilized with palmityl-nitroDOPA.[123] The particles are incorporated in the lipid membrane, and using alternating magnetic fields the control release and dose from the vesicles can be achieve by locally heating the membrane.

Using the concept of local heating by magnetic fields as well as global thermal heating reversible aggregate formation can be obtained. Figure 6.14A illustrates the synthesis the anchor group nitrodopamine which is the key attachment to the Fe_3O_4 surface as well as end-grafted PNIPAM brushes with facilitate grafting density independent LCST.[124] The unexplored uses of thermally stimuli-responsive materials in devices and other separation and extraction applications in the presence of magnetic field can be envision (Figure 6.14B). Combining dopamine affinity towards iron oxide along with thermally, pH, or electromagnetic-responsive monomers will provide numerous opportunities for remote stimuli responsiveness.

One of the attractive applications of stimuli-responsive magnetic nanoparticles for diagnostic target capture and concentration control

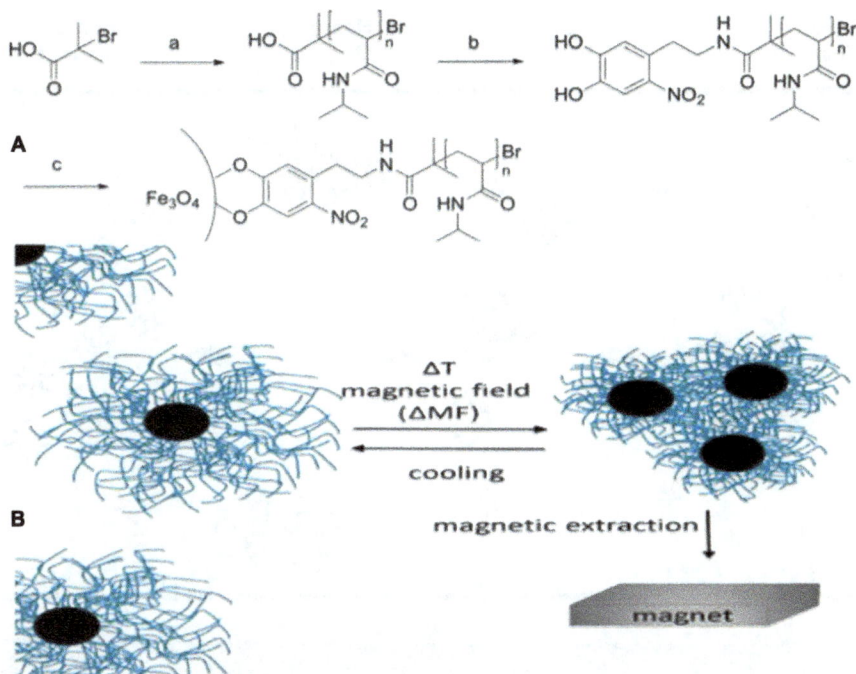

Figure 6.14 (A) Synthesis and magneto-thermal actuation of iron oxide core–PNIPAM shell nanoparticles. (B) Using oscillating magnetic field, aggregation or deaggregation can be achieved. Reprinted with permission from ref. 124. Copyright (2015) American Chemical Society.

has been developed for microfluidic lab card settings.[125] This can be accomplished by synthesizing telechelic poly(*N*-isopropylacrylamide) (PNIPAAm) polymer chains with dodecyl tails at one end and a reactive carboxylate on the other using RAFT polymerization. When PNIPAAm chains self-associate into nanoscale micelles, they can be used as dimensional confinements to synthesize the magnetic nanoparticles. The resulting superparamagnetic nanoparticles exhibit a γ-Fe_2O_3 magnetic nanoparticle (mNP) core with a layer of carboxylate-terminated PNIPAAm chains as a surface corona. The general description of the PNIPAAm mNP separation device is depicted in Figure 6.15. The mNPs are soluble and free flowing in the PEGylated channels when the temperature is below the LCST of PNIPAAm. The small size of these PNIPAAm mNPs facilitates low magnetophoretic mobility, thus preventing from capturing by an applied magnetic field under flow conditions below the LCST. The mNPs diffuse and capture targets as isolated small particles below the LCST. As they flow into

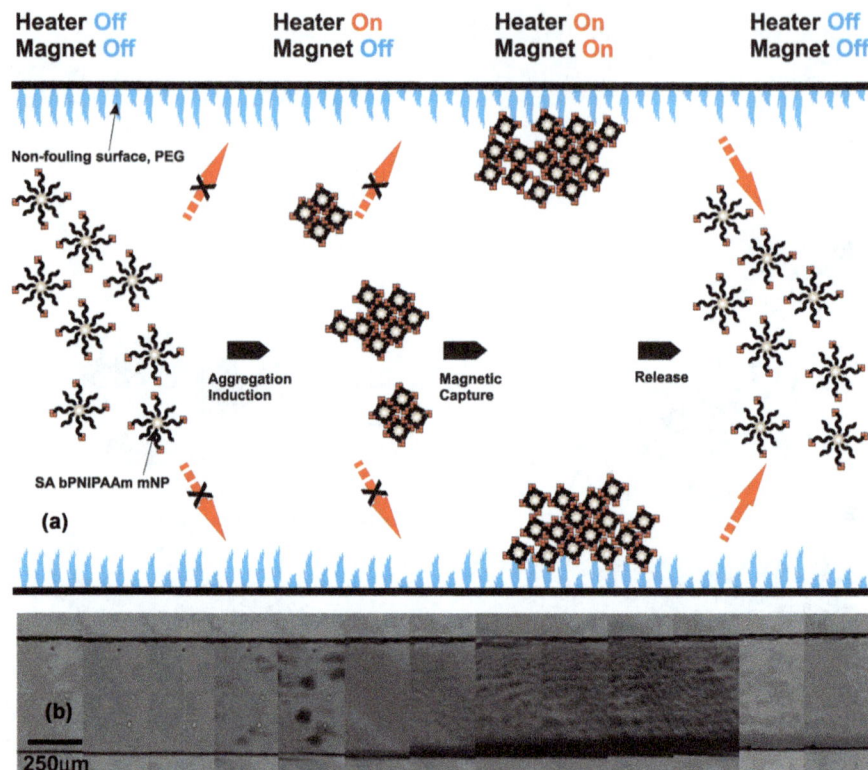

Figure 6.15   Particle captures and releases in dual-responsive system. Reprinted with permission from ref. 125. Copyright (2007) American Chemical Society.

the heated region of a microchannel, the temperature is raised above the LCST of the PNIPAAm and the mNPs aggregate, but do not stick to the non-fouling, PEGylated channel walls in the absence of an applied magnetic field. The mNPs are attracted by the applied magnetic field and accumulated at the PEGylated channel walls above the LCST. The reversal of both the temperature and applied magnetic field results in the redissolution of the aggregated mNPs and their diffusive reentry into the flow stream.

Stable mesoporous structures, large surface areas, tunable pore sizes and volumes, and well-defined surface properties of the organically functionalized mesoporous silica materials make these materials attractive for hosting molecules of various sizes, shapes, and functionalities.[126] Thus, mesoporous silica-based carrier systems for controlled-release delivery of drugs was a logical application. However, due to high toxicity, many site-selective deliveries require 'zero release' before reaching the targeted cell or tissues. Unfortunately, in many release systems utilizing encapsulated compounds the drug delivery occurs immediately upon dispersion of the drug/polymer in water. In the systems where the drug release relies on hydrolysis of biodegradable polymer-based drug delivery systems, other site effects such as protein denaturalization or aggregation may occur. Meso-porous silica-based controlled-release delivery system with built-in stimuli responsiveness and chemically inert to the matrix entrapped compounds consists of a 2-(propyldisulfanyl) ethylamine functiona-lized mesoporous silica nanosphere (MSN) material with a particle size of ~ 200 nm and an average pore diameter of ~ 2.3 nm. Figure 6.16 illustrates the mesopores of the MSN material serving as reservoirs to soak up aqueous solutions of various pharmaceutical drug molecules and neurotransmitters, such as vancomycin and adenosine triphosphate (ATP).[127] The openings of the mesopores of the drug/neurotransmitter-loaded MSN material will cap *in situ* by allowing the pore surface-bound 2-(propyldisulfanyl) ethylamine functional groups to covalently capture the water-soluble mercap-toacetic acid-derivatized cadmium sulfide (CdS) nanocrystals[128] using well-known amidation reaction.[129] The resulting disulfide linkages between the MSNs and the CdS nanoparticles are chemically labile in nature and can be cleaved with various disulfide-reducing agents, such as dithiothreitol (DTT) and mercaptoethanol (ME). Using this approach, the release of the CdS nanoparticle caps from the drug/neurotransmitter-loaded MSNs can be regulated by introducing vari-ous release triggers (Figure 6.16).

Fluorescence resonance energy transfer (FRET) is a non-radiative process during which the electronic excitation energy of a donor chromophore is transferred to a nearby acceptor molecule via long-range dipole–dipole interactions.[130] While excellent mono-graphs[131–135] provide an extensive coverage of this topic, Figure 6.17 outlines the basic principles of FRET.[136] The main factors that impact the efficiency of the FRET process include (i) the extent of overlap between the emission spectrum of the donor and the absorption spectrum of the acceptor, (ii) the relative orientation of the transition dipoles, and (iii) the distance between the donor and acceptor mol-ecules. Due to the ability of biological detection, simplicity and sen-sitivity towards organic dyes and fluorescent proteins, FRET-based analytical methods became quite attractive.[137,138] However, the lim-iting factors are short observation times and many dyes exhibit nar-row absorption and broad emission spectra. And this is how quantum dots (QDs) come to the rescue. Their strong photoluminescence, high

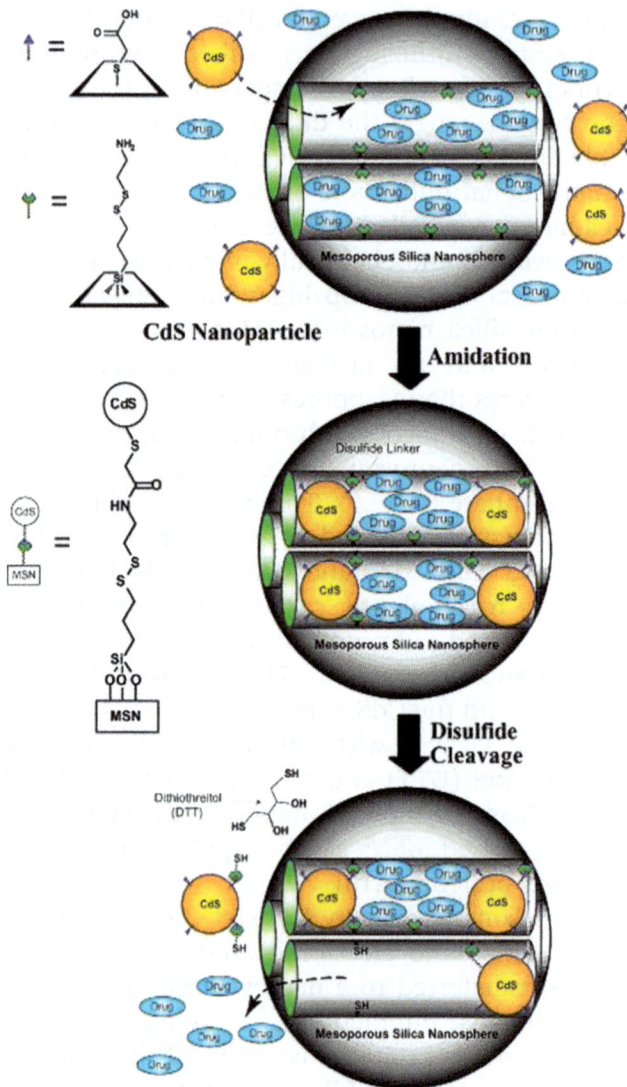

Figure 6.16 A schematic representation of the CdS nanoparticle-capped MSN-based drug/neurotransmitter delivery system. The controlled-release mechanism of the system is based on chemical reduction of the disulfide linkage between the CdS caps and the MSN hosts.
Reprinted with permission from ref. 116. Copyright (2002) American Chemical Society.

quantum yields, and good photostability, large Stokes shift, narrow emission, and broad excitation spectrum make QDs quite attractive.[139,140] However, aside from the concerns resulting from the QD

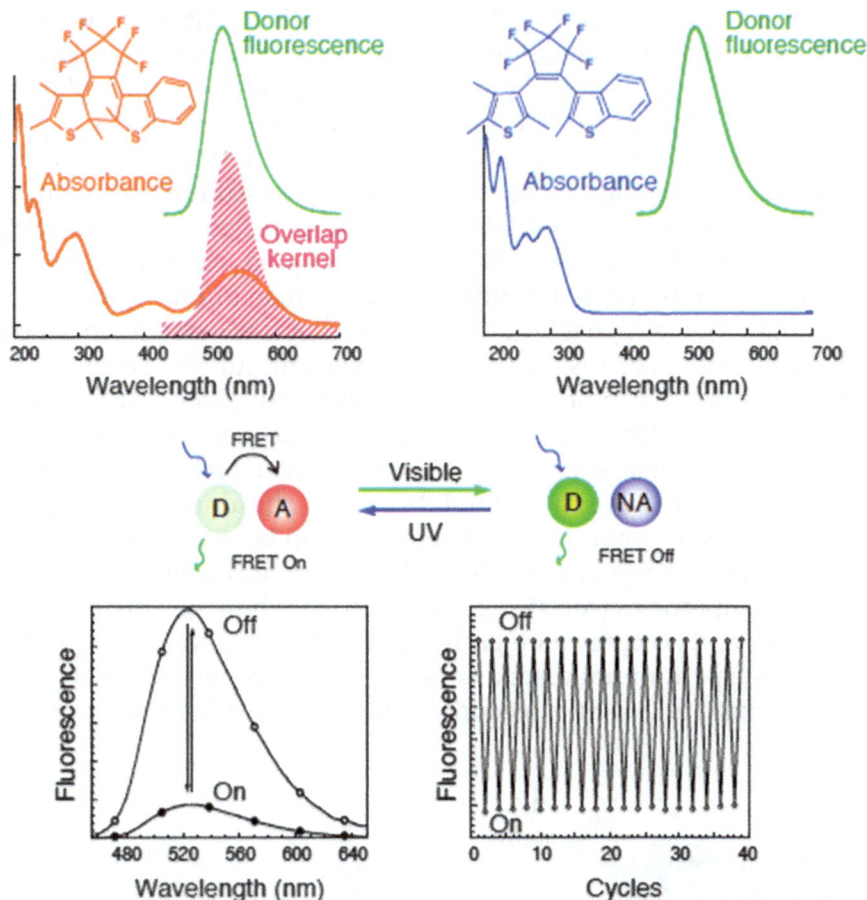

Figure 6.17 Photochromic FRET (pcFRET). The chemical structures depicted correspond to the photochromic dithienylethene in the colorless open form (right upper) and colored closed form (left upper). The absorption spectrum of the latter overlaps well with the emission spectrum of the donor; the kernel of the overlap integral (striped) corresponds to the lucifer yellow donor selected for a model compound 27. Ultraviolet light induces the photochromic transition to the closed form (On), and visible (green) light reverses the process to the open form (Off). Bottom: corresponding donor spectra and multiple cycles between the two states of the system.
Reprinted by permission from Macmillan Publishers Ltd: Nature (ref. 127), copyright (2010).

potential toxicity and chemical stability,[141–143] the excitation of QDs often requires the use of UV radiation resulting in low signal-to-noise ratio caused by auto-fluorescence.[144,145] One of the fluorescent bio-labels are near-infrared (NIR)-to-visible up-conversion nanoparticles

(UCNPs). The uniqueness of this approach is the ability of converting a longer wavelength radiation (*e.g.* NIR light) to shorter wavelength fluorescence (*e.g.* visible light) *via* a two-photon or multiphoton mechanism.[146] Because NIR light can penetrate biological samples and tissues without damage while avoiding auto-fluorescence a luminescence resonance energy transfer (LRET)-based immunoassay, it can also be excited by NIR irradiation. As depicted in Figure 6.18, amino-modified NaYF4:Yb, Er UCNPs and gold nanoparticles (Au NPs) can be prepared and conjugated with human immunoglobulin G (IgG) and rabbit antigoat IgG, respectively.[147] A LRET system, in which the biofunctionalized NaYF4:Yb, Er UCNPs and Au NPs serve as energy donor and acceptor, respectively. This UCNP-based LRET system can be used to determine the amount of goat antihuman IgG in a sandwich-type bioassay. This detection system facilitates the two conditions that are critical for LRET from the donor to acceptor nanoparticles to occur: (a) the spectral overlap between the emission of the donor nanoparticles (UCNPs) and the absorption of the acceptor nanoparticles (Au NPs) as well as (b) the short spacing between the donor and acceptor nanoparticles, which is achieved through the sandwich-type immunoreactions between the primary antibody and the proteins antigen and secondary antibody chemically conjugated

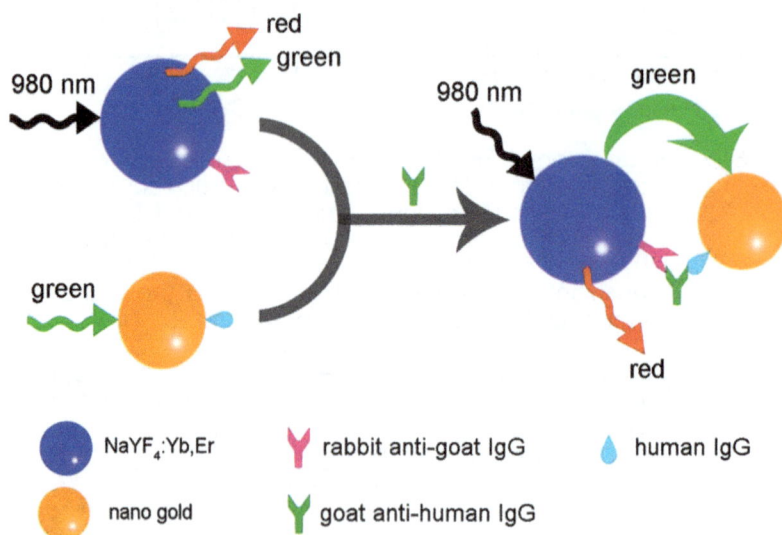

Figure 6.18   Schematic illustration of the LRET process between NaYF4:Yb, Er UCNPs (donor) and Au NPs (acceptor).
Reprinted with permission from ref. 147. Copyright (2009) American Chemical Society.

to the surface of the donor and acceptor nanoparticles, respectively (Figure 6.18). This is the first example of using a UCNPs-based LRET system for stimuli-responsive antibody detection.

6.3.1 Stimuli-responsive Vesicles

Similar to many other studies, the primary interest in vesicles arose from their potential use in drug delivery, coatings and nanoreactor applications.[148–150] It was immediately recognized that in order to make use of the vesicles, stimuli responsiveness would be the critical component. Thus, several studies used temperature and/pH responsiveness to achieve triggered delivery and release.[151,152] One of the classical examples is the formation of a three-layer polymeric vesicle consisting of polystyrene-*b*-poly(2-vinylpyridine)-*b*-poly(ethylene oxide) (PS-*b*-P2VP-*b*-PEO) triblock copolymer.[153] The pH sensitivity of the P2VP shell can be used to tune the micelle size from a hydrodynamic diameter (DH) of 75.4 nm at pH > 5 to 135.2 nm at pH < 5. This attributed to the electrostatic repulsion between the charged P2VP blocks and is not completely reversible because of the formation of salt with each pH cycle. Because stability of these vesicles is critical, a common strategy is to use crosslinking to improve the stability of self-assembled structures. Another approach to obtain pH sensitivity and to be able to manipulate the size and shape of the supramolecular structures formed by self-assembly of block copolymers in aqueous medium is to utilize the secondary structure of polypeptides.[154] One example of self-organization is a polybutadiene *b*-poly(L-glutamic acid) (PB-*b*-PGA) diblock copolymer in aqueous solution. Another strategy of controlling the size and structural features is to prepare diblock copolypeptides that self-assemble into spherical vesicular assemblies with the size and structures dictated by the ordered conformations of the polymer segments.[155] The key is to incorporate the functionality that reply to external stimuli. The vesicle assemble in an aqueous suspensions consists of $K^P_{160}(L_{0.3}/K_{0.7})_{40}$, where $K^P = $ poly($N\varepsilon$-2-(2-(2-methoxy-ethoxy)ethoxy)acetyl-L-lysine), $L = $ poly(L-leucine) and the subscript numbers refer to the average number of residues within each polymer segment. At pH > 9 vesicles are formed, but protonation the amino side chains on the lysine residues in $K^P_{160}(L_{0.3}/K_{0.7})_{40}$ will enhances the hydrophilicity by destabilizing the α-helical structure of the leucine-rich domain due to electrostatic repulsions. As a result, helix-to-coil conformation transitions within this domain make the vesicle pH responsive.

6.3.2   Nanotubes and Nanowires

It is well established that carbon nanotubes (CNTs) exhibit unusual electrical, mechanical, optical, thermal, and chemical properties. Since their 1991 discovery, they have attracted scientists as well as have led to technological advances across many disciplines.[156] Their applications range from nanoelectronics to drug delivery, composites, optical sensors, to name just a few. In spite of these advantages, CNTs exhibit one inherent problem: during their synthesis, due to inherently built-in hydrophobic interactions they agglomerate to form bundles. A simple solution to this problem is to modify interfacial interactions by using covalent bonding or physical adsorption. Numerous modifications were offered, which may or may not adversely impact useful properties. Single-walled carbon nanotubes (SWNTs) can be effectively modified using phospholipids. Using a simple surface modification of SWNTs, followed by deposition of 1,2-bis(10,12-tricosadiynoyl)-*sn*-glycero-3-phosphoethanolamine (DCPE) phospholipid, results in stable water-dispersible SWNTs with highly uniform thickness.[157] The use of copolymer stimuli-responsive stabilizer is even more attractive. As shown in Figure 6.19, using copolymer stabilizer, temperature and ionic strength responsive

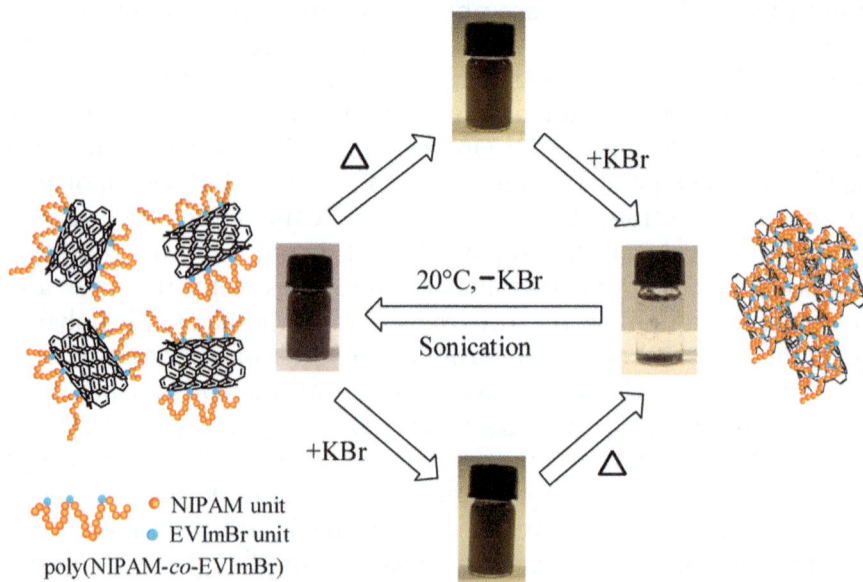

Figure 6.19   Cycle process of using temperature and ionic strength responsive MWCNT dispersion using a copolymer stabilizer. Reprinted with permission from ref. 158. Copyright (2012) American Chemical Society.

MWCNT dispersions can be effectively achieved.[158] The copolymer stabilizer can be prepared *via* free radical polymerization of NIPAM and an ionic liquid monomer, 1-ethyl-3-vinylimidazolium bromide (EVImBr, content: 2.2–12.8 mol%). The uniqueness of this approach is the ability to control the stimuli responsiveness and tune in and out MWCNT dispersity.

Iron oxides are of particular interest and because shape also contributes to magnetic properties,[159,160] different shapes of α-Fe_2O_3, including nanorods, nanospheres, nanowires and a spindle-type shape have been prepared. Attempts to make hollow tubular structures have been made because this shape may exhibit unique magnetic properties, but the highest obtained magnetization values were 2.7–2.9 emu g^{-1} or smaller. The first example of using phospholipids to produce ferromagnetic nanotubes is shown in Figure 6.20.[161] Using 1,2-bis(10,12-tricosadiynoyl)-*sn*-glycero-3-phosphocholine (DC8,9PC) phospholipid nanotubes will facilitate polymerization in the presence of Fe^{2+}/H_2O_2 redox agents, leading to amorphous basic iron(III) sulfate concentric nanotubes separated by crosslinked phospholipid (PL)

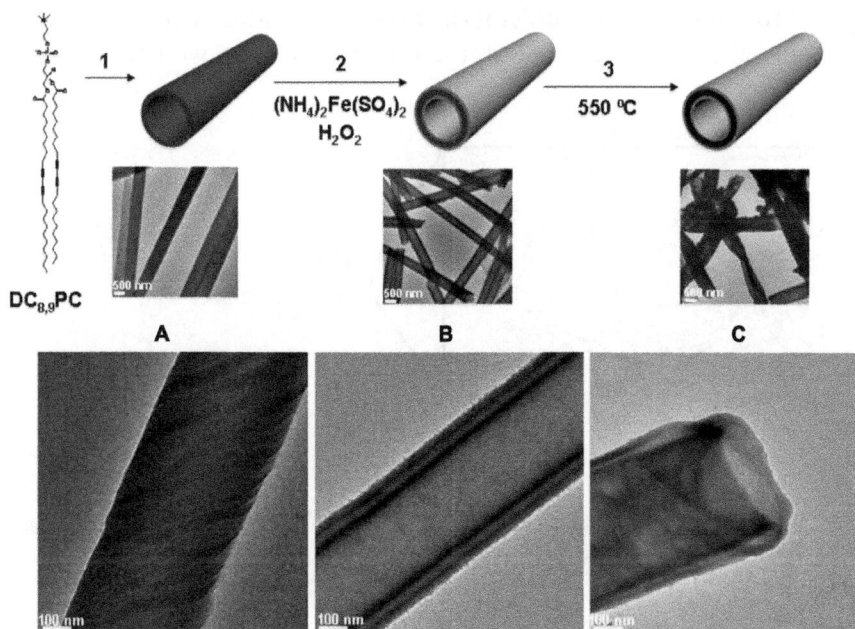

Figure 6.20   TEM images of nanotubes: (A) Step 1: self-assembly of DC8,9 PC into nanotubes. (B) Step 2: reactions of $(NH_4)_2Fe(SO_4)_2$ and H_2O_2. (C) Step 3: thermal exposure to 550 °C for 0.5 h. Reproduced from ref. 161 with permission from The Royal Society of Chemistry.

layers. Further exposure of the nanotubes to 550 °C resulted in the formation of ferromagnetic iron oxide concentric nanotubes separated by carbon sheet interlayers with the remanent magnetization (Mr) of 4.62 emu g^{-1}, the saturation magnetization (Ms) of 46.12 emu g^{-1}, the squareness (Mr/Ms) of 0.1002, and the coercivity (Hc) of 51.35 Oe.

One of the limiting factors for the development of stimuli-responsive materials is their ability of dimensional changes on demand and to reversibly adopt by responding to environmental changes. New technological platforms can be envisioned if selected polymers meet the criteria of reversible shape and size changes.[162] One example of shape changes in nanotubes is illustrated in Figure 6.21.[163] These materials can be prepared by polymerization of N-isopropylacrylamide (NIPAAM) in the presence of biologically active 1,2-bis(tricosa-10,12-diynoyl)-sn-glycero-3-phosphocholine (DC8,9PC) diacetylenic phospholipids (PL). As a result, thermally responsive poly-NIPAM phospholipid nanotubes (PNNTs) can be prepared. Polymerization reactions occur within hydrophilic regions of PL bilayers, whereas PL hydrophobic zones facilitate transport and supply of the monomer for polymerization. The unique feature of PNNTs is that, above 37 °C, the outer diameter (OD) as well as the wall

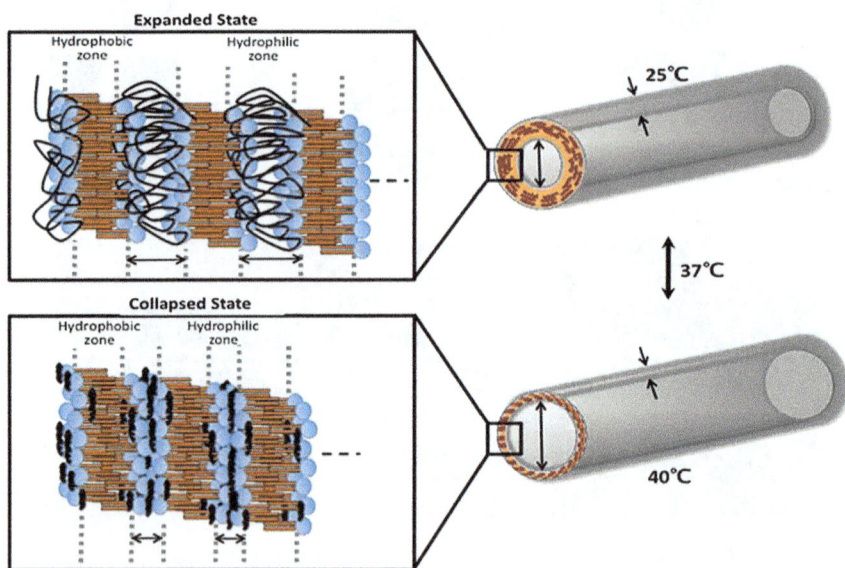

Figure 6.21   Schematic diagram of temperature-sensitive PNNTs.
Reprinted with permission from ref. 163. Copyright (2012) American Chemical Society.

thickness (WT) shrink by 20 and 55%, respectively, whereas the inner diameter (ID) increases by $\sim 16\%$. This behavior is attributed to the PNIPAM backbone buckling induced by local rearrangements within PL bilayered morphologies. The presence of acetylenic moieties along the PL bilayers in PNNTs provides an opportunity for irreversible 'locking' of designable dimensions facilitated by the formation of crosslinked PNNTs (CL-PNNTs). It is also worth mentioning that miscible micelles formed from surfactants and phospholipids may serve as stable loci for lipophilic monomers and nanostructured templates[164] and is capable of providing a stimuli-responsive environment during film formation through which individual surface stabilizing components can be driven to the film–air (F–A) or film–substrate (F–S) interfaces.

6.4 Length Scales and Stimuli-responsiveness

Perhaps one of the most intriguing challenges in designing stimuli-responsive materials is the responsiveness across various length scales. Beginning with a single atom, building a molecule from which macromolecules and nanotubes or nanosheets are formed, or creating molecular devices at larger scale lengths is an exciting challenge. However, to achieve stimuli-responsiveness at micro- and larger length scales, higher volumes are required as well as the ability of coordinated movements. Using an analogy of moving electrons across a conductive wire that results in an electric current, we need to realize that the size really matters. When electric potential is applied across a conductive wire, electrons move, thus generating electric current, but due to their small size (~ 10–18 m), they do not need much space. Being significantly smaller that the surroundings, electrons can get around very easily. As we go to the larger length and volume scales depicted in Figure 6.22, A and B, from the Angstroms to nano, micro, millimeters and meters, we begin to realize that significantly greater volumes are needed to execute responsiveness. Furthermore, 'connectivity' of individual responsive elements will be essential at the larger length scales. For example, the heart impulse begins with the electrical pulse of specialized cells referred to as sinoatrial node (SA), thus causing the walls of the atria to contract, and forcing blood into ventricles. This sets the rhythm and rate of our heart through SA nodes. This process is regulated by another cluster of cells referred to as atrioventricular (AV) node, which serves as a gate, slowing down the electrical pulses before they approach

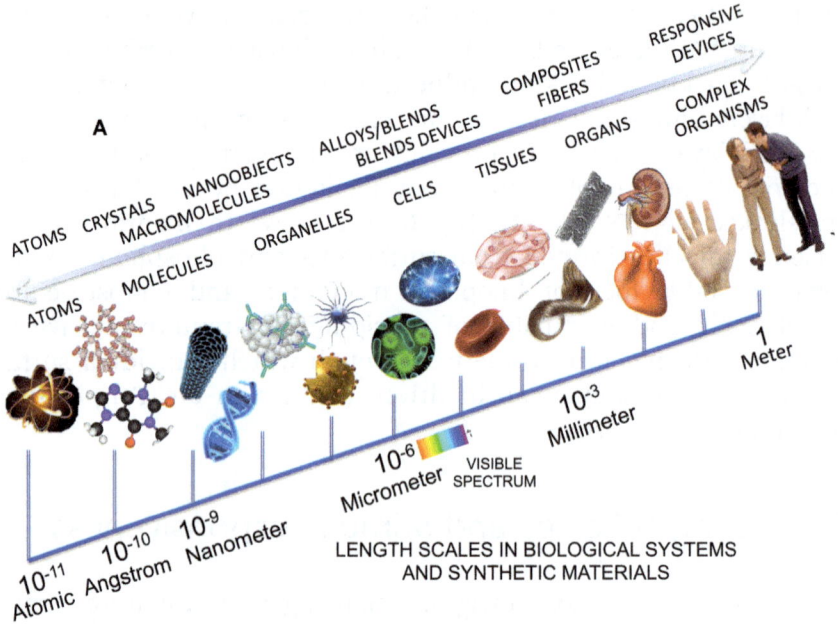

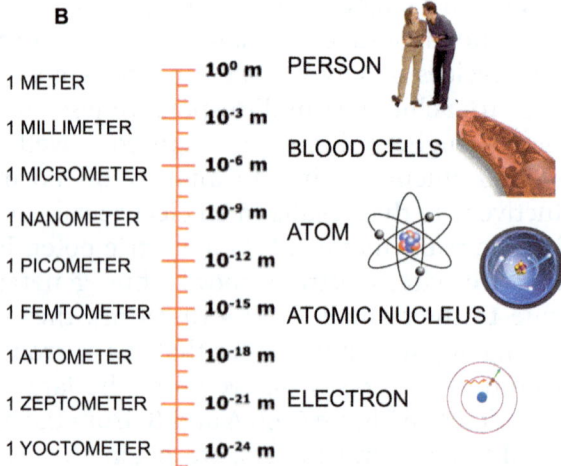

Figure 6.22 (A) Comparison of length scales between biological and synthetic materials. (B) Nomenclature and comparison of various scale lengths ranging from 10^{-24} to 1 meter.

ventricles. In essence, this delay gives atria sufficient time to contract before the ventricles respond. Lastly, so-called His-Purkinje networks fibers send the impulse to the muscular walls of the ventricles, thus causing them to contract, forcing blood out of the heart to the body. As this process is repeated many times during the life-cycle, this abbreviated description of the heart beat required chemical and electrical energy. Yet, for the outside observer, it is autonomous. Furthermore, there are 'connected' processes that are highly coordinated and the development of synthetic devices that would be able to offer similar outcomes is remarkably challenging, but it will be achievable in not distant future. Perhaps the good indicators of the future trends are computational approaches utilized to predict the mechanics and physics of dynamic materials at these length scales. They may range from atomistic methods such quantum mechanics, molecular dynamics, and to the largest length scale where continuum mechanics and the finite element methods are utilized. However, each method is appropriate for a particular, limited range of length scale. One of the major challenges is that many theoretical approaches fall into a gap between length scales depicted in Figure 6.22A and B, but none of them are through the entire process. In materials, there are numerous examples, which include nanoporous or nanocrystaline ceramics, metals or polymers, with microstructural features on the order of a few nanometers, thus requiring atomistic resolution, but the volume changes may be too large to be modeled by molecular dynamics. Typically, at the quantum level, *ab initio* quantum mechanics/molecular mechanics and density functional theory approaches can be used. Molecular dynamics (MD) simulations, with both reactive and nonreactive force fields can be expanded to understand the behavior of interfacial regions between soft and hard materials, and the quantum simulations can be utilized to develop the required force fields, thus serving as an input to another level. However, using dissipative particle dynamics (DPD) one can envision modeling at the mesoscale levels, thus offering the interface between MD and continuum. An excellent example that accounts for the microscopic dynamics of the membranes and coatings upon environmental changes (pH, temperature) is the structural evolution of gel membranes. [165,166] To accurately capture the interactions between the functional groups embedded into a surrounding medium at the nanoscale will lead to the development of transitions between different scale lengths.

Autonomous dynamic behavior characteristics of living systems, such as heart beating or repulsive neuronal signals, are perhaps one of the most intriguing phenomena that fascinate synthetic materials chemists. As mentioned above, although often perceived by an external viewer as autonomous, these processes also require chemical, electrical, and/or thermal energy input, which is converted to visible mechanical responses. Inspired by these oscillatory behaviors, 'self-oscillating' polymer gels were developed, which exhibit spontaneous and periodic volume changes without external intervention.[167] They utilize known oscillatory Belousov–Zhabotinsky (BZ) reactions.[168,169] The overall process of the BZ reaction includes oxidation of organic substrates, such as malonic acid oxidized by sodium bromate catalyzed by metals under acidic conditions. Oscillation is achieved by the concentration changes of several reaction intermediates and the redox states of metal catalysts, thus causing swelling (oxidation) and deswelling (reduction) of the surrounding gel. This is schematically depicted in Figure 6.23A. The chemical structure of the oscillating gel consists of PNIPAAm containing ruthenium tris(2,2′-bipyridine) (Ru(bpy)$_3$) side chains (Figure 6.23B). Since Ru(bpy)$_3$ is the metal catalyst for the BZ reactions, when p(NIPAAm-r-Ru(bpy)$_3$) gel is placed in the aqueous solution containing BZ substrates, such as nitric acid (HNO$_3$), malonic acid (MA) and sodium bromate (NaBrO$_3$), the redox state of Ru(bpy)$_3$ periodically changes as BZ reactions are catalyzed by the Ru(bpy)$_3$ incorporated into a polymer network gel. As a result of the hydrophilicity changes resulting for the redox state changes from Ru(bpy)$_3{}^{3+}$ (more hydrophobic) to Ru(bpy)$_3{}^{2+}$, the volume phase transition temperature and swelling ratio of the gel changes. Thus, at constant temperature, volume changes occur with the redox changes, and periodic volume changes can be driven by the BZ reaction. Therefore,

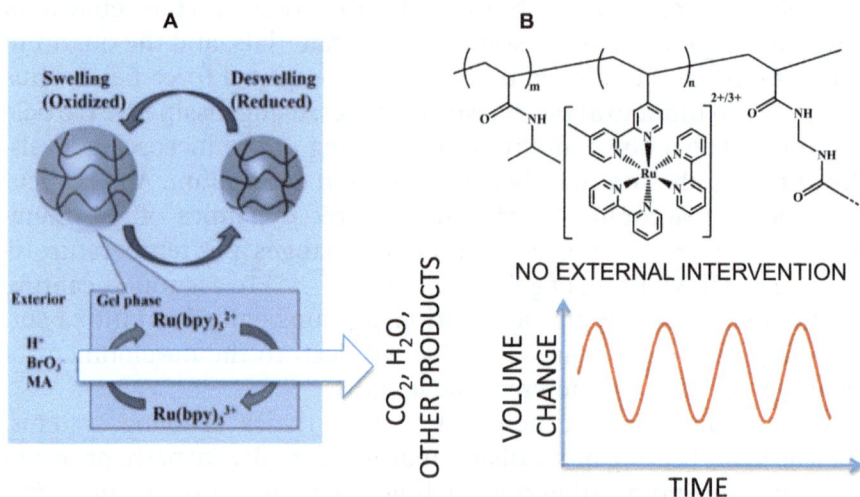

Figure 6.23 (A) Swelling and deswelling of a polymer gels resulting from oxidation–reduction along the polymer backbone. (B) The polymer backbone of swelling–deswelling gels consists of PNIPAAM containing Ru(BPY)$_3$ side chains that change volume without external intervention. Adapted from ref. 168 and 169.

the p(NIPAAm-r-Ru(bpy)$_3$) gel immersed in a catalyst-free aqueous solution with BZ substrates exhibits autonomous periodic volume changes without external intervention in an oscillatory manner. This system inspired by oscillatory processes observed in living systems has been expanded to numerous applications which resulted in the development of oscillating polymer gels, tubular oscillating gels capable of mass transport, self-oscillating brushes,[170] block copolymers, micelles,[171] vesicles, and artificial cells.[172] These self-oscillating supramolecular self-assemblies can be manipulated to form higher ordered structural features, including rhythmic sol–gel conversions.

References

1. T. W. Martinek, Apparatus for preparing colloidal dispersions, *US Pat.*, 2 905 448, 1959.
2. E. B. Pierre, J. S. P. Francoi and P. Bebin, Process for preparing a colloid oxychloride of copper, *US Pat.*, 2 201 928, 1940.
3. E. N. Savariar, S. Ghosh and S. Thayumanavan, Disassembly of noncovalent amphiphilic polymers with proteins and utility in pattern sensing, *J. Am. Chem. Soc.*, 2008, **130**(16), 5416–5417.
4. N. Nishiyama and K. Kataoka, Current state, achievements, and future prospects of polymeric micelles as nanocarriers for drug and gene delivery, *Pharmacol. Ther.*, 2006, **112**(3), 630–648.
5. R. Savić, A. Eisenberg and D. Maysinger, Block copolymer micelles as delivery vehicles of hydrophobic drugs: micelle–cell interactions, *J. Drug Targeting*, 2006, **14**(6), 343–355.
6. A. V. Kabanov and V. Y. Alakhov, Pluronic® block copolymers in drug delivery: From micellar nanocontainers to biological response modifiers, *Crit. Rev. Ther. Drug Carrier Syst.*, 2002, **19**(1), 72–144.
7. K. Kataoka, A. Harada and Y. Nagasaki, Block copolymer micelles for drug delivery: design, characterization and biological significance, *Adv. Drug Delivery Rev.*, 2001, **47**(1), 113–131.
8. Y. Kakizawa and K. Kataoka, Block copolymer micelles for delivery of gene and related compounds, *Adv. Drug Delivery Rev.*, 2002, **54**(2), 203–222.
9. V. P. Torchilin, Structure and design of polymeric surfactant-based drug delivery systems, *J. Controlled Release*, 2001, **73**(2), 137–172.
10. N. Rapoport, Physical stimuli-responsive polymeric micelles for anti-cancer drug delivery, *Prog. Polym. Sci.*, 2007, **32**(8), 962–990.

11. S. Ghosh, V. Yesilyurt, E. N. Savariar, K. Irvin and S. Thayumanavan, Redox, ionic strength, and pH sensitive supramolecular polymer assemblies, *J. Polym. Sci. Part A: Polym. Chem.*, 2009, **47**(4), 1052–1060.

12. Z. L. Tyrrell, Y. Shen and M. Radosz, Fabrication of micellar nanoparticles for drug delivery through the self-assembly of block copolymers, *Prog. Polym. Sci.*, 2010, **35**(9), 1128–1143.

13. A. Klaikherd, C. Nagamani and S. Thayumanavan, Multi-stimuli sensitive amphiphilic block copolymer assemblies, *J. Am. Chem. Soc.*, 2009, **131**(13), 4830–4838.

14. R. L. McCarley, Redox-responsive delivery systems, *Annu. Rev. Anal. Chem.*, 2012, **5**, 391–411.

15. L. Sambe, K. Belal, F. Stoffelbach, J. Lyskawa, F. Delattre and M. Bria *et al.*, Multi-stimuli responsive supramolecular diblock copolymers, *Polym. Chem.*, 2014, **5**(3), 1031–1036.

16. D. G. Hall, *Boronic Acids: Preparation, Applications in Organic Synthesis and Medicine*, John Wiley & Sons, 2006.

17. M. Sana, G. Leroy and C. Wilante, Enthalpies of formation and bond energies in lithium, beryllium, and boron derivatives. A theoretical attempt for data rationalization, *Organometallics*, 1991, **10**(1), 264–270.

18. J.-S. Wang and K. Matyjaszewski, Controlled/"living" radical polymerization. Atom transfer radical polymerization in the presence of transition-metal complexes, *J. Am. Chem. Soc.*, 1995, **117**(20), 5614–5615.

19. M. Kato, M. Kamigaito, M. Sawamoto and T. Higashimura, Polymerization of methyl methacrylate with the carbon tetra-chloride/dichlorotris-(triphenylphosphine) ruthenium (II)/ methylaluminum bis (2, 6-di-tert-butylphenoxide) initiating system: possibility of living radical polymerization, *Macromolecules*, 1995, **28**(5), 1721–1723.

20. A. P. Vogt and B. S. Sumerlin, An efficient route to macro-monomers via ATRP and click chemistry, *Macromolecules*, 2006, **39**(16), 5286–5292.

21. J. Chiefari, Y. Chong, F. Ercole, J. Krstina, J. Jeffery and T. P. Le *et al.*, Living free-radical polymerization by reversible addition-fragmentation chain transfer: the RAFT process, *Macromolecules*, 1998, **31**(16), 5559–5562.

22. S. Perrier and P. Takolpuckdee, Macromolecular design via reversible addition–fragmentation chain transfer (RAFT)/xanthates (MADIX) polymerization, *J. Polym. Sci., Part A: Polym. Chem.*, 2005, **43**(22), 5347–5393.

23. G. Moad, E. Rizzardo and S. H. Thang, Living radical polymerization by the RAFT process, *Aust. J. Chem.*, 2005, **58**(6), 379–410.
24. P. De, S. R. Gondi and B. S. Sumerlin, Folate-conjugated thermoresponsive block copolymers: highly efficient conjugation and solution self-assembly, *Biomacromolecules*, 2008, **9**(3), 1064–1070.
25. Y. Ren, X. Jiang and J. Yin, Poly (ether tert-amine): A novel family of multiresponsive polymer, *J. Polym. Sci., Part A: Polym. Chem.*, 2009, **47**(5), 1292–1297.
26. R. Wang, X. Jiang, G. Yin and J. Yin, Well-defined multi-stimuli responsive fluorinated graft poly (ether amine) s (fgPEAs), *Polymer*, 2011, **52**(2), 368–375.
27. W. Yuan, J. Wang, L. Li, H. Zou, H. Yuan and J. Ren, Synthesis, Self-Assembly, and Multi-Stimuli Responses of a Supramolecular Block Copolymer, *Macromol. Rapid Commun.*, 2014, **35**(20), 1776–1781.
28. W. Xu, P. A. Ledin, Z. Iatridi, C. Tsitsilianis and V. V. Tsukruk, Multiresponsive Star-Graft Quarterpolymer Monolayers, *Macromolecules.*, 2015, **48**(10), 3344–3353.
29. J. Dong, Y. Wang, J. Zhang, X. Zhan, S. Zhu and H. Yang *et al.*, Multiple stimuli-responsive polymeric micelles for controlled release, *Soft Matter*, 2013, **9**(2), 370–373.
30. X. J. Loh, Js del Barrio, P. P. C. Toh, T.-C. Lee, D. Jiao and U. Rauwald *et al.*, Triply triggered doxorubicin release from supramolecular nanocontainers, *Biomacromolecules*, 2011, **13**(1), 84–91.
31. O. Bertrand, C.-A. Fustin and J.-Fo. Gohy, Multiresponsive Micellar Systems from Photocleavable Block Copolymers, *ACS Macro Lett.*, 2012, **1**(8), 949–953.
32. G. Wu, S.-C. Chen, C.-L. Liu and Y.-Z. Wang, Direct Aqueous Self-Assembly of an Amphiphilic Diblock Copolymer toward Multi-stimuli-Responsive Fluorescent Anisotropic Micelles, *ACS Nano*, 2015, **9**(4), 4649–4659.
33. M. Motornov, Y. Roiter, I. Tokarev and S. Minko, Stimuli-responsive nanoparticles, nanogels and capsules for integrated multifunctional intelligent systems, *Prog. Polym. Sci.*, 2010, **35**(1), 174–211.
34. E. Fleige, M. A. Quadir and R. Haag, Stimuli-responsive polymeric nanocarriers for the controlled transport of active compounds: concepts and applications, *Adv. Drug Delivery Rev.*, 2012, **64**(9), 866–884.

35. S. Mura, J. Nicolas and P. Couvreur, Stimuli-responsive nano-carriers for drug delivery, *Nat. Mater.*, 2013, **12**(11), 991–1003.
36. O. J. Cayre, N. Chagneux and S. Biggs, Stimulus responsive core-shell nanoparticles: synthesis and applications of polymer based aqueous systems, *Soft Matter*, 2011, **7**(6), 2211–2234.
37. Z. Tian, W. Wu and A. D. Li, Photoswitchable fluorescent nanoparticles: preparation, properties and applications, *ChemPhysChem*, 2009, **10**(15), 2577–2591.
38. *Stimuli-Responsive Polymeric Films and Coatings*, ed. A. M. Urban and M. W. Urban, American Chemical Society, Washington, DC, 2005.
39. T. T. Min, A. Klein, M. S. El-Aasser and J. W. Vanderhoff, *J. Polym. Sci., Polym. Chem.*, 1983, **21**, 2845.
40. R. U. Dreher and M. W. Urban, Stable nonspherical fluroine conating colloidal dispersion, *Macromolecules*, 2005, **38**, 2205–2212.
41. M. Okubo, A. Yamada and T. Matsumoto, *J. Polym. Sci., Polym. Chem.*, 1980, **18**, 3219.
42. Y. Zhao and M. W. Urban, Novel STY/nBA/GMA and STY/nBA/MAA Core-Shell Latex Blends: Film Formation, Particle Morphology, and Cross-Linking. 20. A Spectroscopic Study, *Macromolecules*, 2000, **33**, 8426.
43. D. J. Lestage and M. W. Urban, Cocklebur-shaped colloidal dispersions, *Langmuir*, 2005, **21**(23), 10253–10255.
44. D. J. Lestage and M. W. Urban, Hollow colloidal particles obtained by nano-extrusion in the presence of phospholipids, *Langmuir*, 2005, **21**(10), 4266–4267.
45. S. Blomberg, S. Ostberg, E. Harth, A. W. Bosman, B. Van Horn and C. J. Hawker, Production of crosslinked, hollow nanoparticles by surface-initiated living free-radical polymerization, *J. Polym. Sci., Part A: Polym. Chem.*, 2002, **40**(9), 1309–1320.
46. V. Franklin and R. W. Burhans, Process of producing hollow particles and resulting product, *US Pat.*, 2 797 201, 1957.
47. A. Kowalski and M. Vogel, Sequential heteropolymer dispersion and a particulate material obtainable therefrom, useful in coating compositions as an opacifying agent, *US Pat.*, 4 469 825, 1984.
48. C. J. McDonald and M. J. Devon, Hollow latex particles: synthesis and applications, *Adv. Colloid Interface Sci.*, 2002, **99**(3), 181–213.
49. V. Pavlyuchenko, O. Sorochinskaya, S. Ivanchev, V. Klubin, G. Kreichman and V. Budtov *et al.*, Hollow-particle

latexes: Preparation and properties, *J. Polym. Sci., Part A: Polym. Chem.*, 2001, **39**(9), 1435–1449.

50. K. H. Roh, D. C. Martin and J. Lahann, Biphasic Janus particles with nanoscale anisotropy, *Nat. Mater.*, 2005, **4**(10), 759–763.
51. A. Walther and A. H. E. Müller, Janus Particles: Synthesis, Self-Assembly, Physical Properties, and Applications, *Chem. Rev.*, 2013, **113**(7), 5194–5261.
52. J. Du and R. K. O'Reilly, Anisotropic particles with patchy, multicompartment and Janus architectures: preparation and application, *Chem. Soc. Rev.*, 2011, **40**(5), 2402–2416.
53. J. Hu, S. Zhou, Y. Sun, X. Fang and L. Wu, Fabrication, properties and applications of Janus particles, *Chem. Soc. Rev.*, 2012, **41**(11), 4356–4378.
54. Z. Nie, W. Li, M. Seo, S. Xu and E. Kumacheva, Janus and ternary particles generated by microfluidic synthesis: Design, synthesis, and self-assembly, *J. Am. Chem. Soc.*, 2006, **128**(29), 9408–9412.
55. T. Nisisako, T. Torii, T. Takahashi and Y. Takizawa, Synthesis of monodisperse bicolored janus particles with electrical anisotropy using a microfluidic co-flow system, *Adv. Mater.*, 2006, **18**(9), 1152–1156.
56. R. Erhardt, A. Böker, H. Zettl, H. Kaya, W. Pyckhout-Hintzen and G. Krausch *et al.*, Janus micelles, *Macromolecules*, 2001, **34**(4), 1069–1075.
57. S. C. Glotzer and M. J. Solomon, Anisotropy of building blocks and their assembly into complex structures, *Nat. Mater.*, 2007, **6**(8), 557–562.
58. A. Walther and A. H. E. Mueller, Janus particles, *Soft Matter*, 2008, **4**(4), 663–668.
59. D. Dendukuri and P. S. Doyle, The synthesis and assembly of polymeric microparticles using microfluidics, *Adv. Mater.*, 2009, **21**(41), 4071–4086.
60. L. Hong, S. Jiang and S. Granick, Simple method to produce Janus colloidal particles in large quantity, *Langmuir*, 2006, **22**(23), 9495–9499.
61. H. W. Gu, Z. M. Yang, J. H. Gao, C. K. Chang and B. Xu, Heterodimers of nanoparticles: Formation at a liquid-liquid interface and particle-specific surface modification by functional molecules, *J. Am. Chem. Soc.*, 2005, **127**(1), 34–35.
62. Y. Lu, H. Xiong, X. Jiang, Y. Xia, M. Prentiss and G. M. Whitesides, Asymmetric dimers can be formed by dewetting half-shells of gold deposited on the surfaces of spherical oxide colloids, *J. Am. Chem. Soc.*, 2003, **125**(42), 12724–12725.

63. L. Nie, S. Y. Liu, W. M. Shen, D. Y. Chen and M. Jiang, One-pot synthesis of amphiphilic polymeric Janus particles and their self-assembly into supermicelles with a narrow size distribution, *Angew. Chem., Int. Ed*, 2007, **46**(33), 6321–6324.

64. H. Gu, R. Zheng, X. Zhang and B. Xu, Facile one-pot synthesis of bifunctional heterodimers of nanoparticles: a conjugate of quantum dot and magnetic nanoparticles, *J. Am. Chem. Soc.*, 2004, **126**(18), 5664–5665.

65. S. Reculusa, C. Poncet-Legrand, A. Perro, E. Duguet, E. Bourgeat-Lami and C. Mingotaud *et al.*, Hybrid dissymmetrical colloidal particles, *Chem. Mater.*, 2005, **17**(13), 3338–3344.

66. N. Zhao and M. Gao, Magnetic Janus particles prepared by a flame synthetic approach: synthesis, characterizations and properties, *Adv. Mater.*, 2009, **21**(2), 184–187.

67. A. Misra and M. W. Urban, Acorn-Shape Polymeric Nano-Colloids: Synthesis and Self-Assembled Films, *Macromol. Rapid Commun.*, 2010, **31**(2), 119–127.

68. C. C. Corten and M. W. Urban, Shape evolution control of phase-separated colloidal nanoparticles, *Polym. Chem.*, 2011, **2**(1), 244–250.

69. D. Ramachandran, C. C. Corten and M. W. Urban, Color- and shape-tunable colloidal nanoparticles capable of nanopatterning, *RSC Adv.*, 2013, **3**(24), 9357–9364.

70. S. Berger, A. Synytska, L. Ionov, K.-J. Eichhorn and M. Stamm, Stimuli-Responsive Bicomponent Polymer Janus Particles by "Grafting from"/"Grafting to" Approaches, *Macromolecules*, 2008, **41**(24), 9669–9676.

71. T. Tanaka, M. Okayama, H. Minami and M. Okubo, Dual Stimuli-Responsive "Mushroom-like" Janus Polymer Particles as Particulate Surfactants, *Langmuir*, 2010, **26**(14), 11732–11736.

72. S. Jiang, Q. Chen, M. Tripathy, E. Luijten, K. S. Schweizer and S. Granick, Janus particle synthesis and assembly, *Adv. Mater.*, 2010, **22**(10), 1060–1071.

73. Q. Chen, J. K. Whitmer, S. Jiang, S. C. Bae, E. Luijten and S. Granick, Supracolloidal Reaction Kinetics of Janus Spheres, *Science*, 2011, **331**(6014), 199–202.

74. B. P. Binks and P. D. I. Fletcher, Particles adsorbed at the oil-water interface: A theoretical comparison between spheres of uniform wettability and "Janus" particles, *Langmuir*, 2001, **17**(16), 4708–4710.

75. S. Jiang and S. Granick, Controlling the geometry (Janus balance) of amphiphilic colloidal particles, *Langmuir*, 2008, **24**(6), 2438–2445.
76. Q. Chen, S. C. Bae and S. Granick, Directed self-assembly of a colloidal kagome lattice, *Nature*, 2011, **469**(7330), 381–384.
77. C.-C. Ho, W.-S. Chen, T.-Y. Shie, J.-N. Lin and C. Kuo, Novel fabrication of Janus particles from the surfaces of electrospun polymer fibers, *Langmuir*, 2008, **24**(11), 5663–5666.
78. X. Y. Ling, I. Y. Phang, C. Acikgoz, M. D. Yilmaz, M. A. Hempenius and G. J. Vancso *et al.*, Janus particles with controllable patchiness and their chemical functionalization and supramolecular assembly, *Angew. Chem., Int. Ed.*, 2009, **48**(41), 7677–7682.
79. A. H. Gröschel, A. Walther, T. I. Löbling, J. Schmelz, A. Hanisch and H. Schmalz *et al.*, Facile, solution-based synthesis of soft, nanoscale janus particles with tunable janus balance, *J. Am. Chem. Soc.*, 2012, **134**(33), 13850–13860.
80. K.-H. Roh, D. C. Martin and J. Lahann, Biphasic Janus particles with nanoscale anisotropy, *Nat. Mater.*, 2005, **4**(10), 759–763.
81. M. Lattuada and T. A. Hatton, Synthesis, properties and applications of Janus nanoparticles, *Nano Today*, 2011, **6**(3), 286–308.
82. A. Walther, M. Hoffmann and A. H. Müller, Emulsion polymerization using Janus particles as stabilizers, *Angew. Chem.*, 2008, **120**(4), 723–726.
83. A. Misra and M. W. Urban, Acorn-Shape Polymeric Nano-Colloids: Synthesis and Self-Assembled Films, *Macromol. Rapid Commun.*, 2010, **31**(2), 119–127.
84. C. Lu and M. W. Urban, Tri-phasic size-and janus balance-tunable colloidal nanoparticles (JNPs), *ACS Macro Lett.*, 2014, **3**(4), 346–352.
85. J. C. Love, B. D. Gates, D. B. Wolfe, K. E. Paul and G. M. Whitesides, Fabrication and wetting properties of metallic half-shells with submicron diameters, *Nano Lett.*, 2002, **2**(8), 891–894.
86. J. Choi, Y. Zhao, D. Zhang, S. Chien and Y.-H. Lo, Patterned fluorescent particles as nanoprobes for the investigation of molecular interactions, *Nano Lett.*, 2003, **3**(8), 995–1000.
87. H. Takei and N. Shimizu, Gradient sensitive microscopic probes prepared by gold evaporation and chemisorption on latex spheres, *Langmuir*, 1997, **13**(7), 1865–1868.

88. X. Y. Ling, I. Y. Phang, C. Acikgoz, M. D. Yilmaz, M. A. Hempenius and G. J. Vancso *et al.*, Janus particles with controllable patchiness and their chemical functionalization and supramolecular assembly, *Angew. Chem.*, 2009, **121**(41), 7813–7818.

89. B. Li and C. Y. Li, Immobilizing Au nanoparticles with polymer single crystals, patterning and asymmetric functionalization, *J. Am. Chem. Soc.*, 2007, **129**(1), 12–13.

90. B. Li, C. Ni and C. Y. Li, Poly (ethylene oxide) single crystals as templates for Au nanoparticle patterning and asymmetrical functionalization, *Macromolecules*, 2008, **41**(1), 149–155.

91. B. Wang, B. Li, B. Zhao and C. Y. Li, Amphiphilic Janus gold nanoparticles via combining "solid-state grafting-to" and "grafting-from" methods, *J. Am. Chem. Soc.*, 2008, **130**(35), 11594–11595.

92. L. Liu, M. Ren and W. Yang, Preparation of polymeric Janus particles by directional UV-induced reactions, *Langmuir*, 2009, **25**(18), 11048–11053.

93. M. D. McConnell, M. J. Kraeutler, S. Yang and R. J. Composto, Patchy and multiregion janus particles with tunable optical properties, *Nano Lett.*, 2010, **10**(2), 603–609.

94. J. Gong, X. Zu, Y. Li, W. Mu and Y. Deng, Janus particles with tunable coverage of zinc oxide nanowires, *J. Mater. Chem.*, 2011, **21**(7), 2067–2069.

95. A. Walther and A. H. Müller, Janus particles: synthesis, self-assembly, physical properties, and applications, *Chem. Rev.*, 2013, **113**(7), 5194–5261.

96. C.-C. Lin, C.-W. Liao, Y.-C. Chao and C. Kuo, Fabrication and characterization of asymmetric Janus and ternary particles, *ACS Appl. Mater. Interfaces*, 2010, **2**(11), 3185–3191.

97. T. Nisisako, T. Torii, T. Takahashi and Y. Takizawa, Synthesis of Monodisperse Bicolored Janus Particles with Electrical Anisotropy Using a Microfluidic Co-Flow System, *Adv. Mater.*, 2006, **18**(9), 1152–1156.

98. T. Tanaka, M. Okayama, Y. Kitayama, Y. Kagawa and M. Okubo, Preparation of "Mushroom-like" Janus Particles by Site-Selective Surface-Initiated Atom Transfer Radical Polymerization in Aqueous Dispersed Systems†, *Langmuir*, 2010, **26**(11), 7843–7847.

99. T. Tanaka, M. Okayama, H. Minami and M. Okubo, Dual Stimuli-Responsive "Mushroom-like" Janus Polymer Particles as Particulate Surfactants†, *Langmuir*, 2010, **26**(14), 11732–11736.

100. A. Walther, X. André, M. Drechsler, V. Abetz and A. H. Müller, Janus discs, *J. Am. Chem. Soc.*, 2007, **129**(19), 6187–6198.

101. W. R. Dreher, W. L. Jarrett and M. W. Urban, Stable nonspherical fluorine-containing colloidal dispersions: synthesis and film formation, *Macromolecules*, 2005, **38**(6), 2205–2212.
102. A. Singh, W. R. Dreher and M. W. Urban, Phospholipid-assisted synthesis of stable F-containing colloidal particles and their film formation, *Langmuir*, 2006, **22**(2), 524–527.
103. D. Ramachandran, C. C. Corten and M. W. Urban, Color- and shape-tunable colloidal nanoparticles capable of nanopatterning, *RSC Adv.*, 2013, **3**(24), 9357–9364.
104. A. R. Tao, S. Habas and P. Yang, Shape control of colloidal metal nanocrystals, *Small*, 2008, **4**(3), 310–325.
105. V. N. Manoharan, M. T. Elsesser and D. J. Pine, Dense packing and symmetry in small clusters of microspheres, *Science*, 2003, **301**(5632), 483–487.
106. M. Okubo, Y. Murakami and T. Fujiwara, Formation mechanism of anomalous "golf ball-like" composite polymer particles by seeded emulsion polymerization, *Colloid Polym. Sci.*, 1996, **274**(6), 520–524.
107. S. H. Im, U. Jeong and Y. Xia, Polymer hollow particles with controllable holes in their surfaces, *Nat. Mater.*, 2005, **4**(9), 671–675.
108. C. Lu and M. Urban, Rationally Designed Gibbous Stimuli-Responsive Colloidal Nanoparticles, *ACS Nano*, 2015, **9**(3), 3119–3124.
109. F. Romano and F. Sciortino, Colloidal self-assembly: Patchy from the bottom up, *Nat. Mater.*, 2011, **10**(3), 171–173.
110. Y. Lalatonne, J. Richardi and M. Pileni, Van der Waals versus dipolar forces controlling mesoscopic organizations of magnetic nanocrystals, *Nat. Mater.*, 2004, **3**(2), 121–125.
111. M. G. Warner and J. E. Hutchison, Linear assemblies of nano-particles electrostatically organized on DNA scaffolds, *Nat. Mater.*, 2003, **2**(4), 272–277.
112. A. M. Kalsin, M. Fialkowski, M. Paszewski, S. K. Smoukov, K. J. Bishop and B. A. Grzybowski, Electrostatic self-assembly of binary nanoparticle crystals with a diamond-like lattice, *Science*, 2006, **312**(5772), 420–424.
113. S. Sacanna, W. Irvine, P. M. Chaikin and D. J. Pine, Lock and key colloids, *Nature*, 2010, **464**(7288), 575–578.
114. Y. Wang, Y. Wang, D. R. Breed, V. N. Manoharan, L. Feng and A. D. Hollingsworth *et al.*, Colloids with valence and specific directional bonding, *Nature*, 2012, **491**(7422), 51–55.

115. Y. Min, M. Akbulut, K. Kristiansen, Y. Golan and J. Israelachvili, The role of interparticle and external forces in nanoparticle assembly, *Nat. Mater.*, 2008, **7**(7), 527–538.

116. Y. Wang, Y. Wang, X. Zheng, G.-R. Yi, S. Sacanna and D. J. Pine *et al.*, Three-dimensional lock and key colloids, *J. Am. Chem. Soc.*, 2014, **136**(19), 6866–6869.

117. L. Feng, B. Laderman, S. Sacanna and P. Chaikin, Re-entrant solidification in polymer–colloid mixtures as a consequence of competing entropic and enthalpic attractions, *Nature Mater.*, 2014, **14**(1), 61–65.

118. L. Cademartiri and K. J. Bishop, Programmable self-assembly, *Nat. Mater.*, 2015, **14**(1), 2–9.

119. E. Amstad, M. Textor and E. Reimhult, Stabilization and functionalization of iron oxide nanoparticles for biomedical applications, *Nanoscale*, 2011, **3**(7), 2819–2843.

120. J. H. Waite and M. L. Tanzer, Polyphenolic substance of Mytilus edulis: novel adhesive containing L-dopa and hydroxyproline, *Science*, 1981, **212**(4498), 1038–1040.

121. H. Gu, Z. Yang, J. Gao, C. Chang and B. Xu, Heterodimers of nanoparticles: Formation at a liquid-liquid interface and particle-specific surface modification by functional molecules, *J. Am. Chem. Soc.*, 2005, **127**(1), 34–35.

122. E. Amstad, T. Gillich, I. Bilecka, M. Textor and E. Reimhult, Ultrastable iron oxide nanoparticle colloidal suspensions using dispersants with catechol-derived anchor groups, *Nano Lett.*, 2009, **9**(12), 4042–4048.

123. E. Amstad, J. Kohlbrecher, E. Müller, T. Schweizer, M. Textor and E. Reimhult, Triggered release from liposomes through magnetic actuation of iron oxide nanoparticle containing membranes, *Nano Lett.*, 2011, **11**(4), 1664–1670.

124. S. Kurzhals, R. Zirbs and E. Reimhult, Synthesis and magneto-thermal actuation of iron oxide core–PNIPAM shell nanoparticles, *ACS Appl. Mater. Interfaces*, 2015, **7**, 19342.

125. J. J. Lai, J. M. Hoffman, M. Ebara, A. S. Hoffman, C. Estournès and A. Wattiaux *et al.*, Dual magnetic-/temperature-responsive nanoparticles for microfluidic separations and assays, *Langmuir*, 2007, **23**(13), 7385–7391.

126. A. Stein, B. J. Melde and R. C. Schroden, Hybrid inorganic-organic mesoporous silicates—nanoscopic reactors coming of age, *Adv. Mater.*, 2000, **12**(19), 1403–1419.

127. C.-Y. Lai, B. G. Trewyn, D. M. Jeftinija, K. Jeftinija, S. Xu and S. Jeftinija *et al.*, A mesoporous silica nanosphere-based

carrier system with chemically removable CdS nanoparticle caps for stimuli-responsive controlled release of neuro-transmitters and drug molecules, *J. Am. Chem. Soc.*, 2003, **125**(15), 4451–4459.

128. V. Colvin, A. Goldstein and A. Alivisatos, Semiconductor nano-crystals covalently bound to metal surfaces with self-assembled monolayers, *J. Am. Chem. Soc.*, 1992, **114**(13), 5221–5230.
129. W. C. Chan and S. Nie, Quantum dot bioconjugates for ultra-sensitive nonisotopic detection, *Science*, 1998, **281**(5385), 2016–2018.
130. K. E. Sapsford, L. Berti and I. L. Medintz, Materials for fluo-rescence resonance energy transfer analysis: beyond traditional donor–acceptor combinations, *Angew. Chem. Int. Ed*, 2006, **45**(28), 4562–4589.
131. B. W. Van Der Meer, G. Coker and S.-Y. S. Chen Resonance en-ergy transfer: theory and data. 1994.
132. E. A. Jares-Erijman and T. M. Jovin, FRET imaging, *Nat. Bio-technol.*, 2003, **21**(11), 1387–1395.
133. D. M. Jameson, J. C. Croney and P. D. Moens, [1] Fluorescence: Basic concepts, practical aspects, and some anecdotes, *Methods Enzymol.*, 2003, **360**, 1–43.
134. R. M. Clegg, Fluorescence resonance energy-transfer and nu-cleic-acids, *Methods Enzymol.*, 1992, **211**, 353–388.
135. A. Kirsch, V. Subramaniam, A. Jenei and T. Jovin, Fluorescence resonance energy transfer detected by scanning near-field op-tical microscopy, *J. Microsc.*, 1999, **194**(2–3), 448–454.
136. L. Giordano, T. M. Jovin, M. Irie and E. A. Jares-Erijman, Dihe-teroarylethenes as thermally stable photoswitchable acceptors in photochromic fluorescence resonance energy transfer (pcFRET), *J. Am. Chem. Soc.*, 2002, **124**(25), 7481–7489.
137. S. Wang, B. S. Gaylord and G. C. Bazan, Fluorescein provides a resonance gate for FRET from conjugated polymers to DNA intercalated dyes, *J. Am. Chem. Soc.*, 2004, **126**(17), 5446–5451.
138. M. Zimmer, Green fluorescent protein (GFP): applications, structure, and related photophysical behavior, *Chem. Rev.*, 2002, **102**(3), 759–782.
139. A. R. Clapp, I. L. Medintz and H. Mattoussi, Förster Resonance Energy Transfer Investigations Using Quantum-Dot Fluor-ophores, *ChemPhysChem*, 2006, **7**(1), 47–57.
140. T. Jamieson, R. Bakhshi, D. Petrova, R. Pocock, M. Imani and A. M. Seifalian, Biological applications of quantum dots, *Bio-materials*, 2007, **28**(31), 4717–4732.

141. E. Chang, N. Thekkek, W. W. Yu, V. L. Colvin and R. Drezek, Evaluation of quantum dot cytotoxicity based on intracellular uptake, *Small*, 2006, **2**(12), 1412–1417.
142. J. Lovrić, S. J. Cho, F. M. Winnik and D. Maysinger, Unmodified cadmium telluride quantum dots induce reactive oxygen species formation leading to multiple organelle damage and cell death, *Chem. Biol.*, 2005, **12**(11), 1227–1234.
143. J. M. Tsay and X. Michalet, New light on quantum dot cyto-toxicity, *Chem. Biol.*, 2005, **12**(11), 1159–1161.
144. G. S. Yi and G. M. Chow, Synthesis of Hexagonal-Phase NaYF4: Yb, Er and NaYF4: Yb, Tm Nanocrystals with Efficient Up-Conversion Fluorescence, *Adv. Funct. Mater.*, 2006, **16**(18), 2324–2329.
145. O. Ehlert, R. Thomann, M. Darbandi and T. Nann, A four-color colloidal multiplexing nanoparticle system, *ACS Nano*, 2008, **2**(1), 120–124.
146. F. Auzel, Upconversion and anti-stokes processes with f and d ions in solids, *Chem. Rev.*, 2004, **104**(1), 139–174.
147. M. Wang, W. Hou, C.-C. Mi, W.-X. Wang, Z.-R. Xu and H.-H. Teng *et al.*, Immunoassay of goat antihuman immuno-globulin G antibody based on luminescence resonance energy transfer between near-infrared responsive NaYF$_4$: Yb, Er upconversion fluorescent nanoparticles and gold nanoparticles, *Anal. Chem.*, 2009, **81**(21), 8783–8789.
148. H. A. Klok and S. Lecommandoux, Supramolecular materials via block copolymer self-assembly, *Adv. Mater.*, 2001, **13**(16), 1217–1229.
149. J.-M. Lehn, *Supramolecular chemistry*, Vch, Weinheim, 1995.
150. I. W. Hamley, *The Physics of Block Copolymers*, Oxford University Press, New York, 1998.
151. M. Sauer and W. Meier, Responsive nanocapsules, *Chem. Commun.*, 2001, **1**, 55–56.
152. G. B. Sukhorukov, A. A. Antipov, A. Voigt, E. Donath and H. Möhwald, pH-controlled macromolecule encapsulation in and release from polyelectrolyte multilayer nanocapsules, *Macromol. Rapid Commun.*, 2001, **22**(1), 44–46.
153. J. F. Gohy, N. Willet, S. Varshney, J. X. Zhang and R. Jérôme, Core–shell–corona micelles with a responsive shell, *Angew. Chem.*, 2001, **113**(17), 3314–3316.
154. F. Chécot, S. Lecommandoux, Y. Gnanou and H. A. Klok, Water--soluble stimuli-responsive vesicles from peptide-based diblock copolymers, *Angew. Chem., Int. Ed*, 2002, **41**(8), 1339–1343.

155. E. G. Bellomo, M. D. Wyrsta, L. Pakstis, D. J. Pochan and T. J. Deming, Stimuli-responsive polypeptide vesicles by conformation-specific assembly, *Nat. Mater.*, 2004, **3**(4), 244–248.
156. S. Iijima, Helical microtubules of graphitic carbon, *Nature*, 1991, **354**(6348), 56–58.
157. P. He and M. W. Urban, Controlled phospholipid functionalization of single-walled carbon nanotubes, *Biomacromolecules*, 2005, **6**(5), 2455–2457.
158. S. Soll, M. Antonietti and J. Yuan, Double stimuli-responsive copolymer stabilizers for multiwalled carbon nanotubes, *ACS Macro Lett.*, 2011, **1**(1), 84–87.
159. A. Wachowiak, J. Wiebe, M. Bode, O. Pietzsch, M. Morgenstern and R. Wiesendanger, Direct observation of internal spin structure of magnetic vortex cores, *Science*, 2002, **298**(5593), 577–580.
160. S.-B. Choe, Y. Acremann, A. Scholl, A. Bauer, A. Doran and J. Stöhr *et al.*, Vortex core-driven magnetization dynamics, *Science*, 2004, **304**(5669), 420–422.
161. M. Yu and M. W. Urban, Formation of concentric ferromagnetic nanotubes from biologically active phospholipids, *J. Mater. Chem.*, 2007, **17**(44), 4644–4646.
162. M. A. C. Stuart, W. T. Huck, J. Genzer, M. Müller, C. Ober and M. Stamm *et al.*, Emerging applications of stimuli-responsive polymer materials, *Nat. Mater.*, 2010, **9**(2), 101–113.
163. S. Kawano and M. W. Urban, Expandable temperature-responsive polymeric nanotubes, *ACS Macro Lett.*, 2011, **1**(1), 232–235.
164. D. J. Lestage, M. Yu and M. W. Urban, Stimuli-responsive surfactant/phospholipid stabilized colloidal dispersions and their film formation, *Biomacromolecules*, 2005, **6**(3), 1561–1572.
165. M. Dutt, O. Kuksenok, M. J. Nayhouse, S. R. Little and A. C. Balazs, Modeling the Self Assembly of Lipids and Nanotubes in Solution: Forming Vesicles and Bicelles with Transmembrane Nanotube Channels, *ACS Nano*, 2011, **5**, 4769–4782.
166. J. S. Amelang, G. N. Venturini and D. M. Kochmann, Summation rules for a fully-nonlocal energy-based quasicontinuum methods, *J. Mech. Phys. Solids*, 2015, **82**, 378–413.
167. R. Yoshida, T. Takahashi, T. Yamaguchi and H. Ichijo, Self-Oscillating Gel, *J. Am. Chem. Soc.*, 1996, **118**, 5134–5135.
168. R. J. Field, E. Koros and R. M. Noyes, Oscillations in chemical systems. II. Thorough analysis of temporal oscillation in the bromate-cerium-malonic acid system, *J. Am. Chem. Soc.*, 1972, **94**(25), 8649–8664.

169. I. R. Epstein and J. A. Pojman, *An introduction to nonlinear chemical dynamics: oscillations, waves, patterns, and chaos*, Oxford University Press, New York, NY, 1998.
170. T. Ueki, M. Shibayama and R. Yoshida, Self-Oscillating Micelles, *Chem. Commun.*, 2013, **49**, 6947–6949.
171. R. Tamate, T. Ueki and R. Yoshida, Self-Beating Artificial Cells: Design of Cross-Linked Polymersomes Showing Self-Oscillating Motion, *Adv. Mater.*, 2015, **27**, 837–842.
172. T. Ueki, Y. Takasaki, K. Bundo, T. Ueno, T. Sakai, Y. Akagi and R. Yoshida, Autonomous viscosity oscillation via metallo-supramolecular terpyridine chemistry of branched poly(ethylene glycol) driven by the Belousov–Zhabotinsky reaction, *Soft Matter*, 2014, **10**, 1349.

7 Biologically Responsive Polymers

7.1 Introduction

Polymers that respond to external stimuli are of great interest in medicine. Owing to the fact that they can be used as controlled drug release vehicles and in other therapeutic applications, these materials will shape the future of human existence. One has to be aware, however, that certain biological processes may induce external stimuli; for example, it is well known that infections usually result in pH changes, which in turn will induce other responses. Thus, of particular interest are synthetic or modified biological materials that can undergo conformational or phase changes not only in response to classic stimuli, *i.e.* temperature and/or pH or electromagnetic radiation, but also enzymatic, glucose, vasoactive, peptide or microbial changes. Ever since nanotechnology saturated various fields and our vocabularies, responsive polymers have become a major focus of many emerging nanoscale studies, moving right along with biological responses. Aside from diversified stimuli, the fundamental difference between the 'regular' polymers and biologically responsive polymers is the non-linear responsiveness to the stimuli. Although there are many elements that may be incorporated into a polymer backbone, the real question is whether these multi-level elements will be able to trigger timely and synchronized responsiveness. The last decade taught us that the responses of the polymers are usually due to multiple cooperative interactions, including orchestrated supramolecular and covalent bonding, hydrophobic and hydrophilic

Stimuli-Responsive Materials: From Molecules to Nature Mimicking Materials Design
By Marek W. Urban
© Marek W. Urban, 2016
Published by the Royal Society of Chemistry, www.rsc.org

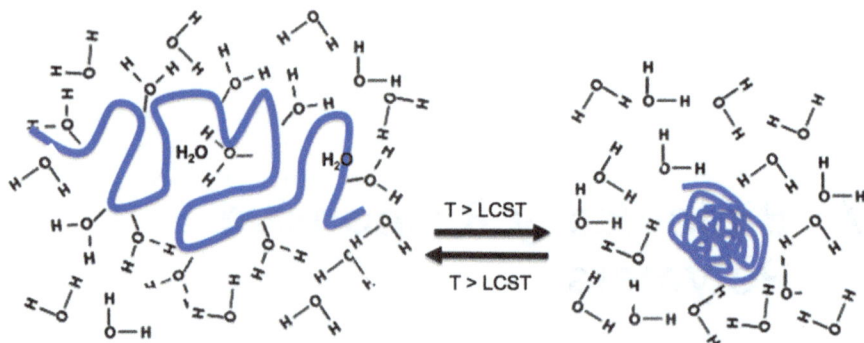

Figure 7.1 Polymer conformational changes as a function of lower critical solution temperature (LCST).

interactions and the combinations thereof. Ironically, the majority of the biomaterials review articles begin with the cartoon depicted in Figure 7.1 and the short description that ... the most studied synthetic polymer is poly(*N*-isopropylacrylamide) (PNIPAm), which undergoes a coil–globule transition in aqueous environments at 32 °C, changing from a hydrophilic state below to a hydrophobic state above this temperature.[1] Depicted many times, this transition is believed to arise from the entropic gain as water molecules associated with the side-chain isopropyl moieties are released back into the aqueous environment.

Indeed, the temperature at which this transition occurs (the lower critical solution temperature or LCST) corresponds to the region in the phase diagram at which the enthalpic contributions of hydrogen-bonded water to the polymer chain becomes smaller than that of the entropic origin. Another polymer that exhibits temperature and pH responsiveness is poly(2-*N*,*N*-dimethylamino)ethyl methyl-acrylate or p(DMAEMA) where temperature sensitivity results from (dimethylamino)ethyl groups and generates the LCST at around 40 °C,[2] and pH responsiveness is attributed to protonation/deprotonation of the tertiary amine side groups.[3] The most signifi-cant aspect of these homopolymers is that their respective LCSTs are close to the physiological temperature of 37 °C. A simple strategy to tune into this exact temperature is to copolymerize another monomer that would increase (for p(NIPAm)) or decrease (for p(DMAEMA)) the LCST. Other common temperature and pH responsive monomers are listed in Chapter 3.

With this approach one can design the combination of hydrophilic, hydrophobic and charged groups within one polymer chain, which coupled with the ability to alter these properties *via* temperature or pH switching, will facilitate the resemblance to biological systems that can be classified as responsive polymeric nanoparticles (sometimes incorrectly called polymeric micelles), polymeric gels, and polymer–biopolymer conjugates. They are schematically depicted in Figure 7.2.[4]

7.2 LCST-based Stimuli-responsive Bioactive Polymers

Within the first category, complex multi-block responsive micellar-like materials have been prepared containing poly(propylene oxide) (PPO) as the hydrophobic component with p(DMAEMA) as a cross-linkable unit and poly(oligoethyleneglycolmethacrylate) (POEGMA) as a solubilizing block.[5–7]

Responsive polymer hydrogels exhibit enormous potential applications in a variety of drug-loading and release formats. The objective in designing these materials is to correlate the gel responsiveness manifested by the gel collapse point, structural feature of the matrix and drug release. Thus, significant synthetic efforts are necessary in designing pH- and temperature-sensitive gels for therapeutic applications. For drug delivery applications polymer response should be non-linear, *i.e.* with distinct and selective 'on' and 'off' switches. An alternative approach is to design gels with very sharp response, capable of responding with a very minute stimulus from the environment. Tuning the time responses is extremely critical to many applications. For instance, Figure 7.2B illustrates an example of grafting of p(NIPAm) to crosslinked hydrogels, which leads to quick aggregation of the non-crosslinked oligomeric components. These aggregates are often responsible for the nucleation at this hydrophobic site, causing adhesion of other network components.

The field for responsive polymer–biopolymer conjugates deals with the targeting of gene expression *via* switchable polymers. Figure 7.2C shows an example of antisense nucleotide binding behavior.[8] In this case p(NIPAm) conjugates with pendant oligodeoxynucleotides (ODNs) were prepared *via* direct copolymerization of NIPAm with a methacryloyl-terminated ODN. This particular conjugate exhibits the temperature-induced coil–globule transition at 33 °C in physiological-like buffers (pH 7.4, 100 mM NaCl). The activity of the conjugate, which contained the antisense sequence for the ribosomal binding site of mRNA encoding enhanced green fluorescent protein, was assessed in *E. coli*. The uniqueness of the translation of a plasmid

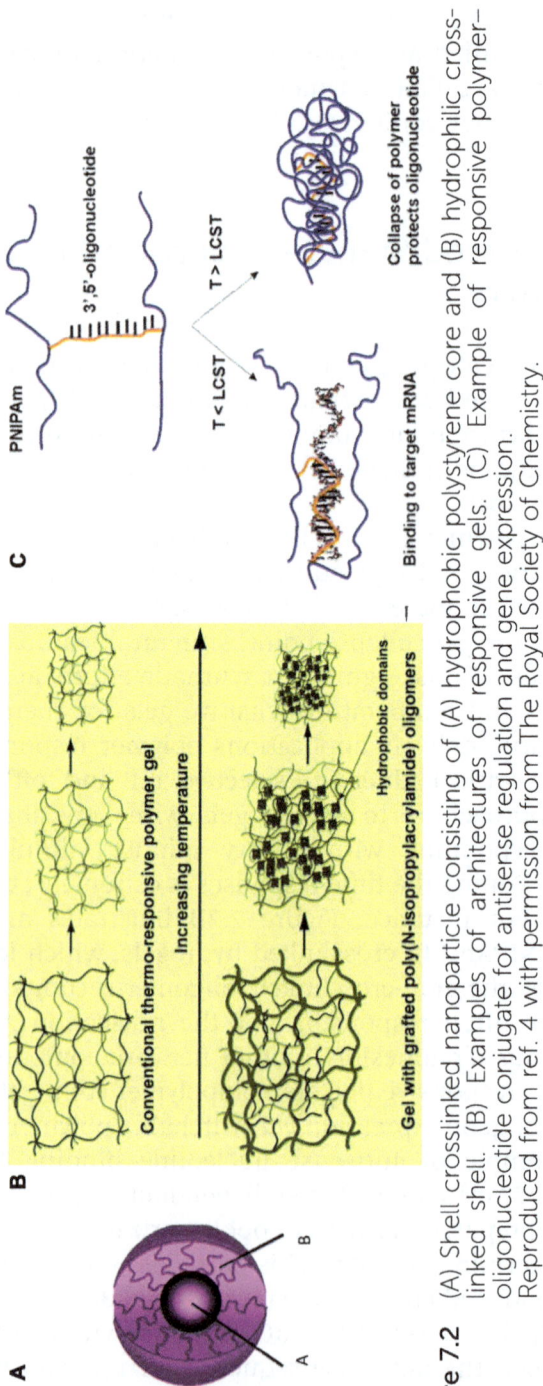

Figure 7.2 (A) Shell crosslinked nanoparticle consisting of (A) hydrophobic polystyrene core and (B) hydrophilic cross-linked shell. (B) Examples of architectures of responsive gels. (C) Example of responsive polymer–oligonucleotide conjugate for antisense regulation and gene expression. Reproduced from ref. 4 with permission from The Royal Society of Chemistry.

encoding EGFP was manifested by the ability to suppress with a dose-dependent experiment by the p(NIPAm)–antisense ODN conjugates in which no translational repression was observed for p(NIPAm) alone.

7.3 Glucose-responsive Polymers

Driven by many potential applications of glucose sensing and insulin delivery systems, polymers that respond to glucose have received considerable attention. The most common glucose-responsive polymeric systems are typically based on enzymatic oxidation of glucose by glucose oxidase (GOx), binding of glucose with concanavalin A (ConA), or reversible covalent bond formation between glucose and boronic acids. Glucose-responsive polymers are based on the GOx-catalyzed reaction of glucose with oxygen. In general, glucose sensitivity is not caused by direct interaction of glucose with the responsive polymer, but by the response of the polymer to by-products that result from the oxidation of glucose. Reactions that lead to glucose-responsive systems are shown in Figure 7.3.

This is the specificity of the enzymatic action of GOx on glucose that leads to by-products of gluconic acid and hydrogen peroxide (H_2O_2). In a typical scenario, a pH-responsive polymer is loaded or conjugated with GOx, and the gluconic acid by-product that results from the reaction with glucose induces a response in the

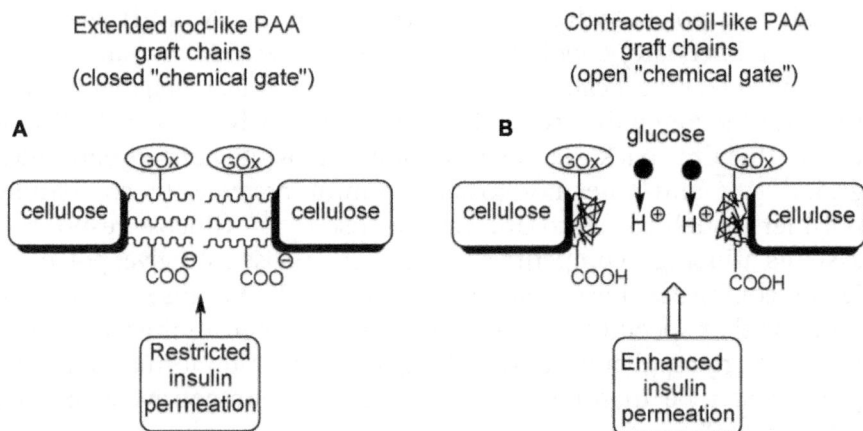

Figure 7.3 (A) When glucose is not present, the PAA chains are extended, thus lowering the permeability of the membrane. (B) The presence of glucose leads to a lowering of local pH and chain collapse due to a reduction in electrostatic repulsion.
Reprinted from ref. 9. Copyright (1989) with permission from Elsevier.

pH-responsive macromolecule. When a cellulose film is covalently modified with GOx-conjugated poly(acrylic acid) (PAA),[9] at neutral and high pH levels, the carboxylate units of the PAA chains are negatively charged and extended due to electrostatic repulsive forces. As a consequence, occlusion of the pores in the cellulose membrane occurs. The gluconic acid that forms from the addition of glucose led to a local pH reduction, protonation of the PAA carboxylate moieties, and concomitant collapse of the chains, obscuring the membrane pores and facilitating the release of entrapped insulin. Similar concepts can be used in the preparation of glucose-responsive hydrogels.[10] In this case, poly(methacrylic acid (PMAA)-*graft*-ethylene glycol) gels can be synthesized in the presence of GOx. Again, at neutral and high pH values, these gels are swollen by repulsion between negative charges on methacrylate units. When pH is reduced upon oxidation of glucose by GOx, the gel networks collapse. Under these conditions the efficient response can be achieved not only by a reduction in electrostatic repulsion, but the enhanced hydrogen bonding between the carboxyl and ether groups of the ethylene glycol units. Similar hydrogels can be synthesized by copolymerization of *N*-isopropylacrylamide (NIPAM) with methacrylic acid[11] or a sulfadimethoxine monomer with *N,N*-dimethylacrylamide (DMA).[12] In contrast to gluconic acid resulting in the chain collapse of carboxylate-containing polymers, the lowering of pH may also lead to chain expansion in the presence of a polybase.[13] Glucose-responsive crosslinked poly(2-hydroxyethyl methacrylate-*co-N,N*-dimethylamino-ethyl methacrylate) copolymers (p(HEMA-*co*-DMAEMA)) that contain entrapped GOx and catalase,[14,15] can be used, but the challenge is to increase biocompatibility in GOx-based responsive materials. To alleviate this drawback the incorporation of poly(ethylene glycol) (PEG) grafts[13,16,17] and other non-toxic, non-immunogenic, biocompatible polymers, such as chitosan, can be used.[17,18] Glucose-responsive systems utilizing a competitive binding of glucose with glycopolymer–lectin complexes have also been designed.[19] Because lectins are proteins that specifically bind carbohydrates, their multivalency may facilitate glycopolymer crosslinking or aggregate formation. This process can be disrupted by introducing a competitively binding saccharide.[20]

Synthetically, monosubstituted conjugates of glucosyl-terminal PEG (G-PEG) and insulin can be produced[21] and conjugated to ConA along a PEG-poly(vinylpyrrolidone-*co*-acrylic acid) backbone. However, elevated concentrations of glucose will lead to competitive binding of glucose with ConA, thus forcing the displacement and release of the G-PEG–insulin conjugates. ConA–glycogen gel also exhibits a gel–sol transition in the presence of glucose as a result of preferred binding of ConA with free glucose over the glycogen-containing gel.[22] Glucose-responsive hydrogels *via* copolymerization of ConA vinyl macromonomers with a monomer modified with pendant glucose units[23] can also be prepared, but the addition of free glucose will result in the glycopolymer–ConA complex dissociation, thus causing the gel to swell. One of the important concerns is that the entrapment of ConA in the hydrogel often leads to leakage of the protein and irreversible swelling, which can be prevented by copolymerization of reversibly responsive hydrogels.[24,25]

It has been recognized that the ability of boronic acids to reversibly complex with sugars may offer an opportunity for the development of glucose sensors as well as ligand binding moieties in chromatographic experiments.[26] As depicted in Figure 7.4, the uniqueness of boronic acids comes from their water solubility that can be tuned by pH changes or diol concentration.[27] In aqueous systems boronic acids exist in equilibrium between an undissociated neutral form (**1**) and a dissociated anionic form (**2**),[28–32] but in the presence of 1,2- or 1,3-diols, cyclic boronic esters between the neutral boronic acid and a diol are generally considered hydrolytically unstable.[29] In contrast, the anionic form (**2**) will reversibly bind with diols to form a boronate ester (**3**), shifting the equilibria to the anionic forms (**2** and **3**).[32] Polymers containing neutral boronic acid groups are generally hydrophobic, whereas the anionic boronate groups impart water solubility.[30] At increased concentrations of glucose, the ratio of the anionic forms (**2** and **3**) to the neutral form (**1**) is also elevated, and the hydrophilicity of the system is enhanced. The solubility of boronic acid-containing polymers is dependent not only on pH, but also on the concentration of compatible diols in the surrounding medium.[28]

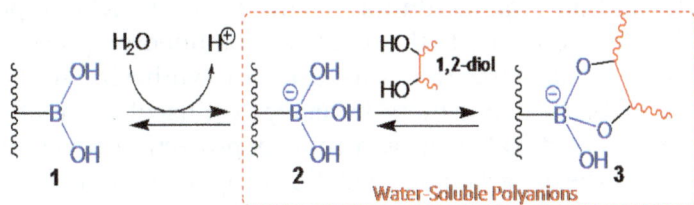

Figure 7.4 Aqueous ionization equilibria of boronic acids. As the concentration of diol increases, the equilibria shift toward the anionic boronate forms of the boronic acid.
Reprinted from ref. 27. Copyright (1989) with permission from Elsevier.

It should be noted that the majority of boronic acid-based responsive polymers are in the form of hydrogels. One example is the synthesis of a glucose-responsive hydrogel composed of terpolymers of 3-acrylamidophenylboronic acid (APBA), (*N,N*-dimethylamino)-propylacrylamide (DMAPA) and DMA.[33] Under these conditions boronic acid groups of the terpolymer were complexed with poly(vinyl alcohol) (PVA) at physiological pHs, but addition of glucose results in competitive displacement of PVA, and decreased crosslink density. The latter will lead to swelling. Similarly, responsive hydrogels can be prepared by complexing PVA with poly(*N*-vinyl-2-pyrrolidone-*co*-APBA)[34] or poly(DMA-*co*-3-methacrylamidophenylboronic acid-*co*-D-MAPA-*co*-butyl methacrylate).[35] Along the same lines, glucose-responsive polymeric hydrogels prepared *via* copolymerization of NIPAM with APBA and *N,N*-methylene-bis-acrylamide (MBA) as a crosslinker can be readily obtained.[31] Glucose-responsive polymer gel particles were reported using a similar system,[36] and PNIPAM-based comb-type grafted hydrogels with rapid response to glucose concentration were recently reported by Xie *et al.*[37] In addition to conventional hydrogels, several glucose-responsive microgels[38-41] and fluorescent nanospheres based on poly(NIPAM-*co*-APBA)[42] have been reported. Kataoka *et al.* recently reported the synthesis of a glucose-responsive copolymer of 4-(1,6-dioxo-2,5-diaza-7-oxamyl) phenylboronic acid (DDOPBA) ($pK_a \approx 7.8$) and NIPAM.[43,44]

Most glucose-responsive boronic acid-based (co)polymers have been synthesized by conventional radical polymerization to yield random copolymers,[45,46] gels[39] or other crosslinked entities.[47,48] However, precise synthetic efforts are required to fully capitalize on boronic acid-containing polymer properties. This can be achieved by controlling molecular weights, narrow molecular weight distributions, and retained chain end functionalities.[49,50] As pointed out in Chapter 2, atom transfer radical polymerization (ATRP)[51-53] and reversible addition-fragmentation chain transfer (RAFT) polymerization[54-56] should facilitate this need for organoboron polymers.[57-61] In fact, RAFT polymerization was used to synthesize well-defined boronic acid block copolymers by using a pinacol ester of 4-vinyl-phenylboronic acid followed by a mild deprotection procedure[59] as well as the synthesis of well-defined block copolymers *via* direct RAFT polymerization of unprotected APBA.[60,61] Synthesized block co-polymers of hydrophilic DMA and responsive boronic acid monomers are usually hydrophilic in nature and fully soluble above the pK_a of the boronic acid units. At $pH < pK_a$ these block copolymers self-assemble to form polymeric micelles. However, the presence of boronic acid functionality within the hydrophobic core may lead to dissociation of these micelles under elevated pH or higher glucose concentration levels. Replacing the hydrophilic poly(DMA) block with temperature-responsive pNIPAM[62] will facilitate the formation of triply-responsive block copolymers responsive to pH alterations, glucose concentration and temperature.[61] Depending on the combination of stimuli applied, these block copolymers were capable of forming micelles or reverse micelles.

7.4 Enzyme-responsive Polymers

Design of materials that undergo macroscopic property changes when triggered by the selective catalytic actions of enzymes are of significant importance.[63,64] Because enzymes are highly selective in their reactivity, and are vital components in many biological pathways, their catalytic properties may lead to changes in supramolecular architectures manifested by macroscopic transitions.[64] As was discussed in the previous chapters, block copolymers can assemble into various morphological confinements, ranging from spherical polymeric micelles to rods, lamellas, and other structures. However, it is the triggering motif within a block copolymer that is responsible for transitions from one morphology to another. Among numerous stimuli that have been utilized, enzymes and enzyme-catalyzed reactions offer a very promising avenue for biological applications. This is primarily attributed to their involvements in biological processes. However, from the chemical point of view, enzymes exhibit unique selectivity and specificity, and responsiveness to pH, temperature, or electromagnetic radiation. Equally important, is their ability to form a variety of assemblies driven by electrostatic, H-bonding, van der Waals or hydrophobic interactions. Although enzymes offer many diverse properties, typically they can be classified into three main categories: enzyme-triggered polymer assemblies, enzyme triggered structural reorganization, and enzyme triggered sol–gel and gel–sol transitions.

Enzyme-triggered polymer assemblies often contain phosphate functionalities to trigger self-assembly and aggregation. For example, Figure 7.5 illustrates an example of enzyme-triggered self-assembly from an initially water-soluble diblock copolymer made of polyethylene glycol (PEG) and a phosphorylated poly(4-vinylphenol) (PVPh) block was synthesized by nitroxide-mediated polymerization (NMP) utilizing a PEG-based mediating agent.[65]

A   **Water soluble**

Red = Hydrophilic
Black= Hydrophobic

B

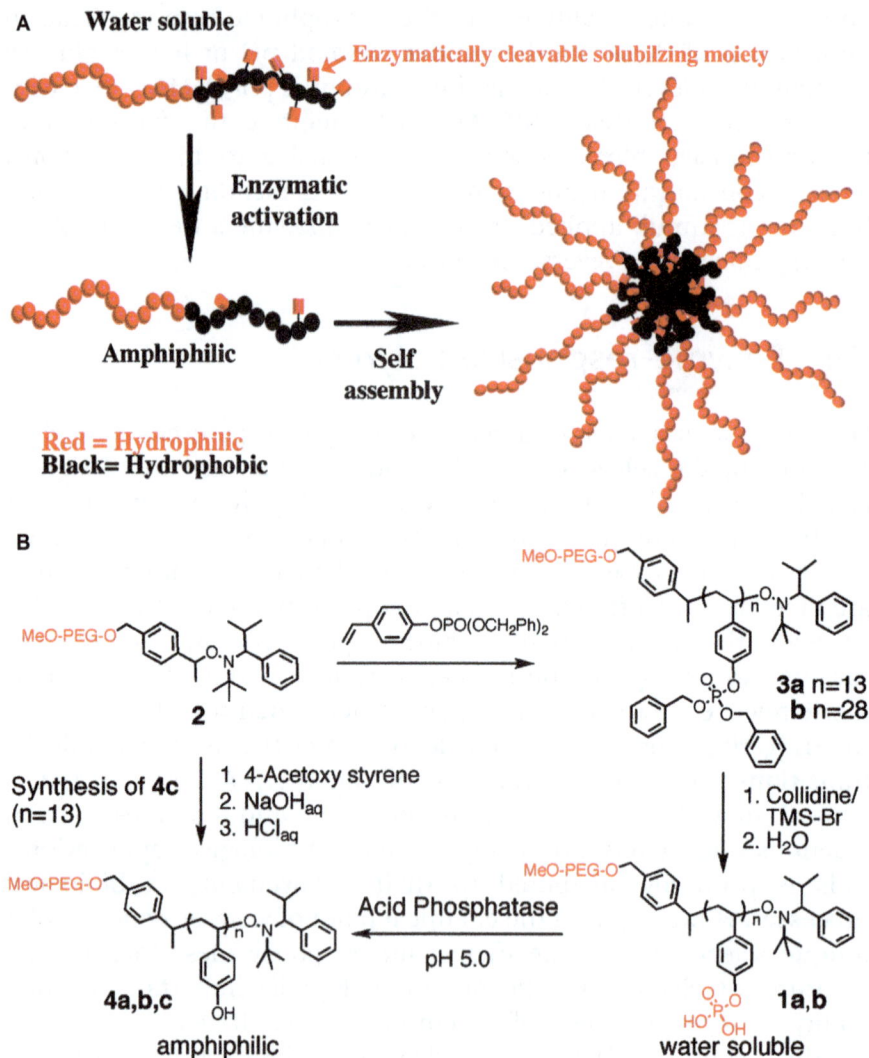

Figure 7.5   (a) Synthetic process employed in the formation of enzyme-responsive diblock copolymers. (b) Schematic illustration of enzyme triggered self-assembly resulting from the conversion of a water-soluble diblock copolymer into an amphiphilic diblock copolymer.
Reproduced from ref. 65 Copyright 2009, American Chemical Society.

The ability of peptides to form spontaneous secondary structures represented by α-helix and β-sheets opened up another opportunity for designing hybrid copolymers that consist of a synthetic solubilizing block and a peptide or protein block. An illustrative example is

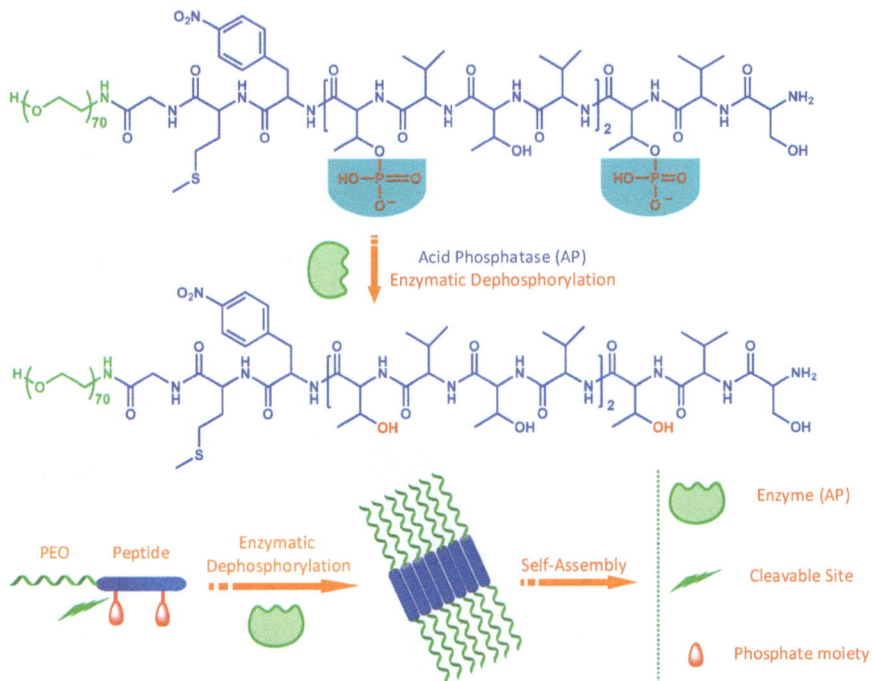

Figure 7.6 BioSwitch produced from a PEG–peptide hybrid diblock co-polymer that utilized enzyme-triggered dephosphorylation reactions.
Reproduced from ref. 66.

depicted in Figure 7.6, which shows a synthesized PEG–peptide conjugate containing a (TV)5-peptide aggregator domain, with five repeats of alternating threonine (T) and valine (V) diads.[66] The self-organization of the hybrid block is suppressed by site-specific phosphorylating hydroxyl functionalities of three threonine residues. At the same time, AP-catalyzed hydrolysis of threoninephosphate ester moieties turn on the self-assembling process due to the presence of the (TV)5-peptide domain, forming highly organized nanostructures.

There is also another group of enzyme–polyion complexes that form micellar structures and can be produced from block polyelectrolyte and enzyme-reactive molecules with opposite charges. One example are positively charged PEG-b-P-(L-lysine hydrochloride) and negatively charged adenosine 5'-triphosphate (ATP) which initially forms micellar morphologies, but in the presence of phosphatase resulting in selective hydrolysis of ATP disintegrates.[67] Along the same lines, acetylcholinesterase (AChE)-responsive polymeric supra-amphiphiles can be produced from a mixture of poly(ethylene

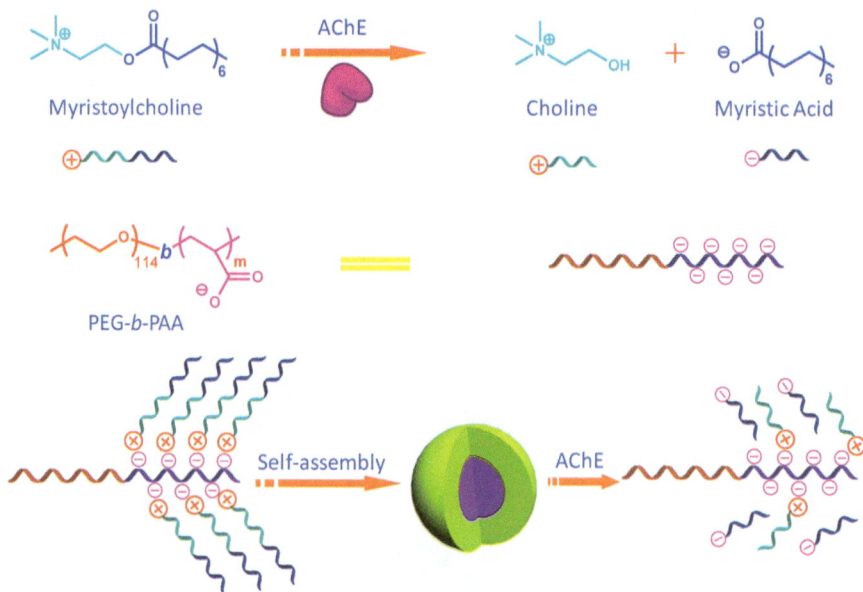

Figure 7.7   Self-assembly and acetylcholinesterase (AChE)-triggered disassembly of polymeric supra-amphiphiles. Reproduced from ref. 68, Copyright 2012, American Chemical Society.

glycol)-b-poly(acrylic acid) (PEG-b-PAA) diblock copolymers and myristoylcholine chloride in aqueous environments.[68] The uniqueness of these assemblies is the formation of noncovalently linked supra-amphiphiles that self-assemble into spherical 40–150 nm aggregates, but the presence of AChE cleaves the ester linkage of myristoylcholine chloride, thus leading to the micellar morphology disintegration. This is schematically depicted in Figure 7.7.

Perhaps one of the most appealing and fascinating features of using enzymes as triggering components is their ability to program the evolution of polymeric assemblies composed of amphiphilic peptide hybrid copolymers.[69] Figure 7.8 illustrates an elegant example of synthesis of amphiphilic peptide hybrid copolymer containing hydrophobic backbones grafted with peptide side chains containing PKA- and MMP-reactive sequences. Amphiphilic polymer-peptides were designed to contain substrates of four different cancer-associated enzymes: protein kinase A (PKA),[70] protein phosphatase-1 (PP1),[71] and matrix-metalloproteinases MMP-2 and MMP-9.[72–74] These enzyme substrates were incorporated into the polar head groups of the copolymers, resulting in the micelle morphology and aggregation which can be modified by the following

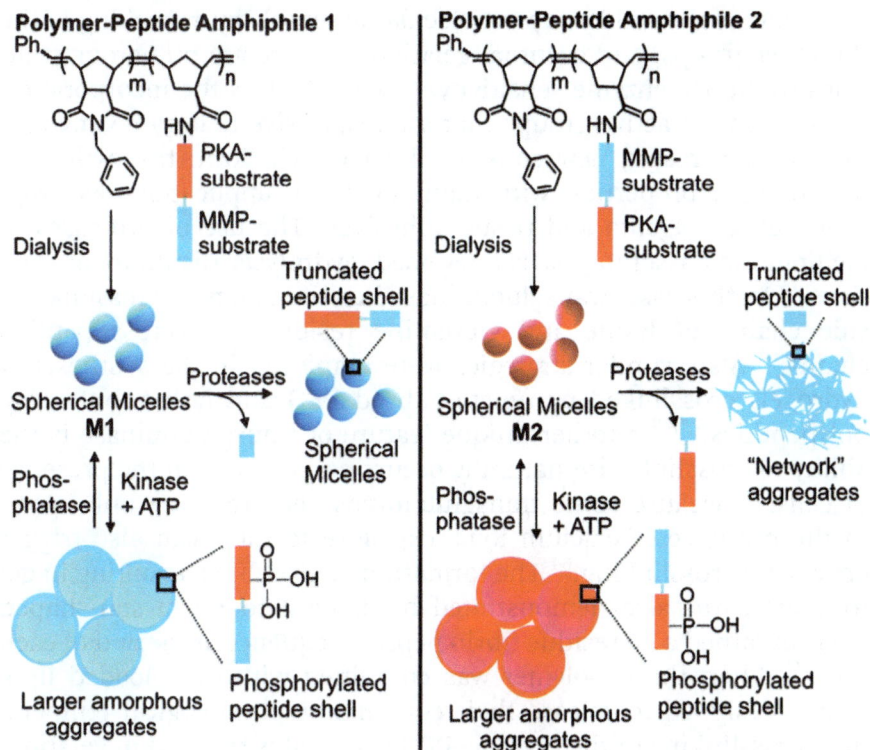

Figure 7.8 Schematic representation of peptide-substrate polymeric amphiphiles assemble into spherical micelles. The peptide moieties within the micelle corona are able to interact with enzymes to generate a variety of morphologies of polymeric amphiphile aggregates which depend upon the design of the peptide and the type of enzyme.
Reproduced from ref. 69. Copyright 2011, American Chemical Society.

mechanisms: (1) phosphorylation by PKA at serine residues, (2) dephosphorylation by PP1 at serine residues, (3) peptide cleavage by MMPs at Gly-Leu peptide bonds. These studies are an excellent example of multi-enzyme responsiveness that can lead to materials with signaling capabilities with predefined molecular patterns known in biological systems, which are often viewed as ubiquitous.

Enzymes may cause swelling/collapse of gels, or the transformation of surface properties.[64] These attributes resulted in the development of the *in situ* non-invasive self-growing hydrogels. One example is the use of enzymatic dephosphorylation to induce a sol–gel transition using fluorenylmethyloxycarbonyl (FMOC)-tyrosine phosphate. These materials, upon exposure to phosphatase, will cause the removal of phosphate groups, thus leading to a reduction in electrostatic repulsions, followed by supramolecular assembly by π-stacking of the fluorenyl groups, and eventual gelation.[75] There are numerous strategies to induce enzyme sensitivity. One of them is the incorporation of functional reactive groups that are responsive to enzyme environments. As a result, new covalent linkages will form that will alter macroscopic properties, with many potential applications as drug/gene delivery agents and tissue adhesives. The use of proteases to facilitate self-assembly is *via* reversed hydrolysis (ligation) of peptides.[76] In this case, transglutaminase has the ability to crosslink the side chains of lysine and glutamine residues or across peptide chains.[64] Also, similar strategies were employed in the synthesis of hydrogels crosslinked by functionalized PEG and lysine-containing polypeptides.[77,78] Another unique feature of transglutaminase is the ability to crosslink with naturally occurring polymers in the presence of cells.[79] Because some transglutaminase enzymes are only active in the presence of calcium ions, exposure to Ca^{2+} can also trigger enzymatic crosslinking.[64] The formation of crosslinks resulting in gel formation was also demonstrated by using a four-arm star-shaped PEG contained a 20-residue fibrin peptide sequence at the end of each arm.[80] When the copolymer was combined with Ca^{2+}-loaded liposomes designed to release their contents at physiological temperature, crosslinking of the peptide–PEG conjugates resulted in gelation.

The use of proteases in hydrogels has resulted in a variety of strategies for synthesizing hydrogels. The idea is to create gels sensitive to proteases in such a way that when the hydrogel is exposed to a protease enzyme, hydrolysis of protein or peptide-based crosslinkers leads to gel degradation and subsequent release of encapsulated contents. Using a Michael-type reaction between vinyl sulfone-functionalized multi-armed telechelic polyethylene glycol (PEG) macromers and mono-cysteine adhesion peptides or bis-cysteine matrix metalloproteinases, hydrogels can be produced in the presence of cells.[81] Perhaps the most intriguing part is that the local degradation of hydrogels caused by response of cell-surface proteases may facilitate the forming of paths for cell growth or formation. Furthermore, when the gels are subjected to the α-chymotrypsin solution, they dissolve. It was suggested that protease-responsive hydrogels may also facilitate the removal of toxins or entrapment of drug molecules.[63] In this case the response was caused by osmotic pressure changes, not degradation. Copolymers composed of acrylamide and PEG macromonomers can also be modified by an enzyme-cleavable tripeptide composed of glycine, phenylalanine and positively charged arginine residues and due to electrostatic repulsions, swelling is imparted.

However, in the presence of proteases, tripeptides can be cleaved, primarily losing arginine. As a consequence, protective repulsive interactions are diminished, thus causing the collapse of the gel.

7.5 Antigen-responsive Polymers

Antigen–antibody interactions are also highly specific and are associated with complex immune responses that facilitate the recognition and neutralization of infection-causing entities. Because the binding between antigens and antibodies relies on a variety of non-covalent interactions, such as hydrogen bonding, van der Waals forces, and electrostatic and hydrophobic interactions, antibodies are highly desirable in immunological assays. Therefore, they serve as sensors and detectors of biological and non-biological substances.[82] High affinity and specificity of the interactions with antigens can be yoked in responsive synthetic polymeric systems. Therefore, antigen–antibody binding can be used to generate responses in hydrogels prepared by physically entrapping antibodies or antigens, chemical conjugation of the antibody or antigen to the gel network, or forming reversible crosslinkers from antigen–antibody within the gel networks.[83] Antigen-sensitive hydrogels can be prepared by coupling immunoglobulin G (like rabbit IgG) with *N*-succinimidylacrylate (NSA). When such modified monomer is polymerized in the presence of goat anti-rabbit IgG as an antibody, acrylamide and MBA, the formation of a hydrogel crosslinked both covalently and by antigen–antibody interactions can be achieved. Due to competitive binding after addition of rabbit IgG as a free antigen with that of the goat anti-rabbit IgG antibodies causes a loss of the antigen–antibody crosslinkers and simultaneous swelling of the hydrogel.[84] This is schematically depicted in Figure 7.9A. Similar hydrogels can exhibit reversible responsiveness[85] when the antigen (rabbit IgG) and antibody (goat anti-rabbit IgG) are independently functionalized with vinyl groups, followed by their polymerizations with acrylamide (antibody) and acrylamide and MBA (antigen). As a result, semi-interpenetrating network hydrogel (semi-IPN) was produced. Interestingly, the addition of free rabbit IgG antigen results in competitive binding with the antibodies, thus causing disruption of antigen–antibody crosslinks and subsequent hydrogel swelling. However, due to the presence of covalent bonds the swelling process was reversible. This is illustrated in Figure 7.9B.

It should be also noted that antigen-responsive hydrogels can be prepared by copolymerization of antibody Fab′ fragments[73] with

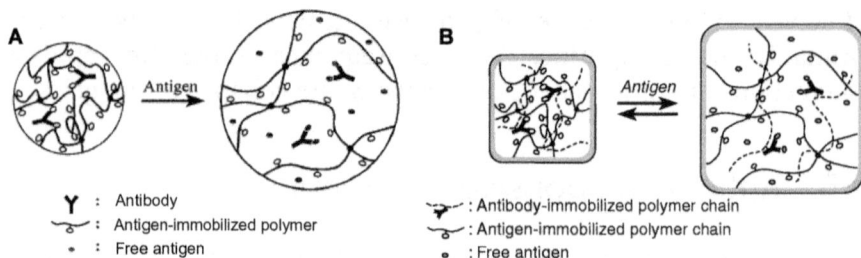

A

Y : Antibody
⌐ : Antigen-immobilized polymer
• : Free antigen

B

: Antibody-immobilized polymer chain
: Antigen-immobilized polymer chain
• : Free antigen

Figure 7.9   Non-reversible (A) and reversible (B) antigen-responsive hydro-
gels. Non-reversible response (A) results from the loss of the
antibody crosslinks upon the addition of antigen. Reversibility
(B) can be achieved due to covalent immobilization of antigen
in the network.
Reprinted with permission from ref. 84. Copyright (1999)
American Society of Chemistry. Reprinted by permission from
Macmillan Publishers Ltd: [*Nature*] (ref. 85), © 1999.

NIPAM and MBA. Again, binding of antigens to the Fab' results in
reversible volume changes which is Fab' content, temperature, and
pH dependent. These antigen-responsive hydrogels can be used in
membranes and it has been shown that crosslinked dextran backbone
grafted with both a fluorescein isothiocyanate (FITC) antigen and
a sheep anti-FITC IgG antibody can serve this purpose.[86] These
materials, upon addition of free sodium fluorescein, will lose antibody–
antigen crosslinks by competitive binding of the free antigen and
again, reversible hydrogel swelling.

7.6   Polymer Surface Modifications

Polymers that exhibit tunable physical and chemical properties offer
a number of unique applications when used in biological systems.
Notable applications range from medical implants,[87–89] microarrays,[90]
biosensors and bio-actuators[91] to tissue engineering,[92–94] or gene and
drug delivery systems[93,95–97] to name just a few. The most commonly
used polymers are poly(tetrafluoroethylene) (PTFE), poly(ethylene
terephthalate) (PET), ultra-high molecular weight poly(ethylene)
(UHMWPE), poly(polypropylene) (PP), polyether ether ketone (PEEK),
poly(methylmethacrylate) (PMMA), polypyrrole (PPy), polythiophene
(PT), poly(lactide-*co*-glycotide) (PLGA), poly(*N*-isopropylacrylamide)
(PNIPAAm), poly-L-lactic acid (PLLA), poly(*N*,*N*'-(dimethylamino)ethyl
methacrylate) (pDMAEMA), poly(ethylene glycol) (PEG), poly(ε-caprolactone) (PCL), poly(acrylic acid) (PAA), poly(allylamine) (PAH), and

polyaniline (PANI). There are others, and each polymer offers specific and unique attributes that are pertinent to their functions.

Many of us do not realize that the lifetime of a pacemaker is not determined by a battery, but biocompatibility and durability of polymeric coatings that cover platinum leads going to a heart muscle. Premature degradation will shorten an electrical circuit and lead to the dysfunction of a pacemaker. The consequences are obvious. For these and other reasons modification of polymeric surfaces to create biocompatible or bioactive materials has been explored utilizing an array of approaches ranging from physisorption to creating covalent bonds. This is, however, not stimuli responsiveness; these are the properties that are necessary to function in a given environment. In this context, numerous studies involving attachments of biologically active species which exhibit antimicrobial[98-100] and anticoagulant[98] properties have been conducted in various biomedical application areas.[101-104] Reactions leading to polymer surface modifications may range from the simple addition of bioactive molecules to a polymer matrix during processing[102,103] or non-covalent physisorption using layer-by-layer deposition,[97,105,106] and self-assembled monolayers (SAMs),[104,107,108] as well as covalent bonding using grafting-to[102,109] and grafting-from[102,103,110] approaches. The latter may offer greater control over surface chemistry and morphologies,[98] as well as the use of chemical spacers to impart mobility and accessibility to the bioactive species of interest.[98] Although this chapter focuses on stimuli-responsive polymer surfaces, it is almost impossible not to mentioned selected surface modifications, as they paved the way to stimuli responsiveness, and Inertness, mechanical strength, or biocompatibility are those attributes that allow many polymeric materials to be successfully integrated into biological systems, but not without adverse effects. And this is why stimuli responsiveness is critical.

7.6.1 Selected Surface Modifications

Because the main characteristics of biocompatible polymeric systems within a biological environment are surface properties, approaches to immobilize bioactive molecules on polymeric surfaces by either non-covalent or covalent bonding have been of significant interest. Non-covalent bonding methods include physisorption,[87,97] electrostatic interactions,[111] and ligand-receptor recognition,[108] while covalent bonding[87] is achieved by chemical reactions of functional groups for form stable surface entities. Figure 7.10 depicts physical

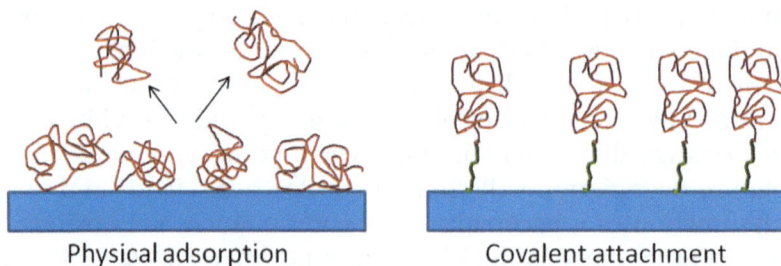

Physical adsorption   Covalent attachment

Figure 7.10   Schematic illustration of physical adsorption (non-covalent) and covalent attachment.

adsorption and covalent attachment methods to immobilize molecules onto surfaces that may or may not be beneficial in long range surface stability.

Physisorption of bioactive molecules onto polymer surfaces is the simplest method of temporarily altering biopolymer surface properties. For example, gentamicin[102] buffer solutions are being used to soak PP mesh implants before surgery in order to prevent infections. Similarly, by combining mammalian cells with polymers such as PLA, it is possible to create new skin for burn patients.[93]

Another short term stability surfaces are achieved using layer-by-layer (lbl) approach which takes advantage of electrostatic interactions by dipping of a charged substrate into dilute aqueous solutions of polyelectrolyte with opposite charges, thus allowing adsorbing of alternating charges on the substrate.[97,112,113] Multilayer films can be produced by sequential adsorptions of anionic and cationic polyelectrolytes, which result in heterogeneous surfaces with limited stability.[113] An example of lbl surface modification is multilayer thromboresistant thin films containing poly(ethylenimine) (PEI), dextran sulfate (DS), and heparin (HP), deposited on poly(vinyl chloride) (PVC) surfaces, which is depicted in Figure 7.11A.[106]

In addition to electrostatic interactions, one of the strongest non-covalent bonding with the disassociation constant K_d of 10^{-15} M,[114] is the ligand-receptor recognition through biotin–streptavidin interactions. Streptavidin exhibits a specific affinity for biotin molecules with four receptor sites leading to specific orientation of immobilized species onto surfaces.[114] Figure 7.11B illustrates the deposition process of biotinylated green fluorescent proteins on SAMs through streptavidin-biotinylated ligands to generate surface nanopatterns.[108] Although non-covalent bonding can effectively bind biomolecules to polymer surfaces, covalent attachments can be used to achieve significantly more stable surfaces.

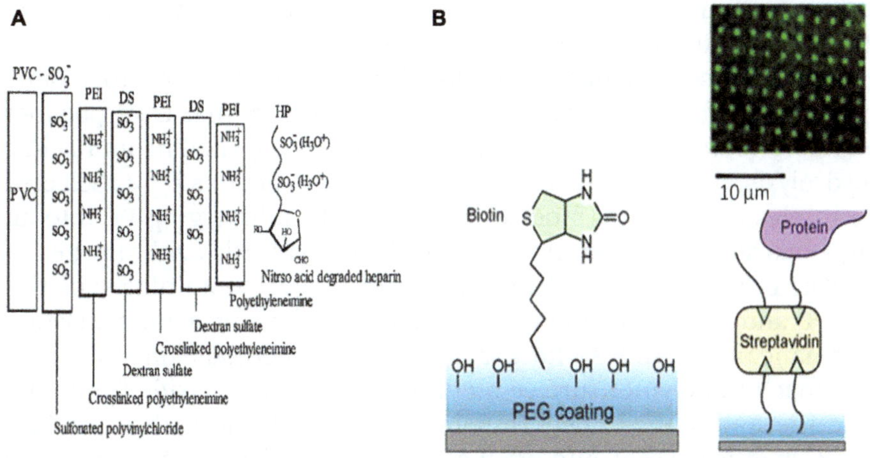

Figure 7.11 (A) Schematic diagram of lbl structure on PVC surface.[96]
(B) Schematic diagram of biotinylated green fluorescent proteins deposited on SAMs through biotin–avidin recognition and fluorescent nanopatterns generated.
Reprinted with permission from ref. 106. Copyright (1998) American Society of Chemistry.

To achieve covalently modified polymer surfaces, the first step is to generate a surface reactive species from which biorelevant molecules can be anchored. One approach is to react maleic anhydride to form acid groups, which can be followed by further reactions of a spacer[100] to provide mobility of the species of interest attached to the other end of the spacer. Well-studied chemical reactions for surface attachment *via* linkages between functional groups include esterification,[90] amidation,[115] and etherification.[116]

Photografting *via* UV radiation has been used to affix bioactive molecules to polymer surfaces using photoinitiators and photo-sensitizers.[117] This approach has been useful for the fabrication of DNA oligonucleotide arrays on PMMA, using UV treatment to generate carboxylic acid functionality.[90,118] In addition to photografting, surface-initiated polymerization, such as controlled living polymerizations including RAFT and ATRP have gained popularity due to their ability to provide controllable chain lengths as well as higher grafting density. Examples of these processes include RAFT surface grafting polymerization of pDMAEMA on cellulose fiber surfaces to produce antimicrobial quaternary ammonium groups.[109] RAFT polymerization was also used to synthesize cell mimicking polymer brushes on the surface of polysulfone.[119] Additionally, surface initiated polymerization of ATRP grafted poly(poly(ethylene glycol) methyl ether monomethacrylate) (PPEGMA) to poly(dimethyl siloxane) (PDMS) was used to impart non-fouling properties.[120]

Solvent-based chemical functionalization can be used to modify polymer surfaces with the use of liquid reagents that treat polymer surfaces to create reactive groups *via* aminolysis, alkaline and acidic hydrolysis, as well as hydrogen peroxide reactions.[90,117] Selected examples of chemical modifications include submerging PE into an aqueous solution of chromium trioxide and sulfuric acid to generate COOH groups,[121] or the introduction of oxygen-containing moieties on PE and PP by reacting with a mixture of chromic acid, potassium permanganate, and sulfuric acid.[90] These methods offer an easy approach to polymer surface modification; however, aside from non-specific and rather uncontrollable functionalization, they produce toxic chemical waste, and often cause etching of the surface resulting in significant morphological alterations.[117]

Other solvent based functionalization methods include click chemistry reactions[122] which are relatively fast, high yielding reactions achieved under mild conditions.[123] Examples of click reactions include Diels–Alder reactions, nucleophilic substitution with exopy and aziridine compounds, thiol-yne reactions and Cu(I) catalyzed Huisgen 1,3-dipolar cycloaddition (azide-alkyne).[124] These reactions have been utilized in the functionalization of polymeric micelles and vesicles,[125] however their use on polymeric substrates has not been explored. For these reactions to be useful in polymer surface reactions, the surfaces must contain moieties necessary for click reactions to occur.

Using high energy radiation is an effective method for modifying the surfaces of biopolymers.[117] They include electron beam radiation[126] and radio or microwave plasma reactions.[127,128] In particular, microwave plasma offers a clean, fast, and solventless route to surface functionalization without affecting bulk properties. Plasma reactions require vacuum conditions in gaseous environments, typically N_2, O_2, CO_2, He, or Ar. Here, excited ionized gas is used to create reactive groups in the presence of a desirable monomer, thus providing controllable chemical conditions for covalently attaching desirable chemical entities which generally occur within 0.5–10 seconds.[127,129] Glow discharge plasma is created in which the gases are ionized to generate highly reactive species within the reactor.[117] These reactive species interact with the polymer surface in the reaction chamber to create in both chemical and physical alterations of the surface. Plasma induced changes at the substrate surface can reach depths of several hundred angstroms,[117] but it may be in the micron range if high energy input per area and extended exposure times are applied.

The first step is the attachment of an anchor molecule through the reaction of maleic anhydride to form acid groups, followed by reactions of a spacer to provide mobility to the active antibiotic molecule attached to the other end of the spacer. Previous studies have employed plasma reactions for the modification of biopolymer surfaces including patterning of PDMS followed by ammonolysis with amoxicillin to create antimicrobial surfaces.[90,130-132] Patterning can be achieved by masking PDMS substrates during microwave plasma reactions.[99] Prevention of undesirable protein adsorption, blood coagulation, or bacterial biofilm formation is critical in many applications and incorporation of these attributes is necessary for polymer substrates in contact with biological environments.

7.6.2   Bioactivity of Modified Polymer Surfaces

Owing to low surface energies, polymers, when in contact with biological systems, are become susceptible to microbial growth and/or undesirable protein adsorption.[98] Microbial growth, thrombosis, and protein absorption are biological events that must be minimized or eliminated in order for a polymer surface to be functional in biological environments. Several specific strategies have been taken to prevent these problems and modified polymer surfaces play an essential role in enhancing antimicrobial, anticoagulant and antifouling properties. The significance of surface modifications is manifested by the fact that each year, the number of deaths resulting from infection continues to increase, with approximately 64% of infections acquired at hospitals being attributed to the attachment of viable bacteria to medical devices and implants.[112] Many of these infections could be avoided by taking the appropriate preventative measures, such as the introduction of antimicrobial agents onto the surfaces of medical devices and implants which would minimize or prevent bacterial attachment and biofilm formation. The summary of typical materials that exhibit anticoagulant, antifouling, and antimicrobial properties is shown in Table 7.1. It should be noted that the majority of these chemicals are utilized in a liquid phase. The challenge is to anchor these materials on polymeric surfaces in such a fashion that they will retain their anticoagulant, antifouling, and antimicrobial properties.

7.6.2.1   Antimicrobial Surfaces

Several methods have been employed in an attempt to introduce antimicrobial agents into biomaterials in order to prevent biofilm formation. A simple approach is adding antimicrobial agents into the

Table 7.1 Examples of materials for preventing coagulant, fouling and microbial processes (adopted from ref. 88).

Anticoagulant		Antifouling	
Name	Chemical structure	Name	Chemical structure
Warfarin		Poly(oligo(ethylene glycol methyl ether) methacrylate)[149]	
Heparin		Zwitterionic polymer	
Hirudin		• Phosphoryl-choline (PC) • Sulfobetaine (SB)	

- Carboxybetaine
  (CB)

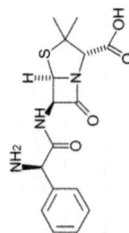

Metals: silver ion,
tin, mercury

Ag$^+$, Sn, Hg

Quaternary
ammonium

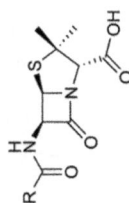

Beta-lactams
- Penicillin

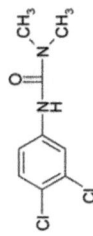

- Ampicillin

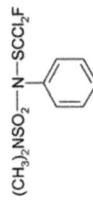

Argatroban

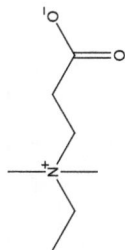

Chlorothalonil

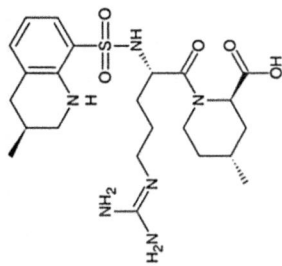

Diuron

Dichlofluanid

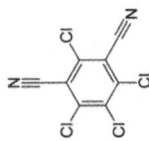

Table 7.1 *(Continued)*

Anticoagulant		Antifouling	
Name	Chemical structure	Name	Chemical structure
Phosphoryl-choline-PDMS[154]		• Amoxicillin	
PEG-fluoropolymer[145]		Aminoglycosides • Gentamicin	
		• Streptomycin	
		Chitosan	

polymer during processing.[133] Their release is controlled by diffusion from a polymer matrix.[102] Many antimicrobial agents have been used for this purpose. They include quaternary ammonium (or phosphonium) salts (QAS),[111,133] chitosan,[134] antimicrobial peptides (AMPs),[133,135] silver ions (Ag$^+$),[136] bacteriophages[137] and antibiotics.[98] As expected, they exhibit different and for the most part unknown mechanisms of inhibition of bacterial growth. While metals cause the displacement of essential Ca^{2+} and Zn$^+$ ions in bacteria,[138] quaternary ammonium salts[111] and chitosan[134] bind to negatively charged bacterial surfaces causing the leakage of intracellular components. The attachment of poly(quaternary ammonium) to PP surfaces using photochemical synthesis and ATRP, were successful against *S. aureus* and *E. coli* bacteria, and molecular weight and the density of the quaternary ammonium (QA) can be controlled.[139] An example of antimicrobial QAS on surfaces is illustrated in Figure 7.12, and involves a lbl assembly of alternating anionic poly(acrylic acid) (PAA) and cationic cetyltrimethylammonium bromide (CTAB) layers that demonstrate antimicrobial activity as the CTAB moiety diffuses to the surface of the assembly.[111] In contrast to the above approaches, Ag$^+$ ions bind to electron donor groups containing sulfur, oxygen, or nitrogen, which are present in biological molecules as thio, amino, imidazole, carboxylate and phosphate groups. Presumably, by

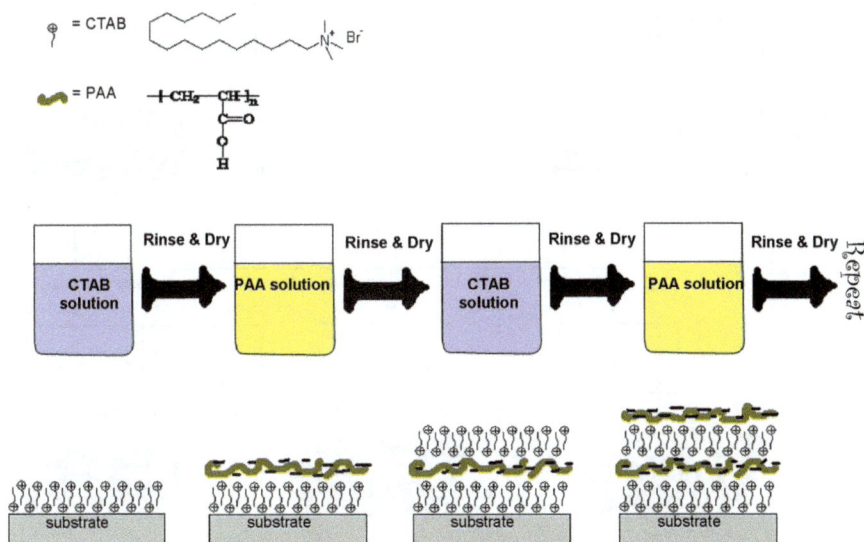

Figure 7.12 Schematic diagram of lbl of PET film consisting of alternating layers of PAA and antimicrobial CTAB. Antimicrobial activity is shown by the zone of inhibition of bacterial growth.
Reprinted with permission from ref. 111. Copyright (2009) American Society of Chemistry.

displacing other ions, such as Ca^{2+} and Zn^+, Ag^+ effectively interrupts a number of cellular transport and oxidation processes.[138] For that reason Ag^+ ions have been incorporated onto PET films by lbl surface modification with PEI and PAA which exhibits biocidal activity against *S. aureus* and *E. coli*,[136] but there a number of poorly understood potential side effects that may prohibit the use of silver.

On the contrary, antibiotics have different modes of action for antimicrobial activity, with the common classes being aminoglycoside and beta-lactam-based antibiotics. Aminoglycosides, such as gentamicin, kanamycin, and streptomycin, prevent bacterial protein synthesis by entering the bacteria and binding to ribosomes within the cell, whereas beta-lactams, such as penicillin (PEN), amoxicillin, and ampicillin (AMP), inhibit bacterial cell wall formation. PEN[100,140] and AMP[141] were successfully attached to the ePTFE surfaces *via* prior microwave plasma reactions leading to grafted carboxylic acid groups. These studies showed high effectiveness against both Gram $(+)$ and $(-)$ bacteria. Microwave plasma reactions and subsequent attachment of antibiotics are illustrated in Figure 7.13. The first step is the attachment of an anchor molecule through the reaction of maleic

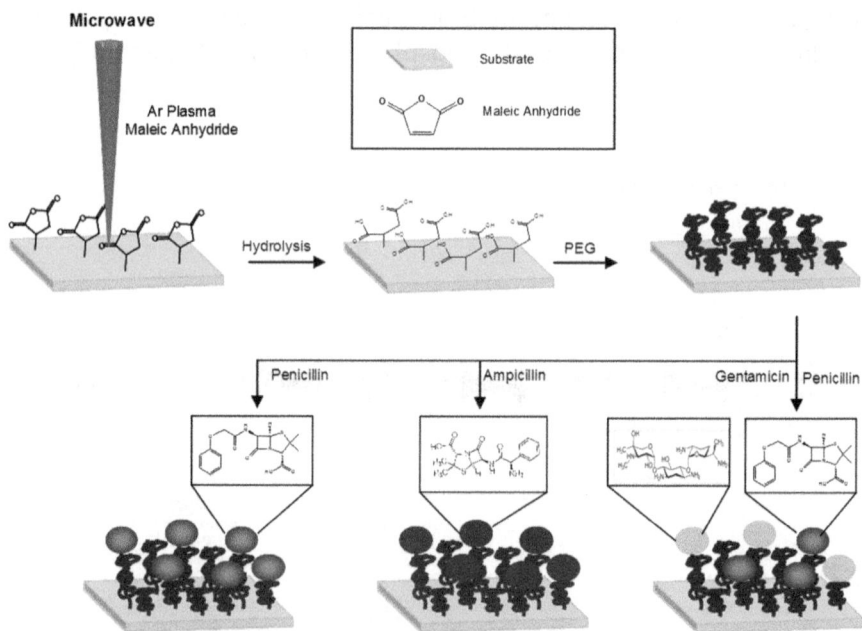

Figure 7.13 Schematic diagram of the covalent attachment of antibiotics to polymer surfaces *via* microwave plasma reactions. From ref. 141 reproduced by permission of John Wiley and Sons. Copyright © 2009 WILEY-VCH Verlag GmbH & Co. KGaA, Weinheim.

anhydride to form acid groups. This solventless and sterilized process takes less than 10 seconds. The second step involves reactions of a spacer to provide mobility to the active antibiotic molecule attached to the other end of the spacer. The covalent attachment of poly(vinyl-*N*-hexylpyridinium) onto HDPE and PET surfaces also indicated bactericidal activity against *S. aureus* and *E. coli*.[142]

7.6.2.2 Anticoagulant Surfaces

Due to low polymer surface energies, most polymeric materials exhibit highly thrombogenic surfaces, and consequently may adsorb fibrin, thrombus, or other proteins resulting in clot formation.[103] Thrombosis, or blood clotting within minor wounds, is necessary for hemostasis to allow healing. After the clot has served its purpose, it is dissolved through a process called fibrinolysis.[143,144] Naturally occurring hirudin and synthetically attained bivalirudin peptides prevent thrombosis.[103] Covalent immobilization of these peptides onto polymer surfaces such as PET,[145] and PLGA[146] have been widely studied in order to inhibit thrombus formation. Another anticoagulant species, heparin (HEP), is a linear polysaccharide consisting of uronic acid-(1,4)-D-glucosamine repeating disaccharide subunits,[147] which acts by binding to a thrombin inhibitor known as antithrombin III (AT-III).[147,148] In this case, HEP molecule dissociates from the complex and can be reused. Several studies have sought to develop hemocompatible devices such as dialysis membranes, catheters, coronary stents, and vascular grafts by immobilizing heparin onto polymer surfaces including PET,[149] PTFE,[149,150] and PU[151] through lbl surface modification in order to prevent blood coagulation.

Acetyl salicylic acid (aspirin) and dipyridamole are also important anti-thrombosis agents. Their anticoagulant activities involve inactivation of clotting enzymes which facilitate thrombus formation. Aspirin has been incorporated into polymer matrix such as PLCA[152] and PVA[153] to enhance blood compatibility. Similarly, dipyridamole has been covalently attached onto PUR surfaces *via* photomodification.[154] Minimization of biofouling and blood coagulation at the surface through various surface modifications greatly enhances biocompatibility of polymeric materials. Selected chemicals offering anticoagulant properties are listed in Table 7.1.

7.6.2.3 Antifouling Surfaces

The adsorption of proteins to a polymer substrate in contact with biological environments may be detrimental to biocompatibility

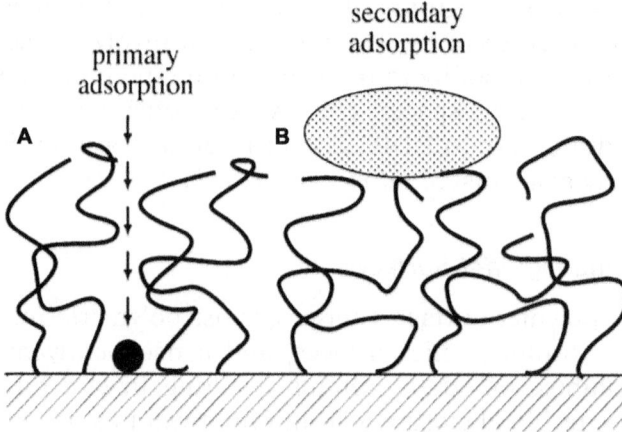

Figure 7.14   Two modes of adsorption of a particle on a polymeric layer grafted to a substrate: (A) primary adsorption in which proteins penetrate the polymer brush and (B) secondary adsorption at the brush–solvent interface.
Reprinted from ref. 157. Copyright (1989) with permission from Elsevier.

because the adsorbed proteins may initiate platelet adhesion and activation.[156,157] The complexity of protein adsorption is due to electrostatic interactions between the surface and protein, protein concentration levels, and surface energy.[117] There are typically two recognized modes of interaction involved in protein adsorption on the surface, which are depicted in Figure 7.14.[157] The primary adsorption (A) occurs at the polymer surface–brush interface, and occurs when small proteins are able to penetrate through the polymer brush and adsorb to the substrate surface while secondary adsorption (B) is due to larger proteins being attracted by van der Waals interactions at the solvent–brush interface.[90,157]

Because adhesion strength of biomolecules to polymer surfaces is a function of surface energy and polymer matrix modulus, attachment of hydrophilic PEG with a low surface energy will result in resistance to biofouling, providing a high activation barrier for protein adsorption as well as steric repulsions which are also important for protein resistance.[98,155] PEG-functionalized polymer brushes synthesized by surface-initiated ATRP have been intensively examined which have resulted in surfaces with significant resistance to protein adsorption as well as cell adhesion.[120,158–161] On the other hand, elastomeric polymers such as PDMS exhibit favorable low modulus, but high surface energy which results in an increase in protein adhesion strength. In order to circumvent these problems, several polymer architectures and surface modification techniques have been explored involving PDMS.[162–164] For example, zwitterionic phosphorylcholine polymer that resembles the structure of natural membrane lipids[155] have been grafted onto PDMS *via* photo-induced polymerization, resulting an greatly improved surface hydrophilicity and antifouling properties.[164] Several zwitterionic containing polymers including phosphorylcholine, polybetaine, carboxybetaine, and sulfobetaine, are used as antifouling polymers due to their hydrophilicity as well as electroneutrality,[165] and are shown in Table 7.1.[98]

Significant advances have been made in surface modifications of commodity polymers in order to achieve antimicrobial, anticoagulant, and antifouling properties. In spite of the efforts there are major unresolved scientific and technological problems associated with the safe use of polymers in biological environments. Understanding surface modifications of polymers and the development of reproducible technologies is timely due to devastating effects of microbial infections and thrombosis.

7.7 Stimuli-responsive Surface Modifications

The majority of physicochemical approaches ranging from layer-by-layer deposition,[166] chemical etching,[167] or radiation grafting,[127] appear to have short life span effectiveness. Although in the layer-by-layer approach multiple layers are deposited onto surfaces by dipping, and multilayers are held together by opposite electrostatic charges, the resulting films are mechanically unstable and often may exhibit chemically heterogeneous surfaces.[168] Chemical etching alters hydrophobicity *via* surface oxidation and morphology changes, whereas radiation grafting uses radiation sources such as infrared, visible light, ultraviolet, and γ radiation as well as high energy electrons to generate reactive groups on surfaces in the presence of monomer to graft species to/from the surface.[126,167] In contrast, microwave plasma surface reactions offer a fast, solventless, and sterile method for covalently attaching carboxylic acid groups to almost any polymer substrate.

If aliphatic polymer substrates, such as PE and PP, exhibit –COOH surface groups, their presence facilitates an opportunity for further reactions. Although several studies demonstrated that indeed, various surface entities can be covalently attached, it was apparent that lower reaction yields were achieved as more layers were reacted. Because click chemistry[122,123] represents one of the highly efficient (>95%)

synthetic routes of achieving high reaction yields in relatively short time under mild conditions, we will take advantage of combining both approaches to create antimicrobial PE and PP surfaces. Reactions that meet 'click' chemistry criteria, but are not limited to, include Huisgen 1,3-dipolar cycloaddition,[123] Diels–Alder reactions,[122] nucleophilic substitution with epoxy and aziridine compounds,[122] Sharpless dihydroxylation,[122] and thiol-yne reactions.[169] Notably, the Huisgen 1,3-dipolar cycloaddition involves a copper(I)-catalyzed[124] 1,2,3-triazole formation from azides and alkyne functionalities with complete specificity of reactants.[170] In terms of surface reactions, cycloaddition click reactions yielding high efficiencies with Cu,[123] Au,[171,173] Si,[169,173] glass,[174] and even carbon nanotube[175] surfaces have shown promising results, such as the biofunctionalization of Si surfaces with 'clicked' biotin and glucose were successful.[123,125,173] These unique attributes of click chemistry also broke new grounds for selective reactions with complex dendrimers,[176,177] hydrogels,[178] vesicles,[179] and nanoparticles.[172,180]

Although click chemistry has been used in a variety of model studies, their practicality has been limited by one major roadblock: the ability to react to typically inert polymeric surfaces. In view of the previous studies using microwave plasma reactions leading to –COOH functionalization of aliphatic polymer surfaces, simple and clean surface reactions were developed and formulated a platform for almost any polymer surface modifications without adverse effects on polymer bulk properties.[181] Figure 7.15 schematically depicts a sequence of surface reactions using two simple steps: microwave plasma reactions that lead to –COOH formation, followed by covalent attachment of alkyne moieties resulting in polymeric surfaces to

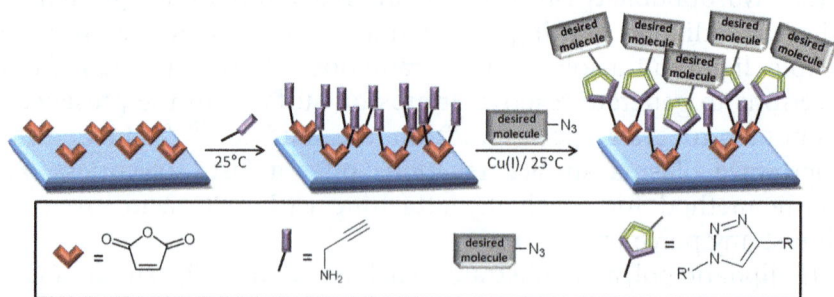

Figure 7.15 Schematic diagram of 'clickable' polymeric surfaces exhibiting alkyne functionalities for 'clicking' any azide-containing molecule.
Reproduced from ref. 181 with permission from The Royal Society of Chemistry.

which any azide containing molecule may be 'clicked'. Selected examples of 'desired molecules' include peptides used in stem cell adhesion,[182] fluorophores for labeling hydrogels,[183,184] polyhedral oligomeric silsequioxane (POSS),[169] biotin,[174,185] bacteriophages,[186] or DNA,[187] to name just a few. These studies focused on attachment of antibiotics, such as ampicillin (AMP) which exhibits Gram (+) and (−) antimicrobial functions.

Figure 7.16 illustrates the two-step process in which MA was reacted to PE and PP (A)[128,188] to obtain –COOH groups, followed by their conversion to acid chloride. The second step relies on reactions of propargylamine (B) to obtain alkyne functionalized polymeric surfaces to which any azide containing molecule may be reacted.

Figure 7.17 shows the actual surface modifications resulting from surface reactions conducted on PE surfaces. The same reactions can be utilized to modify silicone elastomers as well as fluoropolymers. Analysis of the data indicates that clicked AMP on both PE and PP facilitates a major enhancement of the antimicrobial activity manifested by a drop of CFUs from 4000–5000 for PE-MA-PPA and PP-MA-PPA specimens to 10–100 for PE-MA-PPA-AMP and PP-MA-PPA-AMP specimens. These AMP clicked surfaces exhibit highly efficient antimicrobial activity against *S. aureus* with a 97–99.8% decrease of bacterial growth.

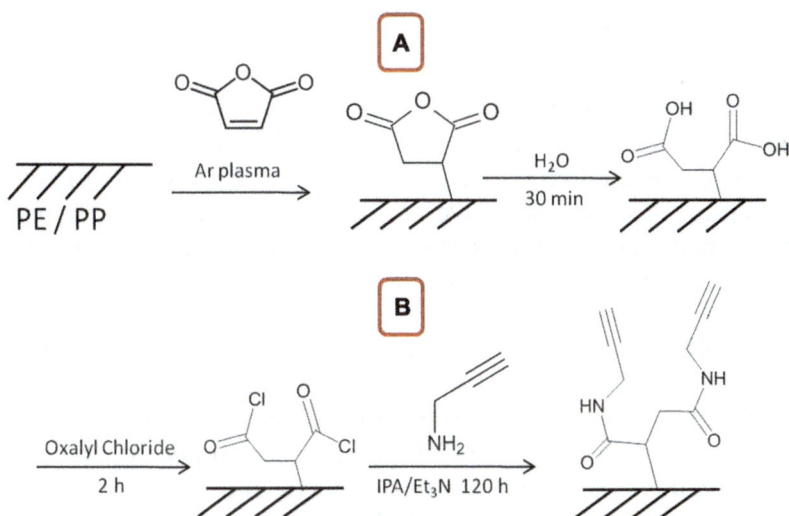

Figure 7.16 Surface reactions leading to the formation of alkyne surface groups. Microwave plasma reactions in the presence of MA (A) and alkyne functionalization using propargylamine (B). Reproduced from ref. 181 with permission from The Royal Society of Chemistry.

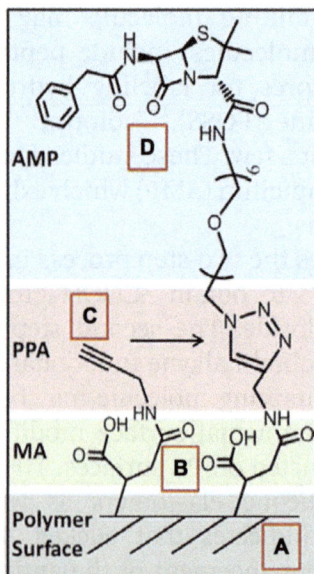

Figure 7.17   A illustrates structural features that develop from reactions depicted in steps A and B of Figure 7.16. To examine the effectiveness of AMP covalently attached to polymer surfaces against *S. aureus* bacteria, PE-MA-PPA-AMP and PP-MA-PPA-AMP surfaces were tested for antimicrobial activity against *Staphylococcus aureus*.
Reproduced from ref. 181 with permission from The Royal Society of Chemistry.

Although numerous studies have been conducted on polymer surface modifications, recent advances[189] in surface and interfacial reactions have resulted in the development of a new class of materials that exhibit stimuli-responsive characteristics. For example, by creating a switchable surfaces through the use of polyelectrolytes,[190] followed by functionalizing the end groups of these polyelectrolytes, a given surface may adapt itself to environmental pH changes. Taking this concept a step further allows creating not only functional surfaces, but engineering surface responsiveness to an array of external stimuli. The majority of surface reactions have focused on silicon and gold surfaces,[191] but reactions on polymer substrates are challenging due to the inert nature of polymer surfaces and their morphological heterogeneities. This is particularly important if responses to an external stimulus, such as temperature,[192–195] ionic strength,[196] UV radiation,[195] or pH[190,192,193] are required. In the recent studies stimuli-responsive polymeric surfaces 'decorated' with cationic and anionic polyelectrolytes terminated by –COOH or –NH$_2$ groups were

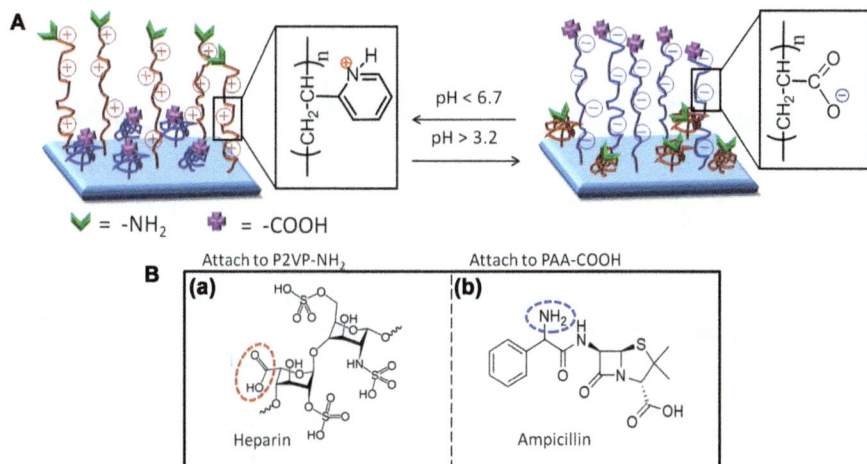

Figure 7.18 (A) Schematic diagram of dual stimuli-responsive polyelectro-lyte surfaces terminated with –NH₂ and –COOH moieties. (B) Structures of (a) HEP and (b) AMP, circled moieties are available for attachment.
Reproduced from ref. 187 with permission from The Royal Society of Chemistry.

developed.[181] The advantages of oppositely charged polyelectrolytes not only facilitates the formation of hydrophilic surfaces on hydrophobic polymer substrates, but also introduces surface dynamic properties as a function of pH. Specifically when cationic segments extend, anionic tethers collapse, and *vice versa*. Figure 7.18A illustrates an overall scheme of the coupling reactions which facilitate the attachment of poly(2-vinyl pyridine) (P2VP) and poly(acrylic acid) (PAA). Because P2VP and PAA chain ends are terminated with –NH₂ and –COOH functionalities, respectively, the scope of these studies is further expanded to the attachment of heparin (HEP) (on P2VP) and ampicillin (AMP) (on PAA) (Figure 7.18B) to achieve antimicrobial and anticoagulant pH responsive polymeric surfaces with antimicrobial and anticoagulant moieties was explored.

While Figure 7.18 depicts the general theme to achieve pH responsive anticoagulant and antimicrobial surfaces, Figure 7.19 illustrates reaction sequences leading to the formation of covalently attached multilayers (CAM) tethered to Si, PE, and PTFE surfaces.[197]

Similarly to the previous studies (step 1), MA can be reacted to a substrate,[128,188] followed by hydrolysis to obtain –COOH groups, which is then converted to acid chloride. In the step 2, EDN is reacted with –COCl groups resulting in –NH₂ terminated surfaces. The next step (3) involves the attachment of dicarboxy-terminated P*t*BA to

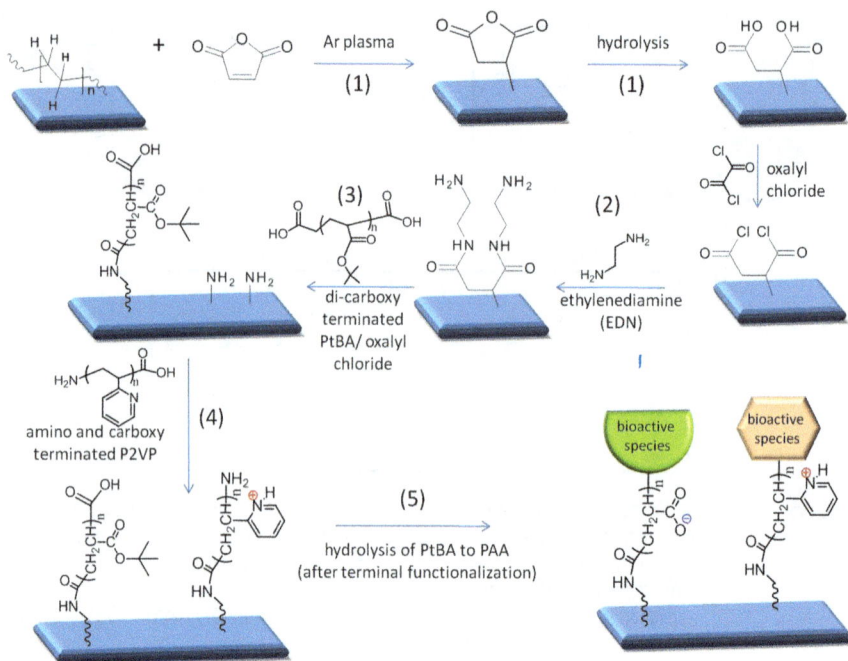

Figure 7.19 Reaction sequences on polymeric substrates leading to the formation of stimuli-responsive polyelectrolyte surfaces terminated with –COOH and –NH$_2$ functionalities and ultimately exhibiting bioactive molecules.
Reproduced from ref. 187 with permission from The Royal Society of Chemistry.

the –NH$_2$ terminated surfaces *via* amide linkages, followed by the attachment of P2VP (4), carried out by reacting amide linkages between –COOH groups on P2VP and the –NH$_2$ groups. Upon terminal functionalization with desired bioactive species, P*t*BA can be hydrolyzed under mild conditions to PAA polyelectrolyte. This choice of sequences with polyelectrolyte chain ends containing –NH$_2$ and –COOH terminal groups facilitated to achieve two objectives: the ability to attach bioactive species as well as facilitate pH sensitive expansion/collapse responses. Figure 7.20A illustrates a sequence of CAM reactions leading to –COOH terminal entities. The first step involved MA reactions to a substrate, followed by EDN and P*t*BA (PAA) functionalization. Reactions of P2VP to polymer surfaces facilitates pH responsiveness in the pH range of 1 to 6.7 as well as the opportunity for terminating each tether with –NH$_2$ functionalities for further reactions. Figure 7.20B illustrates a sequence of CAM reactions leading to –NH$_2$ terminal groups.

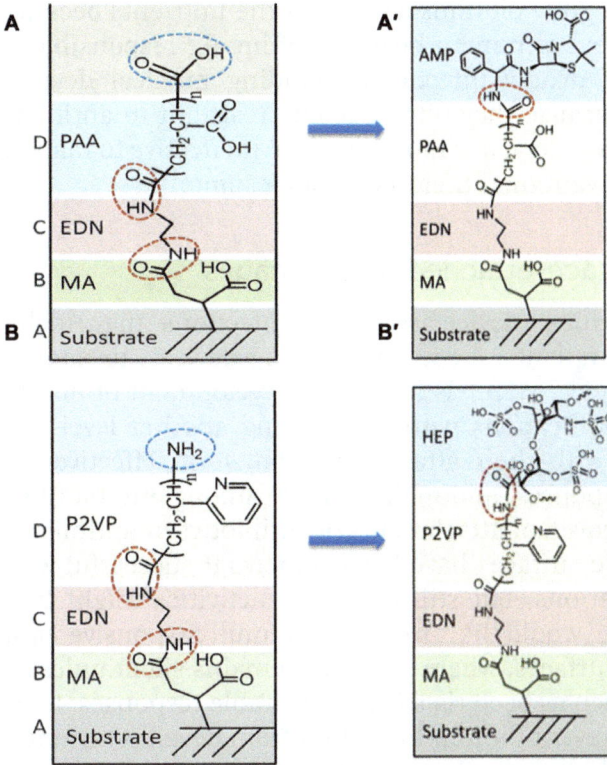

Figure 7.20 Chemical structures resulting from reactions with AMP to: PTFE-MA-EDN-PAA (A/A′) and HEP to PTFE-MA-EDN-P2VP (B/B′) (circles represent reaction sites).
Reproduced from ref. 197 with permission from The Royal Society of Chemistry.

Covalent attachment of ampicillin (AMP) as well as heparin (HEP) to modified polyethylene (PE-MA-EDN-PAA or PE-MA-EDN-P2VP) or polytetrafluoroethylene (PTFE-MA-EDN-PAA/PTFE-MA-EDN-P2VP) surfaces are shown in Figure 7.20, which showed excellent antimicrobial and anticoagulant properties.

The majority of interactions between biologically active species and synthetic materials are inherently non-favorable. However, formation of biofilms represent an important and unwelcome exception resulting from the attachment of bacteria to a synthetic surface, leading to the formation of a complex biofilm community that is often encased on a surface of polymeric materials in contact with blood. Microbial biofilms are very resilient communities that resist removal by chemical or physical means because their residents, bacterial cells, are capable of adhering to a variety of biotic and abiotic surfaces, and continue to grow biofilms as long as the nutrients become available. In the context of human health, biofilm are responsible for the vast majority of deadly infections including medical device-associated diseases. Because they often become resistant to antibiotics or host defenses, the use of antibiotics may be ineffective to many pathogens, making conventional therapies troublesome.[198]

7.7.1 Surface-attached Macrophages

Surface medical device-associated infections may include simple catheters, implants, stents, monitoring devices, to name just a few. The first logical step has led to the development of novel drugs, but surface modifications with drugs bring another level of challenges associated with their attachment, long-term effectiveness, altering immunoresponses, and maintenance. Among anti-biofilm formation strategies, covalent attachments of antibiotics or antimicrobial agents to polymeric surfaces have been somewhat successful with relatively longer durations, but still long-term activities might be limited.[188] Ideally, one would like to create stimuli-responsive properties on polymeric surfaces, where a surface remains silent unless an external stimulus such as bacteria triggers desirable responses. In essence, the goal is to prevent biofilm formation from developing at its inception. Almost entirely unexplored area of inhibition of bacterial film formation is the use of bacteriophages. These living bacterial viruses are capable of selective binding to specific receptors of the target bacteria. Upon binding, they inject their own DNA, which upon reproduction inside the bacteria, kills the host, and release their progeny.[199] The use of these species has been only sporadically explored.[137,200,201] Although control of their progeny is critical, it seems that, under controllable conditions, bacteriophages will offer an alternative and powerful approach for inhibiting bacterial infections by attaching specifically to their target host bacteria, inject their genetic material, reproduce inside the host, kill the host, and release their progeny. Although the first observations of lytic phages to cure infectious diseases go back to the end of the 19th[202] and early 20th centuries,[199,203,204] it was not until 1960s, where prophylaxis and treatment of bacterial infections appeared favorable choice with a remarkable recovery efficacy above 92%.[205] Later on, the use of bacteriophages was exploited by physical mixing with or physisorbed on polymers (Nylon™),[137,206] glass,[200] and gold.[207] The unique attributes of bacteriophages are their host-dependent reproduction – bacteriophages will remain silent until they find a specific bacterium, thus minimizing safety concerns associated with excessive concentration levels. Furthermore, due to evolution with their hosts and host specificity, there are numerous bacteriophages for each bacterium. Taking advantage of the ability of bacteriophages to recognize a host bacterium and their ability to kill specific hosts, T1 and Φ11 phages were covalently attached onto surfaces of polytetrafluoroethylene (PTFE) and ultra-high molecular weight polyethylene (PE).[208] Figure 7.21A–D depicts a sequence of steps leading to the formation and subsequent destruction of the bacteria attempting to form biofilms on the phage-modified polymeric substrates. The first step (Figure 7.21A) involves the formation of reactive acid groups on polymeric substrates[128] accomplished by simple and clean microwave plasma reactions in the presence of maleic anhydride, followed by covalent attachment of T1 phages *via* acid–amine reactions leading to amide linkages.

When bacteria attempt to adhere to the surface of the phage-modified polymer substrate, the phage attaches specifically to an external structure of the bacterium (*e.g.* lipopolysaccharide or protein) and injects its genetic material into its target (Figure 7.21C). Upon completion of this process, the phage DNA is replicated by the bacterial host machinery, several capsid proteins are produced and assembled, phage DNA is packaged and the bacteriophage progeny are released by lysis of the bacterial host. Depending on the particular bacteriophage, the progeny can be up to 200 for each individual infection. Perhaps the most significant advantage of using bacteriophages comes from the fact that each member of the progeny is capable of infecting more bacteria and releasing a progeny of its own. This amplification effect continues until all bacteria cells are killed. Of course, one can envision potential challenges associated with control of the phage population, especially when used in therapy, which needs to be carefully adjusted.

In spite of alarming health-related problems associated with antimicrobial, anticoagulant, and antifouling properties, it is quite apparent that, for the most part, relatively minor efforts are given to eliminating and controlling microbial film formation or coagulation/fouling properties at their inception. Infections are not associated with specific diseases, but they are the diseases themselves that may strike under most unpredictable conditions.[209] Yet, compared to the disease-driven programs, marginal efforts focus on understanding of molecular processes leading of the inhibition of infections before it is too late. This is further amplified by antibiotic resistant infections reflected in multi-resistant strains of bacteria. Although the use of bacteriophage therapy has proven promising in several studies, their

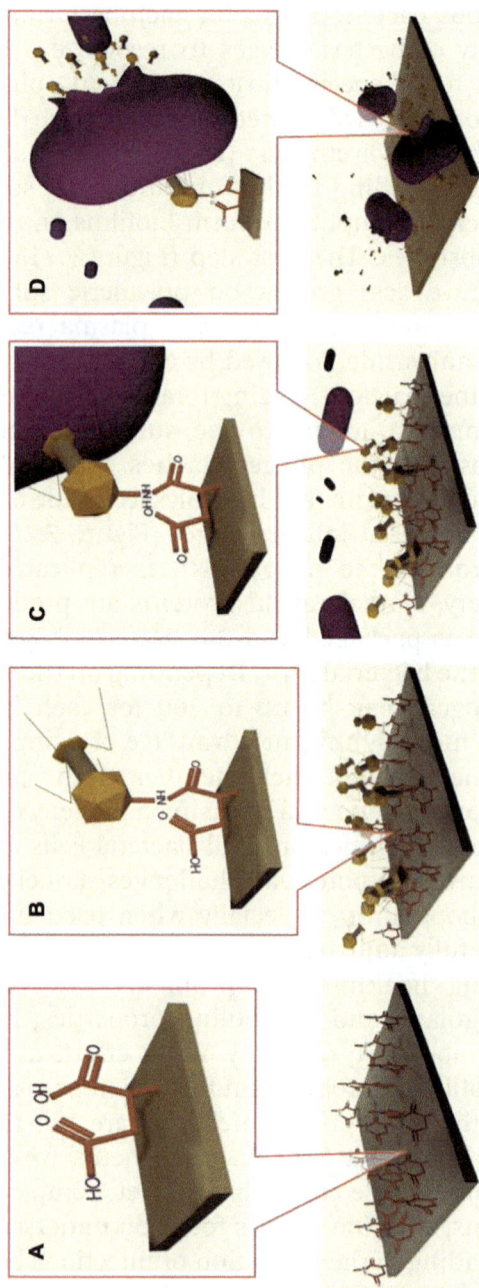

Figure 7.21 (A) Covalent attachment of acid groups to polymeric surfaces. (B) Reactions of NH$_2$ groups of T1 with polymer surface acid groups. (C) Attachment of phages to bacteria. (D) Replication of DNA and destruction of bacteria. Reprinted with permission from ref. 208. Copyright (2013) American Society of Chemistry.

role is not fully understood and innovative approaches are necessary to tackle the uncharted areas in combating antibiotic resistant infections. It seems that, combined with signaling of stimuli-responsive components, may formulate a new paradigm in fighting infections.

References

1. H. G. Schild, Poly (N-isopropylacrylamide): experiment, theory and application, *Prog. Polym. Sci.*, 1992, **17**(2), 163–249.
2. M. Okubo, H. Ahmad and T. Suzuki, Synthesis of temperature-sensitive micron-sized monodispersed composite polymer particles and its application as a carrier for biomolecules, *Colloid Polym. Sci.*, 1998, **276**(6), 470–475.
3. A. S. Lee, V. Bütün, M. Vamvakaki, S. P. Armes, J. A. Pople and A. P. Gast, Structure of pH-dependent block copolymer micelles: charge and ionic strength dependence, *Macromolecules*, 2002, **35**(22), 8540–8551.
4. C. de las Heras Alarcón, S. Pennadam and C. Alexander, Stimuli responsive polymers for biomedical applications, *Chem. Soc. Rev.*, 2005, **34**(3), 276–285.
5. K. S. Murthy, Q. Ma, E. E. Remsen, T. Kowalewski and K. L. Wooley, Thermal shaping of shell-crosslinked (SCK) nanoparticles, facilitated by nanoconfinement of fluid-like cores, *J. Mater. Chem.*, 2003, **13**(11), 2785–2795.
6. J. V. Weaver, Y. Tang, S. Liu, P. D. Iddon, R. Grigg and N. C. Billingham *et al.*, Preparation of shell cross-linked micelles by polyelectrolyte complexation, *Angew. Chem.*, 2004, **116**(11), 1413–1416.
7. S. Liu and S. P. Armes, The facile one-pot synthesis of shell cross-linked micelles in aqueous solution at high solids, *J. Am. Chem. Soc.*, 2001, **123**(40), 9910–9911.
8. M. Murata, W. Kaku, T. Anada, Y. Sato, M. Maeda and Y. Katayama, Temperature-dependent regulation of antisense activity using a DNA/poly (N-isopropylacrylamide) conjugate, *Chem. Lett.*, 2003, **32**(11), 986–987.
9. Y. Ito, M. Casolaro, K. Kono and Y. Imanishi, An insulin-releasing system that is responsive to glucose, *J. Controlled Release*, 1989, **10**, 195–203.
10. C. M. Hassan, F. J. Doyle and N. A. Peppas, Dynamic behavior of glucose-responsive poly(methacrylic acid-*g*-ethylene glycol) hydrogels, *Macromolecules*, 1997, **30**, 6166–6173.

11. H. Y. Huang, J. Shaw, C. Yip and X. Y. Wu, Microdomain pH gradient and kinetics inside composite polymeric membranes of pH and glucose sensitivity, *Pharm Res.*, 2008, **25**, 1150–1157.

12. S. I. Kang and Y. H. Bae, A sulfonamide based glucose-responsive hydrogel with covalently immobilized glucose oxidase and catalase, *J. Controlled Release*, 2003, **86**, 115–121.

13. K. Podual, F. J. Doyle III and N. A. Peppas, Dynamic behavior of glucose oxidase-containing microparticles of poly(ethylene glycol)-grafted cationic hydrogels in an environment of changing pH, *Biomaterials*, 2000, **21**, 1439–1450.

14. T. Traitel, Y. Cohen and J. Kost, Characterization of glucose-sensitive insulin release systems in simulated in vivo conditions, *Biomaterials*, 2000, **21**, 1679–1687.

15. A. Guiseppi-Elie, S. I. Brahim and D. Narinesingh, A chemically synthesized artificial pancreas: Release of insulin from glucose-responsive hydrogels, *Adv. Mater.*, 2002, **14**, 743–746.

16. L. M. Schwartz and N. A. Peppas, Novel poly(ethylene glycol)-grafted, cationic hydrogels: Preparation, characterization, and diffusive properties, *Polymer*, 1998, **39**, 6057–6066.

17. V. Ravaine, C. Ancia and B. Catargi, Chemically controlled closed-loop insulin delivery, *J. Controlled Release*, 2008, **132**, 2–11.

18. N. Kashyap, B. Viswanad, G. Sharma, V. Bhardwaj, P. Ramarao and M. N. Ravi Kumar, Design and evaluation of biodegradable, biosensitive *in situ* gelling system for pulsatile delivery of insulin, *Biomaterials*, 2007, **28**, 2051–2060.

19. M. Brownlee and A. Cerami, A glucose-controlled insulin-delivery system: Semisynthetic insulin bound to lectin, *Science*, 1979, **206**, 1190–1191.

20. E. S. Gil and S. M. Hudson, Stimuli-responsive polymers and their bioconjugates, *Prog. Polym. Sci.*, 2004, **29**, 1173–1222.

21. F. Liu, S. C. Song, D. Mix, M. Baudy and S. W. Kim, Glucose-induced release of glycosylpoly(ethylene glycol) insulin bound to a soluble conjugate of concanavalin A, *Bioconjugate Chem.*, 1997, **8**, 664–672.

22. S. Y. Cheng, J. Gross and A. Sambanis, Hybrid pancreatic tissue substitute consisting of recombinant insulin-secreting cells and glucose-responsive material, *Biotechnol. Bioeng.*, 2004, **87**, 863–873.

23. T. Miyata, A. Jikihara, K. Nakamae and A. S. Hoffman, Preparation of reversibly glucose-responsive hydrogels by covalent

immobilization of lectin in polymer networks having pendent glucose, *J. Biomater. Sci., Polym. Ed.*, 2004, **15**, 1085–1098.

24. T. Miyata, A. Jikihara and A. S. Hoffman, Preparation of poly(2-glucosyloxyethyl methacrylate)-concanavalin A complex hydrogel and its glucose-sensitivity, *Macromol. Chem. Phys.*, 1996, **197**, 1135–1146.

25. K. Nakamae, T. Miyata, A. Jikihara and A. S. Hoffman, Formation of poly(glucosyloxyethyl methacrylate)-Concanavalin A complex and its glucose-sensitivity, *J. Biomater. Sci., Polym. Ed.*, 1994, **6**, 79–90.

26. T. D. James and S. Shinkai, Artificial Receptors as Chemosensors for Carbohydrates, *Top. Curr. Chem.*, 2002, **218**, 159–200.

27. J. N. Cambre and B. S. Sumerlin, Biomedical applications of boronic acid polymers, *Polymer*, 2011, **52**(21), 4631–4643.

28. T. Miyata, T. Uragami and K. Nakamae, Biomolecule-sensitive hydrogels, *Adv. Drug Delivery Rev.*, 2002, **54**, 79–98.

29. J. P. Lorand and J. O. Edwards, Polyol complexes and structure of the benzeneboronate ion, *J. Org. Chem.*, 1959, **24**, 769–774.

30. K. Kataoka, H. Miyazaki, T. Okano and Y. Sakurai, Sensitive glucose-induced change of the lower critical solution temperature of poly[N,N-(dimethylacrylamide)-*co*-3-(acrylamido)-phenylboronic acid] in physiological saline, *Macromolecules*, 1994, **27**, 1061–1062.

31. K. Kataoka, H. Miyazaki, M. Bunya, T. Okano and Y. Sakurai, Totally synthetic polymer gels responding to external glucose concentration: Their preparation and application to on-off regulation of insulin release, *J. Am. Chem. Soc.*, 1998, **120**, 12694–12695.

32. G. Springsteen and B. Wang, A detailed examination of boronic acid-diol complexation, *Tetrahedron*, 2002, **58**, 5291–5300.

33. I. Hisamitsu, K. Kataoka, T. Okano and Y. Sakurai, Glucose-responsive gel from phenylborate polymer and poly(vinyl alcohol): Prompt response at physiological pH through the interaction of borate with amino group in the gel, *Pharm Res.*, 1997, **14**, 289–293.

34. S. Kitano, Y. Koyama, K. Kataoka, T. Okano and Y. Sakurai, A novel drug delivery system utilizing a glucose responsive polymer complex between poly(vinyl alcohol) and poly(N-vinyl-2-pyrrolidone) with a phenylboronic acid moiety, *J. Controlled Release*, 1992, **19**, 162–170.

35. A. Kikuchi, K. Suzuki, O. Okabayashi, H. Hoshino, K. Kataoka and Y. Sakurai *et al.*, Glucose-sensing electrode coated with

polymer complex gel containing phenylboronic acid, *Anal. Chem.*, 1996, **68**, 823–828.

36. A. Matsumoto, T. Kurata, D. Shiino and K. Kataoka, Swelling and shrinking kinetics of totally synthetic, glucose-responsive polymer gel bearing phenylborate derivative as a glucose-sensing moiety, *Macromolecules*, 2004, **37**, 1502–1510.

37. S. B. Zhang, L. Y. Chu, D. Xu, J. Zhang, X. J. Ju and R. Xie, Poly(-N-isopropylacrylamide)-based comb-type grafted hydrogel with rapid response to blood glucose concentration change at physiological temperature, *Polym. Adv. Technol.*, 2008, **19**, 937–943.

38. Y. Zhang, Y. Guan and S. Zhou, Synthesis and volume phase transitions of glucose-sensitive microgels, *Biomacromolecules*, 2006, 7, 3196–3201.

39. H. Ge, Y. Ding, C. Ma and G. Zhang, Temperature-controlled release of diols from N-isopropylacrylamide-co-acrylamidophenylboronic acid microgels, *J. Phys. Chem. B*, 2006, **110**, 20635–20639.

40. V. Lapeyre, I. Gosse, S. Chevreux and V. Ravaine, Monodispersed glucose-responsive microgels operating at physiological salinity, *Biomacromolecules*, 2006, 7, 3356–3363.

41. T. Hoare and R. Pelton, Charge-switching, amphoteric glucose-responsive microgels with physiological swelling activity, *Biomacromolecules*, 2008, **9**, 733–740.

42. G. Zenkl, T. Mayr and I. Klimant, Sugar-responsive fluorescent nanospheres, *Macromol. Biosci.*, 2008, **8**, 146–152.

43. A. Matsumoto, S. Ikeda, A. Harada and K. Kataoka, Glucose-responsive polymer bearing a novel phenylborate derivative as a glucose-sensing moiety operating at physiological pH conditions, *Biomacromolecules*, 2003, 4, 1410–1416.

44. A. Matsumoto, R. Yoshida and K. Kataoka, Glucose-responsive polymer gel bearing phenylborate derivative as a glucose-sensing moiety operating at physiological pH, *Biomacromolecules*, 2004, **5**, 1038–1045.

45. J. Pellon, L. H. Schwind, M. J. Guinard and W. M. Thomas, Polymerization of vinyl monomers containing boron. ii. p-vinylbenzeneboronic acid, *J. Polym. Sci.*, 1961, **55**, 161–167.

46. K. Shiomori, A. E. Ivanov, I. Y. Galaev, Y. Kawano and B. Mattiasson, Thermoresponsive properties of sugar sensitive copolymer of N-isopropylacrylamide and 3-(acrylamido)phenylboronic acid, *Macromol. Chem. Phys.*, 2004, **205**, 27–34.

47. W. J. Lennarz and H. R. Snyder, Arylboronic acids. III. Preparation and polymerization of p-vinylbenzeneboronic acid, *J. Am. Chem. Soc.*, 1960, **82**, 2169–2171.

48. R. L. Letsinger and S. B. Hamilton, Organoboron compounds. X. Popcorn polymers and highly cross-linked vinyl polymers containing boron, *J. Am. Chem. Soc.*, 1959, **81**, 3009–3012.
49. K. Matyjaszewski and T. Davis, *Handbook of Radical Polymerization*, Wiley Interscience, Hoboken, NJ, 2002.
50. W. A. Braunecker and K. Matyjaszewski, Controlled/living radical polymerization: Features, developments, and perspectives, *Prog. Polym. Sci.*, 2007, **32**, 93–146.
51. J. S. Wang and K. Matyjaszewski, Controlled/"living" radical polymerization. Atom transfer radical polymerization in the presence of transition-metal complexes, *J. Am. Chem. Soc.*, 1995, **117**, 5614–5615.
52. M. Kato, M. Kamigaito, M. Sawamoto and T. Higashimura, Polymerization of methyl methacrylate with the carbon tetrachloride/dichlorotris(triphenylphosphine)ruthenium(II)/methylaluminum bis(2,6-di-tert-butylphenoxide) initiation system: Possibility of living radical polymerization, *Macromolecules*, 1995, **28**, 1721–1723.
53. A. P. Vogt and B. S. Sumerlin, An efficient route to macromonomers *via* ATRP and click chemistry, *Macromolecules*, 2006, **39**, 5286–5292.
54. J. Chiefari, Y. K. Chong, F. Ercole, J. Krstina, J. Jeffery and T. P. T. Le *et al.*, Living free-radical polymerization by reversible addition-fragmentation chain transfer: The RAFT process, *Macromolecules*, 1998, **31**, 5559–5562.
55. S. Perrier and T. Pittaya, Macromolecular design via reversible addition-fragmentation chain transfer (RAFT)/xanthates (MADIX) polymerization, *J. Polym. Sci., Part A: Polym. Chem*, 2005, **43**, 5347–5393.
56. P. De, S. R. Gondi and B. S. Sumerlin, Folate-conjugated thermoresponsive block copolymers: highly efficient conjugation and solution self-assembly, *Biomacromolecules*, 2008, **9**, 1064–1070.
57. Y. Qin, G. Cheng, O. Achara, K. Parab and F. Jäkle, A New Route to Organoboron Polymers via Highly Selective Polymer Modification Reactions, *Macromolecules*, 2004, **37**, 7123–7131.
58. Q. Yang, C. Guanglou, S. Anand and F. Jäkle, Well-defined boron-containing polymeric lewis acids, *J. Am. Chem. Soc.*, 2002, **124**(43), 12672–12673.
59. J. N. Cambre, D. Roy, S. R. Gondi and B. S. Sumerlin, Facile strategy to well-defined water-soluble boronic acid (co)polymers, *J. Am. Chem. Soc.*, 2007, **129**, 10348–10349.

60. D. Roy, J. N. Cambre and B. S. Sumerlin, Sugar-responsive block copolymers by direct RAFT polymerization of unprotected boronic acid monomers, *Chem. Commun.*, 2008, 2477–2479.

61. D. Roy, J. N. Cambre and B. S. Sumerlin, Triply-responsive boronic acid block copolymers: solution self-assembly induced by changes in temperature, pH, or sugar concentration, *Chem. Commun.*, 2009, **16**, 2106–2108.

62. H. G. Schild, Poly(N-isopropylacrylamide): Experiment, theory, and application, *Prog. Polym. Sci.*, 1992, **17**, 163–249.

63. P. D. Thornton, G. McConnell and R. V. Ulijin, Enzyme responsive polymer hydrogel beads, *Chem. Commun.*, 2005, 5913–5915.

64. R. V. Ulijin, Enzyme-responsive materials: a new class of smart biomaterials, *J. Mater. Chem.*, 2006, **16**, 2217–2225.

65. R. J. Amir, S. Zhong, D. J. Pochan and C. J. Hawker, Enzymatically triggered self-assembly of block copolymers, *J. Am. Chem. Soc.*, 2009, **131**(39), 13949–13951.

66. H. Kühnle and H. G. Börner, Biotransformation on polymer–peptide conjugates: a versatile tool to trigger microstructure formation, *Angew. Chem., Int. Ed.*, 2009, **48**(35), 6431–6434.

67. C. Wang, Q. Chen, Z. Wang and X. Zhang, An enzyme-responsive polymeric superamphiphile, *Angew. Chem.*, 2010, **122**(46), 8794–8797.

68. Y. Xing, C. Wang, P. Han, Z. Wang and X. Zhang, Acetylcholinesterase responsive polymeric supra-amphiphiles for controlled self-assembly and disassembly, *Langmuir*, 2012, **28**(14), 6032–6036.

69. T.-H. Ku, M.-P. Chien, M. P. Thompson, R. S. Sinkovits, N. H. Olson and T. S. Baker *et al.*, Controlling and switching the morphology of micellar nanoparticles with enzymes, *J. Am. Chem. Soc.*, 2011, **133**(22), 8392–8395.

70. J. L. Maller, B. E. Kemp and E. G. Krebs, In vivo phosphorylation of a synthetic peptide substrate of cyclic AMP-dependent protein kinase, *Proc. Natl. Acad. Sci. U. S. A.*, 1978, **75**(1), 248–251.

71. S. Wera and B. A. Hemmings, Serine/threonine protein phosphatases, *Biochem. J.*, 1995, **311**(Pt. 1), 17.

72. D. G. Vartak and R. A. Gemeinhart, Matrix metalloproteases: underutilized targets for drug delivery, *J. Drug Targeting*, 2007, **15**(1), 1–20.

73. J. D. Raffetto and R. A. Khalil, Matrix metalloproteinases and their inhibitors in vascular remodeling and vascular disease, *Biochem. Pharmacol.*, 2008, **75**(2), 346–359.

74. K. Kessenbrock, V. Plaks and Z. Werb, Matrix metalloproteinases: regulators of the tumor microenvironment, *Cell*, 2010, **141**(1), 52–67.
75. Z. Yang, H. Gu, D. Fu, P. Gao, J. K. Lam and B. Xu, Enzymatic formation of supramolecular hydrogels, *Adv. Mater.*, 2004, **16**, 1440–1444.
76. S. Toledano, R. J. Williams, V. Jayawarna and R. V. Ulijin, Enzyme-triggered self-assembly of peptide hydrogels via reversed hydrolysis, *J. Am. Chem. Soc.*, 2006, **128**, 1070–1071.
77. L. G. Griffith and J. J. Sperinde, Synthesis and characterization of enzymatically cross-linked poly(ethylene glycol) hydrogels, *Macromolecules*, 1997, **30**, 5255–5264.
78. J. J. Sperinde and L. G. Griffith, Control and prediction of gelation kinetics in enzymatically cross-linked poly(ethylene glycol) hydrogels, *Macromolecules*, 2000, **33**, 5476–5480.
79. T. Chen, D. A. Small, M. K. McDermott, W. E. Bentley and G. F. Payne, Enzymatic methods for *in situ* cell entrapment and cell release, *Biomacromolecules*, 2003, **4**, 1558–1563.
80. T. J. Sanborn, P. B. Messersmith and A. E. Barron, In situ crosslinking of a biomimetic peptide-PEG hydrogel via thermally triggered activation of factor XIII, *Biomaterials*, 2002, **23**, 2703–2710.
81. M. P. Lutolf, G. P. Raeber, A. H. Zisch, N. Tirelli and J. A. Hubbell, Cell-responsive synthetic hydrogels, *Adv. Mater.*, 2003, **15**, 888–892.
82. T. Miyata and T. Uragami, Biological stimulus-responsive hydrogels, *Polymeric Biomaterials*, in ed. Dumitriu S., 2nd edn, CRC Press, New York, 2001, pp. 959–974.
83. Z. Lu, P. Kopeckova and J. Kopecek, Antigen responsive hydrogels basedon polymerizable antibody Fab fragment, *Macromol. Biosci.*, 2003, **3**, 296–300.
84. T. Miyata, N. Asami and T. Uragami, Preparation of an antigen-sensitive hydrogel using antigen-antibody bindings, *Macromolecules*, 1999, **32**, 2082–2084.
85. T. Miyata, N. Asami and T. Uragami, A reversibly antigen-responsive hydrogel, *Nature*, 1999, **399**, 766–769.
86. R. Zhang, A. Bowyer, R. Eisenthal and J. Hubble, A smart membrane based on an antigen-responsive hydrogel, *Biotechnol. Bioeng.*, 2007, **97**, 976–984.
87. S. Bauer, P. Schmuki, K. von der Mark and J. Park, Engineering biocompatible implant surfaces; part I: materials and surfaces, *Prog. Mater. Sci.*, 2013, **58**, 261.

88. J. S. Temenoff and A. G. Mikos, *Biomaterials: The Intersection of Biology and Materials Science*, Pearson/Prentice Hall, Upper Saddle River, N.J., 2008.
89. E. Wintermantel and H. Suk-Woo, *Medizintechnik mit bio-kompatiblen Werkstoffen und Verfahren*, Springer, Berlin, 2002.
90. J. M. Goddard and J. H. Hotchkiss, Polymer surface modification for the attachment of bioactive compounds, *Prog. Polym. Sci.*, 2007, **32**, 698.
91. N. K. Guimard, N. Gomez and C. E. Schmidt, Conducting polymers in biomedical engineering, *Prog. Polym. Sci.*, 2007, **32**, 876.
92. H. Shen, X. Hu, J. Bei and S. Wang, The immobilization of basic fibroblast growth factor on plasma-treated poly(lacitide-*co*-glycolide), *Biomaterials*, 2008, **29**, 2388.
93. R. Langer and D. A. Tirrell, Designing materials for biology and medicine, *Nature*, 2004, **428**, 487.
94. K. Y. Lee and D. J. Mooney, Hydrogels for tissue engineering, *Chem. Rev.*, 2001, **101**, 1869.
95. T. Jiang, J. Chang, C. Wang, Z. Ding, J. Chen and J. Zhang *et al.*, Adsoprtion of plasmid DNA onto N,N′-(dimethylamino)ethyl-methacrylate graft-polymerized poly-L-lactic acid folm surface for promotion of in-situ gene delivery, *Biomacromolecules*, 2007, **8**, 1951.
96. J. F. Mano, Stimuli-responsive polymeric systems for biomedical applications, *Adv. Eng. Mater.*, 2008, **10**, 515.
97. A. N. Zeilikin, Drug Releasing Polymer Thin Films: New Era of Surface-Mediated Drug Delivery, *ACS Nano*, 2010, **4**, 2494.
98. M. Yu and M. W. Urban, Polymeric surfaces with anticoagulant, antifouling, and antimicrobial attributes, *Macromol. Symp.*, 2009, **283**, 311.
99. W.-S. Bae and M. W. Urban, Creating patterned poly(di-methylsiloxane) surfaces with amoxicillin and poly(ethylene glycol), *Langmuir*, 2006, **22**, 10277.
100. N. Aumsuwan, S. Heinhorst and M. W. Urban, Antibacterial surfaces on expanded poly(tetrafluoroethylene); penicillin attachment, *Biomacromolecules*, 2007, **8**, 713.
101. J. I. Kroschwitz, *Polymers: Biomaterials and Medical Applications*, Wiley-Interscience/John Wiley & Sons, New York, 1989.
102. E. R. Kenawy, S. D. Worley and R. Broughton, The chemistry and applications of antimicrobial polymers: A state-of-the-art review, *Biomacromolecules*, 2007, **8**, 1359.
103. A. G. Kidane, H. Salacinski, A. Tiwari, K. R. Bruckdorfer and A. M. Seifalian, Anticoagulant and antiplatelet agents: Their

clinical and device application(s) together with usages to engineering surfaces, *Biomacromolecules*, 2004, **5**, 798.

104. W. Senaratne, L. Andruzzi and C. K. Ober, Self-assembled monolayers and polymer brushes in biotechnology: Current applications and future perspectives, *Biomacromolecules*, 2005, **6**, 2427.

105. W. S. Bae, A. J. Convertine, C. L. McCormick and M. W. Urban, Effect of sequential layer-by-layer surface modifications on the surface energy of plasma-modified poly(dimethylsiloxane), *Langmuir*, 2007, **23**, 667.

106. H. Kim and M. W. Urban, Reactions of thrombresistant multi-layered thin films on poly(vinyl chloride) (PVC) surface: A spectroscopic study, *Langmuir*, 1998, **14**, 7235.

107. S. Zhang, Fabrication of novel biomaterials through molecular self-assembly, *Nat. Biotechnol.*, 2003, **21**, 1171.

108. T. Tanii, T. Hosaka, T. Miyake, Y. Kanari, G.-J. Zhang and T. Funatsu *et al.*, Hybridization of deoxyribonucleic acid and immobilization of green fluorescent protein on nanostructured organosilane templates, *Jpn. J. Appl. Phys.*, 2005, **44**, 5851.

109. D. Roy, J. S. Knapp, J. T. Guthrie and S. Perrier, Antibacterial cellulose fiber *via* RAFT surface graft polymerization, *Biomacromolecules*, 2008, **9**, 91.

110. S. B. Lee, R. R. Koepsel, S. W. Morley, K. Matyjaszewski, Y. Sun and A. J. Russell, Permanent, nonleaching antibacterial surface. 1. synthesis by atom transfer radical polymerization, *Biomacromolecules*, 2004, **5**, 877.

111. C. M. Dvoracek, G. Sukhonosova, M. J. Benedik and J. C. Grunlan, Antimicrobial behavior of polyelectrolyte surfactant thin film assemblies, *Langmuir*, 2009, **25**, 10322.

112. J. A. Lichter, K. J. V. Vliet and M. F. Rubner, Design of antibacterial surfaces and interfaces: polyelectrolyte multilayers as a multifunctional platform, *Macromolecules*, 2009, **42**, 8573.

113. W. Chen and T. J. McCarthy, Layer-by-Layer deposition: A tool for polymer surface modification, *Macromolecules*, 1997, **30**, 78.

114. F. E. Black, M. Hartshorne, M. C. Davies, C. J. Roberts, S. J. B. Tendler and P. M. Williams *et al.*, Surface engineering and surface analysis of a biodegradable polymer with biotinylated end groups, *Langmuir*, 1999, **15**, 3157.

115. A.-S. Duwez, S. Cuenot, C. Jérôme, S. Gabriel, R. Jérôme and S. Rapino *et al.*, Mechanochemistry: targted delivery of single molecules, *Nat. Nanotechnol.*, 2006, **1**, 122.

116. S. Onard, I. Martin, J.-F. Chailan, A. Crespy and P. Carriere, Nanostructuration in thin epoxy-amine films inducing controlled specific phase ethericfication: effect on the glass transition temperatures, *Macromolecules*, 2011, **44**, 3485.

117. T. Desmet, R. Morent, N. D. Geyter, C. Leys, E. Schacht and P. Dubruel, Nonthermal plasma technology as a varsatilestrategy for polymeric biomaterials surface modification: a review, *Biomacromolecules*, 2009, **10**, 2351.

118. C. Suituma, Y. Wang, M. Hupert, F. Barany, R. L. McCarley and S. A. Soper, Fabrication of DNA microarrays onto poly(methyl methacrylate) with ultravilet petterning and microfluidics for the detection of low-abundant point mutations, *Anal. Biochem.*, 2005, **340**, 123.

119. Q. Ma, H. Zhang, J. Zhao and Y.-K. Gong, Fabrication of cell outer membrane mimetic polymer brush on polysulfone surface *via* RAFT technique, *Appl. Surf. Sci.*, 2012, **258**, 9711.

120. S. Tugulu and H. A. Klok, Stability and nonfouling properties of poly(poly(ethylene glycol) methacrylate) brushes under cell culture conditions, *Biomacromolecules*, 2008, **9**, 906.

121. J. R. Rasmussen, E. R. Stedronsky and G. M. Whitesides, Introduction, modification, and characterization of functional-groups on surface of low-density polyethylene film, *J. Am. Chem. Soc.*, 1977, **99**, 4736.

122. H. C. Kolb, M. G. Finn and K. B. Sharpless, Click chemistry: diverse chemical function from a few good reactions, *Angew. Chem., Int. Ed.*, 2001, **40**, 2004.

123. W. H. Binder and R. Sachsenhofer, 'Click' chemistry in polymer and materials science, *Macromol. Rapid Commun.*, 2007, **28**, 15.

124. V. V. Rostovtsev, L. G. Green, V. V. Fokin and K. B. Sharpless, A stepwise Huisgen cycloaddition process: copper(I)-catalyzed regioselective "ligation" of azides and terminal alkynes, *Angew. Chem., Int. Ed.*, 2002, **41**, 2596.

125. R. K. Iha, K. L. Wooley, A. M. Nystrom, D. J. Burke, M. J. Kade and C. J. Hawker, Applications of orthogonal "click" chemistries in the synthesis of functional soft materials, *Chem. Rev.*, 2009, **109**, 5620.

126. A. Vahdata, H. Bahramia, N. Ansaria and F. Ziaieb, Radiation grafting of styrene onto polypropylene fibres by a 10 MeV electron beam, *Radiat. Phys. Chem.*, 2007, **76**, 787.

127. H. Yasuda, *Plasma Polymerization*, Academic Press, Orlando, 1985.

128. S. R. Gaboury and M. W. Urban, Microwave plasma reactions of solid monomers with silicone elastomer surfaces: a spectroscopic study, *Langmuir*, 1993, **9**, 3225.

129. W. S. Bae and M. W. Urban, Reactions of antimicrobial species to imidazole-microwave plasma reacted poly(dimethylsiloxane) surfaces, *Langmuir*, 2004, **20**, 8372.

130. P. Chevallier, R. Janvier, D. Mantovani and G. Laroche, In vitro biological performances of phosphorylcholine-grafted ePTFE prostheses through RFGD plasma techniques, *Macromol. Biosci.*, 2005, **5**, 829.

131. Q. Zhang, C. Wang, Y. Babukutty, T. Ohyama, M. Kogoma and M. Kodama, Biocompatibility evaluation of ePTFE membrane modified with PEG in atmospheric ppressure glow discharge, *J. Biomed. Mater. Res.*, 2002, **60**, 502.

132. G. Jia, H. F. Wang, L. Yan, X. Wang, R. J. Pei and T. Yan *et al.*, Cytotoxicity of carbon nanomaterials: Single-wall nanotube, multi-wall nanotube, and fullerene, *Environ. Sci. Technol.*, 2005, **39**, 1378.

133. A. Munoz-Bonilla and M. Fernandez-Garcia, Polymeric materials with antimicrobial activity, *Prog. Polym. Sci.*, 2012, **37**, 281.

134. E. L. Rabea, M. E. Badawy, C. V. Stevens, G. Smagghe and W. Steurbaut, Chitosan as antimicrobial agent: applicatiopns and mode of action, *Biomacromolecules*, 2003, **4**, 1457–1465.

135. M. Kazemzadeh-Narbat, J. Kindrachuk, K. Duan, H. Jenssen, R. E. W. Hancock and i. Wang, Antimicrobial peptides on calcium phosphate-coated titanium for the prevention of implant-associated infections, *Biomaterials*, 2010, **31**, 9519.

136. J. C. Grunlan, J. K. Choi and A. Lin, Antimicrobial behavior of polyelectrolyte multilayerfilms containing cetrimide and silver, *Biomacromolecules*, 2005, **6**, 1149.

137. K. Markoishvili, G. Tsitlanadze, R. Katsarava, J. G. J. Morris and A. Sulakvalidze, A novel sustained-release matrix based on biodegradable poly(ester amide)s and impregnated with bacteriophages and an antibiotic shows promice in management in infected venous stasis ulcers and other poorlt healing woulds, *Int. J. Dermatol.*, 2002, **41**, 453.

138. J. M. Schierholz, J. Beuth, G. Pulverer, D.-P. König, R. S. Scharlack and G. Kampf *et al.*, Silver-containing polymers, *Antimicrob. Agents Chemother.*, 1999, **43**, 2819.

139. J. Huang, H. Murata, R. R. Koepsel, A. J. Russell and K. Matyjaszewski, Antibacterial polypropylene *via* surface-initiated

atom transfer radical polymerization, *Biomacromolecules*, 2007, **8**, 1396.

140. N. Aumsuwan, S. Heinhorst and M. W. Urban, The effectiveness of antibiotic activity of penicillin attahed to expanded poly(tetrafluoroethylene) (ePTFE) surfaces: A Quantitative Assessment, *Biomacromolecules*, 2007, **8**, 3525.

141. N. Aumsuwan, R. C. Danyus, S. Heinhorst and M. W. Urban, Attachment of amicillin to expanded poly(tetrafluoroethylene): surface reactions leading to inhibition of microbial growth, *Biomacromolecules*, 2008, **9**, 1712.

142. J. C. Tiller, S. B. Lee, K. Lewis and A. M. Klibanov, Polymer surfaces derivatized with poly(vinyl-N-hexylpyridinium) kill airborne and waterborne bacteria, *Biotechnol. Bioeng.*, 2002, **79**, 465.

143. O. Larm, R. Larsson and P. Olsson, A new non-thrombogenic surface prepared by selective covalent binding of heparin *via* modified reducing terminal residue, *Biomater., Med. Devices, Artif. Organs*, 1983, **11**, 161.

144. M.-C. Bourin and U. Lindahl, Review article: Glycosaminoglycans and the regulation of blood coagulation, *Biochem. J.*, 1993, **289**, 313.

145. M. D. Phaneuf, S. A. Berceli, M. J. Bide, W. C. Quist and F. W. Logerfo, Covalent linkage of recombinant hirudin to poly(ethylene terephthalate) (Dacron): creation of a novel antithrombin surface, *Biomaterials*, 1997, **18**, 755.

146. B. Seifert, P. Romaniuk and T. Groth, Covalent immobilzation of hirudin improves the haemocompatibility of polylactide-polyglycolide in vitro, *Biomaterials*, 1997, **18**, 1495.

147. D. L. Rabenstein, Heparin and heparan sulfate: Structure and function, *Nat. Prod. Rep.*, 2002, **19**, 312.

148. J. Hirsh, S. G. Shaughnessy, J. L. Halperin, C. Granger, E. M. Ohman and J. E. Dalen, Heparin and Low-molecular weight heparin: Mechanisms of action, pharmacokinetics, dosing, monitoring, efficacy, and safety, *Chest*, 2001, **119**, 64s.

149. T. Chandy, G. S. Das, R. F. Wilson and G. H. R. Rao, Use of plasma glow for surface-engineering biomolecules to enhance bloodcompatibility of Dacron and PTFE vascular prosthesis, *Biomaterials*, 2000, **21**, 699.

150. K. Christensen, R. Larsson, H. Emanuelsson, G. Elgue and A. Larsson, Improved blood compatibility of a stent graft by combining heparin coating and abciximab, *Thromb. Res.*, 2005, **115**, 245.

151. J. M. Goddard and J. H. Hotchkiss, Polymer surface modification for the attachment of biactive compounds, *Prog. Polym. Sci.*, 2007, **32**, 698.
152. Y. Tanga and J. Singh, Controlled delivery of aspirin: Effect of aspirin on polymer degradation and in vitro release from PLGA based phase sensitive systems, *Int. J. Pharm.*, 2008, **357**, 119.
153. F.-M. Li, Z.-W. Gu, G.-W. Li, S. WAang and X.-D. Feng, Synthesis of biocompatible polymers with asparin-moieties for asparin delivery, *J. Bioact. Compat. Polym.*, 1991, **6**, 142.
154. Y. B. Aldenhoff and L. H. Koole, Platelet adhesion studies on dipyridamole coated polyurethane surfaces, *Eur. Cells Mater.*, 2003, **5**, 61.
155. S. Krishnan, R. Ayothi, A. Hexemer, J. A. Finlay, K. E. Sohn and R. Perry *et al.*, Anti-biofouling properties of comblike block copolymers with amphiphilic side chains, *Langmuir*, 2006, **22**, 5075.
156. H. Chen, L. Yuan, W. Song, Z. Wu and D. Li, Biocompatible polymer materials: role of protein-surface interactions, *Prog. Polym. Sci.*, 2008, **33**, 1059.
157. E. P. K. Currie, W. Norde and M. A. C. Stuart, Tethered polymer chains: surface chemistry and their impact on colloidal and suface properties, *Adv. Colloid Interface*, 2003, **100**, 205.
158. H. W. Ma, J. H. Hyun, P. Stiller and A. Chilkoti, Non-fouling oligo(ethylene glycol)-functionalized polymer brushes synthesized by surface-initiated atom transfer radical polymerization, *Adv. Mater.*, 2004, **16**, 338.
159. X. W. Fan, L. J. Lin and P. B. Messersmith, Cell fouling resistance of polymer brushes grafted from Ti substrates by surface-initiated polymerization: effect of ethylene glycol side chain length, *Biomacromolecules*, 2006, 7, 2443.
160. V. E. Wagner, J. T. Koberstein and J. D. Bryers, Protein and bacterial fouling characteristics of peptide and antibody decorated surfaces of PEG-poly(acrylic acid) co-polymers, *Biomaterials*, 2004, **25**, 2247.
161. H. Du, P. Chandaroy and S. W. Hui, Grafted poly-(ethylene glycol) on lipid surfaces inhibits protein adsorption and cell adhesion, *Biochim. Biophys. Acta*, 1997, **1326**, 236.
162. J. Thomas, S. B. Choi, R. Fjeldheim and P. Boudjouk, Silicones containing pendant biocides for antifouling coatings, *Biofouling*, 2004, **20**, 227.
163. H. Chen, M. A. Brook and H. Sheardown, Silicone elastomers for reduced protein adsorption, *Biomaterials*, 2004, **25**, 2273.

164. T. Goda, T. Konno, M. Takai, T. Moro and K. Ishihara, Bio-mimetic phosphorylcholine polymer grafting from poly-dimethylsiloxane surface using photo-induced polymerization, *Biomaterials*, 2006, **27**, 5151.

165. S. Chen, J. Zheng, L. Li and S. Y. Jiang, Strong resistance of phosphorylcholine self-assembled monolayers to protein adsorption: Insights into nonfouling properties of zwitterionic materials, *J. Am. Chem. Soc.*, 2005, **127**, 14473.

166. M. Kyomotoa, T. Morob, K. Saigaa, M. Hashimotoe, H. Itoc and H. Kawaguchic *et al.*, Biomimetic hydration lubrication with various polyelectrolyte layers on cross-linked polyethylene orthopedic bearing materials, *Biomaterials*, 2012, **33**, 4451.

167. R. Silva, E. C. Muniz and A. F. Rubira, Multiple hydrophobic polymer ultra-thin layers covalently anchored to polyethylene films, *Polymer*, 2008, **49**, 4066.

168. W.-S. Bae and M. W. Urban, Reactions of Antimicrobial Species to Imidazole-Microwave Plasma Reacted Poly(dimethylsiloxane) (PDMS) Surfaces, *Langmuir*, 2004, **20**, 8372.

169. R. M. Hensarling, V. A. Doughty, J. W. Chan and D. L. Patton, "Clicking" polymer brushes with thiol-yne chemistry: indoors and out, *J. Am. Chem. Soc.*, 2009, **131**, 14673.

170. H. C. Kolb and K. B. Sharpless, The growing impact of click chemistry on drug discovery, *Drug Discovery Today*, 2003, **8**, 1128.

171. E. W. L. Chan and M. N. Yousaf, Immobilization of ligands with precise control of density to electroactive surfaces, *J. Am. Chem. Soc.*, 2006, **128**, 15542.

172. K. Zhu, Y. Zhang, S. He, W. Chen, J. Shen and Z. Wang *et al.*, Quantification of proteins by functionalized gold nanoparticles using click chemistry, *Anal. Chem.*, 2012, **84**, 4267.

173. G. Qin, C. Santos, W. Zhang, Y. Li, A. Kumar and U. J. Erasquin *et al.*, Biofunctionalization on alkylated silicon substrate surfaces via "click" chemistry, *J. Am. Chem. Soc.*, 2010, **132**, 16432.

174. X.-L. Sun, C. L. Stabler, C. S. Cazalis and E. L. Chaikof, Carbohydrate and protein immobilization onto solid surfaces by sequential diels–alder and azide–alkyne cycloadditions, *Bioconjugate Chem.*, 2006, **17**, 52.

175. H. Li, F. Cheng, A. M. Duft and A. Adronov, Functionalization of single-walled carbon nanotubes with well-defined polystyrene by "click" coupling, *J. Am. Chem. Soc.*, 2005, **127**, 14518.

176. P. Wu, A. K. Feldman, A. K. Nugent, C. J. Hawker, A. Scheel and B. Voit *et al.*, Efficiency and fidelity in a click-chemistry route to

triazole dendrimers by the copper(I)-catalyzed ligation of azides and alkynes, *Angew. Chem., Int. Ed.*, 2004, **43**, 3928.

177. B. S. Sumerlin and A. P. Vogt, Macromolecular engineering through click chemistry and other efficient transformations, *Macromolecules*, 2010, **43**, 1.

178. M. Malkoch, R. Vestberg, N. Gupta, L. Mespouille, P. Dubois and A. F. Mason *et al.*, Synthesis of well-defined hydrogel networks using click chemistry, *Chem. Commun.*, 2006, **26**, 2774.

179. B. Li, A. L. Martin and E. R. Gillies, Multivalent polymer vesicles via surface functionalization, *Chem. Commun.*, 2007, **48**, 5217.

180. R. K. O'Reilly, M. J. Joralemon, K. L. Wooley and C. J. Hawker, Functionalization of micelles and shell cross-linked nano-particles using click chemistry, *Chem. Mater.*, 2005, **17**, 5976.

181. H. Pearson, J. Manti and M. W. Urban, Tethered Polyelectrolyte Stimuli-Responsive Chains on Polymeris Surfaces, *Biomater. Sci.*, 2014, **2**, 512.

182. G. A. Hudalla and W. L. Murphy, Using "click" chemistry to prepare SAM substrates to study stem cell adhesion, *Langmuir*, 2009, **25**, 5737.

183. B. J. Adzima, Y. Tao, C. J. Kloxin, C. A. DeForest, K. S. Anseth and C. N. Bowman, Spatial and temporal control of the alkyne-azide cycloaddition by photoinitiated Cu(II) reduction, *Nat. Chem.*, 2011, **3**, 258.

184. C. M. Nimmo and M. S. Shoichet, Regenerative biomaterials that "click": simple, aqueous-based protocols for hydrogel synthesis, suface immobilization, and 3D patterning, *Bioconjugate Chem.*, 2011, **22**, 2199.

185. X. Wang, L. Liu, Y. Luo and H. Zhao, Bioconjugation of biotin to the interfaces of polymeric micelles via *in situ* click chemistry, *Langmuir*, 2009, **25**, 744.

186. D. E. Prasuhn, P. Singh, E. Strable, S. Brown, M. Manchester and M. G. Finn, Plasma clearance of bacteriophage Qβ particles as a function of surface charge, *J. Am. Chem. Soc.*, 2008, **130**, 1328.

187. G. Qing, H. Xiong, F. Seela and T. Sun, Spatially controlled DNA nanopatterns by "click" chemistry using oligonucleotides with different anchoring sites, *J. Am. Chem. Soc.*, 2010, **132**, 15228.

188. N. Aumsuwan, S. Heinhorst and M. W. Urban, Antibacterial surfaces on expanded polytetrafluoroethylene; penicillin at-tachment, *Biomacromolecules*, 2007, **8**, 713.

189. J. M. Goddard and J. H. Hotchkiss, Polymer surface modification for the attachment of bioactive compounds, *Prog. Polym. Sci.*, 2007, **32**(7), 698–725.

190. N. Houbenov, S. Minko and M. Stamm, Mixed Polyelectrolyte Brush from Oppositely Charged Polymers for Switching of Surface Charge and Composition in Aqueous Environment, *Macromolecules*, 2003, **36**, 5897.
191. L. Ionov, N. Houbenov, A. Sidorenko and M. Stamm, Inverse and Reversible Switching Gradient Surfaces from Mixed Poly-electrolyte Brushes, *Langmuir*, 2004, **20**, 9916.
192. F. Liu and M. W. Urban, Recent advances and challenges in designing stimuli-responsive polymers, *Prog. Polym. Sci.*, 2010, **35**, 3.
193. F. Liu and M. W. Urban, Dual Temperature and pH Respon-siveness of Poly(2-(N,N-dimethylamino)ethyl methacrylate-co-n-butyl acrylate) Colloidal Dispersions and Their Films, *Macro-moleucles*, 2008, **41**, 6531.
194. L. Tang, J. Whalen, G. Schutte and C. Weder, Stimuli-Responsive Epoxy Coatings, *ACS Appl. Mater. Interfaces*, 2009, **1**, 688.
195. F. Liu, D. Ramachandran and M. W. Urban, Colloidal Films That Mimic Cilia, *Adv. Funct. Mater.*, 2010, **20**, 3163.
196. Z. D. Hua, Z. Y. Chen, Y. Z. Li and M. P. Zhao, Thermosensitive and Salt-Sensitive Molecularly Imprinted Hydrogel for Bovine Serum Albumin, *Lamgmuir*, 2008, **24**, 5773.
197. H. A. Pearson, J. M. Andrie and M. W. Urban, Covalent attach-ment of multilayers (CAM): a platform for pH switchable antimicrobial and anticoagulant polymeric surfaces, *Biomater. Sci.*, 2004, **2**(4), 512–521.
198. G. F. Webb, E. M. C. D'Agata, P. Magal and S. Ruan, A model of antibiotic-resistant bacterial epidemics in hospitals, *Proc. Natl. Acad. Sci. U. S. A.*, 2005, **102**, 13343.
199. F. W. Twort, An investigation on the nature of ultramicroscopic viruses, *Lancet*, 1915, **ii**, 1241.
200. Z. Hosseinidoust, T. G. M. Van de Ven and N. Tufenkji, Bacterial capture efficiency and antimicrobial activity of phage-functio-nalized model surfaces, *Langmuir*, 2011, **27**, 5472.
201. L.-M. C. Yang, P. Y. Tam, B. J. Murray, T. M. McIntire, C. M. Overstreet and G. A. Weiss *et al.*, Virus electrodes for universal biodetection, *Anal. Chem.*, 2006, **78**, 3265.
202. E. H. Hankin, L'action bactericide des eaux de la Jumma et du Gange sur le vibrion du cholera, *Ann. Inst. Pasteur*, 1896, **10**, 511.
203. F. D'Herelle, Sur un microbe invisible antagoniste des bacilles dysenteriques, *C. R. Acad. Sci.*, 1917, **165**, 373.
204. R. Stone, Stalin's forgotten cure, *Science*, 2002, **298**, 728.

205. A. Sulakvelidze, Z. Alavidze and J. G. J. Morris, Bacteriophage therapy, *Antimicrob. Agents Chemother.*, 2001, **45**, 649.
206. J. D. Chkhaidze and N. E. Imedashvili, The use of a novel biodegradable preparation capable of the sustained release of bacteriophages and ciprofloxacin, in the complex treatment of multidrug-resistant *Staphylococcus aureus*-infected local radiation injuries caused by exposure to Sr90, *Clin. Exp. Dermatol.*, 2005, **30**, 23.
207. L.-M. C. Yang, P. Y. Tam, B. J. Murray, T. M. McIntire, C. M. Overstreet and G. A. Weiss *et al.*, Virus Electrodes for Universal Biodetection, *Anal. Chem.*, 2006, **78**, 3265.
208. H. A. Pearson, G. S. Sahukhal, M. O. Elasri and M. W. Urban, Phage-Bacterium War on Polymeric Surfaces: Can Surface-Anchored Bacteriophages Eliminate Microbial Infections?, *Biomacromolecules*, 2013, **14**(5), 1257–1261.
209. M. W. Urban, *Advances in Molecular Design of Polymer Surfaces with Antimicrobial, Anticoagulant, and Antifouling Properties*, ed. Santambrogio L., Springer, 2015.

8 Stimuli-responsive Materials in Medical Therapy

8.1 Introduction

This is perhaps the most active and fastest growing field of materials for medical applications. This chapter highlights only the key and established developments related to the use of polymeric systems in drug therapy, but the dynamics of the ongoing and still not resolved activities is overwhelming. There are many materials that are used in medicine, ranging from implants to prosthetic devises in which biocompatibility combined with mechanical strength are essential properties, but drug delivery field is one of the emerging technologies that utilize many findings from materials synthesis that are imbedded into biological systems. Drug delivery refers to the approaches and designs of materials capable of transporting a pharmaceutical compound in the body needed specific therapeutic task. The primary concerns that need to be taken into account in designing materials for drug delivery systems are both quantity needed and duration of drug delivery. Most efforts in drug delivery developments focus on the targeted delivery during which the drug is released in the target area. In an effort to achieve this task over a specific time period in a controllable manner with minimal site effects it is necessary to either attach a drug *via* covalent bonding or encapsulate it. As shown in Figure 8.1, materials that are most often used in these applications include micelles, liposomes, nano- and micro-biodegradable particles, gels, films and capsules. Realizing that there are several pathways for drugs to reach a target, today's nanomedicine offers

Stimuli-Responsive Materials: From Molecules to Nature Mimicking Materials Design
By Marek W. Urban
© Marek W. Urban, 2016
Published by the Royal Society of Chemistry, www.rsc.org

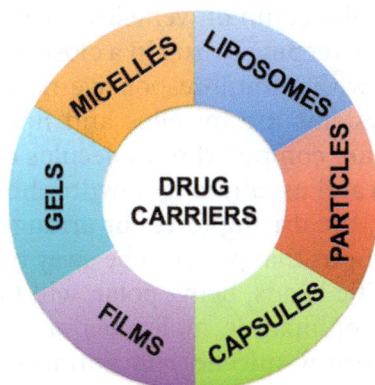

Figure 8.1 Most commonly utilized forms of drug carriers.

nanostructures that can travel though the bloodstream, nanostructures that accumulate in target tissue, enhancing imaging and delivery, such as quantum dots capable of fluorescence emission, and nanocompartments that deliver a drug cargo. As one can imagine, polymeric materials dominate the field and must have temperature and/or pH responsiveness discussed in Chapter 4. Therefore, this chapter will focus on micellar and liposome deliveries as well as selected examples of polymers that are not covered elsewhere, but have great potential.

8.2 Micelles and Liposomes

Due to numerous unacceptable site effects many medications have to be delivered to a target with minimal interactions. These are the side effects that limit our ability to generate optimal medications for many detrimental diseases like cancer, neurodegenerative and infectious diseases. Because drug delivery controls the rate at which a drug is delivered to s specific location in the body, it is logical to administer drugs locally rather than systematically. The latter will affect the whole body, thus leading to site effects. There are two general areas of drug delivery systems and polymers play a key role in each of them: site-specific drug delivery and implantable drug delivery systems. Potential release mechanisms involve: (i) desorption of surface-bound adsorbed drugs; (ii) diffusion through the carrier matrix; (iii) diffusion (nanocapsules) through the carrier wall; (iv) carrier matrix erosion; and (v) combined erosion/diffusion.

In an effort to deliver a drug into a designated area, a drug has to be loaded into a vehicle that will deliver the 'goods'. Aside from the delivery mode that mist common vehicles for drug delivery systems are vesicles capable of keeping a drug in a close cavity. As discussed in the chapter dealing with stimuli-responsive nanoparticles, the most common delivery systems are micelles and liposomes, depicted in Figure 8.2. Micelles are composed of molecules that have polar head groups (hydrophilic) and usually longer hydrophobic tails. Due to this amphipathic nature, the driving force for arrangements is prevailing hydrophobic interactions that prefer to be away from the water phase. In contrast hydrophilic heads, being polar, will face the water phase. As a result, these amphipathic molecules at certain concentration levels, called the critical micelle concentration (CMC), will form thermodynamically stable entities by arranging themselves into a spherical form. Because of their hydrophobic interior, micelles are responsible for many properties, including detergents and soaps. Their action is driven by a hydrophobic interior that will attract insoluble dirt (such as an oil phase) while the hydrophilic face will surround the non-polar dirt. So if instead of dirt, a drug molecule is placed inside a micelle, micelles become vehicles for drug delivery to

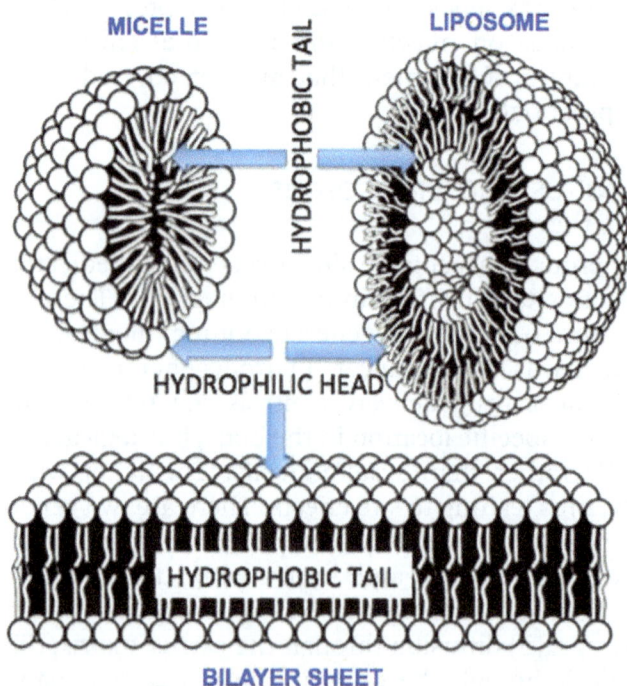

Figure 8.2 Formation of micelles, liposomes and bilayer sheets from amphipathic molecules that have polar head groups (hydrophilic) and hydrophobic tails.

the designated areas of a body. Micelles also facilitate the body absorption of lipids and fat-soluble vitamins. As such, they allow the small intestine to absorb essential lipids and vitamins from the liver and gall bladder. They also deliver lipids such as lecithin and lipid-soluble vitamins (A, D, E and K) to the small intestine. Skin cleaning is also attributed to micellar functions by removing oil and other substances.

Unlike micelles, liposomes contain a lipid bilayer, which is composed of two layers of phospholipids, arranged end to end with the hydrophobic layered buried between the two layers. This is depicted in Figure 8.3. A vesicle is an intracellular membrane capable of transporting and storing substances. Depending upon the shape and the chemical make-up, vesicles may store, transport and digest waste and products from the cell. They can fuse with the plasma membrane to release and absorb substances from the cell, and play an important role in metabolic functions, transport, enzyme storage, and other processes. Taking advantage of the ability to store, liposomes can serve as vehicles for drug delivery. One of the first developed site-specific drug delivery systems using liposomes was developed in the 1960s to 1970s.[1,2] Similarly to surfactants that form micelles, liposomes are formed from phospholipids. Figure 8.3 illustrates an example of a liposome, which may have one or more bilayers that are hydrophobic inside the bilayer, but the outer and inner surfaces are

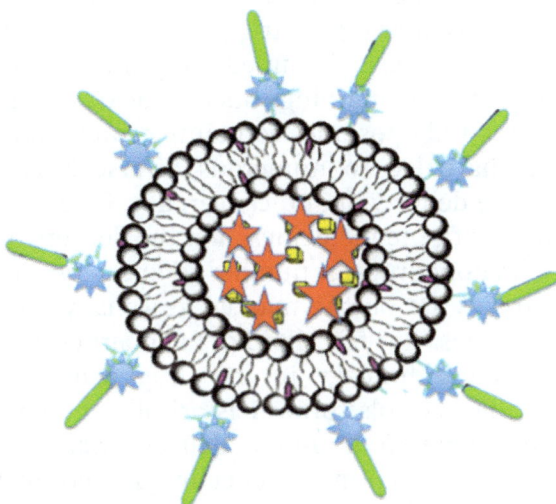

Figure 8.3 A liposome 'loaded' with drug and labeled with recognition markers.

hydrophilic. This polar character of the liposome core enables polar drugs to be encapsulated inside the bilayer cavity. Once encapsulated, a drug can be delivered to a target and is protected from degradative impacts of the environment. If amphiphilic and lipophilic molecules can enter the phospholipid bilayer without destroying it, that's even better, but is often not easily achievable. The stimuli responsiveness of the liposome becomes one of the most critical components of the targeted delivery. As shown in Figure 8.3, modified liposomes can incorporate proteins without loss of their activity within the hydrophobic domain of vesicle membranes, acting as a size-selective filter or storing desired drugs. Drugs that are encapsulated in a nanocage functionalized with channel proteins are effectively protected from premature degradation by proteolytic enzymes. The drug molecule, however, is able to diffuse through the channel, driven by the concentration difference between the interior and the exterior of the nanocage, but with proper stimuli responsiveness release is achieved. The surface can be functionalized by human-derived tumor-targeting antibody that will, upon selective attachment, destroy the liposomal cavity, thus delivering known cancer-killing cytotoxin. Although these approaches sound simple, initially there were a number of problems associated with the *in vivo* use of so-called conventional liposomes. The difficulty of retaining entrapped drugs in the interior, the rapid clearance of liposomes from circulation, how to deliver across cell membranes, or tightening the double layer to reduce the leakage, became significant problems.[3] In essence, decorating liposomes with responsive ligands facilitating the triggered release is the key feature and numerous efforts have included long-circulating (PEGylated) liposomes, liposomes containing nucleic acid polymers, ligand-targeted liposomes and liposomes containing combinations of drugs. These advances have led to numerous diverse biomedical applications such as the delivery of anti-cancer, anti-fungal and antibiotic drugs, the delivery of gene medicines, and the delivery of anesthetics and anti-inflammatory drugs. A number of liposomes in the form of lipidic nanoparticles are on the market, and many more are at the developmental stages. In fact lipidic nanoparticles are the first application of nanotechnology in the drug delivery system that became an established platform. Historically, perhaps the most critical stepping stone in liposome delivery system began when therapeutic outcomes resulting in effective delivery of DNA.[4,5] The delivery of antisense oligodeoxynucleotides (asODN), small interfering or silencing (siRNA), double strength (dsRNA) and microRNA ribonucleic acids using liposomes is critical in many medical applications.[6,7] In general, the principles that govern liposome design for nucleic acid deliveries are somewhat contradictory, but in reality their properties have to be balanced. For example, positive, cationic charge is needed to achieve efficient association with nucleic acids,[8] but to increase circulation half-life of liposomal nucleic acids the surface charge should be nearly neutral. In this context, two approaches have been used: coated cationic liposomes[9] and the use of ionizable lipids.[10] Finally, another critical aspect is efficient endosomal release, followed by internalization,[11] which can be accomplished by ionizable cationic lipids which exhibit bilayer destabilizing capacities achieved by the control of pK_a.[12,13] It is worth mentioning that several mechanisms of stabilization of micelles have been developed which exhibit various types of responses.[14,15]

8.3   Drug Delivery Responsive Polymers

As depicted in Figure 8.1, there are various forms of polymer delivery systems. In general, the mechanism of delivery is pretty straightforward, but to achieve truly drug deliverable materials is not trivial. The general concept is either to attach a drug molecule to a polymer backbone or trap a drug inside a polymer network. In both cases, the presence of stimuli will lead to a drug release. Among various polymer systems, the primary polymers that are used in drug delivery systems are temperature- and pH-sensitive. While Chapter 5 extensively discussed thermally and pH-responsive polymers, in this chapter we focus on volume changes resulting in the release of drug cargo. This process is typically driven by the phase transition and relies of several interactions: van der Waals interactions, hydrophobic interactions, hydrogen bonding, and attractive ionic interactions. Van der Waals interactions will cause a phase transition in hydrophilic gels in mixed solvents, such as an acrylamide gel in an acetone–water mixture. In this case, non-polar solvents can be used to decrease the dielectric constants of a solvent. In contrast, hydrophobic polymers, such poly(*N*-isopropylacrylamide) (NIPAM), will undergo from swollen low temperature state to collapse high temperature state, while supramolecular interactions, such as hydrogen bonding, will exhibit phase transitions induced by the presence of ionic species. For example, in poly(acrylic acid) and poly(acrylamide) gels repulsive ionic interactions will determine the temperature of transitions and the resulting volume changes. Finally, attractive ionic interactions will be responsible for pH-induced volume changes. Typical temperature and/or pH-responsive polymer classes used in drug delivery are those that exhibit a lower critical solution temperature (LCST). They include poly(*N*-alkylacrylamide)s, poly(methyl vinyl ether) (PMVE), poly(*N*-vinyl caprolactam) (PVCa), poly(*N*-ethyl oxazoline) (PEtOx), elastin-like oligo- and polypeptides, poly(acrylic acid-*co*-acrylamide), polycations, polyanions and amphoteric polymers for endosomolytic delivery, and polymer–protein conjugates.

Polyanions and amphoteric polymers for endosomolytic delivery can also be used for drug delivery. Form a chemical point of view, acidification of a polyanion resulting from increase protonation of carboxylate will result in higher hydrophobicity of the polymer backbone. At some degree of the deprotonated polymer will undergo a coil-to-globule transition, thus leading to interactions with a membrane. The pH values at which this will occur will depend on the hydrophilic/hydrophobic balance. For example, poly(acrylic acid) can be protonated at $\sim$pH 3, but hydrophobic groups along the polymer backbone may shift to higher pH values. This simple protonation–deprotonation concept opened up numerous opportunities for endosomolytic delivery. When the coil-to-globule transition occurs at endosomal $\sim$pH 6, the drug load can escape, thus ending its action. Several endosomolytic copolymers are potential candidates, including poly(2-ethylacrylic acid) (PEAA), methacrylic acid (MAA), or octadecylacrylate (ODA). Responsive polymer–protein conjugates have been designed using three general strategies. Typically, a polymer is conjugated to a protein to generate the accessibility of the active site of an enzyme or a receptor binding site. One approach is to attenuate the accessibility, thus enhancing the protein's activity. This can be accomplished by the environmental conditions that alter an enzyme's activity. The use of thermo- and photoresponsive polymer–protein conjugates[16,17] can be accomplished by the synthesis of a photoresponsive pNIPAM-endoglucanase conjugate in site-specific conjugation that becomes inactive for glycoside hydrolysis under irradiation at 350 nm.[17] Another approach involves masking potentially toxic proteins or avoiding an adverse immune response before the protein reaches its target tissue. Upon reaching the target tissue, a trigger is used to irreversibly remove the masking polymer and release the unmasked protein. One example is the employment of polymer–phospholipase conjugates, in which the enzyme activity can be tuned by the polymer–enzyme conjugation.[18] The third approach is to create a polymer–protein conjugate that is insoluble during the transition phase. This approach was successfully demonstrated by retaining 95% activity over many cycles for the conjugation of pNIPAM to trypsin.[19]

Stimuli-responsive polymer–protein conjugates can be also used in drug delivery systems. Three strategies are typically used in this process. The first strategy involved conjugation of a polymer to a protein to impact the accessibility by the active site, for example receptor or enzyme. The idea is to attenuate the accessibility and thus protein activity by environmental changes of pH and temperature. One example of this system is photoresponsive PNIPAM–endoglucanase conjugate. Upon 350 nm irradiation endoglucanase becomes inactive for glycoside hydrolysis.[16,17] A second concept is to mask potentially toxic protein or avoid adverse immune response before the protein reaches its target tissue. Upon reaching the target, it is necessary to trigger the removal of the masking polymer and release the protein. Under these circumstances it is desirable to retain the full activity at the site, which may not be often achievable. For that purpose, polymer–phospholipase conjugates, with tunable enzyme activity were developed.[20,21] Also, conjugated various poly(amido amine)s to melittin were developed which are capable of masking hemolytic activity of melittin.[18] The third concept is to use the polymer–protein conjugate insolubility when the polymer goes through the phase transition, thus facilitating the ability of purification by centrifugation. As demonstrated by conjugation of PNIPAM to trypsin, this concept is quite useful in biotechnology where immobilization of the protein will result in the loss of activity. The field of polymers as therapeutics offers many challenges and opportunities, as described in several review articles.[22,23] This very rapidly growing field will have a significant impact on the treatment of various diseases, the delivery of new drugs in general and anti-cancer drugs in particular. Novel molecular targets and their combinations as well as the development of new stimuli-responsive polymeric materials with defined and responsive architectures are of particular interest.

PNIPAM copolymers have been mainly used in the oral delivery of calcitonin and insulin. In this case, the peptide or hormone is immobilized inside polymeric beads, which retain their stability passing through the stomach. Upon entering the alkaline environment of intestine the beads disintegrate and the drug is released. Poly(methyl vinyl ether) exhibits a transition temperature at 37 °C, which matches physiological temperature and resulted in many biomedical applications. Poly(N-vinyl caprolactam) has not been as intensely utilized, but owing to a transition temperature of 33 °C and water solubility it has many biomedical applications. Poly(N-ethyl oxazoline) has a transition temperature of 62 °C, which is too high for direct use in drug delivery applications, but when copolymerized the temperature may be lowered. The main disadvantage is a lack of tolerance against unprotected functional groups. Polypeptides can also exhibit LCST behavior, when all hydrophilic and hydrophobic components are balanced. The objective is to tune the LCST transition so it is slightly higher than the temperature of the tumor, thus causing the phase transition to occur. As a result, the conjugate becomes insoluble and is being released at the targeted tumor. The formation of inter-penetrating networks consisting of poly(acrylic acid) and poly-acrylamide results in the upper critical solution temperature (UCST) transition at 25 °C. Hydrogen bonding between AAc and AAm units is believed to be responsible for this behavior. Other monomers, in-cluding maleic anhydride (MA) and *N,N*-dimethylaminoethyl metha-crylate (DMAEMA) can also be used. The pH-responsive swelling and collapsing responses are often used to generate the controlled release of caffeine,[24] drugs or cationic proteins (for example lysosome).[25]

Stimuli-responsive hydrogels, especially temperature-sensitive hydrogels that employed PNIPAM as the thermo-responsive units, have been used in many applications;[24,26,27] one example is the cell culture carrier with or without the option of immobilizing bioactive molecules and subsequently releasing them. This approach may be applied in the transplantation of retinal pigment epithelial cell sheets facilitating defect-free recoveries.[28] The immobilization of hydro-lytically sensitive molecules, such as peptides and proteins, can be achieved in PLGA,[29] where pH-sensitive anionic hydrogels can be used for entrapping Ca ions that are released upon thermal stimulus.[30]

In general, the pH variations can be used to trigger the drug release using the following guidelines: (1) In the extracellular tumor tissue the pH may vary from 6.5 to 7.2, which is a notch lower from the normal levels of 7.4.[23] Thus, sensitivity of volume changes polymer should be on a slightly acidic site. (2) After cellular uptake, when the drug conjugates reach the lysosome region with a pH in the 4.5 to 5.0 range, drug release should occur. The aim is to design the volume transitions tuned to this pH range. This is important, especially that pH values in various tissues and cellular compartments vary. Table 8.1 summarizes pH variations in different parts of the body, thus designing polymers for drug delivery systems stimuli-responsive behavior should be adjusted accordingly.[31] Certain cancers as well as inflamed or wound tissues exhibit a pH different from 7.4, as it is in circulation. For example, chronic wounds have been reported to have pH values between 7.4 and 5.4,[32] thus there are significant variations, making the choice in designing delivery systems somewhat ambiguous. It should be also realized that the design of

Table 8.1 pH variations in tissue and cellular components.[33b,34]

Tissue/cellular compartment	pH
Blood	7.35–7.45
Stomach	1.0–3.0
Duodenum	4.8–8.2
Colon	7.0–7.5
Early endosome	6.0–6.5
Late endosome	5.0–6.0
Lysosome	4.5–5.0
Golgi	6.4
Tumor, extracellular	7.2–6.5

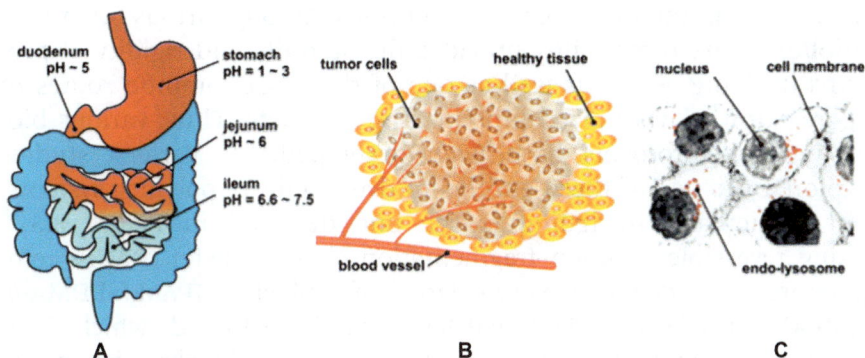

Figure 8.4 Targeting acid-responsive therapies at the organ (A), tissue (B), and cellular levels (C).
Reproduced with permission from Ref. 33, Copyright 2010, American Chemical Society.

stimuli-responsive polymers to acid environments can be facilitated at different levels: (a) organ-targeting levels, (b) tissue-targeting levels, and (c) cellular levels. Figure 8.4 illustrates that at the organ levels, the gastrointestinal track exhibits pH gradients (A), solid tumors have acidic extracellular environments that are different from healthy tissues (B), and endolysosomes are more acidic compared to the cytoplasm (C).[33a]

There are two approaches to utilizing pH changes to control and trigger release of a drug load. In the extracellular tissue, the tumor tissue has extracellular pH_e of 6.5–7.2, thus slightly lower than the normal pH of 7.4.[35] Another approach is after cellular uptake when the drug conjugate reaches the lysosomes in a pH of 4.5–5.0 range. Often hydrolytic enzymes (for example, cathepsin B) can also be used to release a drug load.[36] In both approaches the nanometer size of the drug conjugates or micellar structures facilitate passive targeting.[37]

Redox/thiol sensitive polymers are another alternative used in drug delivery.[24,38–43] As shown below, the reversibility of the reaction is facilitated by the exposure to reducing agents resulting in disulfide exchange in the presence of other thiols, thus making polymers containing disulfide linkages both redox- and thiol-responsive.[44,45]

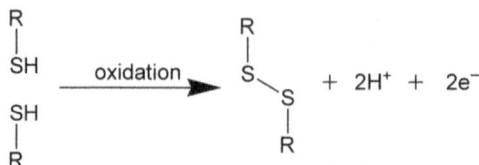

$$
\begin{array}{c}
R \\
| \\
SH \\
\\
SH \\
| \\
R
\end{array}
\quad \xrightarrow{\text{oxidation}} \quad
\begin{array}{c}
R \\
| \\
S{\diagdown}_{\,S} \\
\quad | \\
\quad R
\end{array}
\quad + \; 2H^+ \; + \; 2e^-
$$

Well-documented reversibility of a thiol and disulfide reaction is involved in numerous biological processes. Among various functions, thiol/disulfide reversibility provides the stability and rigidity of proteins in living cells[46] as well as one of the most common routes of aggregation of lyophilized protein drug products. Thus, various bioconjugation protocols have been developed.[44,47] Various studies explored the use of these entities in drug delivery polymer systems, but particularly intriguing is the one depicted in Figure 8.5.[48] Using reversible addition–fragmentation chain transfer (RAFT), temperature and redox-responsive gelation of triblock PNIPAM-*b*-PDMA-*b*-PNIPAM or PDEGA-*b*-PDMA-*b*-PDEGA can be achieved, which form hydrogels upon heating above the LCST. These triblocks exist as molecularly-dissolved unimers in water below the LCST of the respective responsive blocks (Figure 8.5A), but upon heating, as anticipated in the earlier studies,[49] turbid solutions containing micellar aggregates were formed. However, upon heating solutions to 40 °C, freestanding gels will form (Figure 8.5B). Because the products of the aminolysis reaction are thiol-terminated diblock copolymers, they are susceptible to disulfide formation under oxidizing conditions. Therefore, coupling of the thiol-terminated chains leads to reconstruction of the ABA triblock architecture and a transition back to the gel state (Figure 8.5C). Although exposure to air may be sufficient for thiol oxidation to disulfides, the addition of oxidizing agents or catalysts will result in fairly quick gel restoration (Figure 8.5D).

Taking advantage of the fact that a typical intracellular concentration of glutathione (GSH), the most abundant reducing agent in many cells, is about 10 mM compared to ~0.002 mM in the cellular exterior,[50] a delivery system using thiol- and redox-responses can be developed. To accomplish this function, polymeric micelles with

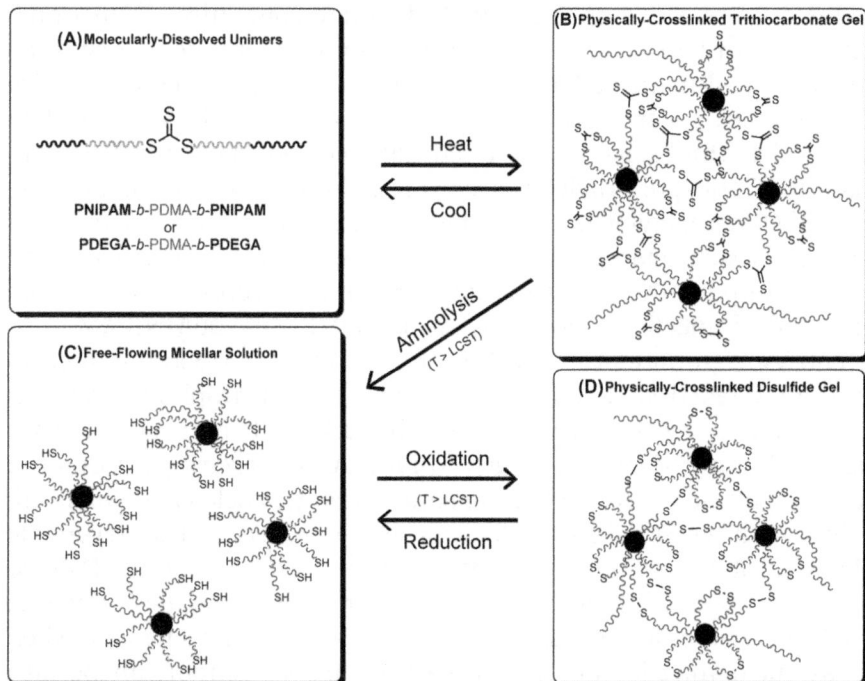

Figure 8.5 Temperature and redox-responsive gelation of triblock co-polymers prepared by RAFT. (A) Molecularly-dissolved unimers of PNIPAM-*b*-PDMA-*b*-PNIPAM or PDEGA-*b*-PDMA-*b*-PDEGA. (B) Hydrogels formed upon heating above the LCST of the responsive PNIPAM or PDEGA blocks. (C) Free-flowing micellar solutions of PNIPAM-*b*-PDMA-SH or PDEGA-*b*-PDMA-SH resulting from trithiocarbonate aminolysis at *T*>LCST. (D) Hydrogels formed from PNIPAM-*b*-PDMA-S-S-PDMA-*b*-PNI-PAM or PDEGA-*b*-PDMA-S-S-PDMA-*b*-PDEGA upon oxidation of the thiol-terminated diblock aminolysis products.
Reproduced from Ref. 48 with permission from The Royal Society of Chemistry.

shells crosslinked *via* thiol-reducible disulfide bonds can be synthesized to serve as biocompatible nanocarriers to release anticancer drugs under the reducing environments of cancer tissues.[51] Using this approach redox-sensitive disulfide groups can be directly attached into side chains or backbones with an appropriate monomer, initiator, or chain transfer agents.[42,52-55]

One approach is to develop a drug carrier by copolymerization of a pyridyl disulfide containing acryloyl monomer with methacrylic acid and butyl acrylate.[42] Because these polymeric systems are thiol- and pH-sensitive, they exhibit membrane-disruptive behavior useful for gene delivery. Using atom transfer radical polymerization (ATRP), redox/thiol-sensitive polymers with disulfide-containing difunctional initiator can be synthesized[56] and employed to produce a disulfide-functional dimethacrylate monomer to generate redox-sensitive nanogels *via* inverse miniemulsion.[45,57] RAFT can be also used to synthesize thiol-sensitive core crosslinked polymer micelles consisting of poly(polyethylene glycol methyl ether methacrylate)-*block*-poly(5'-O-methacryloyluridine) and a dimethacrylate crosslinker.[40] This approach enabled a somewhat encouraging drug release efficiency of up to 70% in several hours. Similarly, redox-responsive core crosslinked micelles of poly(ethylene oxide) (PEO)-*b*-poly(NIPAM-*co*-N-acryloxysuccinimide) *via* RAFT can be produced.[58] Upon the micelle-cystamine reaction, disulfide bonds within the hydrophobic core can be cleaved by treatment with dithiothreitol (DTT) and reform again upon addition of cystamine as a thiol/disulfide exchange promoter. Also, disulfide-based cystamine-containing triblock copolymer micelles with crosslinked shells can be prepared.[59] The reversible cleavage of the micelles can be accomplished with either DTT or tris(2-carboxyethyl)phosphine (TCEP), and the degraded micelles could be re-crosslinked using cystamine as a thiol exchanger. Shell crosslinked micelles of poly(L-cysteine)-*b*-poly(L-lactide) can be also synthesized *via* ring-opening polymerization.[60] The unique feature of this approach is that when treated with DTT, the crosslinks within the micelle shells will vanish, but when DTT is removed by dialysis, the shell crosslinked micelles will reform. Disulfide reduction can also be employed to induce morphological transitions of block copolymer aggregates in solutions.[61] Polymeric micelles with interpolyelectrolyte complexed cores composed of positively and negatively charged chains contain a PEG corona linked by disulfide bonds. Upon reduction with DTT, the PEG segments can be detached from the micelles, forming homopolymer complexes capable of adopting a vesicular structure due to the small curvature, thus not shielded from PEG.

Supramolecular polymer–surfactant complexes can also form micelles sensitive to thiol-induced dissociation.[62] When a polymer decorated with pendant carboxylates attached *via* disulfides is complexed with a cationic surfactant and such supramolecular complexes are treated with GSH, cleavage of the polymer side chains will lead to dissociations of aggregates and release of a hydrophobic part. Thiol-responsive triblocks composed of PNIPAM or PHPMA and PMPC can be prepared using disulfide-based ATRP initiator. Because the integrity of the gels relies on disulfide linkages, controlled degradation induced under physiological conditions in the presence of GSH or DTT will occur.[38] Because dissociation of PHMA-PMC-PHMA triblock copolymer gels occurs under mild reducing conditions, it is believed that materials prepared from such polymers may have potential applications as wound dressings.[38] The ABA triblock copolymers shown in Figure 8.5 are capable of forming hydrogels that are temperature- and redox-responsive.[48] In this case, a difunctional trithiocarbonate RAFT agent was used to synthesize symmetrical triblock copolymers of PNIPAM-*b*-PDMA-*b*-PNIPAM and poly(di(ethylene glycol)ethyl ether acrylate (PDEGA)-*b*-PDMA-*b*-PDEGA. Above the LCST, the outer thermoresponsive block causes gelation. The presence of hydrophilic bridging PDMA segments containing labile trithiocarbonate linkages facilitates aminolysis will lead to a free-flowing solution of thiol-decorated polymeric micelles. Under oxidizing conditions the thiols on the termini of the micelle corona chains couple to form disulfide-containing hydrogels. Notably, aminolysis of RAFT-generated polymers may lead to telechelic thiol polymers potentially capable of the reversibly assemble into responsive cyclic, bicyclic, multiblock, and network architectures.[63,64] Using RAFT, thiol-sensitive biodegradable three-armed star polymers using both 'core-first' and 'arm-first' can be produced.[65] Along the same lines, a disulfide-containing multiblock copolymer of PEO-*b*-poly(-propylene oxide) (PPO-*b*-PEO) was synthesized,[41] in which case a drug release was significantly faster after treatment with GSH. Also, redox-responsive multiblock copolymers of PNIPAM and PDMAEMA can be prepared by disulfide reducing bridges of starting blocks.[24] Similarly, block copolymers with hydrophilic PEG tethered to hydrophobic poly(propylene sulfide) *via* a disulfide bridge are available.[39,66–68] Numerous applications, ranging from cytoplasmic delivery of peptides, proteins, oligonucleotides and DNA, can be envisioned, but the key prerequisite is a sudden vesicular burst within the early endosome.

Gene therapy offers numerous opportunities for treating diseases from genetic disorders, infections and cancer. The achievement of successful gene therapy relies on the development of successful and gene-specific delivery systems. Due to safety, repeatability, and an ease of administration, polymer-based non-viral gene carriers have been used. The use of cationic polymers in non-viral gene therapy is well documented.[69] It is based on the ability of the polycations to complex to nucleotides through electrostatic interaction. The responsiveness of the polymer is critical when the pH drops during cellular uptake as the polymer becomes charged and triggers osmotic or endosomolytic responses. For that purpose many amine-based polymers can be used. PEI has been one of the most commonly employed cationic gene carriers due to its effective transfection efficiency and consistency in transfection in many different types of cells. PEI consists of the primary (25%), secondary (50%) and tertiary amines (25%), of which amines with different pK_a values confer a buffering effect over a wide range of pH. The buffering property gives PEI an opportunity to escape from the endosome (proton sponge effect) [42]. Although tolerable levels of toxicity is a concern, poly(-ethylene imine (PEI) is one type, but there are other candidates. For example, dendrimers, such as polyamidoamines (PAMAM)[70-73] and their derivatives,[74,75] poly(N,N–dimethylaminoethyl methacrylate) (PDMAEMA),[76] poly(L-lysine) (PLL),[77] poly(β-amino esters)[78] and modified chitosan.[79] Selected examples of polymers with a potential of non-viral gene delivery are shown in Figure 8.6.

One example are the chemical structures shown below which are due to show chemical structures of polyethylenimine (PEI)-based gene carriers, water-soluble lipopolymer (WSLP); and a peptide containing Agr–Gly–Asp amino acid sequence (RGD).

Due to diminished cytotoxicity biodegradable gene delivery polymers are quite attractive. Figure 8.7 illustrates several examples of non-viral gene delivery polymers.

Polyelectrolyte complex micelle-based oligonucleotide delivery system offers another alternative. Due to high specificity, antisense oligonucleotides (antisense ODNs) and small interfering RNAs (isRNAs) are able to modulate the gene expressions.[80-82] To overcome antisense ODN and siRNA enzymatic hydrolysis and limited cellular uptake, also polyelectrolyte complex (PEC) micelles were developed.[83-85] This is depicted in Figure 8.8.[86] Due to the interactions between PEG-conjugated oligo nucleic acid (PEGylated oligonucleotide) and polycation (PEI) a colloidal size self-assembly PEC micelles will form. In addition to PEI other polycations can be also used, including fusogenic KALA peptide and protamine. The formation of PEC micelles involves interactions of the negatively charged ODN segment with polycations to form a charge-neutralized hydrophobic inner core. At the same time, the PEG segment is localized outside the core to form a hydrophilic corona.

It would not be fair not to highlight stimuli-responsive polymers in drug delivery without considering many existing technologies on which current medicine relies. These are conventional drug delivery systems, which involve many conventional applications. Over the last five decades, many drug deliveries have been utilized. They include such techniques as compression, spray and dip coating, and encapsulation, which have been used in the pharmaceutical industry

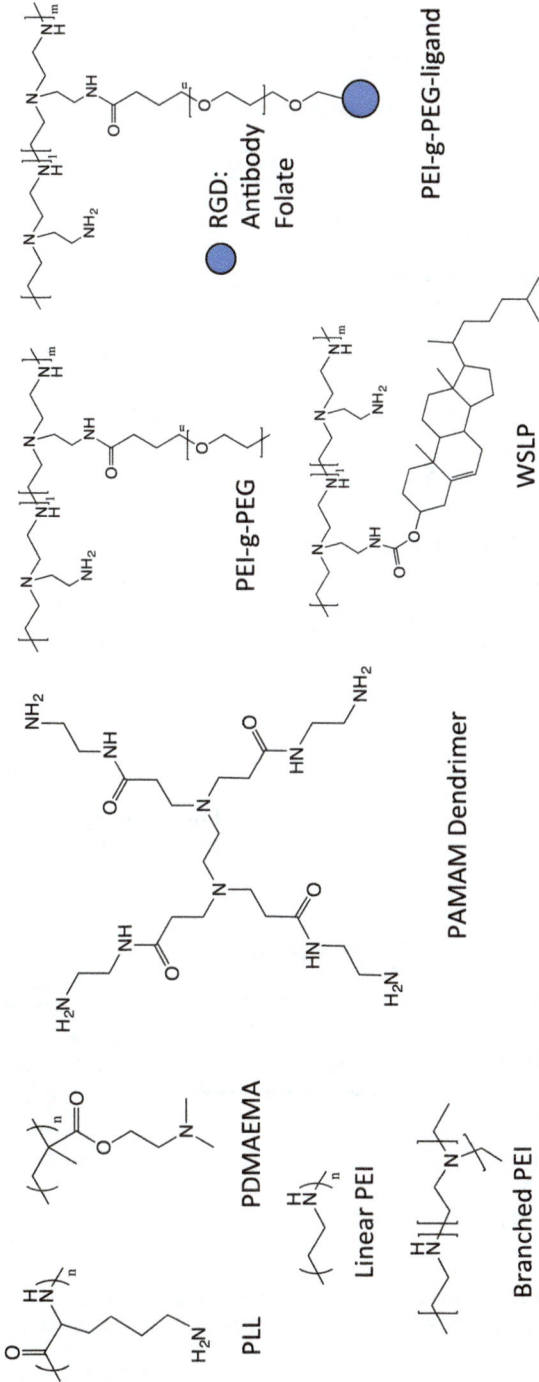

Figure 8.6 Selected examples of non-viral gene delivery polymers.

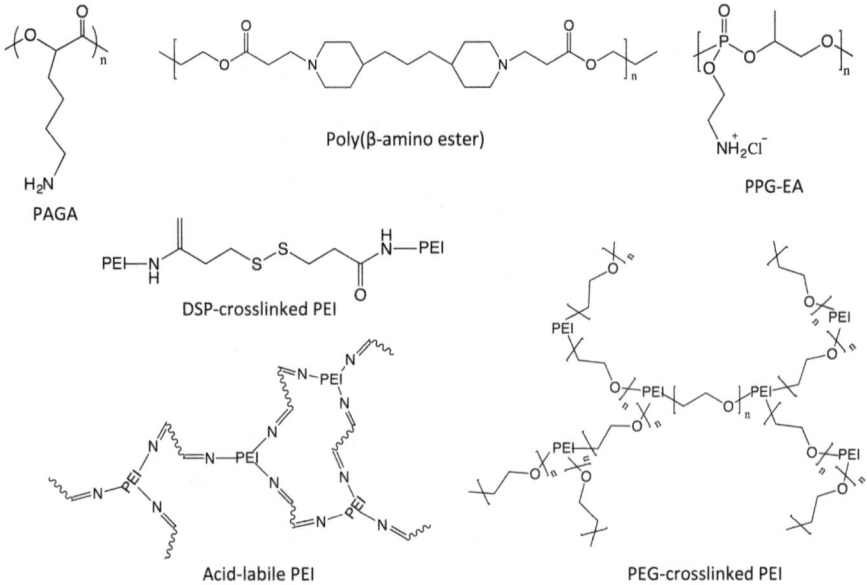

Poly(β-amino ester)

PAGA

PPG-EA

DSP-crosslinked PEI

Acid-labile PEI

PEG-crosslinked PEI

Figure 8.7 Gene carriers based on degradable cationic polymers: poly(α-[4-aminobutyl]-L-glycolic acid) (PAGA); poly(2-aminoethyl propylene phosphate) (PPE-EA); dithiobis(succinimidylpropionate) (DSP).

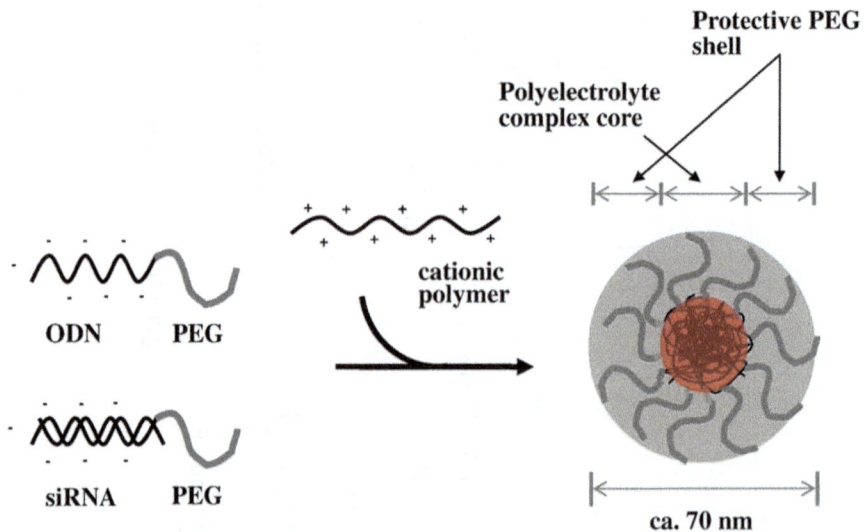

Protective PEG shell

Polyelectrolyte complex core

ODN   PEG

cationic polymer

siRNA   PEG

ca. 70 nm

Figure 8.8 A schematic diagram of the formation of polyelectrolyte complex micelles. ODN, oligodeoxyribonucletide; siRNA, small interfering RNA.
Reprinted from ref. 86. Copyright (2010) with permission from Elsevier.

to incorporate bioactive agents with polymers. The following polymers have been and will continue to be used: cellulose derivatives, poly(ethylene glycol) PEG, and poly(*N*-vinyl pyrrolidone).[87] In terms of the drug delivery prospective, processes that typically control drug deliveries are diffusion-controlled (monolithic devices), solvent-activated (swelling- or osmotically-controlled devices, chemically controlled (biodegradable), or externally-triggered systems (*e.g.* pH, temperature).[88,89] Today, the majority of polymer conjugates are designed as anticancer therapeutics, although other diseases have also been targeted, including rheumatoid arthritis, diabetes, hepatitis B and C, and ischemia.

The high degree of branching, multivalency, globular architecture and well-defined molecular weight of dendrimers make these materials promising as new scaffolds for drug delivery.[90] Over recent years, new advances in the design and synthesis of biocompatible dendrimers have led to many biomedical applications, in particular drug delivery, immunology and the development of vaccines, antimicrobials, antivirals, and more recently in cancer treatment.[91] Their use as delivery of anticancer drugs, including cisplatin and doxorubicin as well as boron neutron capture therapy and photodynamic therapy have been explored. Figure 8.9(a) illustrates commercially available polyamidoamine (PAMAM) dendrimer which was prepared by the divergent growth approach[92] and is one of the commonly used dendrimers for making biological scaffolds.[70] As pointed out,[93] to avoid the toxicity and liver accumulation associated with dendrimers polycationic surfaces the common practice is to modify the surface amine groups with neutral or anionic moieties. Polyaryl ether dendrimers shown in Figure 8.9(b) exhibit limited water solubility,[94] but placing peripheral solubilizing groups will overcome this obstacle.[95] There are also peptide-based dendrimers, such as shown in Figure 8.9(c) which are based on polylysine. It turns out that they are highly suitable for vaccine, antiviral and antibacterial applications. Again, their peripheries can be modified for specific applications.[96,97] Due to multivalent binding capacity these materials exhibit promising immunological applications.[98] Polyester dendrimers based on the monomer 2,2-bis(hydroxymethyl) propionic acid shown in Figure 8.9(d) are good candidates for anticancer drug carriers.[99] Combining these dendrimers with PEO as well as other such monomers as glycerol, succinic acid, phenylalanine and lactic acid shown in Figure 8.9(e) are suitable applications in tissue engineering and other biological applications.[100] Finally, dendritic polymers with incorporated glycerol monomers depicted in Figure 8.9(f) can be also used for biological applications.[101–105]

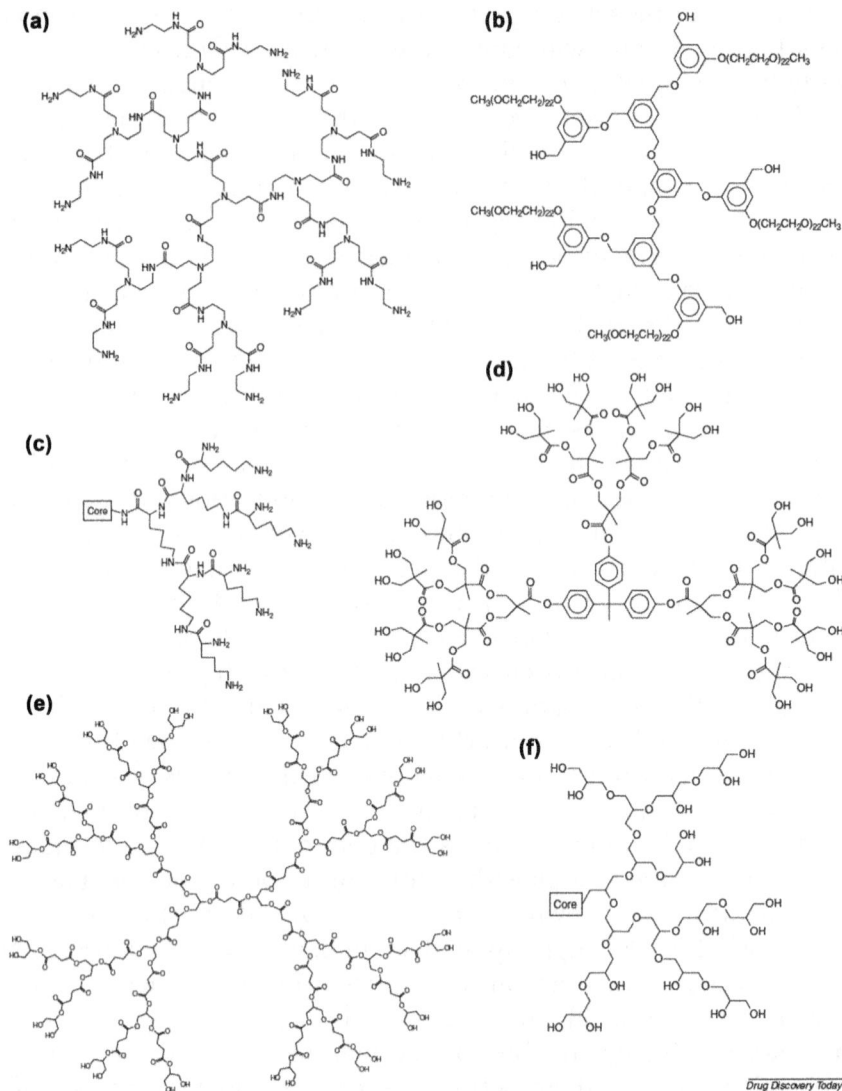

Figure 8.9   Selected structures of biocompatible dendrimers tested for drug delivery applications. (a) PAMAM dendrimer. (b) Polyaryl ether dendrimer. (c) Polylysine dendron. (d) Polyester dendrimer based on 2,2-bis(hydroxymethyl)propionic acid. (e) Polyester dendrimer based on glycerol and succinic acid. (f) Dendritic polyglycerol. Reprinted from ref. 106. Copyright (1989) with permission from Elsevier.

Because dendrimers are highly branched, globular macromolecules with many arms extending and radiating from a central core, they are perfect candidates for drug delivery systems.[92] This is their step-wise synthesis that results in very regular branching patterns with low polydispersity index (PDI) and precise molecular weight that make these polymers so unique. As shown in Figure 8.9,[106] selected structures of dendrimers have been prepared and offer not only bio-compatibility, but also the ability of non-covalent encapsulation as well as the formation of covalent dendrimer–drug conjugates.

As pointed out above, the use of conventional micelles provides an opportunity to encapsulate drugs inside the micellar moiety. Due to relatively weak hydrophobic interactions, the stability of traditional micelles can be a problem in harsh environments. Due to branched structural features, dendritic unimolecular micelles can be readily formed and due to covalent bonding their micellar structure is maintained at all concentration levels. However, the use of dendritic micelles can be limited due to the difficulty of releasing the cargo (drug) from the dendrimer core. On the contrary, the encapsulated drug may not be retained within the cavity of the dendritic core. This is where stimuli-responsive dendritic copolymers came to the rescue. For example, linear–dendritic block copolymers consisting of poly(ethylene oxide) and either a polylysine or polyester dendron can be prepared and hydrophobic groups were attached to the dendrimer periphery by highly acid-sensitive cyclic acetals.[107] The key feature of these materials is that are able to form stable micelles in aqueous environments at around pH 7. However, under mild acidic conditions they disintegrate into unimers, with the loss of hydrophobic groups upon acetal hydrolysis. This is schematically depicted in Figure 8.10. Another approach is to design stimuli-responsive dendritic

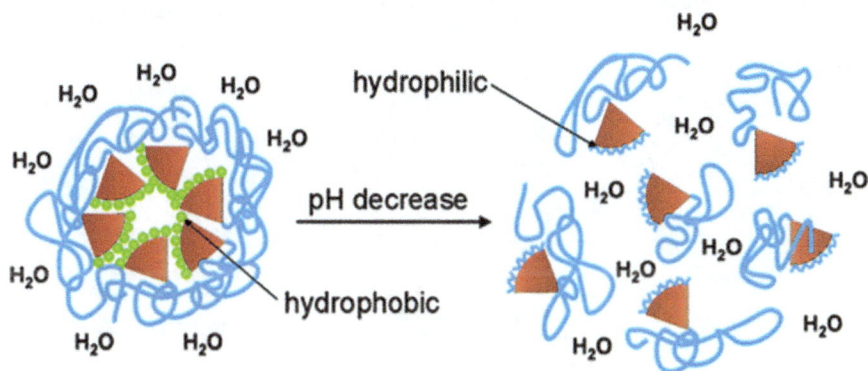

Figure 8.10 Schematic diagram illustrating a drug release from a pH-sensitive dendritic micelle.
Reproduced with permission from ref. 107. Copyright (2004), American Chemical Society.

nanocarriers sensitive to specific stimuli being a component of disease pathology, such as enzyme, protein overexpression, pH, electrolyte levels, and others. This allows the nanocarrier to respond specifically to the pathological triggers and substantially reduce potential side-effects.[108,109] While there are many drug delivery systems that have relied on this approach, the general concept of triggered release is depicted in Figure 8.11.[110] There are two major categories resulting from the nature of interactions between bioactive molecules and dendritic polymer: the release shown in Figure 8.11(a) can be initiated by structural changes with the polymeric scaffold due to cleavage of shell, backbone degradation, conformational changes, or electric charges. In contrast, the mechanism of release shown in Figure 8.11(b) will require the splitting of the molecular linker between a polymer and chemically attached bioactive agent. The simplest approach for pH triggered release of encapsulated guest molecules requires the protonation of the core or the surface of a dendritic structure and the use of amine or carboxylic groups.[111–114] Because there is a

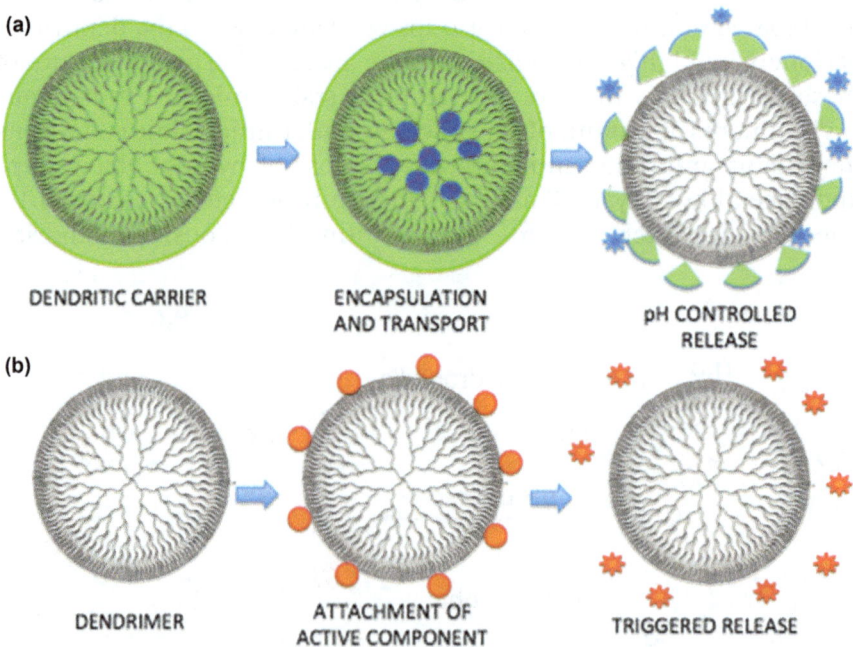

(a)

DENDRITIC CARRIER · ENCAPSULATION AND TRANSPORT · pH CONTROLLED RELEASE

(b)

DENDRIMER · ATTACHMENT OF ACTIVE COMPONENT · TRIGGERED RELEASE

Figure 8.11 Two mechanisms of stimuli-responsive release of bioactive molecules using dendritic nanocarrier: (a) from non-covalent complexes, (b) from covalently conjugated architectures. Adapted from ref. 110. Copyright (2010) with permission from Elsevier.

100–1000 times redox potential difference between the reducing and the oxidizing extracellular space, this creates an opportunity for potential stimuli that will trigger the release of therapeutics agents. There are dendritic delivery systems which contain disulfide linkages which may undergo disulfide cleavage in lysosomal compartments; one example is the glutathione pathway which controls the intracellular redox potential.[115] Using this approach, a drug or gene fragment can be encapsulated or conjugated to dendritic carriers carrying disulfide bonds. Upon reductions of disulfide bonds in the presence of excess of glutathione inside the cell, the drug or gene is released. Enzyme-triggered drug release is highly attractive because involves such functional groups as esters and short polypeptide sequences.[116] These entities are landing substrates for enzymes over-expressed in malignant tissues.[117,118] Also, gene carriers can be prepared in which oligoamine moieties, such as spermidine, spermine and pentaethylenehexamine, are conjugated to hyperbranched polyglycerol using a carbamate linkages.[110] Photo-responsive dendritic architectures rely on the adjustable release of encapsulated or conjugated bioactive components upon irradiation. For example, photolabile *o*-nitrobenzyl,[119] photo-cleavable 2-nitrobenzyl ester and photo-isomerizable azobenzene[120] can be incorporated into the amide dendron in order to produce supramolecular structures with photo-switchable functions. Finally, the main advantage of dendrimers is their ability to reversibly respond to temperature changes within relatively narrow range.[121,122] Hyperbranched polyamidoamine (HPAMAM) polymer with terminal vinyl groups shows useful thermo-responsive properties when end-capped by 1-adamantylamine (ADA). Adjusting the end-capping ratio of ADA can control the lower critical solution temperature.[123]

It should be kept in mind that, although in contrast to passive drug delivery systems stimuli-responsive polymers offer numerous advantages by interacting and responding to the environmental conditions, the choice of stimuli and responsiveness is critical. The ability of tuning into specific targeted areas by adjusting pH or temperature responsiveness offers many opportunities for targeted delivery tailored to specific applications, but in many environments may lack precision. For example, targeting certain types of tissues, such as extracellular cancer, have a relatively narrow margin of parameters to work with. These limitations may cause significant mismatch between the intended and actual drug deliveries. As pointed out in Table 8.1, the accuracy of the pH values, typically measured using a pH electrode is critical. Are there any other tissue parameters that may not only contribute to pH variations? It seems that the energy of water binding in healthy and unhealthy tissues may offer another possibility of triggering a drug release. Incidentally, there is an initial evidence that OH bonds of water[124] may serve this differentiation. Consequently, this may require different sensing approaches and sensing groups in water-soluble stimuli-responsive copolymers.

References

1. G. Gregoriadis, Drug entrapment in liposomes, *FEBS Lett.*, 1973, **36**(3), 292–296.
2. A. Bangham, M. M. Standish and J. Watkins, Diffusion of univalent ions across the lamellae of swollen phospholipids, *J. Mol. Biol.*, 1965, **13**(1), 238–IN27.
3. T. M. Allen and P. R. Cullis, Liposomal drug delivery systems: from concept to clinical applications, *Adv. Drug Delivery Rev.*, 2013, **65**(1), 36–48.
4. R. Fraley, S. Subramani, P. Berg and D. Papahadjopoulos, Introduction of liposome-encapsulated SV40 DNA into cells, *J. Biol. Chem.*, 1980, **255**(21), 10431–10435.
5. R. M. Straubinger and D. Papahadjopoulos, [32] Liposomes as carriers for intracellular delivery of nucleic acids, *Methods Enzymol.*, 1983, **101**, 512–527.
6. A. Kawabata, A. Baoum, N. Ohta, S. Jacquez, G.-M. Seo and C. Berkland *et al.*, Intratracheal administration of a nano-particle-based therapy with the angiotensin II type 2 receptor gene attenuates lung cancer growth, *Cancer Res.*, 2012, **72**(8), 2057–2067.
7. J. Nguyen and F. C. Szoka, Nucleic acid delivery: the missing pieces of the puzzle?, *Acc. Chem. Res.*, 2012, **45**(7), 1153–1162.
8. P. L. Felgner, T. R. Gadek, M. Holm, R. Roman, H. W. Chan and M. Wenz *et al.*, Lipofection: a highly efficient, lipid-mediated DNA-transfection procedure, *Proc. Natl. Acad. Sci.*, 1987, **84**(21), 7413–7417.
9. D. Stuart and T. Allen, A new liposomal formulation for anti-sense oligodeoxynucleotides with small size, high incorporation efficiency and good stability, *Biochim. Biophys. Acta Biomembr.*, 2000, **1463**(2), 219–229.
10. S. C. Semple, S. K. Klimuk, T. O. Harasym, N. Dos Santos, S. M. Ansell and K. F. Wong *et al.*, Efficient encapsulation of antisense oligonucleotides in lipid vesicles using ionizable

aminolipids: formation of novel small multilamellar vesicle structures, *Biochim.Biophys. Acta Biomembr.*, 2001, **1510**(1), 152–166.

11. G. Basha, T. I. Novobrantseva, N. Rosin, Y. Y. C. Tam, I. M. Hafez and M. K. Wong *et al.*, Influence of cationic lipid composition on gene silencing properties of lipid nanoparticle formulations of siRNA in antigen-presenting cells, *Mol. Ther.*, 2011, 1–15.

12. S. C. Semple, A. Akinc, J. Chen, A. P. Sandhu, B. L. Mui and C. K. Cho *et al.*, Rational design of cationic lipids for siRNA delivery, *Nat. Biotechnol.*, 2010, **28**(2), 172–176.

13. M. Jayaraman, S. M. Ansell, B. L. Mui, Y. K. Tam, J. Chen and X. Du *et al.*, Maximizing the potency of siRNA lipid nanoparticles for hepatic gene silencing in vivo, *Angew. Chem., Int. Ed.*, 2012, **51**(34), 8529–8533.

14. K. B. Thurmond, E. E. Remsen, T. Kowalewski and K. L. Wooley, Packaging of DNA by shell crosslinked nanoparticles, *Nucleic Acids Res.*, 1999, **27**(14), 2966–2971.

15. M. Arotçaréna, B. Heise, S. Ishaya and A. Laschewsky, Switching the inside and the outside of aggregates of water-soluble block copolymers with double thermoresponsivity, *J. Am. Chem. Soc.*, 2002, **124**(14), 3787–3793.

16. Z. Ding, R. B. Fong, C. J. Long, P. S. Stayton and A. S. Hoffman, Size-dependent control of the binding of biotinylated proteins to streptavidin using a polymer shield, *Nature*, 2001, **411**(6833), 59–62.

17. T. Shimoboji, E. Larenas, T. Fowler, S. Kulkarni, A. S. Hoffman and P. S. Stayton, Photoresponsive polymer–enzyme switches, *Proc. Natl. Acad. Sci.*, 2002, **99**(26), 16592–16596.

18. N. Lavignac, M. Lazenby, J. Franchini, P. Ferruti and R. Duncan, Synthesis and preliminary evaluation of poly (amidoamine)–melittin conjugates as endosomolytic polymers and/or potential anticancer therapeutics, *Int. J. Pharm.*, 2005, **300**(1), 102–112.

19. Z. Ding, G. Chen and A. S. Hoffman, Unusual properties of thermally sensitive oligomer–enzyme conjugates of poly (N-iso-propylacrylamide)–trypsin, *J. Biomed. Mater. Res.*, 1998, **39**(3), 498–505.

20. E. L. Ferguson and R. Duncan, Dextrin – Phospholipase A2: Synthesis and Evaluation as a Bioresponsive Anticancer Conjugate, *Biomacromolecules*, 2009, **10**(6), 1358–1364.

21. E. Ferguson, D. Schmaljohann and R. Duncan, Polymer-phospholipase conjugates as novel anti-cancer agents: dextrin-phospholipase A2. *Proc 33rd Annual Mtg, Control Release Soc.* 2006, **33**, 660.

22. F. Greco and M. J. Vicent, Combination therapy: opportunities and challenges for polymer–drug conjugates as anticancer nanomedicines, *Adv. Drug Delivery Rev.*, 2009, **61**(13), 1203–1213.
23. D. Schmaljohann, Thermo-and pH-responsive polymers in drug delivery, *Adv. Drug Delivery Rev.*, 2006, **58**(15), 1655–1670.
24. Y.-Z. You, Q.-H. Zhou, D. S. Manickam, L. Wan, G.-Z. Mao and D. Oupicky, Dually Responsive Multiblock Copolymers via Reversible Addition-Fragmentation Chain Transfer Polymerization: Synthesis of Temperature- and Redox-Responsive Copolymers of Poly(N-isopropylacrylamide) and Poly(2-(dimethylamino)ethyl methacrylate), *Macromolecules*, 2007, **40**(24), 8617–8624.
25. F. Tanasa and M. Zanoaga, Polymeric Nanoparticles as Carriers for Stimuli-responsive Drug Delivery Systems, *2nd International Conference on Nanotechnologies and Biomedical Engineering*, Chisinau, Republic of Moldova, April 18–20, 2013.
26. H. Hatakeyama, A. Kikuchi, M. Yamato and T. Okano, Influence of insulin immobilization to thermoresponsive culture surfaces on cell proliferation and thermally induced cell detachment, *Biomaterials*, 2005, **26**(25), 5167–5176.
27. M. Ebara, M. Yamato, T. Aoyagi, A. Kikuchi, K. Sakai and T. Okano, Immobilization of cell-adhesive peptides to temperature-responsive surfaces facilitates both serum-free cell adhesion and noninvasive cell harvest, *Tissue Eng.*, 2004, **10**(7–8), 1125–1135.
28. A. Kubota, K. Nishida, M. Yamato, J. Yang, A. Kikuchi and T. Okano *et al.*, Transplantable retinal pigment epithelial cell sheets for tissue engineering, *Biomaterials*, 2006, **27**(19), 3639–3644.
29. M. van de Weert, R. van't Hof, J. van der Weerd, R. M. Heeren, G. Posthuma and W. E. Hennink *et al.*, Lysozyme distribution and conformation in a biodegradable polymer matrix as determined by FTIR techniques, *J. Controlled Release*, 2000, **68**(1), 31–40.
30. B. Kim, K. La Flamme and N. A. Peppas, Dynamic swelling behavior of pH-sensitive anionic hydrogels used for protein delivery, *J. Appl. Polym. Sci.*, 2003, **89**(6), 1606–1613.
31. A. T. Florence and D. Attwood, *Physicochemical Principles of Pharmacy*, Pharmaceutical Press, 2011.
32. J. Dissemond, M. Witthoff, T. Brauns, D. Haberer and M. Goos, [pH values in chronic wounds, *Evaluation during modern wound therapy*]. *Der Hautarzt; Zeitschrift fur Dermatologie, Venerologie, und verwandte Gebiete*, 2003, **54**(10), 959–965.
33. (a) W. Gao, J. M. Chan and O. C. Farokhzad, pH-Responsive Nanoparticles for Drug Delivery, *Mol. Pharmaceutics*, 2010, 7(6), 1913–1920; (b) P. Watson, A. T. Jones and D. J. Stephens,

Intracellular trafficking pathways and drug delivery: fluorescence imaging of living and fixed cells, *Adv. Drug Delivery Rev.*, 2005, **57**(1), 43–61.

34. M. Grabe and G. Oster, Regulation of organelle acidity, *J. Gen. Physiol.*, 2001, **117**(4), 329–344.

35. L. Y. Qiu and Y. H. Bae, Polymer architecture and drug delivery, *Pharm. Res.*, 2006, **23**(1), 1–30.

36. M. J. Vicent and R. Duncan, Polymer conjugates: nanosized medicines for treating cancer, *Trends Biotechnol.*, 2006, **24**(1), 39–47.

37. Y. Matsumura and H. Maeda, A new concept for macromolecular therapeutics in cancer chemotherapy: mechanism of tumoritropic accumulation of proteins and the antitumor agent smancs, *Cancer Res.*, 1986, **46**(12 Part 1), 6387–6392.

38. J. Madsen, S. P. Armes, K. Bertal, H. Lomas, S. MacNeil and A. L. Lewis, Biocompatible Wound Dressings Based on Chemically Degradable Triblock Copolymer Hydrogels, *Biomacromolecules*, 2008, **9**(8), 2265–2275.

39. S. Cerritelli, D. Velluto and J. A. Hubbell, PEG-SS-PPS: Reduction-Sensitive Disulfide Block Copolymer Vesicles for Intracellular Drug Delivery, *Biomacromolecules*, 2007, **8**(6), 1966–1972.

40. L. Zhang, W. Liu, L. Lin, D. Chen and M. H. Stenzel, Degradable Disulfide Core-Cross-Linked Micelles as a Drug Delivery System Prepared from Vinyl Functionalized Nucleosides via the RAFT Process, *Biomacromolecules*, 2008, **9**(11), 3321–3331.

41. K. H. Sun, Y. S. Sohn and B. Jeong, Thermogelling Poly(ethylene oxide-b-propylene oxide-b-ethylene oxide) Disulfide Multiblock Copolymer as a Thiol-Sensitive Degradable Polymer, *Biomacromolecules*, 2006, **7**(10), 2871–2877.

42. V. Bulmus, M. Woodward, L. Lin, N. Murthy, P. Stayton and A. Hoffman, A new pH-responsive and glutathione-reactive, endosomal membrane-disruptive polymeric carrier for intracellular delivery of biomolecular drugs, *J. Controlled Release*, 2003, **93**(2), 105–120.

43. F. Meng, W. E. Hennink and Z. Zhong, Reduction-sensitive polymers and bioconjugates for biomedical applications, *Biomaterials*, 2009, **30**(12), 2180–2198.

44. P. C. Jocelyn, Chemical reduction of disulfides, *Methods Enzymol.*, 1987, **143**, 246–256.

45. J. K. Oh, D. J. Siegwart, H.-i. Lee, G. Sherwood, L. Peteanu and J. O. Hollinger *et al.*, Biodegradable Nanogels Prepared by Atom

Transfer Radical Polymerization as Potential Drug Delivery Carriers: Synthesis, Biodegradation, in Vitro Release, and Bioconjugation, *J. Am. Chem. Soc.*, 2007, **129**(18), 5939–5945.

46. O. F. Castellani, E. N. Martinez and M. C. Anon, Role of Disulfide Bonds upon the Structural Stability of an Amaranth Globulin, *J. Agric. Food Chem.*, 1999, **47**(8), 3001–3008.

47. G. Saito, J. A. Swanson and K.-D. Lee, Drug delivery strategy utilizing conjugation via reversible disulfide linkages: role and site of cellular reducing activities, *Adv. Drug Delivery Rev.*, 2003, **55**(2), 199–215.

48. A. P. Vogt and B. S. Sumerlin, Temperature and redox responsive hydrogels from ABA triblock copolymers prepared by RAFT polymerization, *Soft Matter*, 2009, **5**, 2347–2351.

49. S. E. Kirkland, R. M. Hensarling, S. D. McConaughy, Y. Guo, W. L. Jarrett and C. L. McCormick, Thermoreversible Hydrogels from RAFT-Synthesized BAB Triblock Copolymers: Steps toward Biomimetic Matrices for Tissue Regeneration†, *Biomacromolecules*, 2007, **9**(2), 481–486.

50. D. P. Jones, J. L. Carlson, P. S. Samiec, P. Sternberg Jr., V. C. Mody Jr. and R. L. Reed *et al.*, Glutathione measurement in human plasma evaluation of sample collection, storage and derivatization conditions for analysis of dansyl derivatives by HPLC, *Clin. Chim. Acta*, 1998, **275**(2), 175–184.

51. A. N. Koo, H. J. Lee, S. E. Kim, J. H. Chang, C. Park and C. Kim *et al.*, Disulfide-cross-linked PEG-poly(amino acid)s copolymer micelles for glutathione-mediated intracellular drug delivery, *Chem. Commun.*, 2008, **48**, 6570–6572.

52. N. V. Tsarevsky and K. Matyjaszewski, Reversible Redox Cleavage/Coupling of Polystyrene with Disulfide or Thiol Groups Prepared by Atom Transfer Radical Polymerization, *Macromolecules*, 2002, **35**(24), 9009–9014.

53. L. Wong, C. Boyer, Z. Jia, H. M. Zareie, T. P. Davis and V. Bulmus, Synthesis of Versatile Thiol-Reactive Polymer Scaffolds via RAFT Polymerization, *Biomacromolecules*, 2008, **9**(7), 1934–1944.

54. J. Liu, V. Bulmus, C. Barner-Kowollik, M. H. Stenzel and T. P. Davis, Direct synthesis of pyridyl disulfide-terminated polymers by RAFT polymerization, *Macromol. Rapid Commun.*, 2007, **28**(3), 305–314.

55. A. Klaikherd, S. Ghosh and S. Thayumanavan, A Facile Method for the Synthesis of Cleavable Block Copolymers from ATRP-Based Homopolymers, *Macromolecules*, 2007, **40**(24), 8518–8520.

56. N. V. Tsarevsky and K. Matyjaszewski, Combining Atom Transfer Radical Polymerization and Disulfide/Thiol Redox Chemistry: A Route to Well-Defined (Bio)degradable Polymeric Materials, *Macromolecules*, 2005, **38**(8), 3087–3092.

57. J. K. Oh, C. Tang, H. Gao, N. V. Tsarevsky and K. Matyjaszewski, Inverse Miniemulsion ATRP: A New Method for Synthesis and Functionalization of Well-Defined Water-Soluble/Cross-Linked Polymeric Particles, *J. Am. Chem. Soc.*, 2006, **128**(16), 5578–5584.

58. J. Zhang, X. Jiang, Y. Zhang, Y. Li and S. Liu, Facile Fabrication of Reversible Core Cross-Linked Micelles Possessing Thermo-sensitive Swellability, *Macromolecules*, 2007, **40**(25), 9125–9132.

59. Y. Li, B. S. Lokitz, S. P. Armes and C. L. McCormick, Synthesis of Reversible Shell Cross-Linked Micelles for Controlled Release of Bioactive Agents, *Macromolecules*, 2006, **39**(8), 2726–2728.

60. J. Sun, X. Chen, T. Lu, S. Liu, H. Tian and Z. Guo *et al.*, Formation of Reversible Shell Cross-Linked Micelles from the Biodegradable Amphiphilic Diblock Copolymer Poly(L-cysteine)-block-Poly(L-lactide), *Langmuir*, 2008, **24**(18), 10099–10106.

61. W.-F. Dong, A. Kishimura, Y. Anraku, S. Chuanoi and K. Kataoka, Monodispersed Polymeric Nanocapsules: Spontaneous Evolution and Morphology Transition from Reducible Hetero-PEG PICmicelles by Controlled Degradation, *J. Am. Chem. Soc.*, 2009, **131**, 3804–3805.

62. S. Ghosh, V. Yesilyurt, E. N. Savariar, K. Irvin and S. Thayumanavan, Redox, ionic strength, and pH sensitive supramolecular polymer assemblies, *J. Polym. Sci. Part A: Polym. Chem.*, 2009, **47**(4), 1052–1060.

63. H. Gemici, T. M. Legge, M. Whittaker, M. J. Monteiro and S. Perrier, Original approach to multiblock copolymers via reversible addition-fragmentation chain transfer polymerization, *J. Polym. Sci. Part A: Polym. Chem.*, 2007, **45**(11), 2334–2340.

64. M. R. Whittaker, Y.-K. Goh, H. Gemici, T. M. Legge, S. Perrier and M. J. Monteiro, Synthesis of monocyclic and linear poly-styrene using the reversible coupling/cleavage of thiol/disulfide groups, *Macromolecules*, 2006, **39**, 9028–9034.

65. J. Liu, H. Liu, Z. Jia, V. Bulmus and T. Davis, An approach to biodegradable star polymeric architectures using disulfide coupling, *Chem. Commun.*, 2008, 6582–6584.

66. A. Napoli, M. Valentini, N. Tirelli, M. Müller and J. A. Hubbell, Oxidation-responsive polymeric vesicles, *Nat. Mater.*, 2004, **3**(3), 183–189.

67. A. Napoli, N. Tirelli, E. Wehrli and J. A. Hubbell, Lyotropic Behavior in Water of Amphiphilic ABA Triblock Copolymers Based on Poly(propylene sulfide) and Poly(ethylene glycol), *Langmuir*, 2002, **18**(22), 8324–8329.

68. A. Napoli, N. Tirelli, G. Kilcher and A. Hubbell, New Synthetic Methodologies for Amphiphilic Multiblock Copolymers of Ethylene Glycol and Propylene Sulfide, *Macromolecules*, 2001, **34**(26), 8913–8917.

69. B. Twaites, C. de las Heras Alarcón and C. Alexander, Synthetic polymers as drugs and therapeutics, *J. Mater. Chem.*, 2005, **15**(4), 441–455.

70. R. Esfand and D. A. Tomalia, Poly (amidoamine)(PAMAM) dendrimers: from biomimicry to drug delivery and biomedical applications, *Drug Discovery Today*, 2001, **6**(8), 427–436.

71. R. Duncan and L. Izzo, Dendrimer biocompatibility and toxicity, *Adv. Drug Delivery Rev.*, 2005, **57**(15), 2215–2237.

72. C. Dufès, I. F. Uchegbu and A. G. Schätzlein, Dendrimers in gene delivery, *Adv. Drug Delivery Rev.*, 2005, **57**(15), 2177–2202.

73. S. C. Richardson, N. G. Pattrick, Y. Stella Man, P. Ferruti and R. Duncan, Poly (amidoamine) s as potential nonviral vectors: ability to form interpolyelectrolyte complexes and to mediate transfection in vitro, *Biomacromolecules*, 2001, **2**(3), 1023–1028.

74. D. G. Anderson, D. M. Lynn and R. Langer, Semi-automated synthesis and screening of a large library of degradable cationic polymers for gene delivery, *Angew. Chem.*, 2003, **115**(27), 3261–3266.

75. K. C. Wood, S. R. Little, R. Langer and P. T. Hammond, A Family of Hierarchically Self-Assembling Linear-Dendritic Hybrid Polymers for Highly Efficient Targeted Gene Delivery, *Angew. Chem., Int. Ed.*, 2005, **44**(41), 6704–6708.

76. F. Verbaan, I. van Dam, Y. Takakura, M. Hashida, W. Hennink and G. Storm *et al.*, Intravenous fate of poly (2-(dimethylamino) ethyl methacrylate)-based polyplexes, *Eur. J. Pharm. Sci.*, 2003, **20**(4), 419–427.

77. M. D. Brown, A. I. Gray, L. Tetley, A. Santovena, J. Rene and A. G. Schätzlein *et al.*, In vitro and in vivo gene transfer with poly (amino acid) vesicles, *J. Controlled Release*, 2003, **93**(2), 193–211.

78. D. G. Anderson, W. Peng, A. Akinc, N. Hossain, A. Kohn and R. Padera *et al.*, A polymer library approach to suicide gene therapy for cancer, *Proc. Natl. Acad. Sci. U. S. A.*, 2004, **101**(45), 16028–16033.

79. T. Kean, S. Roth and M. Thanou, Trimethylated chitosans as non-viral gene delivery vectors: cytotoxicity and transfection efficiency, *J. Controlled Release*, 2005, **103**(3), 643–653.
80. M. Simons, E. R. Edelman, J.-L. Dekeyser, R. Langer and R. D. Rosenberg, *Antisense c-myb Oligonucleotides Inhibit Intimal Arterial Smooth Muscle Cell Accumulation In vivo*, 1992.
81. C. Stein and Y. Cheng, Antisense oligonucleotides as therapeutic agents–is the bullet really magical?, *Science*, 1993, **261**(5124), 1004–1012.
82. S. M. Elbashir, J. Harborth, W. Lendeckel, A. Yalcin, K. Weber and T. Tuschl, Duplexes of 21-nucleotide RNAs mediate RNA interference in cultured mammalian cells, *Nature*, 2001, **411**(6836), 494–498.
83. R. W. Wagner, Gene inhibition using antisense oligodeox-ynucleotides, *Nature*, 1994, **372**(6504), 333–335.
84. J. H. Jeong and T. G. Park, Novel polymer-DNA hybrid polymeric micelles composed of hydrophobic poly (D, L-lactic-co-glycolic acid) and hydrophilic oligonucleotides, *Bioconjugate Chem.*, 2001, **12**(6), 917–923.
85. D. A. Braasch, Z. Paroo, A. Constantinescu, G. Ren, O. K. Öz and R. P. Mason *et al.*, Biodistribution of phosphodiester and phosphorothioate siRNA, *Bioorg. Med. Chem. Lett.*, 2004, **14**(5), 1139–1143.
86. T. G. Park, J. H. Jeong and S. W. Kim, Current status of polymeric gene delivery systems, *Adv. Drug Delivery Rev.*, 2006, **58**(4), 467–486.
87. N. A. Peppas, J. Z. Hilt, A. Khademhosseini and R. Langer, Hydrogels in biology and medicine: from molecular principles to bionanotechnology, *Adv. Mater.-Deerfield Beach then Weinheim*, 2006, **18**(11), 1345.
88. N. Peppas, Y. Huang, M. Torres-Lugo, J. Ward and J. Zhang, Physicochemical foundations and structural design of hydrogels in medicine and biology, *Annu. Rev. Biomed. Eng.*, 2000, **2**(1), 9–29.
89. W. B. Liechty, D. R. Kryscio, B. V. Slaughter and N. A. Peppas, Polymers for drug delivery systems, *Annu. Rev. Chem. Biomol. Eng.*, 2010, **1**, 149.
90. C. Kresge, M. Leonowicz and W. Roth, *Dendrimers and Dendrons. Concepts, Syntheses, Applications*, VCH, Weinheim, 2001.
91. D. A. Tomalia and J. M. Frechet, *Introduction to the Dendritic State*, Wiley Online Library, 2002.
92. D. A. Tomalia, H. Baker, J. Dewald, M. Hall, G. Kallos and S. Martin *et al.*, A new class of polymers: starburst-dendritic macromolecules, *Polym. J.*, 1985, **17**(1), 117–132.

93. N. Malik, R. Wiwattanapatapee, R. Klopsch, K. Lorenz, H. Frey and J. Weener *et al.*, Dendrimers:: Relationship between structure and biocompatibility in vitro, and preliminary studies on the biodistribution of 125I-labelled polyamidoamine dendrimers in vivo, *J. Controlled Release.*, 2000, **65**(1), 133–148.

94. C. J. Hawker and J. M. Frechet, Preparation of polymers with controlled molecular architecture. A new convergent approach to dendritic macromolecules, *J. Am. Chem. Soc*, 1990, **112**(21), 7638–7647.

95. M. Liu, K. Kono and J. M. Fréchet, Water-soluble dendrimer-poly (ethylene glycol) starlike conjugates as potential drug carriers, *J. Polym. Sci. Part A: Polym. Chem.*, 1999, **37**(17), 3492–3503.

96. G. R. Newkome, X. Lin and C. D. Weis, Polytryptophane terminated dendritic macromolecules, *Tetrahedron: Asymmetry*, 1991, **2**(10), 957–960.

97. K. Sadler and J. P. Tam, Peptide dendrimers: applications and synthesis, *Rev. Mol. Biotechnol.*, 2002, **90**(3), 195–229.

98. R. Roy and M.-G. Baek, Glycodendrimers: novel glycotope isosteres unmasking sugar coding. Case study with T-antigen markers from breast cancer MUC1 glycoprotein, *Rev. Mol. Biotechnol.*, 2002, **90**(3), 291–309.

99. S. M. Grayson, M. Jayaraman and J. M. Fréchet, Convergent synthesis and 'surface'functionalization of a dendritic analog of poly (ethylene glycol), *Chem. Commun.*, 1999, **14**, 1329–1330.

100. M. W. Grinstaff, Biodendrimers: new polymeric biomaterials for tissue engineering, *Chem. – Eur. J.*, 2002, **8**(13), 2838–2846.

101. M. M. Boysen, K. Elsner, O. Sperling and T. K. Lindhorst, Glycerol and glycerol glycol glycodendrimers, *Eur. J. Org. Chem.*, 2003, **2003**(22), 4376–4386.

102. H. Frey and R. Haag, Dendritic polyglycerol: a new versatile biocompatible material, *Rev. Mol. Biotechnol.*, 2002, **90**(3), 257–267.

103. R. Haag, A. Sunder and J.-F. Stumbé, An approach to glycerol dendrimers and pseudo-dendritic polyglycerols, *J. Am. Chem. Soc.*, 2000, **122**(12), 2954–2955.

104. S. Fuchs, T. Kapp, H. Otto, T. Schöneberg, P. Franke and R. Gust *et al.*, A Surface-Modified Dendrimer Set for Potential Application as Drug Delivery Vehicles: Synthesis, In Vitro Toxicity, and Intracellular Localization, *Chem. – Eur. J.*, 2004, **10**(5), 1167–1192.

105. M. F. Neerman, W. Zhang, A. R. Parrish and E. E. Simanek, In vitro and in vivo evaluation of a melamine dendrimer as a vehicle for drug delivery, *Int. J. Pharm.*, 2004, **281**(1), 129–132.

106. E. R. Gillies and J. M. Frechet, Dendrimers and dendritic polymers in drug delivery, *Drug Discovery Today*, 2005, **10**(1), 35–43.
107. E. R. Gillies, T. B. Jonsson and J. M. Fréchet, Stimuli-responsive supramolecular assemblies of linear-dendritic copolymers, *J. Am. Chem. Soc.*, 2004, **126**(38), 11936–11943.
108. S. Kommareddy and M. Amiji, Preparation and evaluation of thiol-modified gelatin nanoparticles for intracellular DNA delivery in response to glutathione, *Bioconjugate Chem.*, 2005, **16**(6), 1423–1432.
109. D. Shenoy, S. Little, R. Langer and M. Amiji, Poly (ethylene oxide)-modified poly (β-amino ester) nanoparticles as a pH-sensitive system for tumor-targeted delivery of hydrophobic drugs: part 2. In vivo distribution and tumor localization studies, *Pharm. Res.*, 2005, **22**(12), 2107–2114.
110. M. Calderón, M. A. Quadir, M. Strumia and R. Haag, Functional dendritic polymer architectures as stimuli-responsive nanocarriers, *Biochimie*, 2010, **92**(9), 1242–1251.
111. B. Devarakonda, R. A. Hill and M. M. de Villiers, The effect of PAMAM dendrimer generation size and surface functional group on the aqueous solubility of nifedipine, *Int. J. Pharm.*, 2004, **284**(1), 133–140.
112. O. Milhem, C. Myles, N. McKeown, D. Attwood and A. D'Emanuele, Polyamidoamine Starburst® dendrimers as solubility enhancers, *Int. J. Pharm.*, 2000, **197**(1), 239–241.
113. C. Yiyun and X. Tongwen, Solubility of nicotinic acid in polyamidoamine dendrimer solutions, *Eur. J. Med. Chem.*, 2005, **40**(12), 1384–1389.
114. B. Devarakonda, R. A. Hill, W. Liebenberg, M. Brits and M. M. de Villiers, Comparison of the aqueous solubilization of practically insoluble niclosamide by polyamidoamine (PAMAM) dendrimers and cyclodextrins, *Int. J. Pharm.*, 2005, **304**(1), 193–209.
115. A. Meister and M. E. Anderson, Glutathione, *Annu. Rev. Biochem.*, 1983, **52**(1), 711–760.
116. R. Satchi, T. Connors and R. Duncan, PDEPT: polymer-directed enzyme prodrug therapy, *Br. J. Cancer*, 2001, **85**(7), 1070.
117. M. A. Azagarsamy, P. Sokkalingam and S. Thayumanavan, Enzyme-triggered disassembly of dendrimer-based amphiphilic nanocontainers, *J. Am. Chem. Soc.*, 2009, **131**(40), 14184–14185.
118. R. Duncan, Designing polymer conjugates as lysosomotropic nanomedicines, *Biochem. Soc. Trans.*, 2007, **35**(1), 56.

119. M. Smet, L.-X. Liao, W. Dehaen and D. V. McGrath, Photolabile dendrimers using o-nitrobenzyl ether linkages, *Organic Lett.*, 2000, **2**(4), 511–513.
120. C. Park, J. Lim, M. Yun and C. Kim, Photoinduced Release of Guest Molecules by Supramolecular Transformation of Self-Assembled Aggregates Derived from Dendrons, *Angew. Chem.*, 2008, **120**(16), 3001–3005.
121. W. Li, A. Zhang, Y. Chen, K. Feldman, H. Wu and A. D. Schlüter, Low toxic, thermoresponsive dendrimers based on oligoethylene glycols with sharp and fully reversible phase transitions, *Chem. Commun.*, 2008, **45**, 5948–5950.
122. H.-i. Lee, J. A. Lee, Z. Poon and P. T. Hammond, Temperature-triggered reversible micellar self-assembly of linear–dendritic block copolymers, *Chem. Commun.*, 2008, **32**, 3726–3728.
123. Z. Guo, Y. Zhang, W. Huang, Y. Zhou and D. Yan, Terminal Modification with 1-Adamantylamine to Endow Hyperbranched Polyamidoamine with Thermo-/pH-Responsive Properties, *Macromol. Rapid Commun.*, 2008, **29**(21), 1746–1751.
124. J. Surmacki, J. Musial, R. Kordek and H. Abramczyk, Raman imaging at biological interfaces: applications in breast cancer diagnosis, *Mol. Cancer*, 2013, **12**(1), 48.

9 Photochromic Materials

9.1 Introduction

Just like many stimuli-responsive polymers have been inspired by Nature, photochromic systems are no exception. One example are reversible shape changes of the photoreceptor molecule rhodopsin in eyes that produces a cascade of molecular rearrangements responding to light illumination[1] as discussed in Chapter 1. As we have seen, upon photoexcitation, all-*trans*-retinal undergoes isomerization to 11-*cis*-retinal across the double bond, thus causing shape changes. Although this event occurs at the Angstrom (Å) level, there are many molecular interconnected events and responses that are not fully understood. In materials chemistry somewhat similar processes have been observed in certain molecules, although on much less orchestrated scales. They are known as photochromic molecules and were discovered over 150 years ago in an organic crystal of tetracene, and later on, in inorganic and organometallic materials.[2–4] In spite of a long research history, a quest for new photochromic materials continues and of particular interest is color changes induced by ultraviolet (UV) radiation caused by *trans* ⇌ *cis* isomerization, ring opening–closing reactions, inter- or intramolecular charge transfer, or bimolecular cycloaddition reactions. To make use of their photochromic behavior in various applications, photochromic entities are often incorporated into polymer matrices.

While creating individual molecules and understanding their electronic properties facilitated further advances in understanding molecular mechanisms governing photochromism, useful applications require fabrication into films, sheets, fibers, or beads. This is

Stimuli-Responsive Materials: From Molecules to Nature Mimicking Materials Design
By Marek W. Urban
© Marek W. Urban, 2016
Published by the Royal Society of Chemistry, www.rsc.org

typically achieved by incorporating photochromic molecules to macromolecular matrices by doping or dispersing, or covalently attaching them onto a polymer backbone. Because photoisomerization of chromophores alter equilibrium conformations, dimensions of the surrounding polymer matrix play an essential role in spatial rearrangements. Dimensional classifications of the matrix effect has been extensively examined,[5] with notable applications in ophthalmic lens, optical storage media, photosensors, and biomedical devices. Furthermore, kinetics of conformational changes of polymers with photochromic moieties depends upon their physical or chemical incorporation into the matrix.[6] As one might anticipate, photochromic transitions are slower in polymers with a lower free volume content due to the limited space reducing segmental motions, but also may be altered by local polarity, intermolecular interactions, or steric restrictions.[7-9] Another limiting factor is molecular aggregations of a photochromic entity that may also influence kinetics of responses.[10]

9.2   Photochromic Polymeric Systems

As pointed out earlier, photochromic chromophores can be incorporated into polymer matrices either by dispersing them or covalently attaching them to a polymer backbone and many studies focused on fundamental understanding of mechanisms governing photochromic polymers as well as on applications ranging from photoswitching to optical data storage devices, sensors or light-driven reactors, or artificial muscles. Although we realized that there are many outstanding photochromic materials, and scientific and patent literature provides many detailed and excellent overviews, this chapter focuses on selected examples that influence molecular structures, conformational changes, environmental effects, and kinetics on photochromic responses in polymeric matrices.

9.2.1   Influence of Molecular Structures

The effect of molecular properties such as the length and mobility of polymer chains as well as the nature of the substituents play a significant role on responses of photochromic systems to electromagnetic radiation. The quantum yield (Φ) of photo-induced isomerization, which quantifies the efficiency with which the conformational change (or ring cyclization) takes place, appear to be enhanced for systems with less bulkier substituents.[11,12,13] As *trans*-isomers absorb light, photoinduced *trans* → *cis* conversion at the photostationary state may be as high as 70%, depending upon associated functional groups.[14] For *cis* → *trans* isomerization to occur effectively the molecules require enough space around them, and for bulkier molecules, to be able to exhibit rotational mobility, higher free volumes are required. Optical storage studies conducted on poly[(4-nitrophenyl)[4-[[2-(methacryloyoxy)ethyl]ethylamino]phenyl]diazene] (pDR1M) and poly[(4-nitronaphthyl)[4-[-2-(methacryloyloxy)ethyl]-ethylamino]phenyl]diazene] (pNDR1M) showed that the dynamics of writing depends on the size of the photochromic moiety, while the maximum achievable modulation and the rate of relaxation are size independent.[15] The decrease of the rate of writing, which chemically represents *trans–cis–trans* conformation changes resulted from the reduced quantum yields of photoisomerization attributed to the presence of bulkier naphthyl groups. In contrast, phenyl rings enhanced quantum yields due to smaller critical volume. Similar trends were observed in other photochromic systems as well.[16] Most of the applications of photochromic materials make use of photoinduced reactions in the solid state, which take place in the nano-range free volume of the matrix, where the free volume occupied by the van der Waals envelope of molecular groups is sufficient. Interestingly enough, photochromic probes have also been used to estimate microstructure and free volume distribution inside the polymer networks.[17,18]

For fulgides, where reversible photo-induced ring cyclization is responsible for photochromic reactions, an important requirement is that the carbon atoms that form C–C single bonds come in close contact upon photoirradiation. Thus, conformational freedom of the neighboring molecular groups affects photoisomerization quantum yields, such as in diarylethene derivatives with heterocyclic aryl groups, where photochemical ring cyclization and ring opening reactions occur.

The effect of polymer matrices on photochromic response of diarylethene type species has been investigated by incorporating *cis*-1,2-bis(2-methylbenzothiophene-3-yl)-1,2-dicyanoethene (BTCN), 1,2-bis(2-methylbenzothiophene-3-yl)maleic anhydride (BTMA) and 1,2-bis(2-methylbenzothiophene-3-yl)perfluorocyclopropene (BTF6) into polyethylene (PE), poly(methylmethacrylate) (PMMA) and poly-ethylene grafted polysiloxane (PEG-*g*-PS).[19] The quantum yield (Φ) of photochromic ring cyclization was highest for PEG-*g*-PS, which is the matrix with the highest polarity. The observed behavior was contrasted to that of a solution phase, where a higher polarity of the solvent resulted in lower quantum yields. Also, higher Φ values in PEG-*g*-PS for lowest glass transition temperature (T_g) networks was observed, indicating again that a higher free volume is significant in photochromic responses and appear to be a prerequisite for effective photochromic responses over the polarity of the matrix.

Along the same lines, increased flexibility of side polymer chains also reduces the movement restrictions resulting in a greater conformational freedom. Addition of flexible oligomer plasticizers will lower the T_g, thus increasing their switching speeds.[20] Although the use of plasticizing species may be beneficial, they are usually not chemically attached to the matrix, which may result in phase separation or other adverse effects leading to diffusion or undesirable stratification. One of the permanent structural modifications to increase the flexibility of the polymer chains is attachment of alkyl spacers between the chromophore and the polymer backbone.[21] Covalent attachment of photochromic molecules is advantageous, such as shown in polyurethane elastomers that exhibit mechanochromic properties, achieved by chemically incorporating azobenzene oligomer into a polyurethane copolymer.[22]

Among other factors that may influence the photochromic effect, covalent insertion of inert spacers may minimize the steric hindrance of pendant chromophores, thus providing enhanced mobility to move away responsive species from a bulky polymer backbone.[14] Again, enhanced free volume by the incorporation of alkyl spacers in poly(methyl methacrylate) was shown to be beneficial to enhance the responses.[23] More complex behavior is shown in liquid crystalline polymers, where in addition to the length of the spacer, light induced orientation behavior depends on the enthalpic stability of the mesophase.[24]

Molecular weight of photochromic polymers is another factor which influences the rate of photoisomerization and thermal relaxations,[21] but there are systems with marginal molecular weight dependence.[14] As photochromic responses result from light induced conformational changes of molecular segments that may be preceded by photochemical ring opening or closing reactions, azobenzene functionalized poly(methyl phenylsilane) (PMPS) is an example in which light induced conformational changes occur, but reversible reactions are induced by heating.[25] The efficiency of light-induced conformational changes in PMPS matrix containing pendant 4-nitroazobenzene group increases with the decreasing (–Si–O–) units with respect to azobenzene moiety. Also, conformational changes of polymer chains affect the rate of isomerization reaction of photochromic molecules, as evidenced by kinetic analysis of the isomerization of spiropyran modified poly(L-glutamic acid) (PSLG) coupling reaction.[26] Compared to that the random coil structure, isomerization of the spiropyran side chains in PSLG in α-helix structure was significantly restrained.

9.2.2 Environmental Effects

Photochromic reactions in polymers also depend on the polarity of the polymer matrix. For dispersed in or covalently attached to polymer matrices, the stability of polar merocyanine isomers in a polar matrix will be enhanced. Consequently, the reverse reaction rates in polar matrices will be reduced. Interactions between polar polymer matrices and photochromic 1,3,3-trimethylindoline-6′-nitrobenzopyrylospiran (SP) molecule resulted in a blue shift of the absorption spectrum of SP. As shown in Figure 9.1, for the poly(n-butyl methacrylate) (PnBMA) matrix with a lower T_g, larger blue shifts are observed compared to that of poly(methyl methacrylate) (PMMA) and

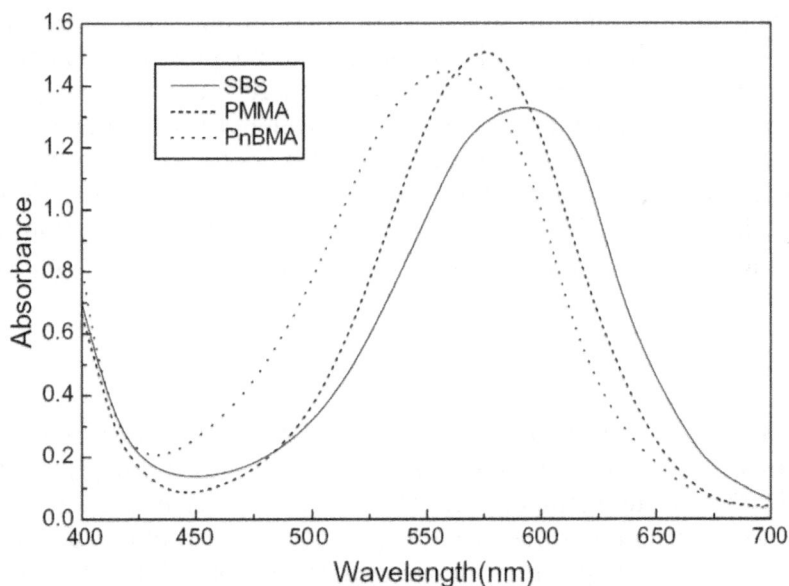

Figure 9.1   Effect of polymer matrix on the spectra of spiropyran. The absorption peak shift towards lower wavelength in polar matrix. Larger blue shifts are observed for the poly(*n*-butyl methacrylate) (PnBMA) compared to poly(methyl methacrylate) (PMMA) and styrene–butadiene–styrene (SBS) matrices.
Reprinted from ref. 27. Copyright (2010) with permission from Elsevier.

styrene–butadiene–styrene (SBS) matrices. This behavior was attributed to polar and ionic interactions between P*n*BMA matrices and SP.[27] The magnitude of these interactions, however, will vary depending upon the T_g of the matrix.

As one would anticipate, the dependence of electronic properties of photochromic chromophores upon environment is more pronounced in solvents, where hypsochromic or bathochromic shifts of the absorption and fluorescence spectra were observed and are strongly influenced by the polarity of solvent molecules.[28] Chromophore properties are also influenced by matrix physical properties, such as viscosity, conductivity, pH, solubility or phase transition temperatures.[29,30] For example, the reversible light induced insolubility of polystyrene with pendant azobenzene in cyclohexane induced by irradiation was attributed to the reduction in solvent–polymer interactions, as the azobenzene chromophore transforms from the less polar *trans*-form to the more polar *cis*-form.[30]

9.2.3 Kinetics of Photoreactions

Understanding mechanistic aspects of photoinduced reactions is necessary for the optimization of photochromic polymers. The rate of light induced reactions and thermal reverse reactions, the magnitude of activation energy barriers, and the wavelength dependence are of particular importance as they govern the response of the system.

The activation energies involved in thermal reversible reactions is typically estimated by monitoring the rate of the reverse reactions as a function of temperature. For solutions, this follows the Arrhenius equation:[31]

$$k = \nu \exp\left(-\frac{E}{RT}\right)$$

where k is the thermal decay rate, ν is a weakly temperature-dependent frequency factor, E is the activation energy, R the universal gas constant and T the absolute temperature.[31] The thermal reverse reaction usually follows a first-order kinetics, although there are exceptions to this behavior.[32] In many polymer matrices, thermal relaxation reactions follow a non-exponential kinetics of the form given by equation:

$$k(t) = k_0 \exp[-(t/\tau)^\beta]$$

where τ is the relaxation time, and β $(0 < \beta \leq 1)$ quantifies the extent of deviation from an exponential behavior. It has been proposed that the difference in decay kinetics between polymer matrices and solutions lies in the fact that in polymer matrices, isomerization processes must thermally overcome a distribution of potential barriers owing to the spatially distributed polymer segments with variable dynamics.[33] Although the kinetics of relaxation reactions has been investigated, no satisfactory model describing the observed behaviors is available.[34]

The kinetics of light-induced forward reactions in photochromic spiropyran derivative, l-β-(4-trifluoromethyl benzoyloxy)ethyl-3,3-dimethyl-spiro[indoline-2,3'-[3*H*]-naphtho[2,1-*b*-1,4]-oxazine], (SO) showed that the forward reaction rates are faster in poly(ethylene terephthalate) (PET) compared to polycaprolactam (Nylon-6). As the polarities of both matrices are similar, those differences were attributed to the increased crystallinity of the latter, which retarded the rotation of the active molecules after the light induced cleavage of the C_{spiro}–O bond. (35) The forward reaction rates also exhibit temperature dependence, as observed for 1,3,3-trimethylindoline-6'-nitrobenzopyrylospiran (SP) dispersed in PMMA and SBS matrices.[27] The rate of photo-isomerization decreases at higher temperatures and, at 100 °C, the reverse and forward reaction rates are equal, resulting in no change of color upon light irradiation. The reverse reactions, however, do not follow the first order kinetics, and instead, a mechanism involving two parallel first-order reactions can be applied.[35]

The presence of undesirable side reactions associated with photochromic transitions may also reduce the number of cycles of forward and reverse reactions. This concept known as a 'fatigue problem' is associated with the types of photochromic molecules and is represented by the following reaction:

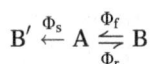

$$B' \xleftarrow{\Phi_s} A \underset{\Phi_r}{\overset{\Phi_f}{\rightleftharpoons}} B$$

where a photochromic molecule A upon irradiation converts to isomer B with a quantum yield Φ_f, where by associated side reactions exhibit quantum yield Φ_s, and B reverts back to A, with a quantum yield Φ_r.

In order to achieve an acceptable usage cycle greater than 10^3 times, the value of Φ_s must be less than 0.0001 (assuming that conversion to B is almost 100%).[36] To increase the photochemical stability of photochromic molecules, that will also increase the fatigue factor, substituents with electron donating and withdrawing groups can be used. For example, as shown in Figure 9.2, the addition of methyl groups at the 4 and 4' positions of the thiophene ring in 1,2-di-(2-dimethyl-5-phenylthiophen-3-yl)perfluorocyclopentene (B) increases the number of repeatable cycles in hexane from 80 to 200 in the absence of oxygen.[37] Notably, synthetic modifications of photochromes

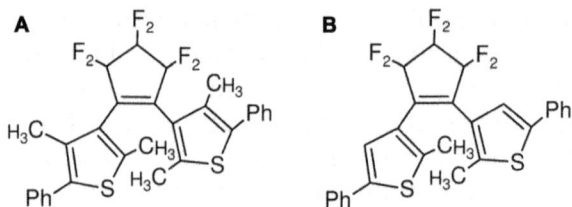

Figure 9.2 Molecular structures of (A) 4, 4′-methyl-1,2-di-(2-dimethyl-5-phenylthiophen-3-yl) perfluorocyclopentene and (B) 1,2-di-(2-dimethyl-5-phenylthiophen-3-yl)perfluorocyclopentene.

can yield enhanced fatigue factors.[36] When 2-(1-(2,5-dimethyl-3-furyl)ethylidene)-3-(2-adamantylidene) succinic anhydride is incorporated in polymer matrices, increased fatigue resistance results from the increased rigidity of the matrix preventing the conversion of the molecules to the inactive isomeric form.[38] It appears that the presence of sufficient free volume to accommodate ring-opening reactions is essential.

9.3 Photochromic Systems

Performance of physically dispersed and covalently attached photochromic molecules in polymer matrices depends on the molecular structure, conformational changes, environment, and kinetic processes. Because there is no single molecule matrix pair that possesses all these attributes, the search for newer materials with higher response speeds and fatigue factor continues[39] which are primarily driven by potential applications. While matrices specificity and complexity are the main road blocks, individual photochromic entities and their molecular and electronic properties typically are grouped based on their scientific and technological importance into azobenzenes, spiropyrans, diarylethenes, and fulgides. More details regarding structural and electronic properties responsible for photochromicity is provided below.

9.3.1 Azobenzenes

Azobenzenes are aromatic compounds containing two phenyl groups separated by an azo (N=N) group which is responsible for reversible *trans* ⇌ *cis* isomerization induced by light, and irreversible *cis* → *trans* induced by heat. The absorption spectrum of azobenzene exhibits a high intensity π–π* transition in the UV region and a low intensity n–π* transition in the visible region. The n–π* transitions are responsible for colors which can be manipulated from the UV to the visible regions *via* modifying the ring substituents. Based on the substitution pattern and the relative energies between n–π* and π–π* states, the azo aromatics are generally categorized into three classes: azobenzene, amino azobenzene, and pseudo stilbene type molecules. While unsubstituted azobenzene represents the simple form, the second class consists of azobenzenes substituted with electron donating groups, particularly the amino groups at *ortho-* and *para-* positions. The third class represents azobenzenes substituted with an electron donor and electron acceptor groups at 4- and 4′-positions generally known as pseudo-stilbene. The n–π* and π–π* transitions in amino substituted azobenzenes are almost completely overlapped, whereas pseudo stilbene type molecules, due to a strong asymmetric electron distribution between the groups of opposite charges exhibit π–π* absorption shift to the lower energy and n–π* to higher. It should be noted that the factors influencing the relative energies of n–π* and π–π* states result from π-conjugation and the nature of substitution pattern on the aromatic ring. The greater the extent of conjugation and the presence of strong electron donating and electron withdrawing substituents, the lower the energy gap, hence shifting absorption bands towards higher wavelength. Reversible isomerization between *trans-* and *cis-*conformations as well as molecular and structural changes resulting from dispersing and/or covalently attaching azobenzenes are probably the most challenging and, at the same time, difficult to control processes.

The mechanism of *trans–cis* photoisomerization in azobenzenes proceed either *via* rotation about the azo bond with the rupture of the π-character under π–π* excitation, or *via* inversion mechanism under n–π* excitation, where the π bond remains intact. Both mechanisms are shown in Figure 9.3. On the other hand, thermally induced reversible *cis–trans* back-reaction proceeds through a rotational mechanism. The *trans–cis* photoconversion is almost completely reversible in picoseconds without any side reactions, whereas thermal *cis–trans* conversion usually has timescales ranging from nano to milliseconds, or even hours, depending upon the substituent and the local environment. A variety of azobenzene systems can be synthesized, including cyclodextrins,[40–42] adamantanes,[43] polycyclics,[44] bacteriorhodopsin[45] and crown ethers.[46]

One of the drawbacks of doping polymer matrices with azobenzenes that exhibit photoinduced motions is that at higher concentration levels, phase separation and microcrystallization may occur. One approach to eliminate this adverse effect is to covalently

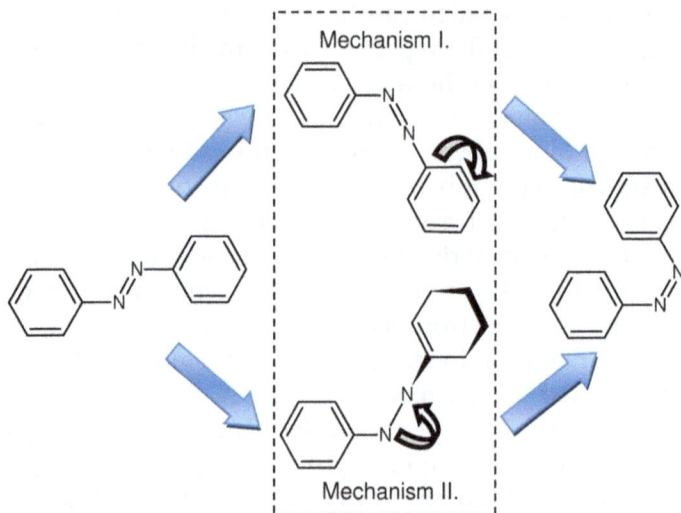

Figure 9.3 Mechanism of *trans* ⇌ *cis* isomerization in azobenzene.

attach azobenzene chromophores to a polymer matrix. Both side and main chain azobenzene substituted polymers with good stability, rigidity and ease of processability involve synthetic routes either using direct polymerization of azobenzene-functionalized monomers or post-functionalization, where the chromophores are introduced to reactive precursor polymers without harsh polymerization conditions. For side and main chain azobenzene polymers, imides,[47] esters,[48] isocyanates,[49] acrylates,[50] methacrylates,[51] urethanes,[52] ethers,[53] organometallic ferrocene polymers,[54] dendrimers[55] and conjugated polydiacetylenes[56] can be used. Selected examples of azobenzene-containing polymer repeating units of (a) side chain, (b) epoxy, (c) organometallic ferrocene, (d) main chain and (e) conjugated polydiacetylene are illustrated in Figure 9.4.

9.3.2 Spiropyrans

Spiropyrans are photochromic materials that exist in two forms: spiropyran (SP, colorless) and merocyanine (MC, colored), and these structures are shown in Figure 9.5. The MC form exhibits a characteristic absorption band due to extended π-electron conjugation in the visible region. In polar solvents, it may exist in any or all of the following four complicated Zwitterionic forms by changing the conformation about the conjugated bond: *trans*-transoid-*trans* (TTT), *trans*-transoid-*cis* (TTC), *cis*-transoid-*cis* (CTC, and *cis*-transoid-*trans*

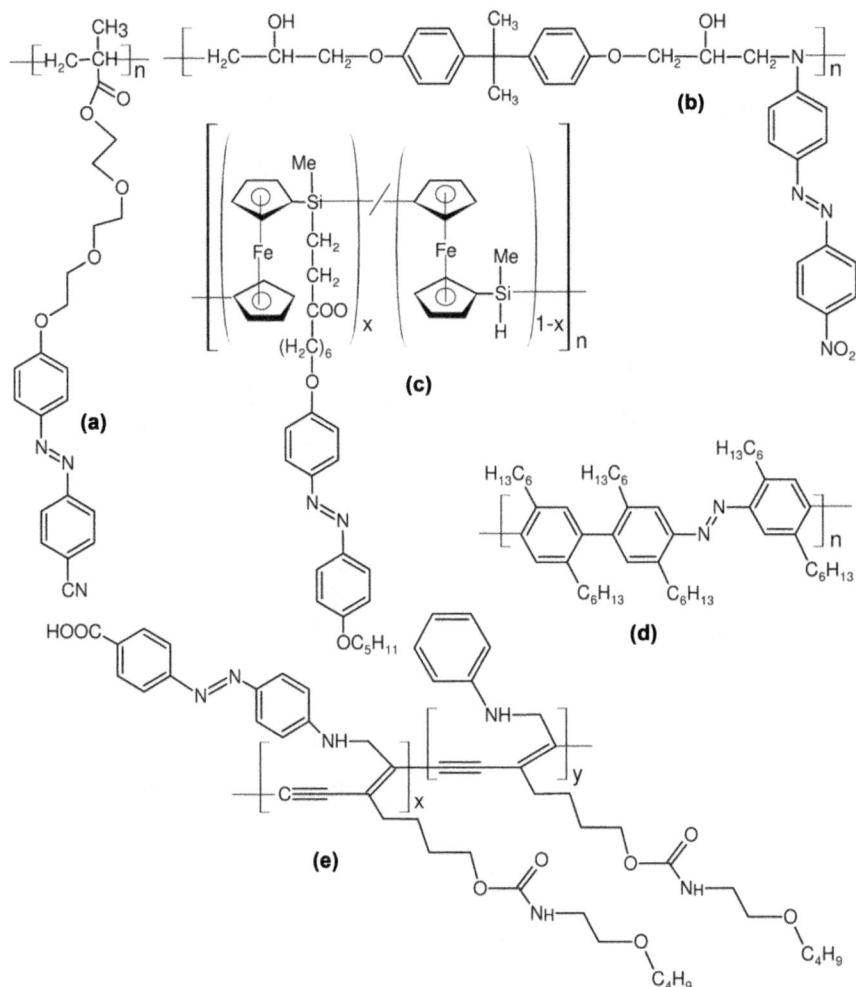

Figure 9.4 Selected examples of azobenzene-based polymers.

(CTT). Furthermore, MC exhibits a strong tendency to aggregate in stacks, giving rise to parallel (head-to-head, J-aggregates) and anti-parallel (head-to-tail, H-aggregates) molecular dipoles.[57] The open MC form converts back to the closed SP form either by irradiation in the visible region or by thermal exposure. Contrary to azobenzenes, where a number of reversible *trans–cis* isomerization cycles are possible without any degradation, in spiropyans, the number of cycles is limited, thereby limiting the applicability. Knowledge of photochemical and thermal processes in the ring opening and closure pathways led to systems with better photochemical and thermal stability. The techniques used include protonation of the MC-phenoxide

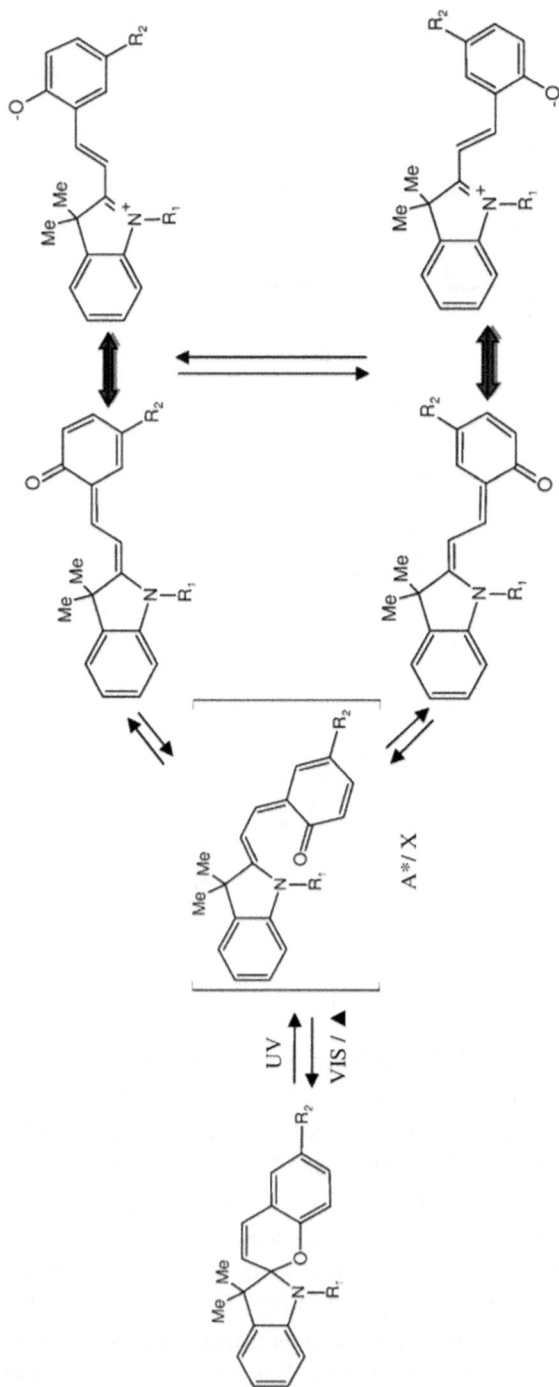

Figure 9.5 Mechanism of ring-opening and closing in spiropyrans.

moiety,[58] synthetic modification of the SP with crown ethers,[59] or a 7-trifluoromethylquinoline group,[60] and intramolecular bidentate metal ion chelation.[61]

Time-resolved spectroscopy provides useful information on the dynamics and nature of intermediates involved in reversible photochemical and thermal ring openings as well as closure reactions in spiropyran compounds. Figure 9.5 depicts these reactions, where the first step involves cleavage of the C–O bonds between the spiro carbon and the oxygen, followed by orthogonal SP to planar MC conversions. The C–O bond cleavage occurs on a timescale of picoseconds or even faster. It has been proposed that the mechanism of the photochemical ring-opening reaction involves the immediate formation of a singlet excited state, intermediate species, or a 'metastable' species in less than 100 fs. A part of the metastable species restores the initial C–O bond back-formation in a few picoseconds, and the remaining portion forms a mixture of transient merocyanine conformers with a decay time constant of 100 ps. The mechanism of photochromic transformation of spiropyrans is shown in Figure 9.5.

The most common synthetic methods for designing spiropyrans involve condensation of a heterocyclic quaternary salt with an alkyl group at the vicinal position to the heteroatom or condensation of 2-hydroxyarenealdehyde with the corresponding heterocyclic methylene base. As a majority of bases have the tendency to dimerize, their use is undesirable during the synthesis. Benzothiazoline, benzodithiole and benzopyrans can be synthesized by using alkylimmonium, oxonium and thionium salts as precursors, respectively. One of the characteristic features of spiropyrans is a strong absorption maximum around the 290–380 nm regions due to the presence of a benzopyran moiety. A number of attempts to incorporate spiropyran molecules into polymer systems have been reported, which include spiropyrans incorporated in polymer matrices,[62] liquid crystal polymers containing spiropyran as mesophases,[63] and spiropyran-grafted dextran,[64] pullulan,[65] polydimethylsiloxane[66] and other synthetic polymers.[67] Selected examples are illustrated in Figure 9.6.

9.3.3 Diarylethenes

Diarylethenes with heterocyclic aryl groups undergo photochemical electrocyclization cycloreversion reactions between 1,3,5-hexatriene (open form) and 1,3-cyclohexadiene (closed form) entities. This is illustrated in Figure 9.7. These materials are the thermally irreversible

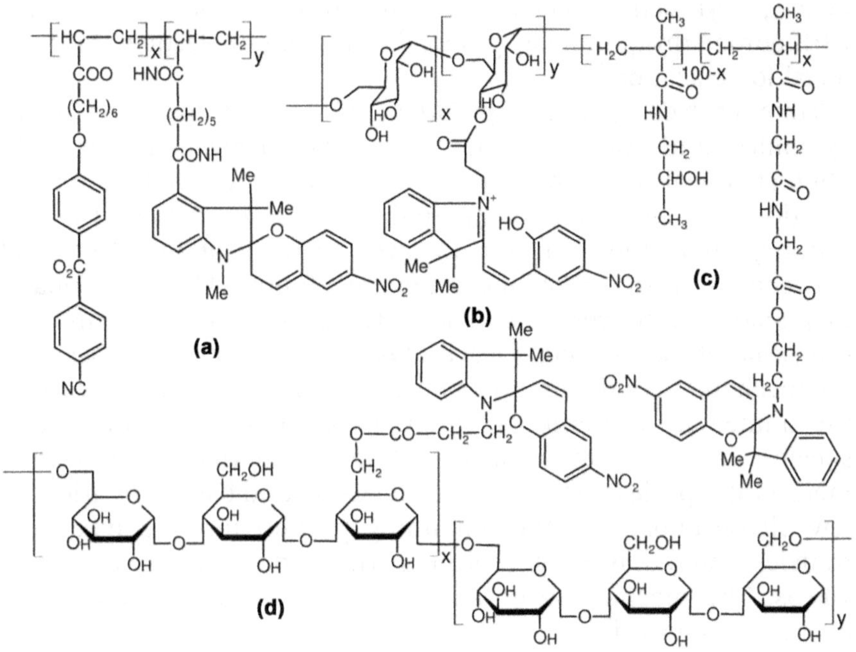

Figure 9.6　Selected examples of spiropyran-based polymers.

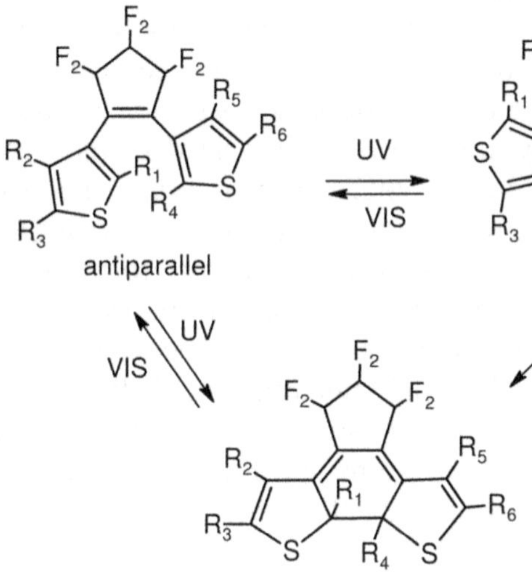

Figure 9.7　UV/vis-induced parallel and anti-parallel conformations in diarylethenes.

organic photochromic compounds that are resistant to fatigue. Their thermal reversibility depends upon the aromatic stabilization energies of the aryl groups. For low aromatic stabilization energy aryl groups, such as thiophene, thiazole and furans, colored closed ring structures are thermally stable and do not return to the corresponding open ring structures. In contrast to aryl groups having high aromatic stabilization energy, such as indoles, phenyls and pyrroles, the colored ring structures are thermally unstable.[68] One example is 1,2-bis(2-cyano-1,5-dimethyl-4-pyrrolyl)perfluorocyclopentene, which upon UV irradiation turns blue. This unstable blue colored closed ring returns to the colorless open ring form with a half-life time of 37 s, which requires a thermal fading activation energy of 32.5 kJ mol^{-1}.[69] The introduction of bulky substituents at the reacting positions of dithenylethenes decrease thermal stability of the closed ring isomers at higher temperatures.[70] An extraordinarily high thermal stability of $t_{1/2} = 4.7 \times 10^5$ years exhibits a closed-ring isomer of 1,2-bis(5-methyl-2-phenylthiazol-4-yl)perfluorocyclopentene at 30 °C.[71] The *ortho*-isomers of dithia-(dithienylethena)phane derivatives with dithienylethene moieties and benzene bridges also exhibit good thermal stability.[72] Thermal stability can be enhanced by oxidation of sulfide bonds in tetrarylene derivatives,[73] but the presence of electron withdrawing and bulky substituents decreases the thermal stability of the closed ring structures by weakening the central carbon–carbon bonds.[74]

The colorless open form of diarylethenes is a twisted π-system in conjugation with two aromatic rings, whereas the colored closed form is a planar π-system. The absorption spectra of both forms exhibit band shifts resulting from the ring substituents in the closed ring isomer, whereas for the open ring isomer the shift depends on cycloalkane structure. In order to achieve sensitivity in the visible and near infrared regions, diarylethenes can be manipulated by the substitution of electron donating or electron withdrawing groups. Closed isomers of diarylethenes with oligothiophenes and polyene groups showed enhancement of the absorption coefficient and a shift towards longer wavelength regions. As shown in Figure 9.7, the two rings of the diarylethene system exhibit parallel (mirror symmetry) and antiparallel (C_2 symmetry) conformations. However, photocyclization reactions proceed only *via* antiparallel conformation. Also, the quantum yield of diarylethene derivatives depends upon the ratio of two conformers and is expected to increase with the increase of the antiparallel conformation. Incorporation of bulky substituents and forceful confinement of the molecule in the parallel conformation with a cubic or rod-like antiparallel conformation enhances the antiparallel content. The switching dynamics of ring opening and closure investigated by time-resolved spectroscopy showed that the switching occurs at a picosecond timescale.[75]

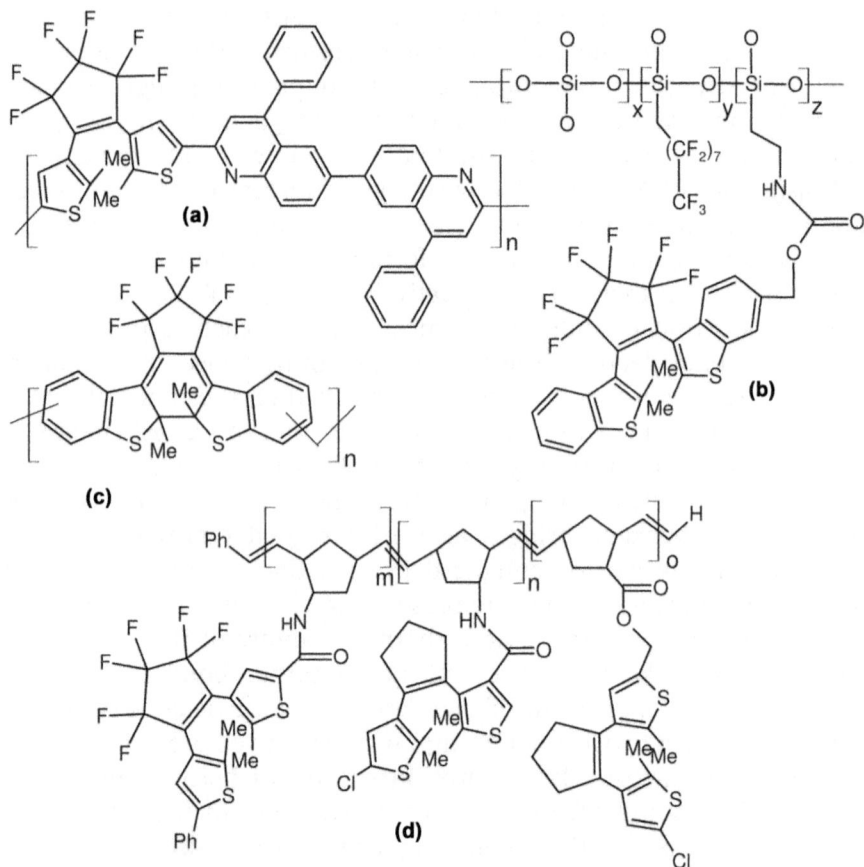

Figure 9.8   Selected examples of diarylethene-based polymers.

Photochromic diarylethenes can be incorporated into polymers by following free radical polymerization,[76] polycondensation,[77] Friedel–Crafts alkylation[78] or oxidative polymerization.[79] The side chain and main-chain homopolymers, copolymers, block copolymers and π-conjugated polymers synthesized by the above mentioned methods have shown good photochromic properties in films and solution phase due to the higher content of diarylethene chromophores in the polymer system compared to simple guest–host systems.[80–82] Selected examples are illustrated in Figure 9.8.

9.3.4   Fulgides

Fulgides are photochromic compounds that exist in colorless open *E*-form and photocyclized colored *C*-form. As shown in Figure 9.9,

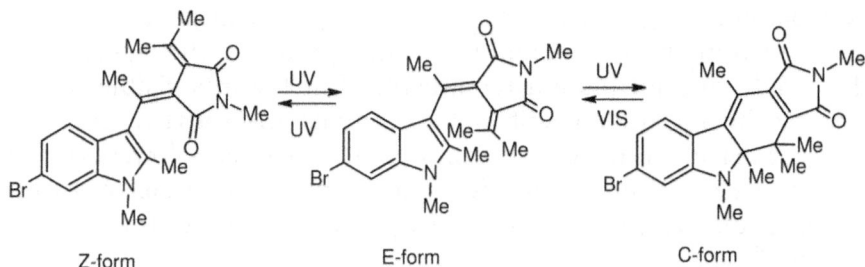

Figure 9.9 UV/vis-induced fulgides in *Z*-, *E*- and C-forms.

they also exist in the *Z*-form, which is a geometrical isomer of the *E*-form that undergoes photochemical *E–Z* isomerization. The C-form undergoes photoisomerization to the *E*-form upon irradiation in the visible region, known as the Woodward–Hoffmann pericyclic photo-reaction, whereas the *E*-form undergoes photoisomerization to the *Z*-form by UV irradiation. Various coloration mechanisms, such as *E–Z* photoisomerization of the double bond,[83] formation of colored radical intermediates,[84] photochemical change between the mesomeric form[85] and photochemical 6π-electrocyclization of the hexatriene moiety[86] are possible.

With a few exceptions, fulgides are mostly thermally irreversible photochromic compounds and the thermal ring opening is usually accompanied with hydrogen rearrangement and dehydrogenative aromatization which are the side reactions.[87,88] In order to prevent thermal side reactions, benzylidene and isopropylidene groups can be replaced with mesitylmethylene groups,[89] and the absence of hydrogen as well as the presence of vicinal methyl groups on the ring closing carbon atom inhibit these reactions.[90] 2,5-dimethyl-3-furyl moiety was the first thermally irreversible photochromic fulgide applied in reversible optical recording data storage media.[91] The presence of a heteroaromatic ring as well as substituents other than hydrogen on the ring resulted in thermally irreversible photochromic systems, also causing steric hindrance that plays an important role in the quantum yield enhancement. The *E*-to-*Z* isomerization quantum yields are smaller for fulgides with larger alkyl groups (*n*-propyl < ethyl < methyl),[92] but the process does not occur in systems with iso-propyl[93] and *tert*-butyl ring substituents.[94] Their good photochromic performance is attributed to thermal stability of both isomers at room temperature, stability against photochemical fatigue, high reaction quantum yields, and high reaction efficiencies. Absorption of fulgides can be shifted toward the visible region by enhancing the electron-donating ability, and by introducing heteroaromatic groups. Electron-donating substituents such as methoxy, dimethylamino, and methylthio groups in the 5-position of the indole ring facilitate a remarkable shift towards visible region, whereas the 6-position results in higher molar absorptivity.[95] Also, the increased strength of the electron donor substituents lowers the UV and visible quantum efficiencies, but the presence of electron withdrawing substituents shifts the absorption towards lower wavelength regions.[96]

Structural features and mechanisms of photoreactions in fulgides[97,98] result in photochemical conversion timescales in the ns range.[99] In polymethyl methacrylate, polystyrene and nitrocellulose polymer matrices, photocyclization processes occur in less than 10 ps.[100] Several synthetic routes, such as Stobbe condensation,[101] oxidative dimerization of 2-lithiocinnamic acid catalyzed by FeCl$_3$,[102,103] substitution of 1,4-butynediol with carbon monoxide in the presence of Pd catalyst, under elevated temperature and high pressure,[104] can be used for the synthesis of fulgides. Copolymers composed of photochromic 2-indolyl fulgides or fulgimides can be also synthesized,[105] and selected examples are shown in Figure 9.10.

Photochromic polymers are an important class of stimuli-responsive materials with many potential applications. These materials can be used in light-activated biological functions, and one example is bis(pyridinium)DTE, a water-soluble diarylethene

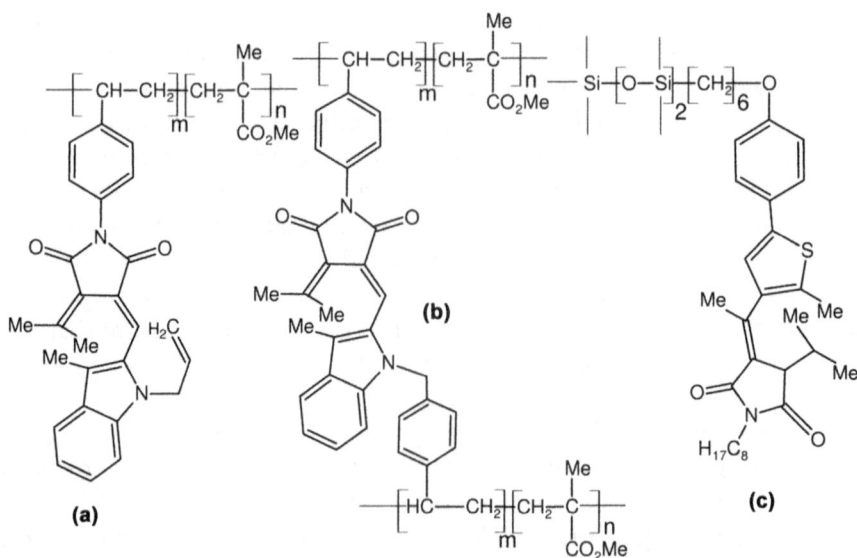

Figure 9.10 Selected examples of fulgide-based polymers.

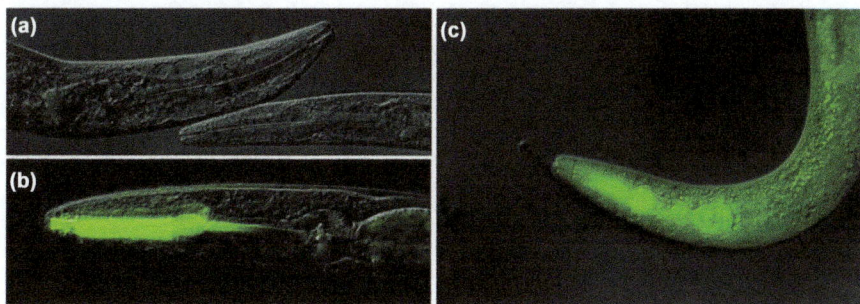

Figure 9.11 Fluorescence microscopy images of *Caonorhabditis elegans* incubated with (a) 10% DMSO, (b) photoswitch within the first 10 min, and (c) after 60 min.
Reprinted with permission from ref. 106. Copyright (2009) American Chemical Society.

derivative, which exhibit significantly enhanced fatigue resistance. This molecule can be used as a light-activated biological switch to induce paralysis in *Caonorhabditis elegans*, a multicellular eukaryotic organism shown in Figure 9.11.[106] Similar functionalized systems can be used in microbial separation or waste water treatments if the functional unit of the photochromic polymer system can be suitably activated by a light triggering ring opening reaction. Of particular interest and importance are copolymerized photochromic molecules that exhibit not only multi-stimuli-responsiveness, but also unique shape and locomotion induced by light. Examples are cilia-like materials created from colloidal particles.[107] Figure 9.12 illustrates the response of cilia-like films to such stimuli as temperature, pH or UV light. Poly(2-(*N,N*-dimethylamino)ethyl methacrylate-*co-n*-butyl acrylate-*co-N,N*-(dimethylamino)azobenzene acrylamide) copolymer particles form films capable of fabricating surfaces with cilia-like features. While film morphological attributes allow the formation of wavy whiskers, chemical composition of the copolymer facilitates chemical, thermal, and electromagnetic responses manifested by simultaneous shape and color changes as well as excitation wavelength dependent fluorescence. Thus, combined thermal, chemical and electromagnetic sensing leading to locomotive and color responses may find numerous applications in sensing devices, intelligent actuators, defensive mechanisms, and others. Another example is the creation of light-driven plastic motors, which enable direct conversion of light to mechanical energy.[108] As illustrated in Figure 9.13, light-induced mechanical motion of azobenzene functionalized liquid crystalline elastomer film is used to run a miniature

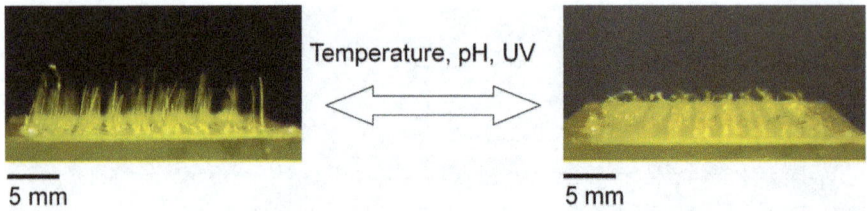

Temperature, pH, UV

5 mm   5 mm

Figure 9.12   Optical images of poly(2-(*N,N*-dimethylamino)ethyl methacry-
late-*co-n*-butyl acrylate-*co-N,N*-(dimethylamino)azobenzene
acrylamide) copolymer that exhibit cilia-like morphology
capable of reversible responses to temperature, pH and
electromagnetic radiation.
Reproduced with permission from ref. 107. Copyright © 2010
WILEY-VCH Verlag GmbH & Co. KGaA, Weinheim.

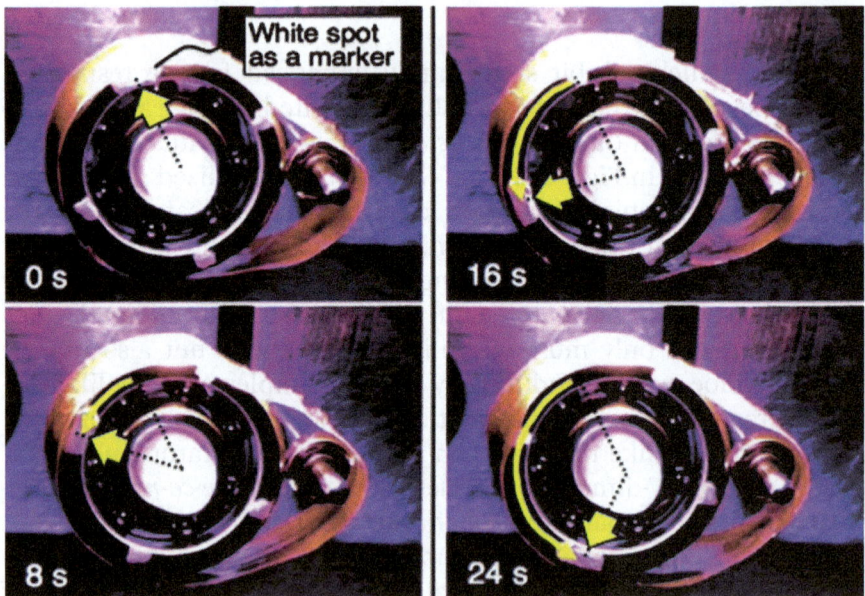

White spot
as a marker

0 s

16 s

8 s

24 s

Figure 9.13   Rotation of the light-driven polymer motor with the liquid
crystal elastomer laminated film induced by simultaneous
irradiation with UV (366 nm) and visible light (>500 nm) at
room temperature.
Reproduced with permission from ref. 108. Copyright © 2008
WILEY-VCH Verlag GmbH & Co. KGaA, Weinheim.

pulley system. This concept of the battery free motions and non-
contact operation mechanical devices is another example of how
incorporating functional and stimuli-responsive entities with photochromic properties may serve unique functions.

References

1. R. Birge, Photophysics and Molecular Electronic Applications of the Rhodopsins, *Annu. Rev. Phys. Chem.*, 1990, **41**, 683–733.
2. M. Fritsche, *C. R. Acad. Sci.*, 1867, **64**, 1035.
3. E. ter Meer, *Ann. Chem.*, 1876, **181**, 1.
4. Y. Hirshberg, *C. R. Acad. Sci.*, 1950, **231**, 903.
5. K. Ichimura, Photoreactive polymeric materials, *Denshi Zairyo*, 1989, 71–76.
6. V. A. Krongauz, Environmental effects on organic photochromic systems, in *Photochromism: Molecules and Systems*, ed. H. Däurr and H. Bouas-Laurent, Elsevier, Amsterdam, 1990, pp. 793–821.
7. Q. Trancong, S. Chikaki and H. Kanato, Relationship between Reorientational Motions of a Photochromic Dopant and Local Relaxation Processes of a Glassy Polymer Matrix, *Polymer*, 1994, **35**, 4465–4469.
8. J. Liu, Q. Deng and Y. Jean, Free-Volume Distributions of Polystyrene Probed by Positron Annihilation Comparison with Free Volume Theories, *Macromolecules*, 1993, **26**, 7149–7155.
9. F. Liu and M. W. Urban, Recent advances and challenges in designing stimuli-responsive polymers, *Prog. Polym. Sci.*, 2010, **35**, 3–23.
10. K. Ichimura, Environmental effects on organic photochromic systems, in *Photochromism: Molecules and Systems*, ed. H. Däurr and H. Bouas-Laurent, Elsevier, Amsterdam, 1990, pp. 903–918.
11. B. Pietro and M. Sandra, Cis-trans photoisomerization of azobenzene. Solvent and triplet donors effects, *J. Phys. Chem.*, 1979, **83**, 648–652.
12. H. Rau and S. Yu-Quan, Photoisomerization of sterically hindered azobenzenes, *J. Photochem. Photobiol., A*, 1988, **42**, 321–327.
13. Y. Morishima, M. Tsuji, M. Kamachi and K. Hatada, Photochromic isomerization of azobenzene moieties compartmentalized in hydrophobic microdomains in a microphase structure of amphiphilic polyelectrolytes, *Macromolecules*, 1992, **25**, 4406–4410.
14. G. Sudesh Kumar, C. Savariar, M. Safran and D. C. Neckers, Chelating copolymers containing photosensitive functionalities. 3. Photochromism of cross-linked polymers, *Macromolecules*, 1985, **18**, 1525–1530.
15. M. Ho, A. Natansohn and P. Rochon, Azo Polymers for Reversible Optical Storage 7. The Effect of the Size of the Photochromic Groups, *Macromolecules*, 1995, **28**, 6124–6127.

16. R. Fernandez, I. Mondragon, M. Galante and P. Oyanguren, Influence of Chromophore Concentration and Network Structure on the Photo-Orientation Properties of Crosslinked Epoxy-Based Azopolymers, *J. Polym. Sci., Polym. Phys.*, 2009, **47**, 1004–1014.

17. K. Anseth, M. Rothenberg and C. Bowman, A Photochromic Technique to Study Polymer Network Volume Distributions and Microstructure during Photopolymerizations, *Macromolecules*, 1994, **27**, 2890–2892.

18. M. Sato and T. Yamashita, Quantitative estimation of free volume distribution of polymers with photochromic reactions, *J. Photopolym. Sci. Technol.*, 2002, **15**, 115–119.

19. D. Kwon, H. Shin, E. Kim, D. Boo and Y. Kim, Photochromism of diarylethene derivatives in rigid polymer matrix: structural dependence, matrix effect, and kinetics, *Chem. Phys. Lett.*, 2000, **328**, 234–243.

20. R. Evans, T. Hanley, M. Skidmore, T. Davis, G. Such, L. Yee, G. Ball and D. Lewis, The generic enhancement of photochromic dye switching speeds in a rigid polymer matrix, *Nat. Mater.*, 2005, **4**, 249–253.

21. A. Altomare, L. Andruzzi, F. Ciardelli, R. Solaro and N. Tirelli, Methacrylic polymers containing permanent dipole azobenzene chromophores spaced from the main chain. ^{13}C NMR spectra and photochromic properties, *Macromol. Chem. Phys.*, 1999, **200**, 601–608.

22. K. Seog-Jun and D. H. Reneker, A mechanochromic Smart Material, *Polym. Bull.*, 1993, **31**, 367–374.

23. T. Yamashita, N. Yoshida and N. Wada, Free volume distribution of poly(alkyl methacrylate)s and the effect of inclusion, *J. Photopolym. Sci. Technol.*, 2005, **18**, 103–108.

24. T. Fischer, L. Laesker, J. Stumpe and S. G. Kostromin, Photo-induced optical anisotropy in films of photochromic liquid crystalline polymers, *J. Photochem. Photobiol., A*, 1994, **80**, 453–459.

25. H. Horiuchi, T. Fukushima, C. Zhao, T. Okutsu, S. Takigami and H. Hiratsuka, Conformational change of poly(methylphenylsilane) induced by the photoisomerization of pendant azobenzene moiety in the film state, *J. Photochem. Photobiol., A*, 2008, **198**, 135–143.

26. I. Katayama, Y. Tezuka, H. Yajima and T. Ishii, Kinetic Study of Conformational Transition Accompanied by Isomerization of Spiropyrans Bound to Poly(L-Glutamic Acid) Side Chains, *J. Photopolym. Sci. Technol.*, 1995, **8**, 65–74.

27. J.-S. Lin, Interaction between dispersed photochromic compound and polymer matrix, *Eur. Polym. J.*, 2003, **39**, 1693–1700.
28. S. Samanta and J. Locklin, Formation of Photochromic Spiropyran Polymer Brushes via Surface-Initiated, Ring-Opening Metathesis Polymerization: Reversible Photocontrol of Wetting Behavior and Solvent Dependent Morphology Changes, *Langmuir*, 2008, **24**, 9558–9565.
29. M. Irie and W. Schnabel, Photoresponsive Polymers – On the Dynamics of Conformational-Changes of Polyamides with Backbone Azobenzene Groups, *Macromolecules*, 1981, **14**, 1246–1249.
30. I. Masahiro and T. Hisami, Photoresponsive polymers. 5. Reversible solubility change of polystyrene having azobenzene pendant groups, *Macromolecules*, 1983, **16**, 210–214.
31. J. Sworakowski, K. Janus and S. Nespurek, Kinetics of photochromic reactions in condensed phases, *Adv. Colloid Interface*, 2005, **116**, 97–110.
32. P. Vandewijer and G. Smets, Photochromic Polymers, *J. Polym. Sci., Part C: Polym. Symp.*, 1968, **22**, 231–245.
33. R. Richert and A. Heuer, Rate-memory and dynamic heterogeneity of first-order reactions in a polymer matrix, *Macromolecules*, 1997, **30**, 4038–4041.
34. G. Such, R. Evans, L. Yee and T. Davis, Factors Influencing Photochromism of Spiro-Compounds within Polymeric Matrices, *J. Macromol. Sci., Polym. Rev.*, 2003, **C43**, 547–579.
35. M. Wang, C. Yeh and A. Hu, Photochromism of Novel Spirooxazine .1. Investigation of the Photocoloration in Polymer Films and Fibers, *Polym. Int.*, 1995, **38**, 101–104.
36. M. Irie, Diarylethenes for memories and switches, *Chem. Rev.*, 2000, **100**, 1685–1716.
37. M. Irie, T. Lifka, K. Uchida, S. Kobatake and Y. Shindo, Fatigue resistant properties of photochromic dithienylethenes: by-product formation, *Chem. Commun.*, 1999, **8**, 747–748.
38. A. Tork, C. Lafond, O. Pouraghajani, M. Bolte, A. M. Ritcey and R. A. Lessard, Matrix effects on the photochemical fatigue resistance of fulgide-doped polymers, *Opt. Eng.*, 2002, **41**, 2310–2314.
39. K. Atsushi, T. Atsuhiro, H. Takeru, O. Toyoji and A. Jiro, Fast Photochromic Polymers Carrying [2.2. Paracyclophane-Bridged Imidazole Dimer, *Macromolecules*, 2010, **43**, 3764–3769.
40. I. Tomatsu, A. Hashidzume and A. Harada, Photoresponsive hydrogel system using molecular recognition of alpha-cyclodextrin, *Macromolecules*, 2005, **38**, 5223–5227.

41. Y. Takashima, T. Nakayama, M. Miyauchi, Y. Kawaguchi, H. Yamaguchi and A. Harada, Complex formation and gelation between copolymers containing pendant azobenzene groups and cyclodextrin polymers, *Chem. Lett.*, 2004, **33**, 890–891.

42. S. Yagai and A. Kitamura, Recent advances in photoresponsive supramolecular self-assemblies, *Chem. Soc. Rev.*, 2008, **37**, 1520–1529.

43. S. Zarwell and K. Ruck-Braun, Synthesis of an azobenzene-linker-conjugate with tetrahedrical shape, *Tetrahedron Lett.*, 2008, **49**, 4020–4025.

44. M. Ai, S. Groeper, W. Zhuang, X. Dou, X. L. Feng, K. Mullen and J. P. Rabe, Optical switching studies of an azobenzene rigidly linked to a hexa-peri-hexabenzocoronene derivative in solution and at a solid-liquid interface, *Appl. Phys. A: Mater. Sci. Process.*, 2008, **93**, 277–283.

45. B. Schmidt, C. Sobotta, B. Heinz, S. Laimgruber, M. Braun and P. Gilch, Excited-state dynamics of bacteriorhodopsin probed by broadband femtosecond fluorescence spectroscopy, *Biochim. Biophys. Acta, Bioenerg.*, 2005, **1706**, 165–173.

46. P. Thuery, M. Nierlich, E. Lamare, J. F. Dozol, Z. Asfari and J. Vicens, Bis(crown ether) and azobenzocrown derivatives of calix[4]arene. A review of structural information from crystallographic and modelling studies, *J. Inclusion Phenom. Macrocyclic Chem.*, 2000, **36**, 375–408.

47. E. L. Aleksandrova, M. Y. Goikhman, L. I. Subbotina, K. A. Romashkova, I. F. Gofman, V. V. Kudryavtsev and A. V. Yakimanskii, Optical and photosensitive properties of comb-shaped polyamide-imides, *Semiconductors*, 2003, **37**, 821–824.

48. C. C. Wu, Q. C. Gu, Y. Huang and S. X. Chen, The synthesis and thermotropic behaviour of an ethyl cellulose derivative containing azobenzene-based mesogenic moieties, *Liq. Cryst.*, 2003, **30**, 733–737.

49. S. Mayer and R. Zentel, A new chiral polyisocyanate: an optical switch triggered by a small amount of photochromic side groups, *Macromol. Chem. Phys.*, 1998, **199**, 1675–1682.

50. Y. Q. Tian, K. Watanabe, X. X. Kong, J. Abe and T. Iyoda, Synthesis, nanostructures, and functionality of amphiphilic liquid crystalline block copolymers with azobenzene moieties, *Macromolecules*, 2002, **35**, 3739–3747.

51. L. Andruzzi, A. Altomare, F. Ciardelli, R. Solaro, S. Hvilsted and P. S. Ramanujam, Holographic gratings in azobenzene side-chain polymethacrylates, *Macromolecules*, 1999, **32**, 448–454.

52. C. K. Park, J. Zieba, C. F. Zhao, B. Swedek, W. M. K. P. Wijekoon and P. N. Prasad, Highly cross-linked polyurethane with enhanced stability of 2nd-order nonlinear-optical properties, *Macromolecules*, 1995, **28**, 3713–3717.
53. X. H. He, H. L. Zhang, D. Y. Yan and X. Y. Wang, Synthesis of side-chain liquid-crystalline homopolymers and triblock copolymers with p-methoxyazobenzene moieties and poly(ethylene glycol) as coil segments by atom transfer radical polymerization and their thermotropic phase behavior, *J. Polym. Sci., Part A: Polym. Chem.*, 2003, **41**, 2854–2864.
54. T. Kondo, T. Kanai and K. Uosaki, Control of the charge-transfer rate at a gold electrode modified with a self-assembled mono-layer containing ferrocene and azobenzene by electro- and photochemical structural conversion of cis and trans forms of the azobenzene moiety, *Langmuir*, 2001, **20**, 6317–6324.
55. D. M. Junge and D. V. McGrath, Photoresponsive dendrimers, *Chem. Commun.*, 1997, **9**, 857–858.
56. M. Sukwattanasinitt, D. C. Lee, M. Kim, X. G. Wang, L. Li, K. Yang, J. Kumar, S. K. Tripathy and D. J. Sandman, New processable, functionalizable polydiacetylenes, *Macromolecules*, 1999, **32**, 7361–7369.
57. V. I. Minkin, Photo-, thermo-, solvato-, and electrochromic spiroheterocyclic compounds, *Chem. Rev.*, 2004, **104**, 2751–2776.
58. A. Fissi, O. Pieroni, N. Angelini and F. Lenci, Photoresponsive polypeptides photochromic and conformational behavior of spiropyran-containing poly(l-glutamate)s under acid conditions, *Macromolecules*, 1999, **32**, 7116–7121.
59. A. Kellmann, F. Tfibel, E. Pottier, R. Guglielmetti, A. Samat and M. Rajzmann, Effect of nitro substituents on the photochromism of some spiro[indoline-naphthopyrans under laser excitation, *J. Photochem. Photobiol., A*, 1993, **76**, 77–82.
60. X. Guo, Y. Zhou, D. Zhang, B. Yin, Z. Liu, C. Liu, Z. Lu, Y. Huang and D. Zhu, 7-trifluoromethylquinoline-functionalized luminescent photochromic spiropyran with the stable merocyanine species both in solution and in the solid state, *J. Org. Chem.*, 2004, **69**, 8924–8931.
61. J. T. C. Wojtyk, E. Buncel and P. M. Kazmaier, Effects of metal ion complexation on the spiropyran-merocyanine interconversion: development of a thermally stable photo-switch, *Chem. Commun.*, 1998, 1703–1704.

62. A. Tork, F. Boudreault, M. Roberge, A. M. Ritcey, R. A. Lessard and T. V. Galstian, Photochromic behavior of spiropyran in polymer matrices, *Appl. Opt.*, 2001, **40**, 1180–1186.

63. G. Berkovic, V. Krongauz and V. Weiss, Spiropyrans and spirooxazines for memories and switches, *Chem. Rev.*, 2000, **100**, 1741–1754.

64. J.-i. Edahiro, K. Sumaru, T. Takagi, T. Shinbo and T. Kanamori, Photoresponse of an aqueous two-phase system composed of photochromic dextran, *Langmuir*, 2006, **22**, 5224–5226.

65. T. Hirakura, Y. Nomura, Y. Aoyama and K. Akiyoshi, Photoresponsive nanogels formed by the self-assembly of spiropyrane-bearing pullulan that act as artificial molecular chaperones, *Biomacromolecules*, 2004, **5**, 1804–1809.

66. R. A. Evans, T. L. Hanley, M. A. Skidmore, T. P. Davis, G. K. Such, L. H. Yee, G. E. Ball and D. A. Lewis, The generic enhancement of photochromic dye switching speeds in a rigid polymer matrix, *Nat. Mater.*, 2005, **4**, 249–253.

67. A. Menju, K. Hayashi and M. Irie, Photoresponsive polymers. 3. reversible solution viscosity change of poly(methacrylic acid) having spirobenzopyran pendant groups in methanol, *Macromolecules*, 1981, **14**, 755–758.

68. M. Irie, Diarylethenes for memories and switches, *Chem. Rev.*, 2000, **100**, 1685–1716.

69. K. Uchida, T. Matsuoka, K. Sayo, M. Iwamoto, S. Hayashi and M. Irie, Thermally reversible photochromic systems photochromism of a dipyrrolylperfluorocyclopentene, *Chem. Lett.*, 1999, **8**, 835–836.

70. K. Morimitsu, K. Shibata, S. Kobatake and M. Irie, Dithienylethenes with a novel photochromic performance, *J. Org. Chem.*, 2002, **67**, 4574–4578.

71. S. Takami, S. Kobatake, T. Kawai and M. Irie, Extraordinarily high thermal stability of the closed-ring isomer of 1,2-bis(5-methyl-2-phenylthiazol-4-yl)perfluorocyclopentene, *Chem. Lett.*, 2003, **32**, 892–893.

72. M. K. Hossain, M. Takeshita and T. Yamato, Synthesis, structure, and photochromic properties of dithia(dithienylethena)phane derivatives, *Eur. J. Org. Chem.*, 2005, **13**, 2771–2776.

73. Y. C. Jeong, C. Gao, I. S. Lee, S. I. Yang and K. H. Ahn, The considerable photostability improvement of photochromic terarylene by sulfone group, *Tetrahedron Lett.*, 2009, **50**, 5288–5290.

74. S. L. Gilat, Dr S. H. Kawai and Prof. J.-M. Lehn, Light-triggered molecular devices: Photochemical switching of optical and

electrochemical properties in molecular wire type diarylethene species, *Chem. - Eur. J.*, 2006, **1**, 275–284.

75. P. Hania, R. Telesca, L. Lucas, A. Pugzlys, J. van Esch, B. Feringa, J. Snijders and K. Duppen, An optical and theoretical investigation of the ultrafast dynamics of a bisthienylethene based photochromic switch, *J. Phys. Chem. A*, 2002, **106**(37), 8498–8507.

76. E. Kim, Y. Choi and M. Lee, Photoinduced refractive index change of a photochromic diarylethene polymer, *Macromolecules*, 1999, **32**(15), 4855–4860.

77. N. Holland, T. Hugel, G. Neuert, A. Cattani-Scholz, C. Renner, D. Oesterhelt, L. Moroder, M. Seitz and H. Gaub, Single molecule force spectroscopy of azobenzene polymers: Switching elasticity of single photochromic macromolecules, *Macromolecules*, 2003, **36**(6), 2015–2023.

78. Y. Jeong, D. Park, E. Kim, S. Yang and K. Ahn, Polymerization of a photochromic diarylethene by Friedel-Crafts alkylation, *Macromolecules*, 2006, **39**(9), 3106–3109.

79. K. Uchida, A. Takata, M. Saito, A. Murakami, S. Nakamura and M. Irie, Synthesis of novel photochromic films by oxidation polymerization of diarylethenes containing phenol groups, *Adv. Funct. Mater.*, 2003, **13**(10), 755–762.

80. L. Shen, C. Ma, S. Z. Pu, C. J. Cheng, J. K. Xu, L. Li and C. Q. Fu, Synthesis and properties of novel photochromic poly(methyl methacrylate-co-diarylethene)s, *New J. Chem.*, 2009, **33**, 825–830.

81. H. Nishi and S. Kobatake, Photochromism and optical property of gold nanoparticles covered with low-polydispersity diarylethene polymers, *Macromolecules*, 2008, **41**, 3995–4002.

82. H. Nakashima and M. Irie, Synthesis of silsesquioxanes having photochromic diarylethene pendant groups, *Macromol. Rapid Commun.*, 1997, **18**, 625–633.

83. D. P. Chakraborty, T. Sleigh, R. Stevenson, G. A. Swoboda and B. Weinstein, Preparation and geometric isomerism of dipiperonylidenesuccinic acid and anhydride, *J. Org. Chem.*, 1996, **31**, 3342–3345.

84. A. Schäonberg, The photochemical formation of organic diradicals. Part III. Investigations on anthracene, the fulgides, thiophosgene and their derivatives, *Trans. Faraday Soc.*, 1936, **32**, 514–521.

85. C. V. Gheorghiu, *Bull. Ec. Polytech. Jassy*, 1947, **2**, 141–155.

86. A. Santiago and R. S. Becker, Photochromic fulgides. Spectroscopy and mechanism of photoreactions, *J. Am. Chem. Soc.*, 1968, **90**, 3654–3658.

87. O. Crescente, H. G. Heller and S. Oliver, Overcrowded molecules. Part 16. Thermal and photochemical reactions of (E,E)-bis(benzylidene)succinic anhydride, *J. Chem. Soc., Perkin Trans. 1*, 1979, **1**, 150–153.

88. H. G. Heller and S. Oliver, Photochromic heterocyclic fulgides. Part 1. Rearrangement reactions of (E)-3-furylethylidene(isopropylidene)succinic anhydride, *J. Chem. Soc., Perkin Trans. 1*, 1981, 197–201.

89. H. G. Heller and R. M. Megit, Overcrowded molecules. Part IX. Fatigue-free photochromic systems involving (E)-2-isopropylidene-3-(mesitylmethylene)succinic anhydride and n-phenylimide, *J. Chem. Soc., Perkin Trans. 1*, 1974, **1**, 923–927.

90. P. J. Darcy, H. G. Heller, P. J. Strydom and J. Whittall, Photochromic heterocyclic fulgides. Part 2. Electrocyclic reactions of (e)-2, 5-dimethyl-3-furylethylidene(alkyl-substituted methylene)-succinic anhydrides, *J. Chem. Soc., Perkin Trans. 1*, 1981, 202–205.

91. B. L. Feringa, W. F. Jager and B. de Lange, Organic materials for reversible optical data storage, *Tetrahedron*, 1993, **49**, 8267–8310.

92. Y. Yokoyama, T. Goto, T. Inoue, M. Yokoyama and Y. Kurita, Fulgides as efficient photochromic compounds. Role of the substituent on furylalkylidene moiety of furylfulgides in the photoreaction, *Chem. Lett.*, 1988, **17**, 1049–1052.

93. Y. Yokoyama, T. Inoue, M. Yokoyama, T. Goto, T. Iwai, N. Kera, I. Hitomi and Y. Kurita, Effects of steric bulkiness of substituents on quantum yields of photochromic reactions of furylfulgides, *Bull. Chem. Soc. Jpn.*, 1994, **67**, 3297–3303.

94. J. Kiji, T. Okano, H. Kitamura, Y. Yokoyama, S. Kubota and Y. Kurita, Synthesis and photochromic properties of fulgides with a t-butyl substituent on the furyl- or thienylmethylidene moiety, *Bull. Chem. Soc. Jpn.*, 1995, **68**, 616–619.

95. Y. Yokoyama, T. Tanaka, T. Yamane and Y. Kurita, Synthesis and photochromic behavior of 5-substituted indolylfulgides, *Chem. Lett.*, 1991, **20**, 1125–1128.

96. A. Tomoda, A. Kaneko, H. Tsuboi and R. Matsushima, Photochromism of heterocyclic fulgides. IV. Relationship between chemical structure and photochromic performance, *Bull. Chem. Soc. Jpn.*, 1993, **66**, 330–333.

97. Y. Yoshioka, T. Tanaka, M. Sawada and M. Irie, Molecular and crystal structures of E- and Z- isomers of 2, 5-dimethyl-3-furylethylidene(isopropylidene) succinic anhydride, *Chem. Lett.*, 1989, **18**, 19–22.

98. V. A. Kumar and K. Venkatesan, Studies on photochromism of a thermally stable fulgide in the crystalline state: X-ray crystallographic investigation of (E)-2-isopropylidene-3-(1-naphthylmethylene)succinic anhydride, *Acta Crystallogr.*, 1993, **B49**, 896–900.

99. C. Lenoble and R. S. Becker, Photophysics, photochemistry, and kinetics of photochromic fulgides, *J. Phys. Chem.*, 1986, **90**, 2651–2654.

100. S. Kurita, A. Kashiwagi, Y. Kurita, H. Miyasaka and N. Mataga, Picosecond laser photolysis studies on the photochromism of a furylfulgide, *Chem. Phys. Lett.*, 1990, **171**, 553–557.

101. M. Fan, L. Yu and W. Zhao, Fulgide Family Compounds: Synthesis, Photochromism, and Applications in Organic Photochromic and Thermochromic Compounds, in *Main Photochromic Families*, ed. J. C. Crano and R. Guglielmetti, Plenum Publishers, New York, 1999, vol. 1, pp. 141–206.

102. A. M. E. Gendy, A. Mallouli and Y. Lepage, Synthesis of 1, 4-diaryl-2, 3-diformylbutadienes, *Synthesis*, 1980, 898–899.

103. H.-L. Elbe and G. Käobrich, *Chem. Ber.*, 1974, **107**, 1654–1666.

104. S. Uchida, Y. Yokoyama, J. Kiji, T. Okano and H. Kitamura, Electronic effects of substituents on indole nitrogen on the photochromic properties of indolylfulgides, *Bull. Chem. Soc. Jpn.*, 1995, **68**, 2961–2967.

105. M. A. Wolak, N. B. Gillespie, C. J. Thomas, R. R. Birge and W. J. Lees, Optical properties of photochromic fluorinated indolylfulgides., *J. Photochem. Photobiol. A*, 2001, **144**, 83–91.

106. U. Al-Atar, R. Fernandes, B. Johnsen, D. Baillie and N. R. Branda, A Photocontrolled Molecular Switch Regulates Paralysis in a Living Organism, *J. Am. Chem. Soc.*, 2009, **131**(44), 15 966–15 967.

107. L. Fang, R. Dhanya and M. W. Urban, Colloidal Films That Mimic Cilia, *Adv. Funct. Mater.*, 2010, **20**(18), 3163–3167.

108. M. Yamada, M. Kondo, J. i. Mamiya, Y. Yu, M. Kinoshita, C. J. Barrett and T. Ikeda, Photomobile polymer materials: Towards light-driven plastic motors, *Angew. Chem., Int. Ed.*, 2008, **47**(27), 4986–4988.

10 Photorefractive Polymers

10.1 Introduction

The photorefractive (PR) effect refers to spatial modulation of the refraction index generated by light-induced charge redistribution in a material with the index of refraction being dependent upon the electric field. The occurrence of this effect arises when photogenerated by a spatially modulated light intensity charge carriers are becoming separated as a result of floating and diffusion to be trapped to produce a non-uniform space-charge distribution area. As a result, internal space-charge electric field is able to modulate the refractive index to form a phase grating or hologram that can diffract a light beam. Thus, strong and intense light such as a laser beam can change the optical properties of such a nonlinear optical material (NLO), which, in turn, alter the properties of the light beam. One example is focusing an infrared laser pulse (wavelength 1064 nm) into a potassium dihydrogen phosphate crystal which can lead to a pulse with a green color at the second harmonic (wavelength 532 nm). For that simple reason the NLO properties of materials can be used to control the phase, the state of polarization, or the frequency of light beams. Thus, they can be used to store and restore information optically. While the growth of doped organic crystals is very attractive, it is also very difficult maintain the concentration of localized dopants, which are often expelled during the crystallization process. In contrast, polymeric materials can be doped with numerous molecules of different sizes and structures, and with today's control of polymerization, the performance of many new polymeric materials will exceed conventional inorganic crystals. The main question this chapter will

Stimuli-Responsive Materials: From Molecules to Nature Mimicking Materials Design
By Marek W. Urban
© Marek W. Urban, 2016
Published by the Royal Society of Chemistry, www.rsc.org

address is how to design the photorefractive (PR) effect in complex polymer structures so when a propagating electromagnetic radiation temporarily alters polymer refractive index, the speed of light will also change. This chapter describes the physics and chemistry of polymeric PR systems.

10.2 Principles of the Photorefractive Effect

Generation of the PR effect requires: (i) photoconductivity, defined as an increase of electrical conductivity, including generation of mobile charges, transport and trapping, upon exposure to electromagnetic radiation of appropriate energy and (ii) electro-optical (EO) responses, which are processes that result in the charge distribution resulting from light perturbation producing internal electric fields locally altering the refractive index.

To develop a PR grating the following steps, illustrated in Figure 10.1, are involved: (a) illumination of the medium with an interference pattern produced by two beams which (b) will create charge carriers (holes and electrons) at brighter regions. (c) The charges will migrate under an external electric field until they are trapped at the energy minimum and (d) an internal space-charge field will be established. The phase-shifted space charge with respect to the incident intensity pattern will be generated due to the migration of photogenerated charges in the media. Because the refractive index grating amplitude phase-shifts by Φ with respect to the initial intensity pattern (d), a modulated index of refraction will be produced. The latter is known as a hologram.

The refractive index of the material can be also altered by other light-triggered phenomena, including thermal effects, photopolymerization, photochromic effects, or transient gratings resulting from nonlinear processes. However, the term photorefractive effect is reserved for a refractive index modulation produced only by the above-described mechanism, which involves light-induced charge generation and a field-induced modulation of the refractive index. With this general background, let us consider specific processes and required materials properties, such as photoconductivity and optoelectrical processes.

The first PR effect was observed in the ferroelectric crystal lithium niobate ($LiNbO_3$) and reported as an unwanted refractive index inhomogeneity caused by light.[1] However, as soon as holographic storage was demonstrated in $LiNbO_3$ using 488 nm incident light,

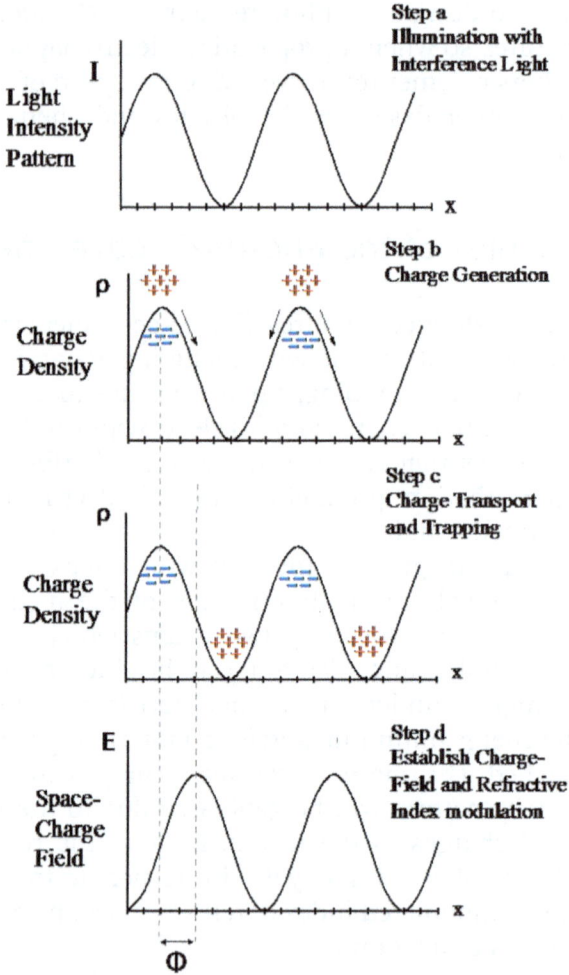

Figure 10.1 Mechanism of formation of photorefractive grating.

a quest for new PR materials begun.[2] The main advantage of PR materials is their capability of storing a replica of the incident non-uniform intensity patterns as a refractive index modulation. More-over, illumination with uniform intensity distribution of appropriate wavelength will erase any grating that has been written inside the material, thus making PR materials uniquely suitable for reversible and real time holography. The first organic crystals and polymers showing this effect were bisphenol-A-diglycidylether 4-nitro-1,2-phenylenediamine doped with the charge-transporting molecule diethylamino-benzaldehyde diphenylhydrazone.[3,4] It should be noted that the refractive index changes as a function of the applied electric

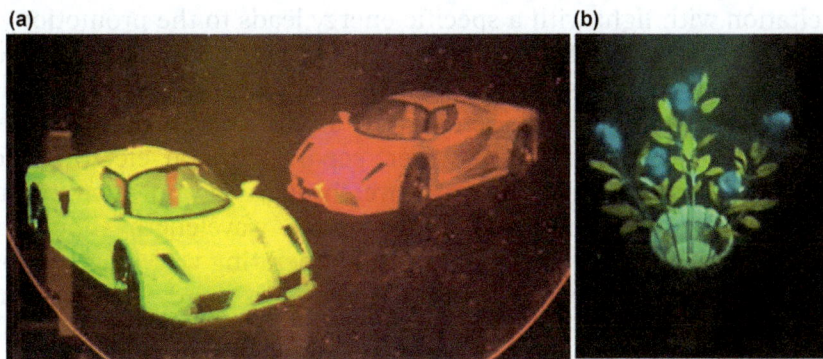

Figure 10.2 (a) Colored hologram of two model cars recorded on a 12-inch diameter photorefractive device. (b) Colored hologram of a vase and flowers.
Reprinted by permission from Macmillan Publishers Ltd: Nature (ref. 10), copyright (2010).

field (voltage) may have many applications, ranging from multiplexed data storage, holographic filters,[5] neural networks,[6] dynamic 3D displays,[7] and others. Although the initial research focused on in-organic crystals, the ease of chemical modifications, processability and cost are clear advantages of polymeric materials. More importantly, many PR polymers exhibit significantly better performance over their inorganic counterparts. For example, diffraction efficiency approaching 100% and two-beam coupling gain coefficient of $\sim 400\,\mathrm{cm}^{-1}$ are achievable, with the gain coefficients reaching $3700\,\mathrm{cm}^{-1}$ in hybrids of liquid crystals or low molecular weight glass.[8,9] An example of the erasable holographic display was achieved by using tetra-phenyldiaminobiphenyl (TPD) and carbaldehyde aniline copolymer.[7] Figure 10.2 illustrates the 3D images recorded on a 12-inch diameter display.[10] There are numerous applications of PR, particularly using infrared (IR) time-gated holographic imaging (TGHI), which can be used in extracting subsurface features of materials.[11] It should also be realized that the complexity of using polymeric systems arises from the difficulty of precisely combining charge-transport and EO properties into a precisely defined and organized polymer matrix.

10.2.1 Photoconductivity

Photoconductive materials are usually insulators in the absence of light, but become conductive upon light irradiation. Therefore, upon exposure, the excess of movable charge carriers is generated.

Excitation with light with a specific energy leads to the promotion of an electron from the highest occupied molecular orbital (HOMO) of molecules to its lowest unoccupied molecular orbital (LUMO), thus leaving a vacancy behind, referred to as 'hole'. The photo-excited electron–hole pair will be, however, separated upon electric field, and the electrons or holes can move through the polymer toward the appropriate electrodes. The absorption wavelength should be chosen so it does not absorb within the operating wavelength. Most polymers are transparent in visible region, and predominant absorption occurs in the ultraviolet regions. It is therefore highly desirable to use polymers merely as charge-transporting media, and use dopant molecules (sensitizers) with the ability of external charge-generating (CG) properties sensitive to the longer wavelength range, thus minimizing the absorptions by all the other components.

Photoconductive polymers can transport either electrons or holes. If the polymer is hole transporting, the CG molecule is an electron acceptor that forms radical cations on photoexcitation. If the polymer is electron transporting, the CG molecule then should be electron donor. Majority of polymers can transport holes only and therefore, the hole in the HOMO of the CG molecule can be filled by an electron from the HOMO of the charge-transporting polymer, thus forming a mobile hole along the polymer backbone. Photogeneration of charge carrier in polymer is a multistep process involving the creation of intermediate states rather than direct generation of carriers following the absorption of light.[12,13] Although photoinduced charge transfer initiated by a sensitizer depends on many factors, the following condition must be satisfied for an efficient charge transfer.[14]

$$I_{D^*} - A_s - U_c \leq 0 \qquad\qquad (10.1)$$

where I_{D^*} is the ionization potential of the excited state of the polymer (donor), A_s is electron affinity of the sensitizer molecule, and U_c denotes the Coulomb attraction between the separated radicals. Factors which may prevent an energetically allowed charge transfer process includes polarizability of the surrounding matrix, the morphology of the polymer which introduce a large separation between the donor and the acceptor, or a potential barrier which prevent the separation of the electron–hole pair.[14,15]

The second requirement for photorefractivity is transport of the photogenerated charge carriers. The charge carriers move within the bulk of the polymer under the influence of an electric field until they are trapped. For example, poly(N-vinylcarbazole) (PVK) sensitized with suitable electron acceptors is photoconductive.[16] This material produced by free radical polymerization of N-vinylcarbazole (NVK) monomer and found applications in first copiers. Its synthesis is show below, and considering aromatic and heteroaromatic features, a variety of other polymers were prepared as organic photoconductors.

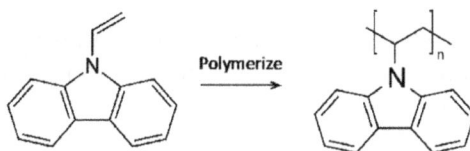

Charge transport in organic semiconductors depends on π-bonding orbitals and quantum mechanical wave-function overlap. Unlike inorganic crystalline materials, there is limited π-bonding overlapping between molecules. Charge carriers pass from one molecule to another in polymers proceeds through a spatially random (positional disorder) and energetically disordered system, this transport process is commonly referred to as hopping transport.[17] The terms positional and energetic disorder imply that the distance between hopping sites and the energy required to hop from one localized to the other vary significantly. Hopping of charge carriers from molecule to molecule depends upon the energy gap between HOMO and LUMO levels. Carrier mobility is dependent on the abundance of similar energy levels for the electrons or holes to move to and hence will experience regions of faster and slower hopping. These can be affected by the temperature and electric fields across the system. In the Bässler model, the mobility is temperature dependent with $\log \mu \propto 1/T^2$, and field dependent according to $\log \mu \propto E$.[18] Until trapped, photo-generated charge carriers will move within the bulk of the polymer under electric field.

10.2.2 EO Response

Optics, the field of interaction of electromagnetic radiation with matter, often uses polarized light that is related to the incident electric field *via*

$$P = \chi E \tag{10.2}$$

where the electrical susceptibility, χ, is a second-rank tensor. In reality, the polarization field is more complicated than the linear relation given above. If the variation is small, the polarization may be expanded in a Taylor series to obtain:

$$P_i = \chi_{ij}^{(1)} E_j + \chi_{ijk}^{(2)} E_j E_k + \chi_{ijkl}^{(3)} E_j E_k E_l + \cdots \tag{10.3}$$

where polarization intensity P_i is the sum over repeated indices. The first coefficient $\chi^{(1)}$ is the linear electric susceptibility discussed previously. The higher order $\chi^{(n)}$ are the nth order nonlinear susceptibilities.

In PR crystals, the refractive index is altered mainly by the linear EO effect, which results in a refractive index modulation of the form given in eqn (10.4):

$$\Delta n(x) = -\frac{1}{2} n^3 r_{\text{eff}} E_{\text{SC}}(x) \tag{10.4}$$

where n is the average refractive index of the material and r_{eff} is the effective EO coefficient. In polymers, a field-dependent refractive index is achieved by the addition of highly polarizable molecules to the host polymer, usually referred to as nonlinear optical (NLO) chromophores. These molecules have an electron-donating-type group linked to an electron-acceptor type group using a π-conjugated system such as benzene, azobenzene, stilbene, biphenyl, heterocycle, polyenes, or tolans. The mismatch in electron affinities leads to a delocalization of the π-electrons resulting in a molecular dipole with the acceptor side denser in electron. At the molecular level, the dependence of the dipole moment of such a molecule on an electric field can be expanded as a Taylor series in the same manner as the polarization field given by eqn (10.5):

$$p_i = \mu_i + \alpha_{ij} E_j + \beta_{ijk} E_j E_k + \gamma_{ijkl} E_j E_k E_l + \cdots \tag{10.5}$$

where α, β and χ are the polarizability, first hyperpolarizability and second hyperpolarizability. Polymers doped with such chromophores have been shown to possess high EO coefficients.[19] The majority of molecules with large first hyperpolarizability will have a strong permanent dipole moment. However, a random ensemble of such molecules will have no macroscopic second-order NLO effect. Therefore, poling techniques have to be employed to break the centrosymmetry of the systems. Polymers of high T_g are poled at temperatures near the T_g by orienting the molecular dipoles along an applied filed. Thus, to preserve the alignment with the electric field applied, the temperature is lowered. Upon poling the dipoles will be permanently fixed and the refractive index modulation is attributed to the EO effect. Polymers with T_g near the room temperature can be poled *in situ* during the PR grating formation, as the dipoles exhibit orientational freedom due to the flexibility of the host resulting in the modulation of the bi-refringence of the material as well as the EO modulation, thereby resulting in a stronger PR effect. This phenomenon is called orientational enhancement.[20]

The role of NLO chromophores in PR system is to provide refractive index modulation in response to an electric field. In systems with low T_g where the molecules are relatively free to rotate due to a higher free volume, the chromophores can reorient in the applied/generated electric field. Due to the anisotropy in polarizability of the chromophores, such a reorientation will induce high refractive index anisotropy termed as orientational birefringence (OB). The ability of a chromophore molecule to induce refractive index change is usually expressed as the figure of merit (FOM) defined by eqn (10.6):[21]

$$\text{FOM} = \frac{1}{M}\left[9\mu_g\beta + \frac{2\mu_g^2\Delta\alpha}{k_BT}\right] \tag{10.6}$$

where M is the molecular mass, μ_g is the ground state dipole moment, β is the first-order hyperpolarizability, $\Delta\alpha$ is the anisotropy in linear polarizability, k_B is the Boltzmann constant and T is the absolute temperature. The first term in eqn (10.6) represents the contribution from the linear EO effect and the second term represents the contribution of OB.

If the molecules are embedded in a polymer host with high T_g and poled, the dominant component to the refractive index modulation is from the linear electro-optic effect known as the Pockel's effect. In this case, the FOM is given by:[22]

$$\text{FOM}_{\text{PEO}} = 9\mu_g\beta_0 \quad \text{and} \quad \beta_0 = \frac{2\mu_{ge}^2\Delta\mu}{E_{ge}^2} \tag{10.7}$$

where μ_{ge} is the transition dipole moment between the ground and excited states, $\Delta\mu$ is the difference between the dipole moments in the excited and ground states, and E_{ge} is the transition energy. Calculated values of FOM of different NLO molecules can be used to predict the refractive index modulation only if the charges are uniformly distributed in space and exhibit equal orientational properties under the internal space-charge field.[23] As anticipated, proper orientation without centro-symmetry must be maintained and the degree of polar order is usually expressed as an order parameter.[13] The change in absorbance induced by poling can be used to estimate the polar order

(φ) in the system, whereas the order parameter can be estimated as $1 - (A_\perp/A_0)$, where A_0 and $A_\perp$ are the absorbance of unpoled and poled specimens measured with the electric field of the probing light perpendicular to the poling direction.

10.2.3 Generation of Photorefractive Gratings

The most common way of writing gratings in a material for PR polymers is four-wave mixing (FWM). Figure 10.3 illustrates the experimental geometry for the photorefractive grating experiment in which two mutually coherent writing beams of the same polarization, beams a and b, enter a polymer film at angles θ_a and θ_b (normal to the surface). At the same time, an electric field is applied along the surface normal to the amplitude E and the NLO chromophores are aligned and rotated to have a non-zero degree angle with respect to the bisector of the writing beams. The grating wave vector K_G and the charge transport occurs *via* hopping from one chromophore to another. Under an external DC field, photogenerated charge carriers will transport in the K_G direction. An external angle of 30° between the writing beams and the tilted specimen at 45° with respect to the specimen normal is typically applied. To read the gratings, a much weaker reading beam c with the same wavelength as the writing beams is applied, and it enters the polymer in a direction that is counter-propagating to one of the writing beams. The read beam is diffracted by the grating with an efficiency η,

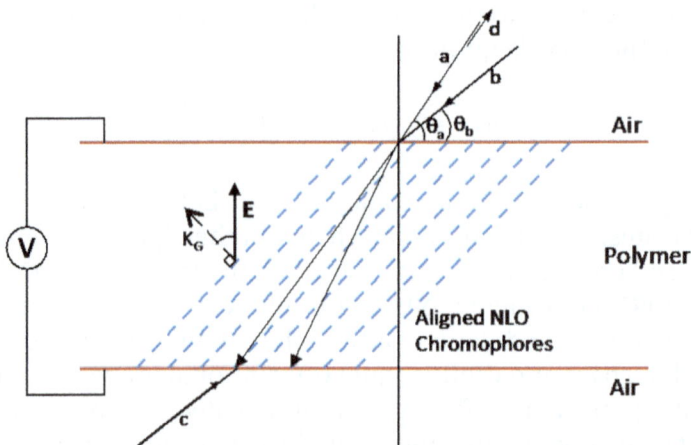

Figure 10.3   Experimental geometry for the PR grating experiment. Adapted from ref. 24.

defined as the intensity ratio of the diffracted beam to the incoming read beam. The diffracted beam is directed into a detector as a background-free signal, which allows very small signals (very weak gratings) to be detected.

10.2.4 The Two-beam Coupling Effect

The phase shifts depicted in Figure 10.1 are a unique property of PR gratings, which results in asymmetric energy exchange between the two mutually coherent light beams that interfere with each other inside a PR material. In this two-beam coupling (TBC) effect two mutually coherent light beams interfere inside a PR medium, whereby one beam gains intensity at the expense of the other. An experimental geometry is shown in Figure 10.4. The writing beams (I_1 and I_2) self-diffract from the grating they created inside the material. This is the case for all types of holographic grating, but for the non-PR case, the outgoing beams $I_{1,\text{Trans}}$, $I_{2,\text{Diffr}}$ and $I_{2,\text{Trans}}$, $I_{1,\text{Diffr}}$ interfere equally in both directions. For phase-shifted gratings, such as described in a PR experiment, the interference is constructive for one beam direction and destructive in the other. Due to the phase differences this asymmetry leads to the amplification of one beam at the expense of the other. Using eqn (10.8) the energy exchange in a photorefractive medium can be quantified by the gain coefficient Γ estimated from the intensities $I_{1,\text{out}}$ and $I_{2,\text{out}}$ measured by the photodiodes PD1 and PD2, respectively.[21]

$$\Gamma = \frac{\cos\theta_1}{d}\ln\left[\frac{\beta_p I_{1,\text{out}}}{I_{2,\text{out}}}\right] \tag{10.8}$$

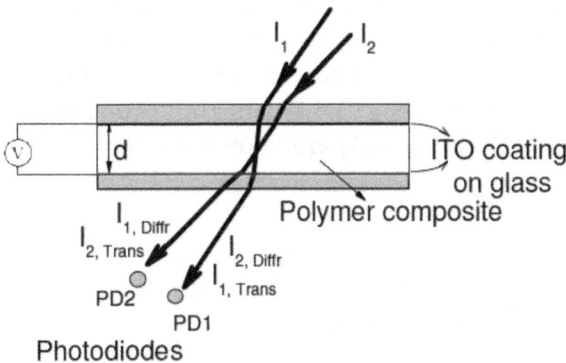

Figure 10.4 Light beam path in a two-beam coupling experiment.

where θ_1 is the angle of incidence of I_1 and $\beta_p = I_2/I_1$. The gain co-efficient is related to the refractive index modulation Δn and the phase shift Φ by the following equation:

$$\Gamma = \frac{4\pi}{\lambda} \frac{\Delta n}{m} \sin(\phi) \tag{10.9}$$

where λ is the wavelength and m is the modulation depth of the interference pattern given by $m = 2\beta_p^{0.5}/(1+\beta_p)$. Using eqn (10.9) the phase difference Φ between the intensity pattern and the resulting grating can be determined which plays a major role in the energy transfer process. If the intensity gain from the TBC effect is greater than the losses by the beam due to absorption and reflection on passing through the PR device, beam amplification can occur. Such an amplifying effect cannot be seen with other types of local gratings unless the light pattern is translated or the medium has large re-fractive index contrast.

PR materials are typically composed of the following components: dopant molecules to generate charges upon illumination of light, photoconducting polymers that can facilitate transport under electric fields, NLO chromophores that generate field-dependent refractive index modulations, and plasticizer molecules that lower the T_g of a polymer matrix.

10.3 Chemistry of Photorefractive Materials

10.3.1 Dopant Molecules

For electron acceptor charge-generating molecules the electron density on the ring system is very low due to the presence of strong electron withdrawing groups such as $-NO_2$ and $-CN$. The chemical structures of some electron acceptors, namely 7,7,8,8-tetra-cyanoquinodimethane (A1), 2,4,7-trinitrofluorenone (TNF) (A2), C_{60} (A3) and squaryllium dye (A4) are shown below.

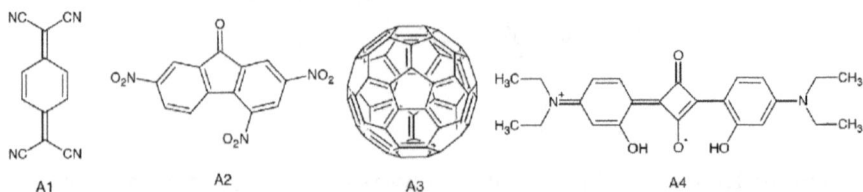

A1   A2   A3   A4

C_{60} molecules as well as the soluble derivative [6,6]-phenyl C_{60}-1-butyric acid methyl ester (PCBM) are among the highly efficient electron acceptor molecules for organic photovoltaic devices. The C_{60} to organic molecules electron transfer occurs through the photo-induced charge transfer (CT) in picoseconds.[25] Electron acceptors may form charge CT complexes with donor type molecules and polymers, giving rise to new absorption band in the visible and near IR region of the spectrum. It turns out that due to large enhancement in steady-state diffraction efficiency and initial rate of grating growth the C_{60} is one the most efficient electron acceptors with relatively small absorption effects due to ionized sensitizer molecules. Also, long wavelength absorption, low reduction potential and high triplet yields make C_{60} a good hole generator for polymer photoconductors with smaller oxidation potential than C_{60}.[25]

Another class of sensitizers are phthalocyanines (Pcs) which exhibit the high absorption in the visible and near IR region and the TBC gain coefficient of $350 \, cm^{-1}$ with response times in the millisecond in a PVK for PR composite sensitized with ZnPc and SiPc derivatives.[26] A comparison of the photoconductivity of PVK based PR composite sensitized with ZnPc, SiPc and C_{60} is shown in Figure 10.5.

Other methods of sensitization include semiconductor quantum dots (CdS or CdSe) and functionalizing polymers with sensitizing moieties such as transition metal complexes.[27] In another example,

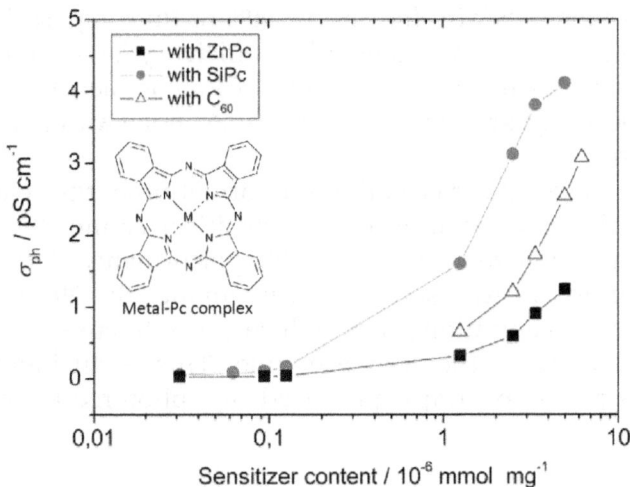

Figure 10.5 Photoconductivity, measured at 22 V μm^{-1}, of composites containing different concentrations of ZnPc (black squares), SiPc (gray circles), or C_{60} (open triangles). Reprinted with permission from ref. 19.

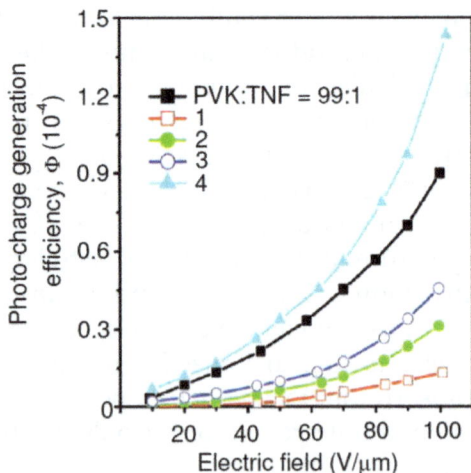

Figure 10.6 Photo-charge generation efficiency of the PVK-Ru complexes 1 (open squares; 0.028 wt%), 2 (closed circles, 0.036 wt%), 3 (open circles; 0.070 wt%), 4 (closed triangles, 0.891 w%), and PVK/TNF composite (closed squares) as a function of external electric field strength.
Reproduced from Ref. 28 with permission from The Royal Society of Chemistry.

the effect of ruthenium (Ru) complex on PVK by covalently attaching Ru(bpy)$_2$(*m*-COOH-4-methyl bpy)(PF$_6$)$_2$ to PVK using Friedel–Crafts acylation showed that the photogeneration efficiency of PVK-Ru complexes increases with increasing wt% of Ru-complex.[28] The field dependence of the photogeneration efficiency of complexes containing 0.028, 0.036, 0.070, 0.891 wt% of Ru complex, labeled 1, 2, 3 and 4 respectively, are shown in Figure 10.6, along with that of a PVK-TNF system.

The role of sensitizer is vital in the formation of the space-charge field, but the concentration of the sensitizer cannot be increased with an intention to increase the PR performance. Typically, the addition of up to 1 wt% sensitizer will enhance the PR properties,[29] but higher concentrations may reduce the refractive index modulation and increase the response time. This is attributed to the increase density of traps produced by photoreduction of the sensitizer.[30]

10.3.2 Photoconducting Polymers

Photoconducting polymers can be broadly classified into two categories: (i) molecularly doped polymers (MDP) in which the dopant molecules give rise to carrier generation and transport, and covalently attached molecules (CAD). As the MDP implies, photoconductivity is achieved by mixing photoactive molecules (dopants) that are dispersed in an inert polymer matrix. Typical polymer matrices are poly(methyl methacrylate), polyvinyl alcohol, polyvinylchloride, polystyrene, bisphenol-A-polycarbonate, and others. If electron acceptor type CG molecules are used, charge-transporting capability is achieved by the addition of molecules with high electron density in the ring system. Examples of charge-transport molecules are *N*-ethylcarbazole (C1), 4-(*N*,*N'*-diethylamino)benzaldehydediphenyl hydrazine (C2) and *N*,*N'*-diphenyl-*N*,*N'*-bis(3-methylphenyl)-[1,1'-biphenyl]-4,4'-diamine (C3) whose chemical structures are illustrated below. It should be realized that the term 'molecular doping' relates to the dopants that retain their identity in the host polymer and their concentration levels may vary from 1–50 wt%. It should be realized that for inorganic semiconductors these levels are significantly smaller, yet the same terminology is applied.[31]

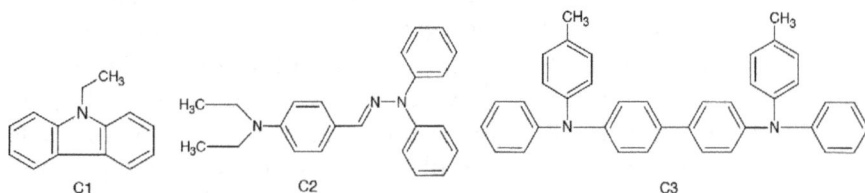

The CAD has the transport unit covalently attached to the polymer main chain. This reduces the passive volume of the host as the polymer itself takes part in the charge-transport process. PVK is the example of polymeric photoconductor with relatively stable radical cations and high charge carrier mobility and its polymerization can be carried out by cationic, free radical, charge transfer, coordination or electrochemical process.[32,33] Photoconductivity in carbazole containing polymers can be enhanced by introducing pendant dimeric carbazole units as the side or main chain using cationic and step growth polymerization of corresponding monomers. The ease of introducing different substituent into the carbazole ring, its low cost, high thermal and photochemical stability make carbazole based compounds attractive as charge-transporting polymers. The scheme below shows selected examples of carbazole-based photoconducting polymers PVK (D1), poly(*N*-vinylcarbazole-*co*-styrene) (D2), poly(*trans*-1-(3-vinyl)carbazolyl)-2-(9-carbazolyl)cyclobutane)

(D3), poly-(*N*-(2-carbazolyl)ethyl acrylate) (D4), and a fully functionalized carbazole containing polymer (D5).

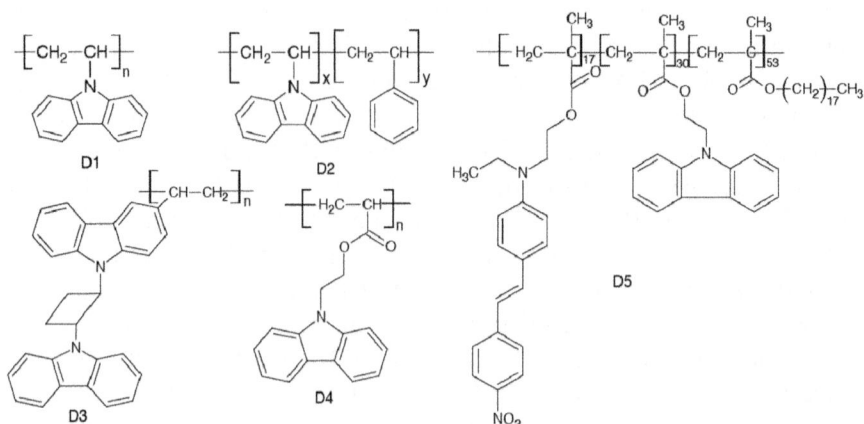

PVK has been used as the photoconducting host in several high-
performance PR composites and demonstrated diffraction efficiencies approaching 100%, TBC gain of $\sim 400\,cm^{-1}$ and refractive
index contrasts of 0.01. However, the carrier mobility in PVK is only
on the order of $10^{-7}\,cm^2\,V\cdot s^{-1}$, which is attributed to the high glass
transition temperature (T_g). One solution of overcome this obstacle
is to use plasticizers. For example, the use of siloxane polymer
with pendant carbazole groups, which has a lower intrinsic T_g and
hole transporting properties similar to that of PVK was used to
prepare PR composites with refractive index modulation of $\sim 10^{-2}$.[34]
Also, PR systems based on poly(methyl-bis-(3-methoxyphenyl)-
(4-propylphenyl)amine)siloxane (MM-PSX-TAA), a charge-transporting
polymer system with siloxane backbone with pendant triarylamine
groups, doped with the nonlinear optical (NLO) chromophore
4-di(2-methoxyethyl) aminobenzylidene malononitrile (AODCST) and
the plasticizer butyl benzyl phthalate (BBP) with response times
in the millisecond range was reported.[35] Other examples of
photoconducting polymers include poly(*N*-vinyldiphenylamine),
poly(vinylarylamines) and polymers containing triphenyldiamine
moieties.[36–38] These polymers have been found to exhibit charge
carrier mobilities that exceed the values of PVK. The chemical
structures of poly(*N*-vinyldiphenylamine) (E1), poly(E,E-[6,2]-para-
cyclophane-1,5-diene) (E2), poly(2-(*N*-ethyl-*N*-3-tolylamino)ethyl methacrylate) (E3) and triphenyldiamine containing condensation polymer
(E4) are shown below.

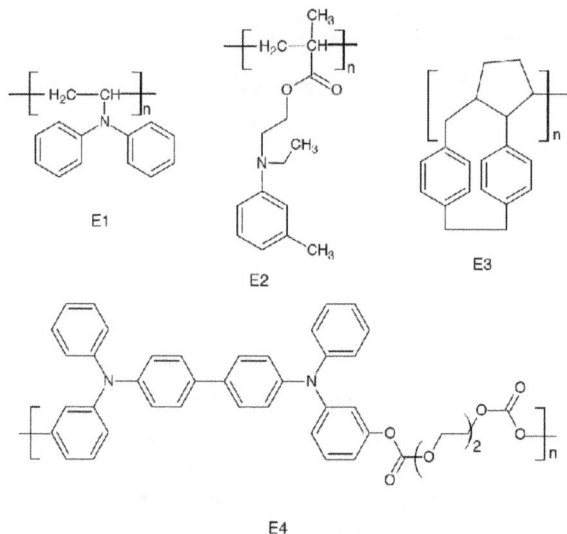

E1

E2

E3

E4

10.3.3 Nonlinear Optical (NLO) Chromophores

As discussed above, a field-dependent refractive index utilizes the addition of molecules possessing an asymmetric charge distribution. The general structure of an electron donating type group linked to an electron acceptor type group using a π-conjugated system such as benzene, azobenzene, stilbene, biphenyl, heterocycle, polyenes or tolans. Selected examples of electron acceptor and donor groups attached to the π-conjugated system are listed in Table 10.1. A typical example of the NLO molecule is para-nitroaniline (pNA) in which the electron donor NH_2 group is attached to the electron acceptor NO_2 group by an aromatic benzene ring. The key property is the relative electron affinities of the donor and acceptor groups as well as the length of the conjugated segment connecting the donor to the acceptor. The chemical structures of selected NLO chromophores, such 4-(dimethyl-amino)-4′-nitrostilbene (DANS), Disperse Red 1 (b), 1-(2′-ethylhexyloxy)-2,5-dimethyl-4-(4′-nitrophenylazo)benzene (EHDNPB) (c), *N,N*′-diethyl-substituted pNA (d), 2,5-dimethyl-4-(*p*-nitrophenylazo)anisole (DMNPAA) (e), 4-piperidin-40ylbenzylidenemalononitrile (f), 4-*N,N*-diethylamino-β-nitro-styrene (h) are shown below.

In designing these materials is it essential that the NLO concentration levels are optimized for an intended application. Specifically, for high dipole moment NLO, excessive concentration levels will adversely impact the carrier mobility. Aside from the problems

Table 10.1 The general structure of non-linear optical chromophores.

General Structure	Donor ——— $\boxed{\pi\text{-conjugated system}}$ ——— Accepter

π-conjugated
   system

Benzene   Azobenzene   Stilbene

Biphenyl   Polyene   Tolane

Electron donor	NH$_2$, –NHCH$_3$, –N(CH$_3$)$_2$, –NHR, –N$_2$H$_3$, –F, –Cl, –Br, –I, –SH, –SR, OR, –CH$_3$, –OH, –NHC(=O)CH$_3$, –OCH$_3$, –SCH$_3$, –OC$_6$H$_5$, –COOCH$_3$
Electron Acceptor	–NO$_2$, –NO, –CN, –COOH, –C(=O)NH$_2$, –C(=O)NHR, –C(=O)NR$_2$, –CHO, –SO$_2$R, –C(=O)R, –CF$_3$, –CH=C(CN)$_2$, –SO$_2$–NH$_2$, –N^{2+}, –NH^{2+}

associated with the phase separation, the formation of dimers will also affect electro-optic responses. The studies have shown that the HOMO energy levels of the NLO molecule relative to the charge-transporting host will also impact the PR performance in a PVK. It turns out that increased concentrations of the NLO molecules with the HOMO levels closest to that of PVK will enhance the PR speed. In contrast, the NLO molecule with the higher HOMO levels may not have an influence on PR properties.[39]

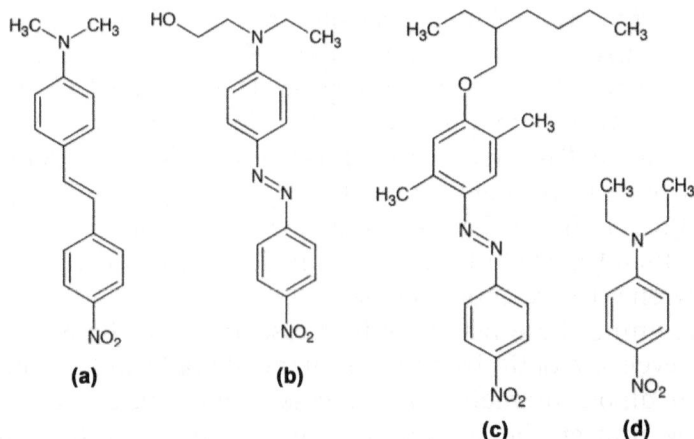

(a)   (b)   (c)   (d)

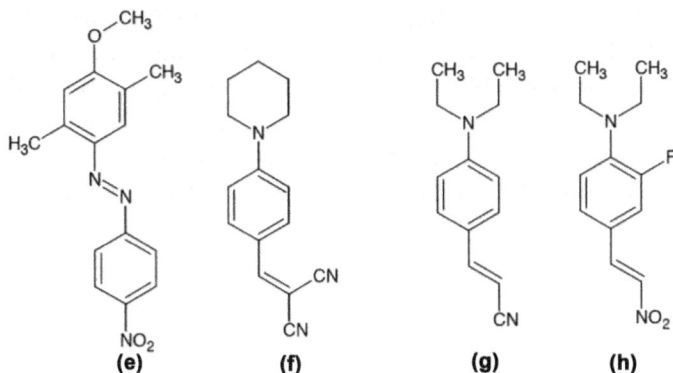

(e) (f) (g) (h)

10.3.4 Plasticizer Molecules

The purpose of using plasticizer molecules is to 'plasticize' high glass transition temperature (T_g) polymers. Plasticizers are small molecules that alter the degree of solvation and, inserted between the polymer chains, push macromolecules apart, resulting in a reduction in the intermolecular cohesive forces. In essence, it is a softening process, which from a practical point of view will lower the processing temperature. As appealing as this may sound, adding a plasticizer may have adverse effects on properties. For example, a common plasticizer for poly(vinylidine fluoride) (PVDF) is vinyl acetate (VA). Unfortunately, with time vinyl acetate will diffuse out of the polymer matrix, which may cause undesirable effects. For example, if this polymer is used to manufacture plastic bags for blood storage, it is anticipated after a certain time vinyl acetate molecules may be found in the blood. Similarly, in food packaging the use of plastics that contain plasticizing agents may cause food contamination. However, the lowering of the T_g in PR materials will facilitate an ease of orientation of the NLO chromophores due to lowered T_g, but will give the rise to higher OB. Interestingly enough, the response time of the PR composite decreased at reduced T_g s for composites based on PVK sensitized with TNF as the photoconductor, 2,5-dimethyl-(4-phenylazo)anisole (DMNPAA) as NLO molecule, and ECZ as the plasticizer.[40] Examples of common plasticizers used in PR systems are illustrated below.

To overcome problems associated with passive plasticizers, and dioctylphthalate is one of them, active plasticizers can be used. These molecules are expected to contribute both the charge-transport or electro-optic effects. For example, *N*-ethylcarbazole (ECZ) is widely used in PR polymers because enhances the charge transport. Thus, there is a tradeoff of properties manifested by the reduced T_g that leads to the enhanced efficiency and reduced CT mobility and PR phase shift. For example, 4-4'-*n*-pentylcyanobiphenyl (5CB) can lower the T_g of a PVK from 230 to 25 °C, while altering the refractive index modulation.[41] Although one can eliminate a plasticizer using a liquid NLO molecule achieving diffraction efficiencies of more than 60% with tens of millisecond of response times, the question of practicality must be raised.[53]

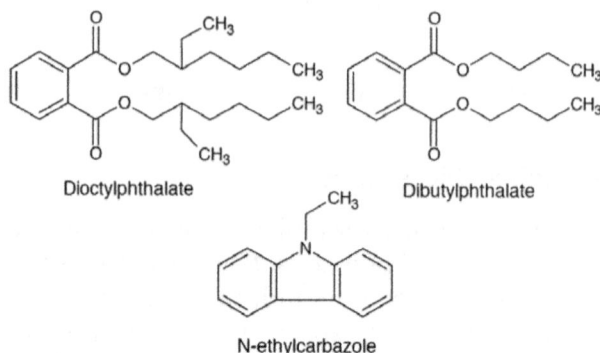

Dioctylphthalate   Dibutylphthalate

N-ethylcarbazole

10.4 Performance of PR Polymers

Typical method to evaluate PR polymer is conducted on polymer sandwiched between two glass plates coated with transparent conducting electrodes such as the indium tin oxide (ITO). Usually, the ITO plate is etched with a mixture of hydrochloric acid and nitric acid at a ratio of 3 : 1, respectively to define conducting regions, and spacers of desired thickness (30–100 μm) are used to assemble a cell. The PR material is filled into the space between the conducting plates of the cell-utilizing capillary forces. Etching helps to define the conducting areas to be covered with the polymer so that device breakdown due to leakage currents can be avoided. Although device preparation depends on the specific PR composite, the main requirements are the application of an electric filed and transparency to light.

10.4.1 Photoconductivity and Electro-optic Responses

Photoconductive polymers are materials that are capable of hole- (p-type) and electron- (n-type) transport. Polymers capable of transporting holes and electrons simultaneously are referred to as bipolar. Typically, bipolarity is achieved by adding electron-transporting molecules such as TNK to donor-like, hole-transporting polymers such as PVK. Basic principles of photoconductivity involve the following processes:

- Absorption of radiation and formation of excitons; photoconductive materials are truly photoconductive in a specific range of wavelength of absorption and to produce wider ranges, for example, in the visible range of the electromagnetic spectrum, electron acceptors forming color charge transfer complexes are added. Upon absorption of light the active are excited to form electron–hole pairs called excitons, which are typically captured at donor or acceptor sites, causing polarization of charges of functional groups. This process competes with the recombination of electron–hole pairs.
- Injection and transport of carriers occur if external photogenerator is used in the presence of charge transporting materials. Dispersed chromophore molecules in conductive polymer matrix serve this function which upon the application of electric field will transport charge carriers. The photogenerated charge carriers will move within the polymer upon applied electric field. In this process photoconductive molecules, such as carbazole in PVK, transport electrons to the electrode and become cation radicals The radicals of PVK will be stabilized by the charge resonance between several chromophores. This transport of carriers is known as the hopping process because the hole hops from one location to another in the direction of the electric field during which moving cation radical may accept an electron from the neighboring neutral carbazole molecule which, in turn, becomes a hole, and the process continues.
- Recombination and trapping. The presence of Coulombic forces will eventually result in recombination of electrons and holes, which completes with the generation of charge carriers. Because the carriers do not move with uniform velocity and, for the most part, remain in localized sites called traps, their mobility in the direction of the electric field can vary. Thus, there are shallow and deep traps, which reflect the time of their release. For example, for PVK at 26 °C, the mobility from 3×10^{-8} to $10^{-6} \, \text{cm}^2 \, \text{V} \, \text{s}^{-1}$ are not unusual.

Photoconductivity can be estimated by a DC photocurrent method that measures the light induced change in the current through the sample under illumination. If a photoconductor is irradiated with appropriate radiation, carrier generation and recombination will take place. If J_{ph} is the photocurrent density $(A \text{ cm}^{-2})$ measured under an electric field E $(V \text{ cm}^{-1})$, the photoconductivity $(1 \ \Omega \ \text{cm}^{-1})$ is given by:

$$\sigma_{ph} = \frac{J_{ph}}{E} \qquad (10.10)$$

The number of generated charges per absorbed photon, called the internal quantum efficiency, is given by:[42]

$$\phi_{int} = \frac{hc}{e\lambda \ln(\alpha)L} \frac{J_{ph}}{I} \qquad (10.11)$$

Here I is the intensity of incident light, L is the thickness of the photoconductor, α is the absorption coefficient and λ is the wavelength of light. If I_{ph} is the photocurrent, P_0 the light power density, A the illuminated area and V the applied voltage, the change in conductivity per incident light intensity, called the photoconductive sensitivity $(S \text{ cm W}^{-1})$ is given by:

$$S = \frac{I_{ph}L}{P_0 A V} \qquad (10.12)$$

The mobility of the carriers can be measured using the time-of-flight (TOF) technique.[43] In these measurements, the molecule under study is sandwiched between two electrodes, of which one is transparent to allow illumination of the sample. If needed, a carrier generation layer can be used to photogenerate and inject carriers to the polymer being studied. Illumination of the sample using a highly absorbed pulse of light through one transparent electrode creates a sheet of charge carriers, which drift under the electric field. The duration of the pulse should be lower than the time taken by the carriers to reach the other electrode. Current through the sample is monitored as the potential drop across a load resistor connected in series with the sample and power supply. The time taken for the arrival of carriers at the other electrode is determined from the knee observed in the current transient. If t_r is the time taken by the carriers under an electric field of magnitude $E = v/d$, where v is the applied voltage and d the thickness, the mobility, defined as the velocity of the carriers under a unit electric field, is calculated using eqn (10.13):

$$\mu = \frac{d}{t_r E} \qquad (10.13)$$

Other methods for the determination of carrier mobility include holographic time of flight (HTOF)[44] and extraction current transients.[45] The density of traps may be estimated with near infrared optical absorption in the case of C60 sensitized PVK systems[45] or can be estimated from thermally stimulated discharge current measurements.[46]

The relevant EO coefficients can be extracted from a measurement of the electric field induced phase shift between the two orthogonal components of electric field vector of polarized light after passing through the sample. In low T_g PR systems the refractive index modulation is due to birefringence, Pockel's effect and Kerr effect.[47] The light induced or field induced birefringence $n_p - n_s$ can be calculated by keeping the sample in between two crossed polarizers and measuring the transmitted intensity as a function of time or applied field as desired.[24] If the contribution from birefringence is dominant, for characterizing the electro-optic response, a quantity called the electro-optic response function is used.[48]

10.4.2 Diffraction Efficiency and Response Times

The diffraction efficiency of a PR polymer system can be measured in the FWM geometry. A grating is written inside the material with two mutually coherent beams and a probe beam is made to counter-propagate one of the writing beams. The probe beam is usually weaker or has a polarization direction orthogonal to that of the writing beams so that it does not interact with the writing beams or affect the grating. The fourth wave, generated by the diffraction of the probing beam, is detected by a photodiode and the diffracted intensity is recorded. The external diffraction efficiency η^{ext} is the ratio of the intensity of the diffracted beam to that of the probing beam. The internal diffraction efficiency is given by $\eta^{int} = I_4/(I_4 + I_3)$. The experimental geometry of a FWM experiment is shown in Figure 10.7.

The diffraction efficiency depends on the thickness of the device as well as the applied voltage. A simplified form the dependence is given by:

$$\eta \propto \sin^2 \left[\frac{\pi \Delta n d}{\lambda (\cos \theta_1 \cos \theta_2)^{\frac{1}{2}}} \right] \quad (10.14)$$

where θ_1 and θ_2 are the writing beam angles in the polymer. The writing beams in a FWM experiment are usually s-polarized due to the fact that beam coupling effects are minimal for this polarization state. The diffraction efficiency is one of the important parameters often quoted with performance figures of all holographic recording materials as it denotes the available intensity for image formation. The PR response speed is usually estimated from the grating build-up

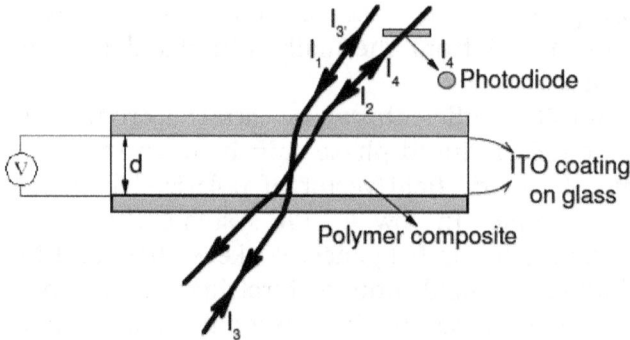

Figure 10.7 Beam paths in the four-wave mixing experiment. The diffracted light beam (J_4) is made to fall on the photodiode
using a semi-silver mirror.

dynamics by fitting the experimental data with a stretched exponential or a biexponential function. It has been reported that if the T_g is
high the grating build-up time is limited by the rotational alignment
of the NLO chromophores under the local electric field. Studies done
on the PR polymer system PVK:TNF:DMNPAA:ECZ with varying
chromophore content showed that irrespective of the chromophore
content the response time of the composites were similar when the T_g
of the systems were near room temperature.[49,50]

It is recognized that the dynamics of grating formation in a
multi-component PR system depend on many factors including the
properties of components such as the charge-transporting polymers,
sensitizer, plasticizer as well as experimental conditions such
as geometry of the experiment, intensity, and wavelength of
writing beams.

10.5   High-performance Photorefractive Polymers

The photorefractive effect, which once was considered an unwanted
phenomenon, is a promising candidate for reversible holographic data
storage. The main challenges are faster response times, low applied
fields, and high stability. As there is no single material that would
fulfill these requirements, variety of materials are being examined and
considered for commercial applications. One of the advantages of PR
polymers over other grating recording materials such as photopolymers
and photo-thermoplastics is the lack of post processing to fix the image
in the medium. PR gratings are reversible, thus enabling erasure of the
stored information by uniform illumination.

10.5.1 Guest–host Polymer Systems

The guest–host approach is the simplest method to prepare a PR polymer system providing a variety of materials combinations. The examples of the guest molecules used in the system with the T_g near room temperature capable of achieving exceptionally high two beam coupling gain coefficients due to the field induced reorientation of the dipolar chromophores are depicted below.[49,51] They are DMNPAA (left), CT complex of PVK with TNF (middle) and ECZ (right).

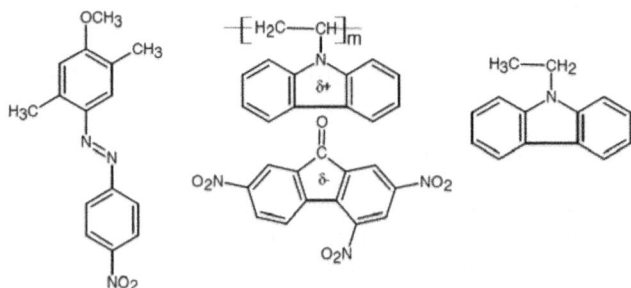

2,5-dimethyl-4-(*p*-nitrophenylazo)synthes (DMNPAA) can be synthesized using diazotization reactions of *p*-nitrobenzenediazonium tetrafluoroborate and 2,5-dimethyl aniline. Similar to other azo dye molecules, this molecule undergoes a *trans–cis–trans* photoisomerization process, which has been used for multilayer polarization-encoded optical data storage.[52]

The electro-optic response functions of the DMNPAA:PVK:ECZ:TNF (50:33:16:1 wt) composite are shown in Figure 10.8. A detailed characterization of this chromophore and some other NLO molecules can be found elsewhere.[23] The T_g of this material is near the room temperature which facilitated the rotational orientation of the DMNPAA molecules, whereas the TBC gain of $220\,\text{cm}^{-1}$ at an electric field of $90\,\text{V}\,\mu\text{m}^{-1}$ exceeded the absorption loss in the material giving a net optical gain of $207\,\text{cm}^{-1}$ for p-polarized light. The large diffraction efficiency observed in the system is usually limited by reflection and absorption losses in the PR device. The high performance of this system was ascribed to the low T_g and the higher loading of the NLO molecule with large $\Delta\alpha$.[49]

Another PR polymer system based on PVK is the composite PVK:PDCST:BBP:C60 in the ratio 49.5:35:15:0.5 wt.%. In electric fields ~120 V μm⁻¹, this PR system exhibits a TBC gain coefficient of $200\,\text{cm}^{-1}$ and a response time of 50 ms at $1\,\text{W}\,\text{cm}^{-2}$ writing intensity.[53] When NLO molecules are substituted with the dicyanostyrene

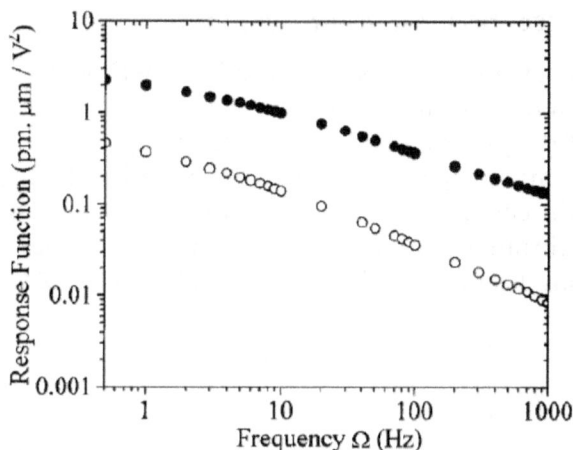

Figure 10.8 Effective electro-optic response functions of the DMNPAA:PVK:ECZ:TNF system estimated with frequency-dependent electro-optic measurements. Response functions measured at the fundamental frequency (solid circles) and its second harmonic (open circles) of different frequencies of the applied voltage.
Reproduced from ref. 47. Copyright 1996, AIP Publishing LLC.

derivatives AODCST and 7-DCST (4-(azepan-1-yl)benzylidenemalononitrile), very fast response times of $\sim$ 5 ms at 1 W cm^{-2} and at 100 V µm^{-1} can be achieved while maintaining the high optical gain.[54] The effect of the NLO molecule on the performance of the PR system PVK:NLO chromophore:BBP:C60 was subjected to a detailed analysis by varying the structure of the NLO molecules. A series of DCST derivatives and a series of cyanoester styrene (CEST) derivatives, both with varying amine donor can be used for this purpose. These systems show very fast response times making them suitable for video rate applications. If NLO molecule is replaced by 2-dicyanomethylen-3-cyano-5,5-dimethyl-4-(4′-dihexylaminophenyl)-2,5-dihydrofuran (DCDHF-6) (below), gain coefficients reaching 400 cm^{-1} can be achieved in an electric field of 100 V µm^{-1}.[55]

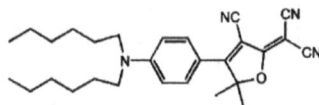

When a sensitizer molecule in the PVK:PDCST:BBP:C60 system is replaced with (2,9,16,23-tetra-*tert*-butyl-phthalocyaninato) zinc (ZnPc) and bis(*p*-methylbenzoate) (2,9,16,23-tetra-*tert*-butyl-phthalocyaninato) silicon (SiPc) (below), light harvesting can be also achieved).[26]

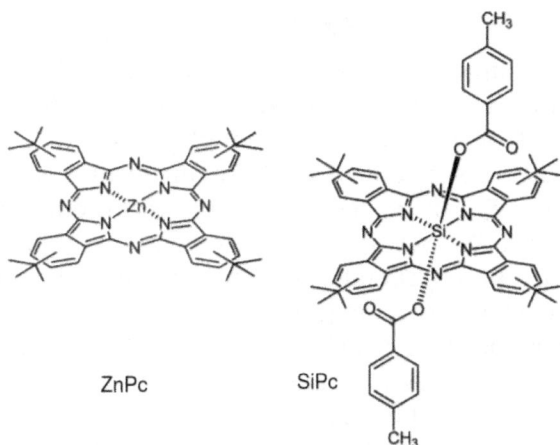

ZnPc   SiPc

One of the concerns in composite systems is the possibility of aggregate formation. This can be avoided by adjusting the solubility during preparation. As Pc's oxidation and reduction potentials can be tuned by changing the central atom and the peripheral substituents, one can envision designing sensitizer requirements of a particular photoconducting host may be designed.

PR polymer systems based on conjugated polymers with delocalized π electron capable of facilitation higher carrier mobilities are shown below. Chemical structures of the plasticizer molecule DPP, NLO molecule DO3 and the polymer constituting and the PR polymer composite is based on the conjugated polymer poly[o(p)-phenylenevinylene-*alt*-2-methoxy-5-(2-ethylhexyloxy)-*p*-phenylenevinylene] (*p*-PMEH-PPV). The prepoled PR system is composed of *p*-PMEH-PPV:DPP:DO3:C60 (74 : 20 : 5 : 1 wt%).[56]

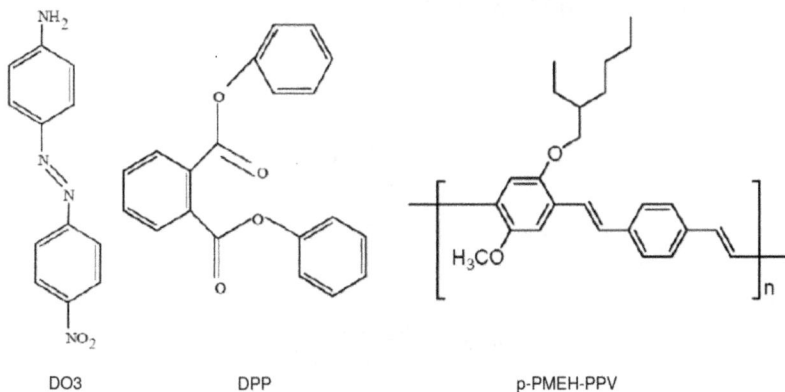

DO3   DPP   p-PMEH-PPV

10.5.2 Functionalized Polymer Systems

In an effort to minimize phase separation or crystallization, covalent functionalization of molecules responsible for PR effect was utilized. However, these strategies are typically employed to low T_g polymers. For example, a NLO moiety linked to the main chain *via* a flexible link (below) and low T_g can be achieved by the incorporation of different alkyl side chains. As shown, this system contains Ru(II)-tri(bispyridyl) complex as the photogenerator moiety,[57] which takes advantage of metal-to-ligand charge transfer (MLCT) taking place on the Ru complex. The conjugated polymer backbone is selected to achieve enhanced carrier mobilities. This system, after poling, exhibits a net optical gain of more than 200 cm^{-1} under zero electric field, but the response time is on the order of several hundreds of seconds. The latter is attributed to undesired structural defects created in the conjugated backbone.

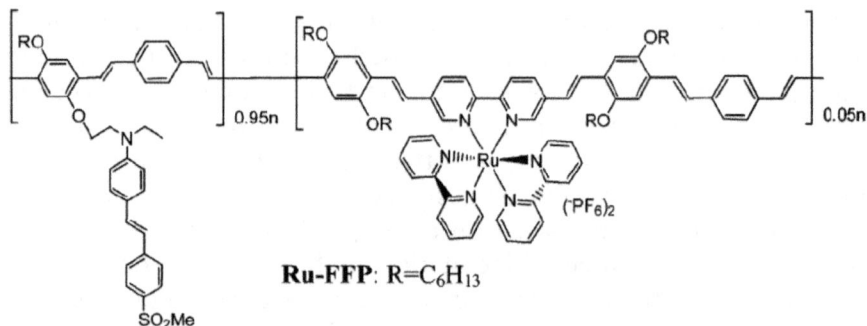

Ru-FFP: R=C$_6$H$_{13}$

The example of another fully functionalized system is shown in below. This material is synthesized following the Stille coupling reaction and is composed of a NLO chromophore attached to conjugated poly(p-phenylene-thiophene) back bone.[58] This fully functionalized system operates at 780 nm wavelength with the gain coefficient of ~ 180 cm^{-1} and external diffraction efficiency of 68% in an electric field of 50 V μm^{-1}.[50]

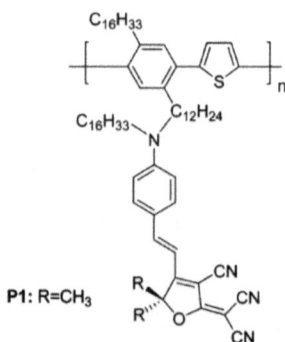

P1: R=CH$_3$

References

1. A. Ashkin, G. Boyd, Ji. Dziedzic, R. Smith, A. Ballman and J. Levinstein *et al.*, Optically-induced refractive index inhomogeneities in LiNbO3 and LiTaO3, *Appl. Phys. Lett.*, 1966, **9**(1), 72–74.
2. F. Chen, J. LaMacchia and D. Fraser, Holographic storage in lithium niobate, *Appl. Phys. Lett.*, 1968, **13**(7), 223–225.
3. K. Sutter, J. Hulliger and P. Günter, Photorefractive effects observed in the organic crystal 2-cyclooctylamino-5-nitropyridine doped with 7, 7, 8, 8-tetracyanoquinodimethane, *Solid State Commun.*, 1990, **74**(8), 867–870.
4. S. Ducharme, J. Scott, R. Twieg and W. Moerner, Observation of the photorefractive effect in a polymer, *Phys. Rev. Lett.*, 1991, **66**(14), 1846.
5. V. Petrov, S. Lichtenberg, J. Petter, T. Tschudi, A. Chamrai and V. Bryksin *et al.*, Optical on-line controllable filters based on photorefractive crystals, *J. Opt. A: Pure Appl. Opt.*, 2003, **5**(6), S471.
6. Y. Frauel, G. Pauliat, A. Villing and G. Roosen, High-capacity photorefractive neural network implementing a Kohonen topological map, *Appl. Opt.*, 2001, **40**(29), 5162–5169.
7. S. Tay, P.-A. Blanche, R. Voorakaranam, A. Tunc, W. Lin and S. Rokutanda *et al.*, An updatable holographic three-dimensional display, *Nature*, 2008, **451**(7179), 694–698.
8. S. Bartkiewicz, K. Matczyszyn, A. Miniewicz and F. Kajzar, High gain of light in photoconducting polymer–nematic liquid crystal hybrid structures, *Opt. Commun.*, 2001, **187**(1), 257–261.
9. R. Angelone, F. Ciardelli, A. Colligiani, F. Greco, P. Masi and A. Romano *et al.*, Unconditionally stable indole-derived glass blends having very high photorefractive gain: the role of intermolecular interactions, *Appl. Opt.*, 2008, **47**(36), 6680–6691.
10. P.-A. Blanche, A. Bablumian, R. Voorakaranam, C. Christenson, W. Lin and T. Gu *et al.*, Holographic three-dimensional telepresence using large-area photorefractive polymer, *Nature*, 2010, **468**(7320), 80–83.
11. B. Lynn, P. A. Blanche and N. Peyghambarian, Photorefractive polymers for holography, *J. Polym. Sci. Part B: Polym. Phys.*, 2014, **52**(3), 193–231.
12. L. M. Hayden, G. F. Sauter, F. R. Ore, P. L. Pasillas, J. M. Hoover and G. A. Lindsay *et al.*, Second-order nonlinear optical measurements in guest-host and side-chain polymers, *J. Appl. Phys.*, 1990, **68**(2), 456–465.

13. D. M. Burland, R. D. Miller and C. A. Walsh, Second-order non-linearity in poled-polymer systems, *Chem. Rev.*, 1994, **94**(1), 31–75.

14. C. Brabec, H. Johannson, F. Padinger, H. Neugebauer, J. Hummelen and N. Sariciftci, Photoinduced FT-IR spectroscopy and CW-photocurrent measurements of conjugated polymers and fullerenes blended into a conventional polymer matrix, *Sol. Energy Mater. Sol. Cells*, 2000, **61**(1), 19–33.

15. R. Lof, M. Van Veenendaal, B. Koopmans, H. Jonkman and G. Sawatzky, Band gap, excitons, and Coulomb interaction in solid C 60, *Phys. Rev. Lett.*, 1992, **68**(26), 3924.

16. H. Hoegl, On Photoelectric Effects in Polymers and Their Sensitization by Dopants1, *J. Phys. Chem.*, 1965, **69**(3), 755–766.

17. W. Moerner, S. Silence, F. Hache and G. C. Bjorklund, Orientationally enhanced photorefractive effect in polymers, *JOSA B*, 1994, **11**(2), 320–330.

18. J. Wu, Birefringent and electro-optic effects in poled polymer films: steady-state and transient properties, *JOSA B*, 1991, **8**(1), 142–152.

19. J. Grazulevicius, P. Strohriegl, J. Pielichowski and K. Pielichowski, Carbazole-containing polymers: synthesis, properties and applications, *Prog. Polym. Sci.*, 2003, **28**(9), 1297–1353.

20. T. Uryu, H. Ohkawa and R. Oshima, Synthesis and high hole mobility of isotactic poly (2-N-carbazolylethyl acrylate), *Macromolecules*, 1987, **20**(4), 712–716.

21. O. Ostroverkhova and W. Moerner, Organic photorefractives: mechanisms, materials, and applications, *Chem. Rev.*, 2004, **104**(7), 3267–3314.

22. A. D. Vannikov AVaG, The photorefractive effect in polymeric systems, *Russ. Chem. Rev.*, 2003, **72**, 471–488.

23. C. R. Moylan, R. Wortmann, R. J. Twieg and I. McComb, Improved characterization of chromophores for photorefractive applications, *JOSA B*, 1998, **15**(2), 929–932.

24. R. Termine, I. Aiello, N. Godbert, M. Ghedini and A. Golemme, Light-induced reorientation and birefringence in polymeric dispersions of nano-sized crystals, *Opt. Express*, 2008, **16**(10), 6910–6920.

25. L. Smilowitz, N. Sariciftci, R. Wu, C. Gettinger, A. Heeger and F. Wudl, Photoexcitation spectroscopy of conducting-polymer–C 60 composites: Photoinduced electron transfer, *Phys. Rev. B*, 1993, **47**(20), 13835.

26. F. Gallego-Gómez, J. A. Quintana, J. M. Villalvilla, M. A. Díaz-García, L. Martín-Gomis and F. Fernández-Lázaro *et al.*, Phthalocyanines as Efficient Sensitizers in Low-T g Hole-Conducting

Photorefractive Polymer Composites, *Chem. Mater.*, 2009, **21**(13), 2714–2720.

27. X. Li, J. W. Chon and M. Gu, Nanoparticle-based photorefractive polymers, *Austr. J. Chem.*, 2008, **61**(5), 317–323.
28. J.-W. Oh, C.-S. Choi, Y. Jung, C. Lee and N. Kim, Photoconducting polymers containing covalently attached ruthenium complexes as a photosensitizer, *J. Mater. Chem.*, 2009, **19**(32), 5765–5771.
29. K. A. Willets, O. Ostroverkhova, M. He, R. J. Twieg and W. Moerner, Novel fluorophores for single-molecule imaging, *J. Am. Chem. Soc.*, 2003, **125**(5), 1174–1175.
30. D. Van Steenwinckel, E. Hendrickx and A. Persoons, Dynamics and steady-state properties of photorefractive poly (N-vinylcarbazole)-based composites sensitized with (2, 4, 7-trinitro-9-fluorenylidene) malononitrile in a 0–3 wt% range, *J. Chem. Phys.*, 2001, **114**(21), 9557–9564.
31. J. Sworakowski and J. Ulański, Electrical properties of organic materials, *Annu. Rep. Prog. Chem., Sect. C: Phys. Chem.*, 2003, **99**, 87–125.
32. Y.-M. Wang, F. Teng, L.-h. Gan, H.-M. Liu, X.-H. Zhang and W.-F. Fu *et al.*, Blue light-emitting bisorthometalated ir (iii) complex: origin of blue emission and application in electrophosphorescent devices, *J. Phys. Chem. C*, 2008, **112**(12), 4743–4747.
33. C. E. Barnes, Stabilized heat polymerizable compositions of nu-vinyl pyrrole compounds, *US Pat.*, 2 483 962, 1949.
34. S. Schloter, U. Hofmann, P. Strohriegl, H.-W. Schmidt and D. Haarer, High-performance polysiloxane-based photorefractive polymers with nonlinear optical azo, stilbene, and tolane chromophores, *JOSA B*, 1998, **15**(9), 2473–2475.
35. D. Wright, U. Gubler, W. Moerner, M. S. DeClue and J. S. Siegel, Photorefractive properties of poly (siloxane)-triarylamine-based composites for high-speed applications, *J. Phys. Chem. B*, 2003, **107**(20), 4732–4737.
36. H. G. Schild, Poly (N-isopropylacrylamide): experiment, theory and application, *Prog. Polym. Sci.*, 1992, **17**(2), 163–249.
37. E. Bellmann, S. E. Shaheen, R. H. Grubbs, S. R. Marder, B. Kippelen and N. Peyghambarian, Organic Two-Layer Light-Emitting Diodes Based on High-T g Hole-Transporting Polymers with Different Redox Potentials, *Chem. Mater.*, 1999, **11**(2), 399–407.
38. J. Kido, G. Harada and K. Nagai, Organic electroluminescent device with aromatic amine-containing polymer as a hole transport layer (II): Poly (arylene ether sulfone)-containing tetraphenylbenzidine, *Polym. Adv. Technol.*, 1996, **7**(1), 31–34.

39. O. Ostroverkhova and K. D. Singer, Space-charge dynamics in photorefractive polymers, *J. Appl. Phys.*, 2002, **92**(4), 1727–1743.
40. R. Bittner, T. K. Däubler, D. Neher and K. Meerholz, Influence of Glass-Transition Temperature and Chromophore Content on the Steady-State Performance of Poly (N-vinylcarbazole)-Based Photorefractive Polymers, *Adv. Mater.*, 1999, **11**(2), 123–127.
41. J. Zhang and K. Singer, Homogeneous photorefractive polymer/ nematogen composite, *Appl. Phys. Lett.*, 1998, **72**, 2948.
42. O. Ostroverkhova, D. Wright, U. Gubler, W. Moerner, M. He and A. Sastre-Santos *et al.*, Recent advances in understanding and development of photorefractive polymers and glasses, *Adv. Funct. Mater.*, 2002, **12**(9), 621–629.
43. D. West, M. Rahn, C. Im and H. Bässler, Hole transport through chromophores in a photorefractive polymer composite based on poly (N-vinylcarbazole), *Chem. Phys. Lett.*, 2000, **326**(5), 407–412.
44. G. Malliaras, V. Krasnikov, H. Bolink and G. Hadziioannou, Control of charge trapping in a photorefractive polymer, *Appl. Phys. Lett.*, 1995, **66**(9), 1038–1040.
45. G. Juška, K. Arlauskas, M. Viliūnas, K. Genevičius, R. Österbacka and H. Stubb, Charge transport in π-conjugated polymers from extraction current transients, *Phys. Rev. B.*, 2000, **62**(24), R16235.
46. C.-J. Chang and W.-T. Whang, Trap characteristics study of photorefractive polymer materials by thermal stimulated current spectroscopy, *J. Polym. Res.*, 1997, **4**(4), 243–252.
47. B. Kippelen, K. Meerholz and N. Peyghambarian, Birefringence, Pockels, and Kerr effects in photorefractive polymers, *Appl. Phys. Lett.*, 1996, **68**(13), 1748–1750.
48. B. Kippelen, K. Meerholz and N. Peyghambarian, Ellipsometric measurements of poling birefringence, the Pockels effect, and the Kerr effect in high-performance photorefractive polymer composites, *Appl. Opt.*, 1996, **35**(14), 2346–2354.
49. K. Meerholz, B. Volodin, B. Kippelen and N. Peyghambarian, *A Photorefractive Polymer with High Optical Gain and Diffraction Efficiency Near 100%*, 1994.
50. R. Bittner, C. Bräuchle and K. Meerholz, Influence of the glass-transition temperature and the chromophore content on the grating buildup dynamics of poly (N-vinylcarbazole)-based photorefractive polymers, *Appl. Opt.*, 1998, **37**(14), 2843–2851.
51. A. V. Vannikov and A. D. Grishina, The photorefractive effect in polymeric systems, *Russ. Chem. Rev.*, 2003, **72**(6), 471–488.

52. X. Li, J. W. Chon, S. Wu, R. A. Evans and M. Gu, Rewritable polarization-encoded multilayer data storage in 2, 5-dimethyl-4-(p-nitrophenylazo) anisole doped polymer, *Opt. Lett.*, 2007, **32**(3), 277–279.

53. A. Grunnet-Jepsen, C. Thompson, R. Twieg and W. Moerner, High performance photorefractive polymer with improved stability, *Appl. Phys. Lett.*, 1997, **70**, 1515–1517.

54. D. Wright, M. Diaz-Garcia, J. Casperson, M. DeClue, W. Moerner and R. Twieg, High-speed photorefractive polymer composites, *Appl. Phys. Lett.*, 1998, **73**(11).

55. D. Wright, U. Gubler, Y. Roh, W. Moerner, M. He and R. J. Twieg, High-performance photorefractive polymer composite with 2-dicyanomethylen-3-cyano-2, 5-dihydrofuran chromophore, *Appl. Phys. Lett.*, 2001, **79**, 4274.

56. D. J. Suh, O. O. Park, T. Ahn and H.-K. Shim, Large two-beam coupling in the p-PMEH-PPV/DPP/DO3/C60, *Jpn. J. Appl. Phys.*, 2002, **41**(4A), L428.

57. Z. Peng, A. R. Gharavi and L. Yu, Synthesis and characterization of photorefractive polymers containing transition metal complexes as photosensitizer, *J. Am. Chem. Soc.*, 1997, **119**(20), 4622–4632.

58. W. You, S. Cao, Z. Hou and L. Yu, Fully functionalized photorefractive polymer with infrared sensitivity based on novel chromophores, *Macromolecules*, 2003, **36**(19), 7014–7019.

11 Self-healing Materials

11.1 Introduction

Due to technological importance and scientific curiosity materials capable of altering their properties in response to chemical or physical stimuli have been of interest for a number of years. The ability of a material to respond to factors such as temperature, pressure, pH, ionic strength, concentration gradients, or electric and magnetic fields opens the door to a range of previously unobtainable applications. Responses to external stimuli may be physical or chemical in nature, or both. Perhaps the most intriguing phenomenon is the ability of materials to self-repair broken bonds, yet being observable. Unquestionably this venue formulated the basis of the 21st century materials and beyond. The challenge in designing stimuli-responsive materials is to create materials by incorporating minute molecular, yet orchestrated changes that lead to significant physicochemical responses upon external or internal stimuli.[1]

11.1.1 Polymeric Materials

Loosely defined in biological systems as a built-in defense mechanism preventing species from losing their living functions, self-healing are quite complex and involve multi-level transient chemico-physical responses in continually changing bioactive environments at different length scales. Self-healing in materials in general, and polymers in particular, is manifested by the ability to regain original properties lost during external damage. If biological systems are the benchmark for self-healing, then a synthetic mimicking should be

Stimuli-Responsive Materials: From Molecules to Nature Mimicking Materials Design
By Marek W. Urban
© Marek W. Urban, 2016
Published by the Royal Society of Chemistry, www.rsc.org

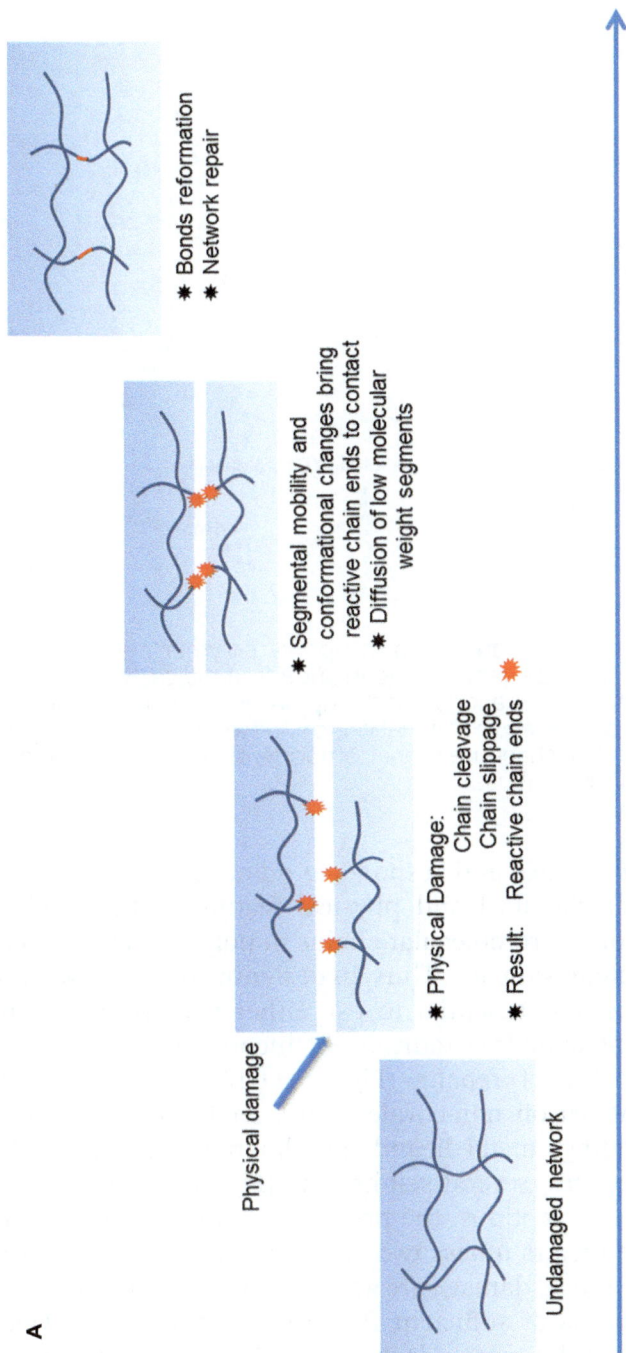

Figure 11.1A (A) Evolution of an ideal damage–repair cycle in polymeric materials; upon physical damage, cleavage of macromolecular chains leads to the formation of reactive end groups, which may be free radicals and/or –C≡C–, –COOH, –NH₂, –OH, –Si–O, SH/S–S, –C=O, or other reactive groups. Segmental mobility and conformational changes bring reactive groups together and bond reformation occurs. Reproduced from ref. 2 with permission from The Royal Society of Chemistry.

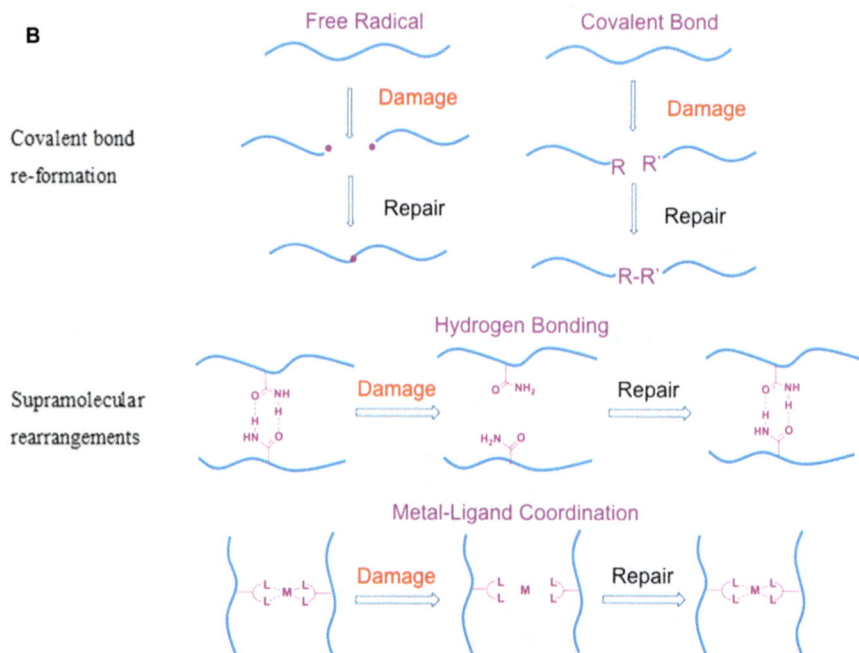

Figure 11.1B (B) Cleavage and re-formation of polymer networks during the damage–repair cycle include free radical, other covalent bond reformations, and supramolecular rearrangements. Synchronized combination of these reactions along with memory shape network components may offer unique self-healing attributes.

able to continually sense and respond to damages over its lifetime, restoring original chemical and physical features without adverse effects. Damage of macromolecular chains in polymers leads to bond cleavage and/or chain slippage. Thus, in designing polymers with self-healing attributes it is essential to use either reactive chain ends, typically represented by free radicals, or incorporate other reactive components capable of repairs (Figure 11.1A) upon mechanical damage. For that reason non-covalent supramolecular interactions, including H-bonding, metal–ligand coordination, or π–π stacking were incorporated into several self-healing (Figure 11.1B) polymers. However, with few exceptions, the presence of reactive species alone will not facilitate repairs unless two spatially separated surfaces created during mechanical damage are able to come in contact with each by mechanical forces of sufficient flow to fill up a wound. Due to limited chain mobility and a lack of bond reformation abilities, common polymers do not exhibit self-repair attributes. The last decade resulted in the development of a few classes of reactions to offer self-repair through the cleavage and reformation of specific bonds. These entities are incorporated into polymer networks, and are typically categorized into reversible covalent and non-covalent bonds. The presence of these reversible bonds introduces molecular level heterogeneities along a polymer backbone, or as pendant groups, which may further facilitate the formation of new phases with extended length scales. Contributions of dynamic bonds and network heterogeneities to self-healing will be discussed in the following sections in the context of reversible covalent bonds, supramolecular chemistry, and encapsulated components.

11.1.2 Metals and Metal Oxides

Self-repairing in metals or metal oxides is another venue that involves entirely different approaches. Unlike polymers which are connected 'mers' by chemical bonds, metals and metal oxides may require different approaches. Because metallic bonding occurs as a result of the strong attractive forces between the positive metal nuclei and the cloud of delocalized electrons, the sharing of free electrons within a lattice will offer a number of unique to metal properties. They include but are not limited to strength, ductility, thermal and electrical properties, electromagnetism, and others. But from the prospective of self-repair, mobility of both the cloud of electrons and metal atoms or ions is the most appealing property. These interactions are dynamic and often described as an array of positive charges in a sea of electrons. As a matter of fact it seems obvious that one would take advantage of this mobility and utilize these properties for the development of self-repairing metals. Yet, relatively little research has been conducted on self-healing metals. Some experimental evidence indicates that hydrogen in the metal substrates will facilitate increased metal-ion transport during oxidation, which may lead to increased oxide growth at the oxide–gas interface. For example, oxides of rare earth metals and Pt will catalyze the dissociation of O_2 and the increased rate of O_2 dissociation can lead to increased transport of oxygen ions in the oxides, thus substantial oxide growth at the substrate–oxide interface. A balanced transport of metal and oxygen ions in metal oxides that leads to oxide growth at both the metal–oxide and at the oxide–gas interface is found to be favorable for the formation of protective oxides with good adherence to the metal substrate. As depicted in Figure 11.2, the proposed mechanism involves simultaneous metal-cation and oxygen-anion transport in order for

Oxide with macro-scale defects

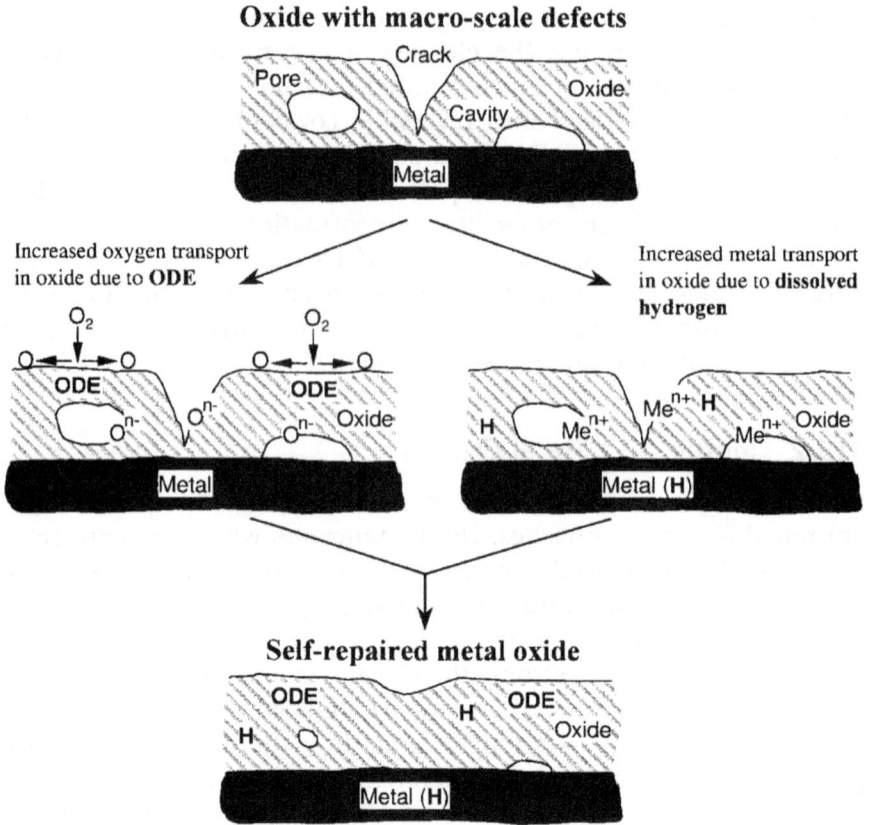

Figure 11.2 Schematic representation of increased transport of dissociated oxygen due to the addition of oxygen dissociating element (ODE) and increased metal transport due to the addition of hydrogen. If well balanced, these additions promote the formation of a self-repaired metal oxide. Reproduced from Ref. 3. With kind permission from Springer Science and Business Media.

protective metal oxides with low densities to form.[3] By addition of H and oxygen dissociating elements (ODEs), such as rare earth metals and Pt, the balance between metal and oxygen transport is favorable to the formation of higher content of protective ODEs and self-repair.

Typical chromium-based coatings are used to prevent so-called white staining (corrosion) of zinc-galvanized steel prior to applications. Recent studies have shown that during this process chromium-based surface layers are formed and consist of chromium ions in dual oxidation states, Cr(III) and Cr(VI). While Cr(III) forms Cr_2O_3 and serves as a protective barrier on the metal, Cr(VI) exists as soluble CrO_4^{2-} and self-repair defects in the surface layer.[4] Although the long-term protection of zinc is not fully understood, recent evidence indicate that corrosion protection is achieved by Zn own corrosion products formed during atmospheric corrosion *via* the formation of a triple layered structure: a thin barrier oxide and two kinds of precipitated layers. The outer precipitated layer is usually porous, whereas the inner layer undergoes densification with time and becomes highly compact and non-permeable.

The concept of incorporating encapsulated reactive components into a materials matrix to be discussed in the later section of this chapter has been also explored in metals and metal oxides. Its origin goes back to hollow fiber reinforced matrix which included inorganic and organic fibers,[5,6] with many applications that followed this original idea.

Another means of self-repair in metals and metal oxides is the use of so-called shape memory alloys. In these materials the shape memory effect (SME) is used in which a material is mechanically deformed, but upon heating, 'remembers' its original shape and returns to it. This behavior is associated with the superelastic or pseudoelastic effect (SE) in which large and recoverable strains are induced by mechanical instead thermal treatments. This is attributed to a first-order displacive transformation manifested in the hardening of steel, called martensite. As shown in Figure 11.3, a steel alloy heated to the temperature where the elevated temperature face-centered cubic (fcc) austenite phase is stable, but rapid cooling produces the hard martensite phase. In SME alloys, the martensite transformation is a thermoelastic process during which the martensite forms disappear on heating and cooling over a relatively small temperature range. The intermetallic phase in these alloys undergoes a displacive, shear-like transformation when cooled below a critical temperature designated as M_S (martensite start) in Figure 11.3. Upon further cooling to temperature M_F (martensite finish), the transformation is complete and the alloy is in the martensitic state. When this martensite is deformed, it undergoes a strain that is recovered upon heating. This recovery results from a reverse transformation to the austenite phase initiated at the temperature A_S (austenite start) and is completed at A_F (austenite finish). This is a first-order phase transformation and the hysteresis associated with the formation of martensite and the reverse transformation to the elevated temperature parent austenite phase. The hysteresis loop of in a typical SME alloy is illustrated in Figure 11.3. Many engineering applications of SME in metals and metal oxides are available which are beyond the scope of this monograph and the interested reader is referred to the literature.[7-9] However, there are numerous opportunities for the development of hybrid materials, combining metals and metal oxides, and fiber- and nano-reinforced composites.

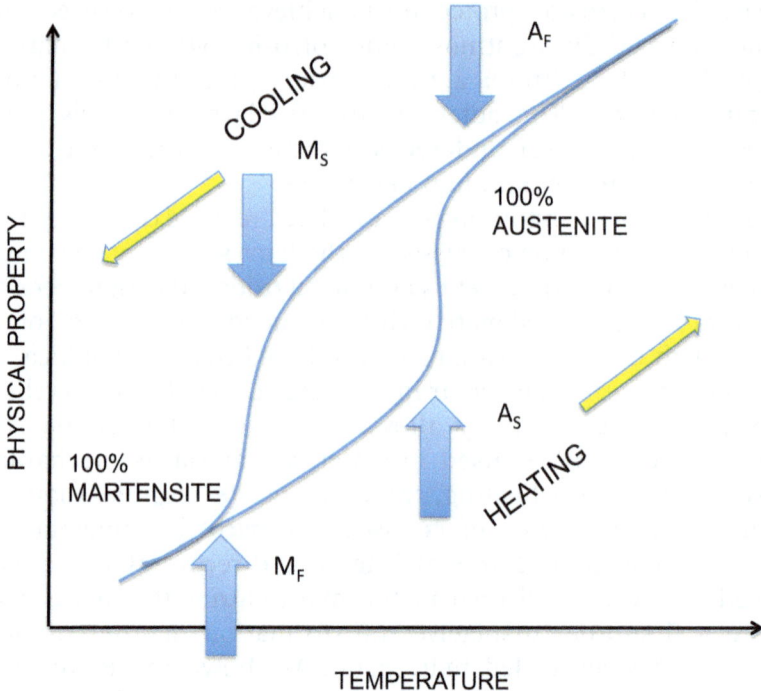

Figure 11.3 Hysteresis loop associated with a SME transformation: MS and MF correspond to the martensite start and finish temperatures, respectively; AS and AF correspond to the start and finish of the reverse transformation of martensite, respectively. The physical property can be volume, electrical conductivity, length, others.
Adapted from ref. 3 with kind permission from Springer Science and Business Media.

11.2 Covalent Bonds

Dissociation and association rates of the dynamic bonds as well as chain mobility are significant factors in designing self-repairing materials. Reversible covalent bonds are good candidates for developing self-repairing polymers due to their high bonding strength, which enhance the mechanical toughness of materials. Four types of self-repairing covalent bonds have been used: (i) reversible cycloaddition reactions, (ii) exchange reactions, (iii) stable free radical-mediated reshuffling reactions, and (iv) heterocyclic compounds/carbohydrates facilitated bond re-formations. One drawback of covalent self-repairing bonds is the high activation energy. For example, for polymers utilizing Diels–Alder (DA) reactions, where activation temperature of the retro-DA reaction responsible for the self-healing is 120 °C,[10] self-healing only occurs by heating the damaged material above this temperature. The advantage is that T_g of the polymer can be as high as 100 °C, but high temperature exposure may not be desirable. Ideally, bond dissociation and reformation should proceed at ambient conditions. Because constrained network mobility becomes the limiting factor, the following section will address how this limitation can be overcome by incorporating heterogeneities, leading to the development of tough materials containing reversible covalent bonds that can repair at ambient conditions.

11.2.1 Reversible Cycloaddition Reactions

Due to their adaptability to various polymers, DA and similar cycloaddition reactions have been extensively investigated. Adduct (*exo* or *endo*) formation *via* DA reaction between furan (diene) and maleimide (dienophile) entities can be used to construct a cross-linked polymer network.[10,11] Retro-DA reactions, typically induced by elevated temperature and/or crack, result in the disconnection of diene and dienophile, and lower temperatures reconstruct the covalent bonds to repair damages. DA pairs have been incorporated in polyesters, epoxies, and polyamides,[12–15] leading to the formation of self-healing networks with various properties. This is summarized in Table 11.1A. As shown in Table 11.1A (a) and (b), the T_g can be tuned as high as 100 °C, but the activation temperature of the retro-DA reaction is above the T_g, which requires heating to high temperatures to activate the self-repair process. Room temperature repairs of polyester containing DA pairs have also been observed (Table 11.1A (c). However, whether DA linkages can dissociate under these conditions is not obvious. The self-healing may be attributed to the material's low T_g or possibly other mechanism. Table 11.1A (d) illustrated a thermally reversible rigid polyamide network whose T_g is higher than the activation temperature of retro-DA reactions. Self-healing at temperatures ~ 100 °C below the T_g resulted in partial repair of the scratch. When chain diffusion is not a limiting factor above the T_g, designed heterogeneity using PEO-*b*-PFGE *block* copolymer, as shown in Table 11.1A (e), did not significantly alter the mechanical properties or self-repairing efficiency compared with the polymer shown in Table 11.1A (a).[15] Self-healing of high T_g materials under ambient conditions are advantageous, but also challenging. DA pairs that are reversible within mild temperature range are rare and require elaborate synthesis, which compromise the practicality. A novel DA

Table 11.1 Cycloaddition reactions responsible for self-repair. The corresponding physical properties and self-repair conditions are listed for corresponding polymer matrices.

Self-repair bonds	Matrix/building block	Phase separation	T_g (°C)	Moduli	Repair conditions	Recovery (%)
A. Thermal Diels–Alder (DA) reactions between furan and maleimide	(a) Multi-diene and multi-maleimide polymer[10]	NA	80–100[11]	4.72 GPa[a]	150 °C 2 h and rt	57
	(b) Thermosetting epoxy[12]	NA	61	NA	140 °C 30 min and 75 °C 5 h	Scratch disappear
	(c) Polyester-containing furan[13]	NA	7.8–8.9	0.014–0.213 GPa[a]	rt 5 d	71.9–16.8
	(d) Polyamide[14]	NA	250	0.566 GPa[a]	120 °C 3 h and 50 °C 5 d	Scratch partially repaired
	(e) PEO-b-PFGE block copolymer[15]	19 nm lamella	NA	5.13 GPa[a]	155 °C 3 h and 65 °C 14 h	Scratch disappear
B. Thermal DA reactions between cyanodithioester and cyclopentadiene	Poly(iBoA-nBA) containing cyclopentadiene end groups[16]	NA	31	NA	120 °C 1 kN for 10 min	>100

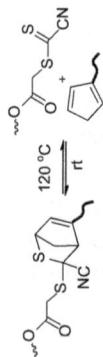

Structure	Material	Phase separation	Modulus[a]	Strength	Healing conditions	Healing efficiency (%)
TCE[17]		NA	Rigid material	42.1 MPa[b]	280 nm UV 10 min	6.4

C. Cinnamoyl [2 + 2] cycloaddition

| | Polyurethane with high M_w PEG (M_w 800 g mol^{-1})[18] | Yes (domain size NA) | 43.7 (−24.9) | 3.5 MPa[c] | 254 and 350 nm UV lights 90 min | 68 |

D. Coumarin [2 + 2] cycloaddition

| | Dendritic macromonomer[19] | Yes (domain size NA) | 75 | NA | 254 nm light 15 min and dark overnight and 366 nm light | Scratch disappear |

E. Anthracene [4 + 4] cycloaddition

[a]Young's modulus.
[b]flexural strength.
[c]Tensile strength; rt: room temperature.

system consisting of cyanodithioester (CDTE) compound and cyclo-pentadiene (Cp), as shown in Table 11.1B, was developed to overcome the limitation of slow self-healing kinetics of the previous systems.[16] Crosslinked network formed between CDTE based tetra-crosslinker and linear poly(iBoA-*n*BA) end-capped with Cp groups offer rapid reversibility and self-healing between 70 and 120 °C without the catalyst.

In order for materials to self-repair without excessive heating, photochemical cycloadditions provide an optically induced self-healing process. These include $[2+2]$ cycloaddition of 1,1,1-tris-(cinnamoyloxy-methyl)ethane (TCE) (Table 11.1C)[17] or coumarin (Table 11.1D),[18] and $[4+4]$ cycloaddition of anthracene derivatives (Table 11.1E).[19] Similar to DA reactions, cycloaddition reactions constructing polymer networks, leading to self-repair *via* four-member ring opening and closure modulated by exposure to light at different wavelengths. Specifically, polymerization of 1,1,1-tris-(cinnamoyloxymethyl)ethane (TCE) monomers *via* photo-stimulated $[2+2]$ cycloaddition gives a rigid polymer.[17] After damage-repair cycle, approximately 6% of the original Young's modulus can be re-covered by photo-radiation to reconstruct the cyclobutane crosslinks (Table 11.1C). Because fragmentations occur at room temperature, self-repair is likely limited by segmental mobility.

The rational design of self-repairable heterogeneous networks should facilitate enhanced segmental mobility. When dihydroxyl coumarin derivatives are reacted with isophorone diisocyanate (IPDI) and polyethylene glycol (PEG) forming linear polyurethane chain, crosslinking occur upon dimerization of coumarin derivatives in-duced by 350 nm UV radiation, forming optical stimuli-responsive polyurethane (IDHPEG).[18] The reversibility of cycloaddition reactions is shown in Table 11.1D. The detection of two T_gs upon crosslinking indicates the coexistence of a rigid phase consisting of coumarin dimer and a mobile phase of PEG-polyurethane segments. Self-repair conditions along with the % recovery for cycloaddition reactions are summarized in Table 11.1. When PEG with lower molecular weight was used, only one T_g is observed owning to good miscibility between soft and hard segments, but % recovery can only reach to 16.2%. This indicates that formation of rubbery domains resulting from micro-phase separation is critical for the reversible photoreactions. Another example is the use of dendritic building block containing anthracene groups on the periphery. Photo-crosslinking through dimerization of anthracene groups increases the T_g from −46 °C for the dendritic macromonomer to 75 °C for crosslinked films (Table 11.1E).

Upon damage, photo-radiation at 254 nm regenerates the building block, allowing reorganization of macromolecular segments, followed by crosslinking upon exposure to 366 nm radiation.[19] The network rearrangements and re-formations occur at room temperature, as shown in Figure 11.4. The recent study also reported that coumarin groups alone are able to facilitate self-healing. When coumarin pendant groups were grafted onto silicone chains, coumarin–coumarin self-associations through π-stacking result in physical crosslinking, and thermoactivated dissociation followed by re-association provide self-healing ability.[20]

11.2.2 Exchange Reactions

Dynamic exchange reactions between reversible covalent bonds that proceed under mild conditions offer another opportunity for the development of self-healing polymers. Different from thermal reversible DA reactions, the activation temperatures of these exchange reactions are lower. Figure 11.5A illustrates the dynamics of acylhydrazone, in which acylhydrazone formation from acylhydrazine and aldehyde groups is reversible in the presence of acid catalyst.[21] Incorporation of acylhydrazone into polyethylene oxide (PEO) networks resulting in a polymer gel that self-heal at room temperature without stimuli.[22] Despite its ability of interchanging bonds even at room temperature, limited studies have been conducted. Another challenge is to achieve quick exchange in solid state without solvent.

Disulfide bonds can undergo metathesis exchange reactions in which two neighboring S–S bonds are disrupted and reformed through free radical or ionic intermediates.[23,24] The cleavage can be activated by photolysis, heating or oxidation.[25] Taking advantage of these conditions, disulfide interchange illustrated in Figure 11.5B was used in self-repairing because S–S entities exhibit good compatibility with polymeric networks and exhibit room-temperature reversibility.[26,27] Mechanical properties and self-repairing abilities vary, depending upon networks morphologies. This is summarized in Table 11.2A though D. Table 11.2A shows an example, in which S–S bonds were introduced into epoxy resin to obtain self-repair, resulting in a crosslinked homogeneous networks with the T_g of approximately $-35\,°C$. No solvent is used in this process, and the tensile stress measurement showed almost complete recovery of properties after heating at 60 °C for 1 h.[26]

When star polymers grafted with poly(n-butyl acrylate) on crosslinked poly(ethylene glycol diacrylate) core were combined with S–S bonds, heterogeneous self-repairable networks can be produced.[27]

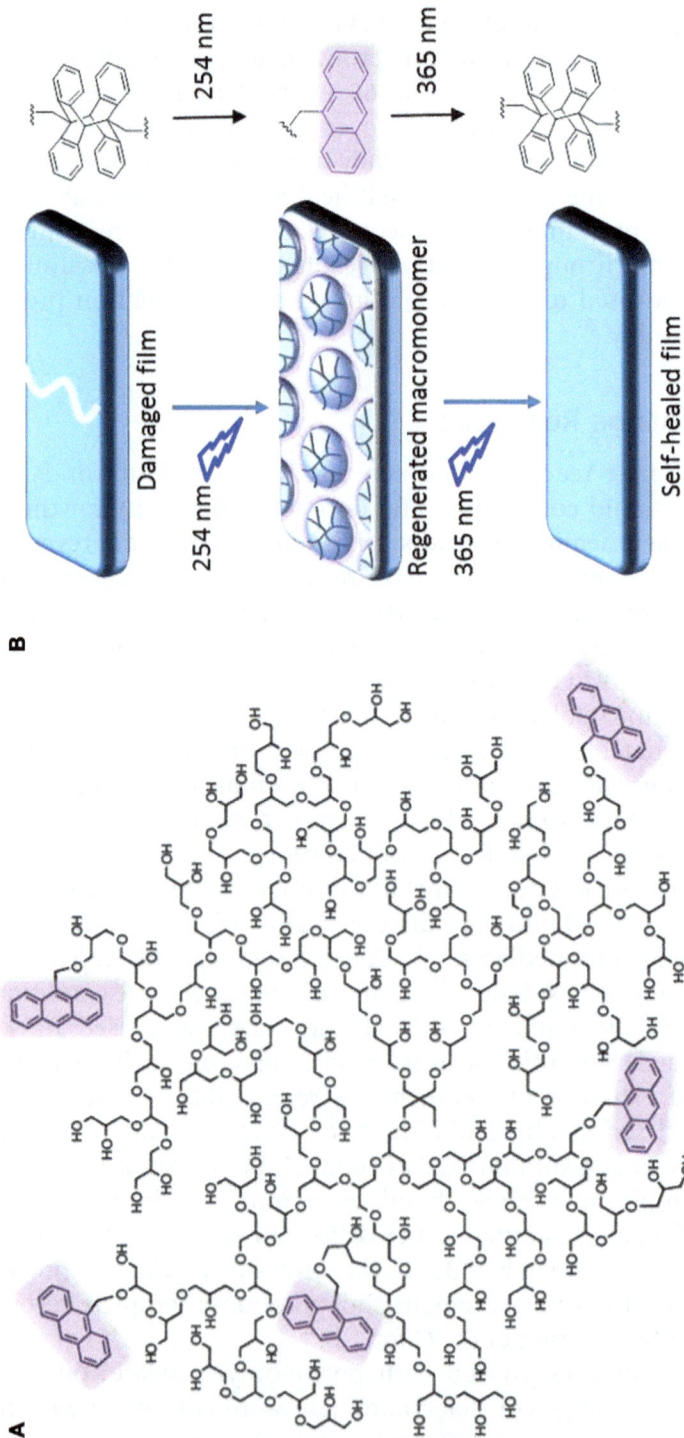

Figure 11.4 (A) Structure of the dendritic building block containing anthracene groups on the periphery. (B) Self-repair process of polymer films synthesized from the dendritic building upon crosslinking *via* anthracene cycloaddition.
Reprinted with permission from ref. 19. Copyright © 2011 WILEY-VCH Verlag GmbH & Co. KGaA, Weinheim.

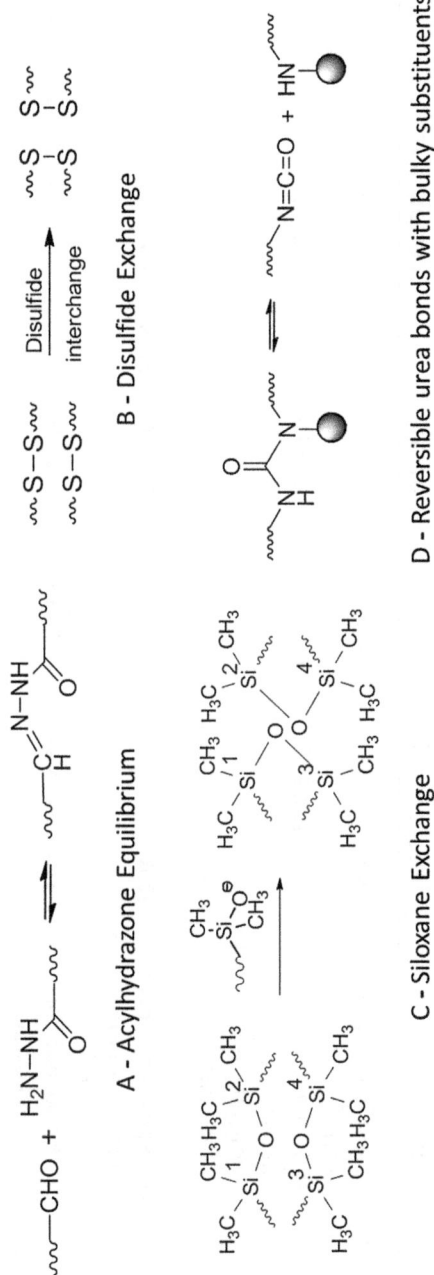

Figure 11.5 Self-repairing exchange reactions responsible for self-healing. Adapted from ref. 21, 26, 30 and 31.

Table 11.2 Physical and self-repairing properties of S–S based homo- and heterogeneous networks.

Self-repairing bond	Matrix	Phase heterogeneity	T_g (°C)	Strength at break	Repair conditions	Recovery (%)
S–S	A. Aliphatic epoxy resin[26]	No	−35	1.3 MPa	60 °C 1 h	96.9
	B. Poly(n-butyl acrylate) grafted poly(ethylene glycol diacrylate) star polymers[27]	Yes (size NA)	NA	20 kPa	rt 20 min to 24 h	NA
	C. Epoxy enhanced with aromatic segments[28]	NA	−3.6	0.18–0.52 MPa	65 °C 240 min	100
	D. Epoxy enhanced with alkoxide inorganic phase[29]	Yes (size NA)	−11 to −9	1.6 MPa	60–70 °C 10 min	100

The presence of the star polymer shown in Figure 11.5A facilitates formation of a network with the high mobility and low intrinsic viscosity, enabling high accessibility of functional groups at ambient healing temperatures. The corresponding properties are shown in Table 11.2B. Significant recovery upon repair is achieved, but at the expense of mechanical integrity. To enhance mechanical properties, rigid aromatic segments were incorporated into epoxy matrices.[28] When aromatic segments are placed along an epoxy backbone, the T_g increase to $-3.6\,°C$ was observed, whereas the T_g of fully aliphatic network is $-46\,°C$. However, longer repair times and lower healing efficiencies were detected (Table 11.2C). Figure 11.5C depicts another heterogeneous inorganic–organic hybrid structure composed of silicone alkoxides inorganic phase and epoxy resin phase containing reversible tetrasulfide bonds.[29] A comparison of the mechanical and self-repair properties of the inorganic–organic hybrid (Table 11.5D) with the aromatic segments enhanced S–S based epoxy networks (Table 11.5C) indicates that incorporation of inorganic phase leads to the formation of a tougher material with higher self-repair efficiency. Thus, enhanced mechanical properties without compromising self-repairing efficiency by introducing heterogeneous phase were achieved.

Exchange reactions in silicone-based materials also facilitated self-repair. Known for their ability to restructure under certain conditions, recently revisited study showed that crosslinked silicone-based polymers containing ethylene bridges and active silanolate end groups synthesized from anionic ring-opening copolymerization of octamethylcyclotetrasiloxane (D_4) and bis(heptamethylcyclotetrasiloxanyl)-ethane (bis-D_4) exhibit remodeling attributes.[30] Figure 11.5C depicts the siloxane exchange reactions responsible for the 'livingness' promoted by the presence of tetramethylammonium dimethylsilanolate anionic end groups, which maintain their activity under ambient conditions, thus providing 'ready-to-respond' active groups. A self-healing polyurea is developed by modifying urea nitrogen with bulky groups, as shown in Figure 11.5D. The modification lowered the dissociation energy of urea C–N bond, resulting in dynamic equilibrium between urea and amine ($-NH_2$)/isocyanate ($-NCO$) reactive groups. When the hindered urea bonds (HUBs) are used to obtain polyurethane-ureas with sub-ambient glass transition temperature, self-repairing under ambient conditions is achieved.[31] When using free NCO groups, even physically entrapped, one needs to realize their toxicity and reactivity toward H_2O (Figure 11.6).

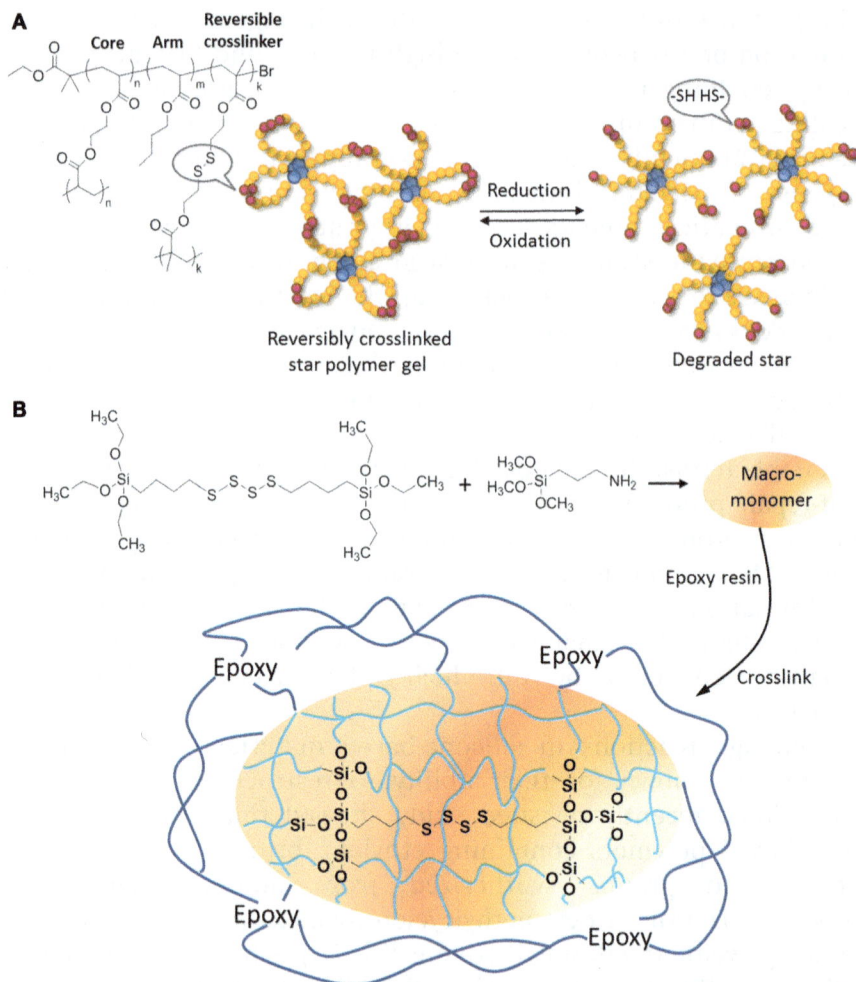

Figure 11.6   Self-healing polymers containing S–S groups that exhibit structural heterogeneities. (A) Formation of star polymers bearing S–S bonds on the periphery. Adapted with permission from ref. 27. (B) Polymer network composed of silicone alkoxides and epoxy resin incorporated with reversible tetrasulfide bonds. Adapted with permission from ref. 29. Copyright (2011) American Society of Chemistry.

11.2.3   Stable Free Radical-mediated Reshuffle Reactions

As pointed out above, it is critical to provide sufficient time for covalent bonds to reform during network remodeling. Mechanical damage of polymers usually generates free radicals as a result of covalent bond cleavages. Recoupling of free radicals from opposite damaged surfaces will reconstruct the network by reforming covalent linkages. If the free radicals are not stabilized, oxidation processes will cause their reactivity loss, thus terminating self-healing. Dynamic reshuffling reactions involving stable free radical formation shown in Figure 11.7 offer an interesting alternative. Taking advantage of dynamic reshuffling of trithiocarbonates (TTC) *via* free radical mechanism shown in Figure 11.7A,[32] where hemolysis of C–S bonds of TTC can be photostimulated by UV radiation at 330 nm generating stable free radicals which can exchange with neighboring TTC groups, self-repair property is attained. Thiuram disulfide moieties exhibiting similar reshuffling characteristics are also used to repair covalently crosslinked polymers.[33] Homolytic cleavage of the disulfide is triggered by visible light, resulting in formation of dithiocarbamate-based free radicals as shown in Figure 11.7B. It was reported that such radicals are able to remain stable for more than two weeks,[34] thus providing sufficient time for network rearrangement before recoupling. TTC free radicals are sensitive to air, it is therefore desirable to seek free radicals that are air tolerant. For example, cleavage of diarylbibenzofuranone (DABBF) groups create arylbenzofuranone (ABF) oxygen insensitive radicals (Figure 11.7C), thus making poly(propylene glycol) (PPG) polymer gels crosslinked by DABBF self-repairable.[35]

When polystyrene (PS) chains are crosslinked by dynamically reversible alkoxyamine (C–ON) groups, self-healing can be achieved by thermal activated dissociation/association of alkoxyamine linkages, such as shown in Figure 11.7D.[36] Crosslinking *via* C–ON groups makes already rigid PS even stiffer. Thus, the challenge is to gain sufficient mobility at the temperatures near the T_g. As shown in Table 11.3, when molar ratio of styrene over the crosslinker C–ON is 7.5/1, two T_gs are detected. While the higher T_g corresponds to crosslinked PS, the lower one is close to linear PS, indicating the existence of a mobile free PS phase arising from dynamic dissociations/associations of C–ON bond within the network. The heterogeneous phase cannot be detected when PS chains between crosslinking points are either shorter (styrene/N–CO = 5/1) or longer (styrene/N–CO = 10/1). As expected, the network containing a mobile phase (styrene/N–CO = 7.5/1) exhibits the highest healing efficiency. When molar ratio of styrene over C–NO is 5/1, free chains generated from scission of N–CO are too short to form a new phase. Thus, self-healing is limited by restricted motion of macromolecular segments. When the ratio is 10/1, the cleaved chains are long, and the chain flexibility will favor re-association, and the healing efficiency is limited by lower concentrations of free radicals. This study demonstrated that the

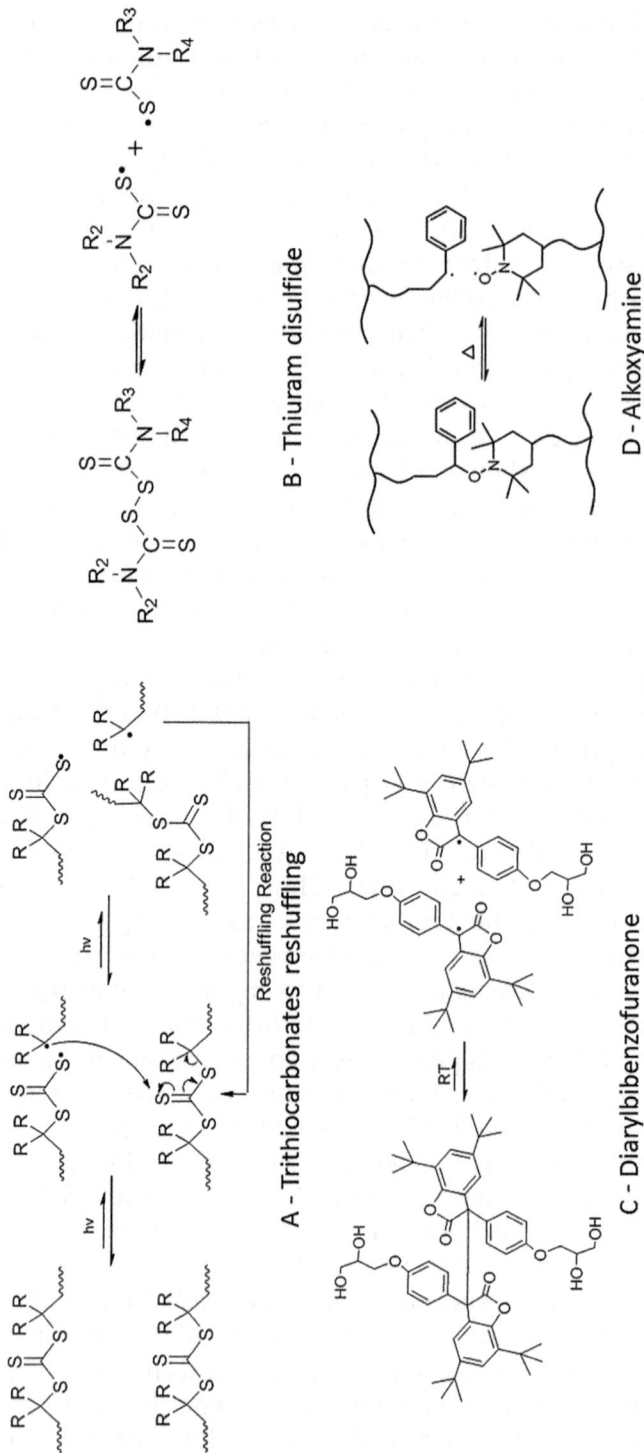

Figure 11.7 Stable free radical mediated reshuffle reactions responsible for self-repair: (A) trithiocarbonates reshuffling (from ref. 32); (B) thiuram disulfide (from ref. 33); (C) diarylbibenzofuranone (from ref. 35); (D) alkoxyamine (from ref. 36).

B - Thiuram disulfide

D - Alkoxyamine

A - Trithiocarbonates reshuffling

C - Diarylbibenzofuranone

Table 11.3 Self-repair and physical properties of alkoxyamine crosslinked polystyrenes in different monomer ratios.

Self-repair bonds	Matrix	T_g (°C)	Phase heterogeneity	Repair conditions	Recovery (%)
Alkoxyamine[36]	Polystyrene: alkoxyamine = 5:1	125	Not detected		26.4
	Polystyrene: alkoxyamine = 7.5:1	(108) 125	Yes (sizes NA)		75.7
	Polystyrene: alkoxyamine = 10:1	125	Not detected	130 °C 2.5 hr	32.6

coexistence of a mobile phase within a rigid network is favorable for self-repairs. Activation temperature of alkoxyamines can be adjusted by placing them in different polymers,[37] and room temperature repair is obtained using linear polyurethane as the matrix.[38] The soft domains originate from PEG segments within polyurethane network facilitate the repair. One limitation of alkoxyamines is that their stability toward air is not clear. The oxidative processes may intercept re-bonding reactions.

11.2.4 Heterocyclic Compounds in Self-healing Polymers

When mechanical forces are evenly distributed within the network, bond rupture typically occurs at positions of lowest bond strength. For polymers, however, external forces are in fact unevenly distributed due to the presence of entanglements and crosslinked points. As a result, not only the weakest bonds, but also other covalent and non-covalent bonds may cleave depending upon the magnitude of the force.[39] While many of the self-healing studies focused only on one type of bond reformation, reactions that reconstruct multiple cleaved bonds are desirable. For example, chain cleavages resulting from mechanical damages of crosslinked polyurethanes will generate $-C^{\bullet}$, $-N^{\bullet}$ and oxygen $-O^{\bullet}$ free radical intermediates. If free radicals remain stable until the chain ends come to close proximity and then recouple, self-repair may occur. For that reason, four and five member heterocyclic compounds, such as oxetanes (OXE) and oxolanes (OXO), which offer low ring opening activation energy producing stable free radicals were used in developing self-repairable polyurethane (PUR) networks.[40,41] The network is shown in Figure 11.8A, where chitosan (CHI) was modified on the C6 position with the OXE, followed by

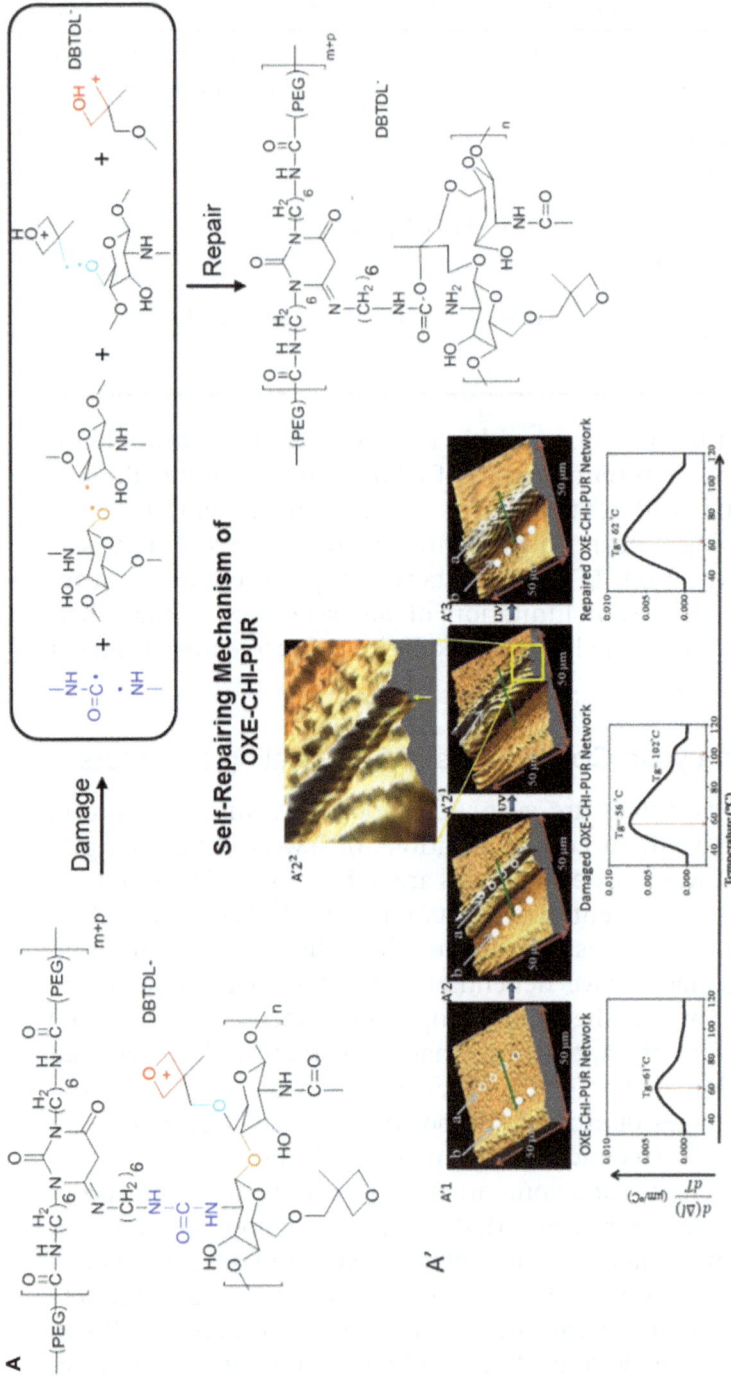

Figure 11.8 Self-repair of polyurethane networks. (A) Oxetane-substituted chitosan polyurethane (OXE-CHI-PUR) network. (A′) Atomic force microscopy (AFM) images of undamaged (A′1), damaged (A′2), UV-exposed (A′2¹, A′2²) and repaired (A′3) OXE-CHI-PUR, and a plot of d(Δl)/dT *vs.* temperature of undamaged, damaged and repaired networks. Reproduced from ref. 40 and 42 with permission of The Royal Society of Chemistry. (B) Methyl α-D-glucopyranoside modified polyurethane (MGP-PUR) network; (B′) Optical images of MGP-PUR self-repair in the presence of CO_2 and H_2O. The Table B′ lists storage modulus (E') of MGP-PUR inside newly created scratch and outside of the scratch measured as a function of time at the oscillation depth of ~200 nm. Reprinted from ref. 43 by permission. © 2014 WILEY-VCH Verlag GmbH & Co. KGaA, Weinheim.

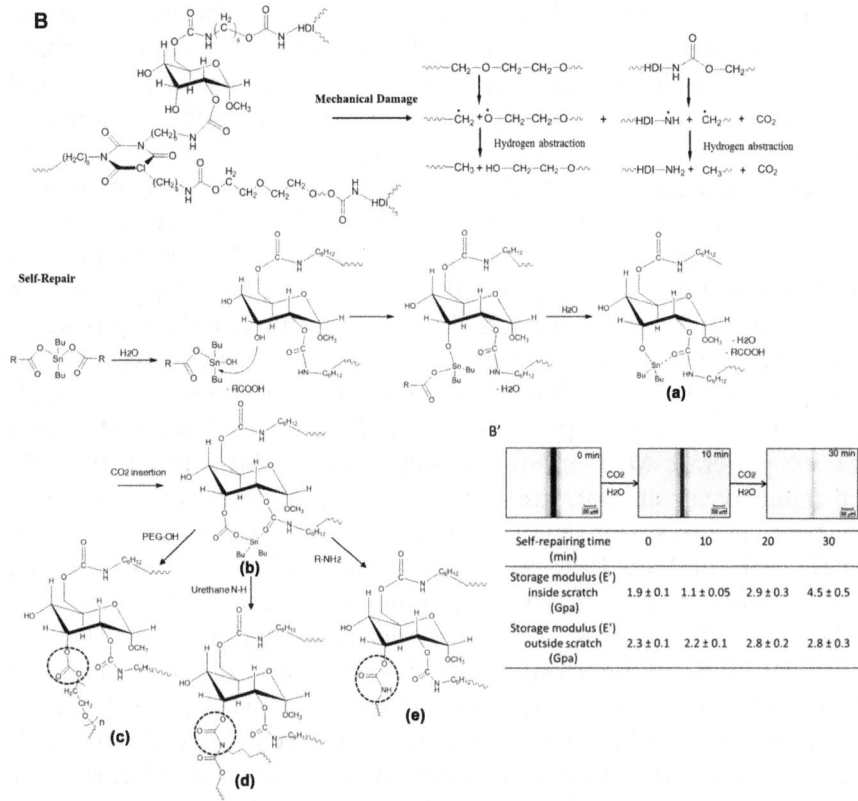

Figure 11.8 Continued.

crosslinking with trifunctional hexamethylene diisocyanate (HDI) in the presence of PEG, forming a heterogeneous crosslinked PUR network (OXE-CHI-PUR). Upon mechanical damage, self-repair occurs by exposure to UV radiation. Specifically, as shown in Figure 11.8A, four labile linkages are affected upon mechanically damage and UV radiation: –NHCONH–PUA linkage, –C–O–C– glycosine units of CHI, –C–O–C– between CHI and OXE, and –C–O–C– of OXE ring. Cleavages of these bonds generate free radicals and a carbocation. In addition, CHI pyranose ring absorb energy from UV, and undergo chair-to-boat conformation changes. Paralleled with OXE cationic ring opening catalyzed by DBTDL and PUA to PUR conversions, these processes lead to self-repair by reformation of urethane and ether linkages. Although mechanistically and kinetically different, the self-healing outcomes with oxolane-chitosan (OXO-CHI) macromonomer in PUR networks are similar. Another interesting aspect is flow of displaced polymer mass.

Mechanical forces applied during damage alter the local free volume. As a result, the increase of localized free volume will favor chain mobility, thus aiding self-repairs. Thus, measurements of the local property changes on nano or micro level are desirable. As shown in Figure 11.8A', following changes of physical properties within the damages during self-repair of OXE-CHI-PUR networks using AFM tips, two T_gs are identified at 61 and 102 °C.[42] The lower T_g corresponds to PEG-polyurethane, while the higher T_g is due to OXE-CHI-urea phase. AFM measurements during repair show a repair front flow from the bottom up, followed by scratch closure upon further UV exposure.

Another approach of obtaining self-repairable PUR networks is by incorporating monosaccharide moieties into thermosetting PURs.[43] When methyl α-D-glucopyranoside (MGP) is reacted with HDI and PEG to generate MGP modified polyurethane (MGP-PUR) network, self-repair occur at room temperature in air. Without CO_2 and H_2O, however, it loses self-repairing property. Analysis of the damage-repair cycle of MGP-PUR revealed cleavages of ether and urethane linkages upon mechanical damage generating nucleophiles including –OH and –NH_2 groups. This is depicted in Figure 11.8B. During repair, MGP-urethane groups first complex with DBTDL assisted by the presence of H_2O, lowering CO_2 insertion activation energy, after which, forming carbonate intermediates. The reactions of –OH/–NH_2 with carbonate intermediates result in the formation of carbonate and urethane linkages, thus repair the networks. This is the first study to use CO_2 and H_2O from the air to self-repair synthetic material, similar to carbon fixation of plants during photosynthesis, making synthetic material living like a plant. It should be noted that localized events measured by nanoindentation revealed initial modulus drop inside the damage, followed by the increase of the modulus until full recovery was reached. This is shown in Figure 11.8B'. In fact, networks with variable morphologies and different T_gs follow different patterns of physical recoveries. For example, thin films made from star polymers containing disulfide crosslinks on a silicon wafer exhibit surface tension driven viscoelastic flow behavior,[27] and self-healing of poly(n-butyl acrylate) network containing alkoxyamine junctions shows enhanced mobility and flow.[44] These *in situ* thermo/mechanical measurements at micro-scale reveal morphology changes of the scratches during repair, thus reveal the processes responsible for damage-repair cycle.

11.3   The Role of Supramolecular Chemistry

Integrity of supramolecular material is achieved by the formation of non-covalent interactions.[45] This is typically accomplished by the presence of associative groups covalently attached to side chain or chain ends of the polymers, which bind the liquid-like polymers into a network that exhibits plastic or rubbery behavior resulting from non-covalent crosslinking. These interactions include hydrogen bonds, metal coordination, ionic interactions, π–π stacking, and host–guest interactions. Several features make supramolecular chemistry particularly attractive when comes to self-healing: reversibility and speed, directionality, and sensitivity. In contrast to covalent bonding, these networks can be remodeled rapidly and reversibly from fluid-like to solid-like states. Due to the distinct nature of secondary bonds which differ from their polymer surroundings, the presence of associative groups usually induce morphology changes such as aggregation and crystallization. Thus, heterogeneities are inherent for many supramolecular networks affecting their physical/chemical properties, and facilitate self-repair. When mechanical force is applied, weaker supramolecular bonds will dissociate, and subsequent regeneration due to their dynamic character will allow self-repair.

11.3.1 Hydrogen Bonding

Although hydrogen bonds (H-bonds) are weaker than common co-valent bonds, due to their directionality and affinity, a wide range of supramolecular polymers with diverse mechanical properties ranging from supramolecular gel to strong rubber-like materials can be prepared. Concentration levels and the strength of H-bonds, length and rigidity of polymer backbone are the main factors determining mechanical strength and network remodeling. Several successful examples of self-repairing materials using H-bonds have been developed and are shown in Figure 11.9.

When fatty diacids and triacids were utilized in a two-step synthetic route involving condensation of acid groups with an excess of diethylenetriamine, followed by reactions with urea, a mixture of oligomers containing H-bonding motifs including amidoethyl imidazolidone, di(amidoethyl) urea and diamido tetraethyltriurea is obtained. Associations of the H-bonding motifs leads to the formation of the self-repairing network shown in Figure 11.9A.[46,50] The resulting supramolecular assembly is a translucent glassy plastic which behaves like a soft rubber at temperatures up to 90 °C, and is able to self-repair at ambient conditions. Although only single T_g just below ambient temperature was observed, NMR measurements revealed substantial heterogeneities consisting of a mobile aliphatic phase and a rigid H-bonded phase.[51] While formation of H-bonds between amide –C=O and amine-functionalized chains is responsible for

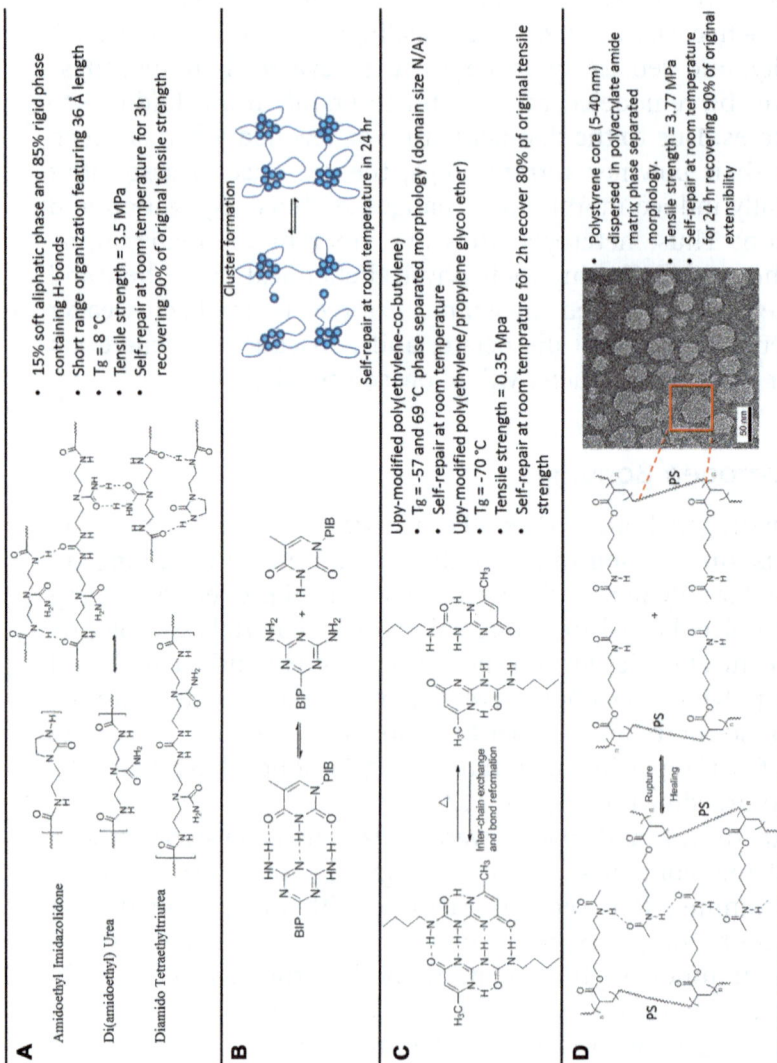

Figure 11.9 Reversible H-bond formations responsible for self-repair and the network's physical and self-repair properties. (A) Diacid and triacid associations from renewable resources. Reproduced from ref. 46 with permission of The Royal Society of Chemistry. (B) Triple H-bond formation between thymine (Thy) and 2,6-diaminotriaine (DAT) (from ref. 47). (C) Dimerization of ureidopyrimidinone (Upy) units by quadruple H-bonds. (D) Copolymers combining hard polystyrene (PS) backbones with soft polyacrylate amide (PA-A) pendant groups carrying multiple H-bonding sites, and TEM image showing heterogeneity of the polymer.
Reprinted by permission from Macmillan Publishers Ltd: Nature (ref. 49) copyright (2012).

network reformation, the presence of ~15% mobile phase enables high healing efficiency at room temperature without any other stimuli. Notably, another interesting aspect of H-bonded networks is that self-healing reactions often compete with external conditions, such as the presence of moisture. For example, extended moisture exposure or heating of freshly cut specimens can significantly diminish the kinetics of self-healing due to the new equilibria,[52] which has been proven for the supramolecular network shown in Figure 11.9A. Not only for supramolecular networks, but for most of the dynamic polymer systems, chemical activity and conformational changes immediately after mechanical damage play a key role in self-healing.

The strength of directional multi-hydrogen bonds, such as triple H-bonds between thymine (Thy) and 2,6-diaminotriaine (DAT), quadruple H-bonds of ureidopyrimidinone (Upy), as well as multi-urea, are comparable to covalent bonds, giving the material soft rubber-like properties. Taking advantage of high segmental mobility of polyisobutylenes (PIB), supramolecular PIB networks were developed containing various directional associative end groups offering variable dynamic behaviors, including self-repair.[53,54] For example, PIBs functionalized with Thy and DAT end groups assemble into strong rubber-like materials by formation of triple H-bonds as shown in Figure 11.9B. Dynamic rheological studies revealed that enhanced mechanical strength is not simply due to the presence of triple-H-bonds, but is also attributed to network heterogeneity. Undirected supramolecular interactions between two or more Thy or DAT lead to complex cluster formation as shown in Figure 11.9B, resulting in stronger crosslinked networks.[55] Taking advantage of such properties, a similar supramolecular PIB network containing barbituric acid moieties were tested for self-repairing.[47] The materials exhibited rubbery plateau in the rheology experiment at high frequency, and is self-repairable at ambient temperature by bringing the two parts into contact. Cluster formation is enhanced by polarity difference between associative chain ends and aliphatic main chains.

Ureidopyrimidinone (Upy) units exhibit high dimerization constant *via* quadruple H-bonds (Figure 11.9C). Supramolecular assemblies containing Upy units exhibit rubber-like behavior like conventional polymers, while exhibiting strong temperature and mechanical property dependence.[56–58] Although homogeneous morphology in the melt state is reported for given polymer backbones,[58] when incorporated within flexible polymer chains such as poly(ethylene-

co-butylene) and PDMS, aggregate formation from multiple Upy-dimers is likely to occur resulting in increased mechanical strength and T_g.[59,60] For example, Upy-modified poly(ethylene-*co*-butylene) is observed to exhibit a low T_g at $-57\,^{\circ}$C and an additional thermo transition at $69\,^{\circ}$C. This indicates a phase separation of aggregated Upy-dimers within amorphous poly(ethylene-*co*-butylene) matrix. Thermal activated dissociation and re-association at lower temperature enables repeatable self-repair, whereas soft polymer chains provide sufficient mobility after dissociation from the aggregate.[48,61]

Due to strong association between Upy units, they exhibit poor solubility in most organic solvents, making it difficult to process, and require high activation temperature. Supramolecular bonds have superior advantages of versatility and tunability of bond strength. If utilizing a multiphase polymer with soft and rigid domains, and placing weaker H-bonds, can the H-bonds within the mobile phase increase self-healing efficiency? Such material can be synthesized from a brush copolymer comprised of rigid polystyrene (PS) backbone and soft polyacrylate amide (PA-A) brush.[49] The brush copolymers collapse into a core-shell nano-structure, where PS chains form the rigid core and PA-A side chains form the soft shell. H-bonds directed assembly through amide groups within the brushes lead to the formation of micro-phase separated microstructure, as shown in the TEM image in Figure 11.9D. Young's moduli of resulting elastomer increased by a factor of three compared to the fatty acid based supramolecular network, while self-repair property at room temperature is maintained. Employing the same concept, different building blocks are used, such as poly(*n*-butyl acrylate) and polystyrene diblock copolymer (PBA-*b*-PS) end capped with Upy motifs at the PBA end yielding cylindrical microphase morphology,[62] as well as an ABA triblock copolymer containing glassy poly(methyl methacrylate) (PMMA) block and two dynamic poly(acrylate amide) (PA-amide) blocks generating core–shell spherical nanoparticles.[63] Both exhibited self-healing attributes.

11.3.2 Metal–Ligand Coordination

Polymers containing metal centers emerged as easily processable materials with distinct optical properties and stimuli responsiveness. The non-covalent binding between metal and macromonomers vary from strong irreversible to highly dynamic.[64–66] The latter has essentially been used in self-repairing metallosupramolecular networks, offering reversibility similar to H-bonds. Their photo responsiveness offers controllable self-repairing, while reversibility and tunability by incorporating different metal ion and ligand substitutes make coordination chemistry particularly attractive. It is distinct from H-bonding that coordination number and conformations of the coordination center are tunable upon stimuli, and various structures can be obtained such as linear, star-shaped, highly branched, or dendritic. These characteristics are preferable for applications in dynamic self-repairing. Typical metal assisted self-repairing networks are illustrated in Figure 11.10.

It is well documented that organometallic conductive polymers comprising of N-heterocyclic carbenes (NHCs) may exhibit corrosion control attributed to conductivity in the order of 10^{-3} S cm^{-1}. Taking advantage of this property and the ability for coordinating with transition metals, formation of complexes utilizing NHCs and transition metals capable of self-repairs was developed.[67] Reversibility of the metal–polymer bonds is shown in Figure 11.10A. When damaged, heating at 200 °C without solvent, or at 150 °C in the presence of DMSO vapor, will repair the damage due to the dynamic equilibrium between metal and the polymer, leading to flow into microcracks and reforming coordination bonds. The high repairing temperature is due to the strength of coordination bond and lack of soft segments.

Similarly, when competitive ligand hydroxyethylethylenediamino triacetic acid sodium salt (HEEDTA) was incorporated into (ruthenium) Ru(II)-bis-terpyridine crosslinked block copolymers, temperature changes decouple Ru metal from the ligand, resulting in bonding and de-bonding.[73] However, the binding strength of Ru(II)-bis-terpyridine is comparable with a typical covalent bond,[74] which requires high energy to dissociate. Metal–ligand coordination can be tuned by choosing from different metal ion and ligand pairs.[75] Similar networks, but bonded through (zinc) Zn(II)-bis-terpyridine[68] and (iron) Fe(II)-bis-terpyridine[69] coordination, exhibited self-repair properties, in which heterogeneities play a critical role during network rearrangements as demonstrated below.

The optically healable supramolecular polymer based on amorphous poly(ethylene-*co*-butylene) core with 2,6-bis(1'-methylbenzimidazolyl)pyridine (Mebip) ligands at the termini is crosslinked *via* Zn(II)-bis-terpyridine coordination, as shown in Figure 11.10B.[68] The presence of a hydrophobic segments and the polar metal–ligand motif results in phase separated morphology. Small-angle X-ray scattering (SAXS) and transmission electron microscopy (TEM) studies both revealed the lamellar morphologies where the metal–ligand complexes form a hard core that physically crosslinks the poly-

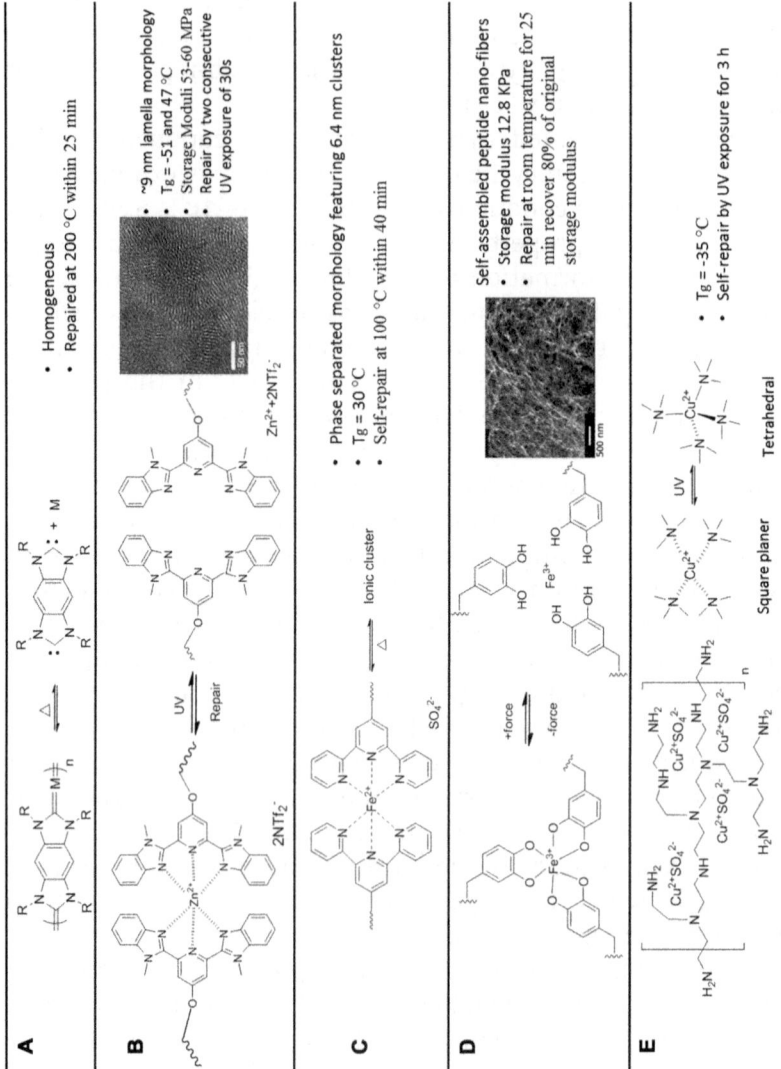

A

- Homogeneous
- Repaired at 200 °C within 25 min

B

$Zn^{2+} + 2NTf_2^-$

$2NTf_2^-$

- ~9 nm lamella morphology
- Tg = −51 and 47 °C
- Storage Moduli 53–60 MPa
- Repair by two consecutive UV exposure of 30s

C

SO_4^{2-}

Ionic cluster

- Phase separated morphology featuring 6.4 nm clusters
- Tg = 30 °C
- Self-repair at 100 °C within 40 min

D

Fe^{3+}

- Self-assembled peptide nano-fibers
- Storage modulus 12.8 KPa
- Repair at room temperature for 25 min recover 80% of original storage modulus

E

$Cu^{2+}SO_4^{2-}$

Square planer

Tetrahedral

- Tg = −35 °C
- Self-repair by UV exposure for 3 h

(ethylene-*co*-butylene) soft domains. The lamella period is character-ized as $\sim 9\,\mathrm{nm}$. Upon exposure to ultraviolet light, the metal–ligand motifs are electronically excited and the absorbed energy is converted into heat, causing temporary disengagement of the metal–ligand motifs and a concomitant reversible decrease in the polymers' mo-lecular mass as well as viscosity. Dissociation of the hard phase gives rise to the mobile macromonomers that rearrange and reform to the original morphology, and reformation of metal–ligand bonds allow-ing healing of mechanical damages.

A somewhat similar metallosupramolecular system, yet chemically different compared to Zn(II)-bis-terpyridine coordination, was de-veloped with terpyridine-Fe^{2+} complexation pair using poly(alkyl methacrylate) as the main chain. This is shown in Figure 11.10C.[69] Polarity difference induced phase separation is not significant in this network. Furthermore, instead of UV exposure, direct thermo treat-ment at $100\,^{\circ}\mathrm{C}$ was applied to repair the damages. It was detected that metal–ligand bonds remain stable at this temperature, indicating that self-repair is not caused by dissociation of the coordination bonds, but was attributed to reversible formation and dissociation of $6.4\,\mathrm{nm}$ ionic clusters triggered by the presence of Fe^{2+}. Although network heterogeneity was not intentionally introduced, it is inherently specific to this network.

There is increasing evidence that metal–ligand coordination play an important role in the dynamics of biological rearrangements. Recent studies took advantage of coordination between Fe(III) and catechol ligands that mimic mussel adhesives in nature, which

Figure 11.10   Reversible metal–ligand coordination reactions responsible for self-repair and the network's physical and self-healing prop-erties. (A) Dynamic equilibrium of N-heterocyclic carbenes and transition metal monomeric species and their polymer. Re-printed by permission from Macmillan Publishers Ltd: Nature (ref. 67) © 2011. (B) Self-repairing supramolecular network containing 2,6-bis(1'-methylbenzimidazolyl)pyridine (Mebip) ligands coordinated with Zn(NTf)$_2$; poly(ethylene-*co*-butylene) is utilized as the main chain, resulting in phase-separated lamella structure as shown in the TEM image (from ref. 68). (C) Terpyridine containing alkyl methacrylates coordinated with (iron) Fe(II) sulfate (from ref. 69). (D) Reversible tris-catechol-Fe^{3+} complex. Reprinted from ref. 70 by permission of John Wiley & Sons, Inc. © 2013. SEM images demonstrate self-assembled peptide nanofibers inter-crosslinked by cate-chol-Fe^{3+} complex (from ref. 71). (E) Polyethylenimine–copper (C_2H_5N–Cu) supramolecular polymer networks and the UV-induced square planar-to-tetrahedral conformational changes (from ref. 72).

resulted in pH-induced crosslinked self-healing polymer with near-covalent elastic moduli.[70,76–78] The networks are formulated from 3,4 dihydroxyphenylalanine (DOPA) modified poly(ethylene glycol) (PEG) or polyallylamine. The uniqueness of this approach is the ability to switch from mono-, bis, to tris-Fe-catechol complexes at different pHs,[70] thus providing the ability of crosslinking control without Fe(III) precipitation. Figure 11.10D depicted reversible tris-Fe(III)-catechol complex formation. Similarly, crosslinked hydrogels were also prepared by complexation of branched catechol derivatized poly(ethyleneglycol) (PEG) with 1,3-benzenediboric acid.[79,80] Later on, a heterogeneous structure inspired by mussel was developed. Peptide amphiphile molecules composed of a hydrophobic lauryl group attached to the N-terminus and DOPA attached to the opposite end of the peptide segment were synthesized.[71] Analogous to mussel byssus, the amphiphilic molecules aggregate into nanofiber structures with hydrophilic peptide sequence containing DOPA exposed to aqueous environment. Interfiber crosslinking by complexation of Fe(III) and DOPA at pH ≈ 10 enhanced the mechanical strength of the network, while dissociation of tris-Fe(III)-catechol bonds create free moving fibers. Self-repair occur upon rearrangement of the nanofibers and reformation of crosslinks. Figure 11.10D shows a TEM image of the nanofibers crosslinked by tris-Fe-catechol. This strategy demonstrated the adaptability of metal binding to various complex supramolecular networks enabling self-repair while enhancing the mechanical strength without disturbing the original morphology.

Facile UV-healable polyethylenimine–copper (C_2H_5N–Cu) supramolecular polymer networks were also developed. The uniqueness of this network is that self-healing is not only induced by reversible formation of C_2H_5N–Cu complexes, but also square-planar-to-tetrahedral conformational changes of the coordination center induced by the charge transfer between $\sigma(N)$ bonding and $dx^2 - y^2$ (Cu) antibonding orbitals upon UV absorption.[72] The structural and conformational changes of the complex center are shown in Figure 11.10E. The potential use of a hybrid polymeric scaffold, in which covalent crosslinked permanent scaffold coexisted with reversible metal–ligand coordinative crosslinks were also investigated.[81] The advantage of the hierarchical design is to create toughness while enabling 'programmable' macroscopic mobility. A self-repairing polydimethylsiloxane/polyethylene glyco-polyurethane (PDMS/PEG-PUR) Cu-catalyzed networks in which Cu–O coordination was embedded into covalently crosslinked polyurethane networks were developed.[82] In addition to UV induced Cu–O bond reformation along with conformational changes, cleavage and reformation of covalent Si–O linkages were detected, indicating a multi-level repair process. Nanoindentation measurements indicated the increase of modulus inside the scratch immediately after damage. This was attributed to the formation of $CuCl_2$ clusters resulting from Cu–O bond cleavage. The interfacial property and heterogeneity at the newly created surfaces upon damage is another area that is critical to self-repair, but has not been extensively studied.

11.3.3 π–π Stacking Interactions

π–π stacking interactions achieved by end-capped π-electron-deficient groups interacting with π-electron-rich aromatic backbone molecules were utilized in the development of thermal-triggered reversible self-healing supramolecular polymer networks. This was achieved by employing chain-folding copolyimide (electron deficient) and pyrenyl (electron rich) end-capped polysiloxane[83] or polyamide[84] chains. The T_g of the network can be tuned to achieve self-healing at relatively wide temperature range ($\sim$50–100 °C) by changing the spacer and adjusting the composition of the blend. Upon heating, the π–π stacking interactions will be interrupted, enabling pyrenyl end-capped chains to disengage from copolyimide and flow due to the presence of a flexible 'soft' spacer. Thus, repair of damage and re-gaining of mechanical strength by reformation of the π–π stacking will occur. Other approaches combined π–π stacking with inter-molecular H-bonding within one supramolecular network to gain efficient thermal healability using polybutadiene containing urethane and urea groups as spacer.[85] Networks with enhanced mechanical properties were produced using tweezer-type bis-pyrenyl end groups instead of mono-pyrenyl groups.[86] Figure 11.11 illustrates the formation of tweezer-type interactions between bis-pyrenyl end groups (compound **1**) and naphthalene-diimide (compound **3**) chains. Non-tweezer type interactions using mono-pyrenyl groups (compound **2**) are also shown. The [1 + 3] blend exhibited higher tensile modulus and elongation compared to [1 + 2] blend, as summarized in Figure 11.11, although requires higher healing temperature and longer time. Nitrobenzoxadiazole (NBD)-containing cholesterol (Chol) derivatives also offer a unique combination of π–π stacking and intermolecular H-bonding interactions,[87] where the length of the spacer connecting NBD and Chol units is critical in controlling gel and healing properties.

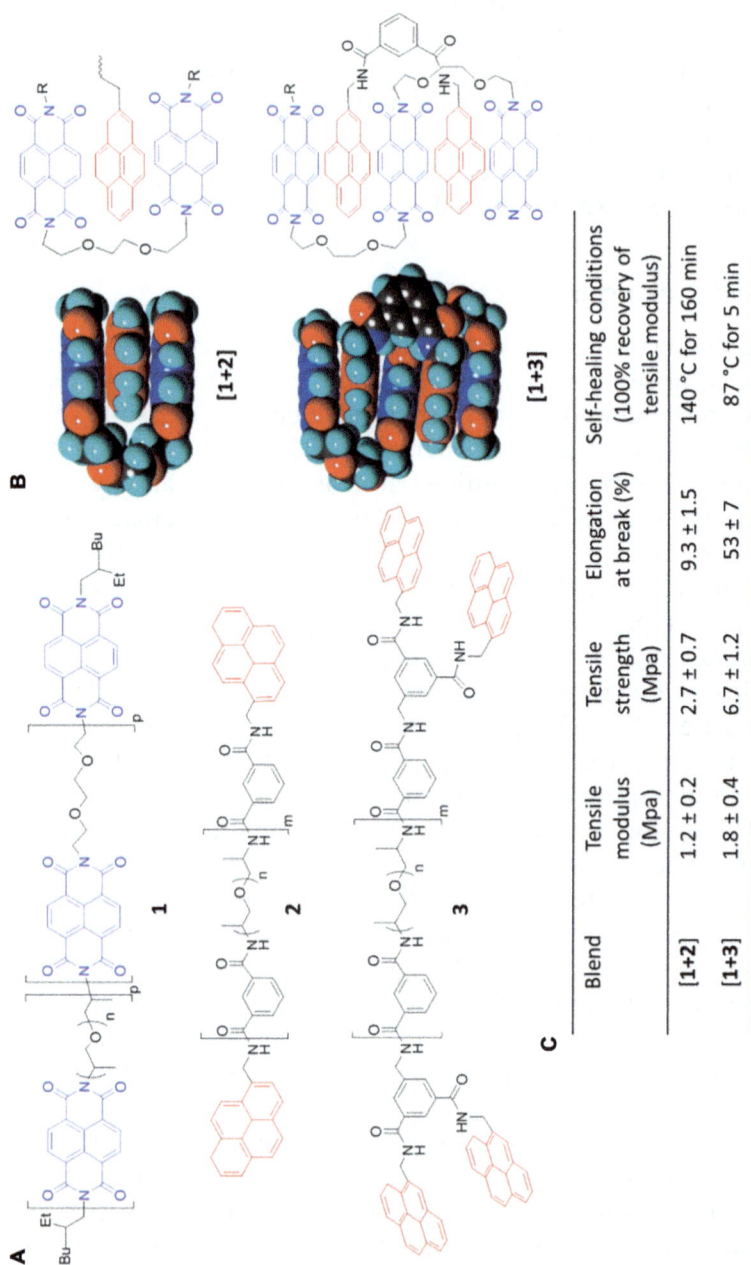

Figure 11.11 (A) Chemical structure of **1** – chain folding units containing naphthalene-diimide receptor, **2** – polyamide end-capped with pyrenyl groups, and **3** – polyamide end-capped with tweeter type bis-pyrenyl groups. (B) Energy-minimized models (molecular mechanics with charge-equilibrium) for complexes formed between compound **1** and **2** ([**1**+**2**]), and between compound **1** and **3** ([**1**+**3**]), in which formations of two or four face-to-face π–π stacking interactions are shown. (C) The table illustrates the results of the tensile measurements and self-healing conditions for blends [**1**+**2**] and [**1**+**3**].
Reprinted with permission from ref. 86 Copyright (2010) American Society of Chemistry.

11.3.4 Ionic Interactions

Ionic interactions in polymers are primarily manifested by the formation of ionomers,[88] and selected polymers demonstrated self-healing attributes. For example, in poly(ethylene-*co*-methacrylic acid) (EMAA) shown in Figure 11.12A, polymer diffusion in a damaged area occurred under ambient and elevated temperatures upon projectile puncture testing.[89] A ballistic puncture in low density polyethylene (LDPE) did not show healing, whereas puncture in EMAA films healed the puncture leaving a scar on the surface. The proposed healing mechanism is a two stage process in which upon projectile impact, ionomeric network is disrupted, and heat generated by the friction during the damage is transferred to the surroundings, generating localized melt state. The molten polymer surfaces fuse together *via* inter-diffusion to seal the puncture, followed by reshuffling of the ionic clustered regions and relaxation.[90] Because self-healing in this case largely relies on thermo activated chain diffusions, increased cluster sizes that immobilize the chains will reduce the healing rate.[91] However, if the clusters are able to dissociate upon stimuli, this process may enhance the healing rate. This was demonstrated by using polyelectrolytes synthesized from poly(acrylic acid) (PAA)/poly(allylamine hydrochloride) (PAH) pairs shown Figure 11.12B. By applying ultracentrifugation and extrusion in the presence of salt yielding compact polyelectrolyte complexes (CoPECs), this systems exhibited self-healing attributes.[92] The self-healing ability becomes stronger with increasing concentrations of NaCl salt, indicating that dynamics of the material originates from interrupting ionic interactions between the polyelectrolyte pairs by salt, which increase the chain mobility. The effect of ionic concentrations on self-healing of EMMA ionomers[93] revealed the relationship between elasticity, diffusion, and the self-healing behavior at various temperatures. Increased ionic content produces higher crosslinking density, but less mobile polymer structures, which yield an increased resilience and elasticity in the melt not observed in non-ionic varieties. It created a much stronger network surrounding the impact site to allow the needed elastic response to repair. The non-ionic variations loose healing ability upon reaching T_m. High concentrations of ionic species are detrimental to low temperature healing due to the rigidity. Such relations between elasticity, diffusivity, healing temperature, and self-healing ability are not limited to ionomers, but also applicable to other self-healing systems.

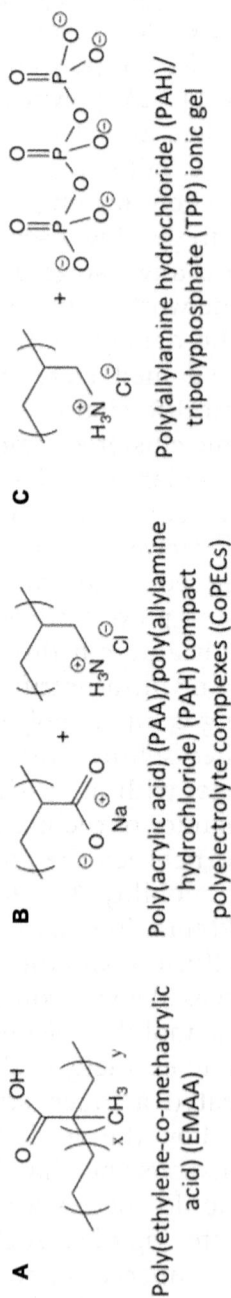

Figure 11.12 Examples of ionomers used in self-healing. (A) Polyethylene-co-methacrylic acid) (EMMA) (from ref. 89). (B) Poly(acrylic acid) (PAA)/poly(allylamine hydrochloride) (PAH) forming compact polyelectrolyte complexes (CoPECs) by ultracentrifuge (from ref. 92). (C) Poly(allylamine hydrochloride) (PAH)/tripolyphosphate (TPP) ionic gel (from ref. 97).

Initiated by the presence of water, polyelectrolyte multilayered assemblies have demonstrated the ability to recover from mechanical deformations.[94,95] Under dry conditions, the Young's modulus is as high as 11.8 GPa, but can be reduced to 0.44 MPa in the presence of H_2O resulting from hydrophilic nature of the ionic pairs. The role of multilayered structures in the presence of ionic interactions was not addressed.[96] Stiff and self-healing gels were also developed from complexation of polyamines with phosphate-bearing multivalent anions, which are shown in Figure 11.12C.[97] They exhibited strong underwater adhesion to both hydrophilic and hydrophobic surfaces as well as self-healing through reversible ionic crosslinks. Another water induced self-healing system, although based on H-bonds instead of ionic interactions, was synthesized from polyacrylate and polymethacrylate polymers surface-functionalized with mussel-inspired catechols.[98] The catechol groups are protected by triethylsilane. Immersing the cut samples in water at pH = 3 removes the triethylsilyl protecting groups at the cut, thus activating catechol groups and subsequent rejoining of the cut surfaces by the formation of intermolecular H-bonds between catechol and oxidized quinones.

In essence, recent trends indicate that all means of self-healing are of interest as long as the dynamics of the system facilitates the process. Even though so-called layer-by-layer (lbl) film deposition process has been known for many years, it turns out that dynamics of charged alternating layers may facilitate sufficient interfacial mobility that is critical during repair. In view of the fact that spraying is an accelerated lbl process,[99] the argument, that the lbl architectures might be responsible for self-healing lacks direct correlation without considering ionic and electrostatic interactions.[100]

11.3.5 Host–Guest Interactions

Host–guest interactions refer to the targeted binding of two structurally dissimilar but dimensionally matched macromolecular units *via* weak interactions including H-bonding, π–π interactions, van de Waals, or hydrophobic interactions. While naturally existing host–guest assemblies exhibit great biological importance,[101] synthetic host–guest pairs capable of reversible guest exchanges[102,103] may facilitate the formation of self-repairable supramolecular gels. Taking advantage of cyclodextrin's (CD) cavity capable of entrapping other molecules, host–guest interactions between CD (host) and polyacrylic acid terminated with ferrocene (guest) were used to create hydrogels.[104] Redox stimuli induced a sol–gel transition, and dissociation/reassociation between CD and ferrocene facilitate the self-repair. Not surprisingly, supramolecular hydrogels prepared from polyacrylamide modified with CD crosslinked with aliphatic guest groups *via* hydrophobic interaction also showed repairing upon rejoining the ruptured surface.[105] Similar supramolecular hydrogels were also prepared from crown ether and bisammonium based poly(methyl methacrylate) (PMMA) polymers. Macrocyclic host molecule cucurbit[8]uril (CB[8]) capable of reversible crosslinking with multivalent copolymers with high binding constants $(K_a > 10^{11}$–$10^{12} \ M^{-2})$ was also shown to have the potential to be used in self-healing materials.[106] In summary, numerous unexplored opportunities in supramolecular hydrogels based on host–guest interactions exist for the development of 3D structures.

Polymer networks that rely on reversible covalent bonds and secondary interactions may offer self-repairs in which bond dynamic and chain mobility affect self-repairing efficiency. A common denominator of many studies is that rationally designed heterogeneities not only facilitate distribution of hard or soft segments and free volume, but also provide bulk integrity. Reversible bonds incorporated into hard-soft domains or their interfaces may enhance dissociation or reposition of soft domains while maintaining local integrity during repair. If designed properly, heterogeneous polymeric materials with glassy or tough rubber-like properties may offer high repair efficiency under ambient conditions. Heterogeneous materials that may also offer self-repair properties are nanocomposites. Owning to large interfacial regions, design of reversible interfacial chemistry may offer unique opportunities for development of new materials. The next sections will examine the role of interfaces in multi-phase composites and their potential impact on self-healing.

11.4   Chemo-mechanical Self-healing

Initial studies on thermoplastic polymers suggested that self-healing of crack involves five stages: segmental surface rearrangements, surface approach, wetting, diffusion, and randomization.[107] In essence, the outcome of this process was to regain mechanical properties at the repaired area *via* chain entanglements. Thus, chain diffusion at the polymer–polymer interface was considered the primary driving force for repairs, and the old reptation model[108] semi-quantified these events by describing the motion of macromolecular segments trapped in a stationary tube. Under these conditions, self-healing was achieved only when the entire chain completely disengaged from the tube, reaching equilibrium state after a given time T_r (where T_r is the time required for complete chain disengagement from the tube). However, self-repairing process can be viewed as chain conformational changes going from non-Gaussian to equilibrated Gaussian state during which each chain escapes from the surroundings, and as expected, longer chains would take longer to escape. However, chain flexibility was considered only recently,[2] which identified the relationship between molecular weight, chain flexibility, and the Gibbs free energy. Intuitively, the ability to flow will facilitate filling damaged areas, but the presence of liquid will obviously undesirable. In an effort to trap a reactive liquid that could potentially reinforce the matrix upon mechanical damage resulted in the concept of encapsulation.

11.4.1 Encapsulation

Initially inspired by other disciplines,[109] introducing intentional heterogeneities into a polymer matrix resulted in the development of self-healing polymers containing hollow fibers containing reactive agents[110–113] and embedded reactive microcapsules.[114] Figure 11.13A, B and C depict these processes. When hollow glass fibers shown in Figure 11.13A containing healing agents such as epoxies, upon mechanical damage were introduced into a polymer matrix, and were mechanically damaged, reactive chemicals 'spilled' from damaged areas and crosslinked, thus serving as an internal glue. The alternative to hollow fibers were embedding reactive liquid-containing microcapsules shown in Figure 11.13B. Using a similar concept, microvascular networks offered the interconnected hollow fibers filled with healing agents.[115–118] Figure 11.13C illustrates the proposed visually resembling microvascular networks which take the advantage of the ability to refilling hollow channels with healing agents upon depletion, thus enabling multiple damage–cycle repairs in the same area.

Critical components of this process are encapsulated reactants. Figure 11.14A–D illustrates chemical reactions involved in encapsulation, hollow fibers, and vascular networks. Initial microencapsulation experiments involved the use of catalyzed reactions of ring-opening metathesis polymerization (ROMP) of dicyclopentadiene (DCPD) microencapsulation in the presence of a ruthenium catalyst shown Figure 11.14A.[119] Later on, multiple DCPD and ruthenium catalysts were employed to adjust the crosslinking reaction rates in

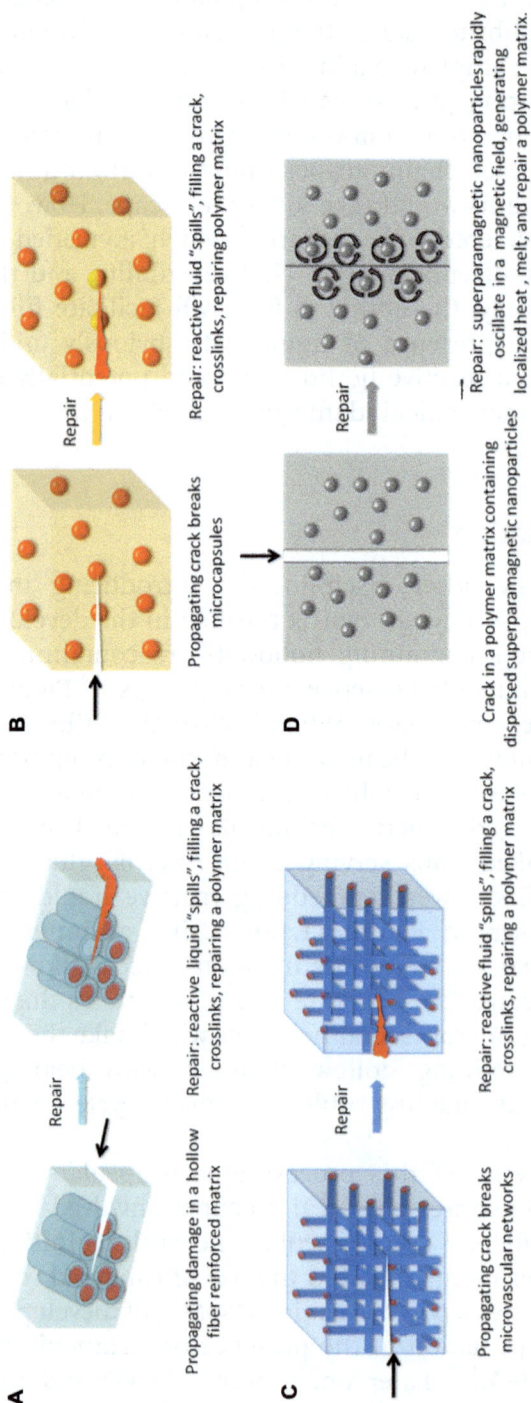

Figure 11.13 Self-healing of polymers using embedded healing agents or nanoparticles. (A) Encapsulation of reactive agents in hollow fibers (adapted from ref. 106). (B) Encapsulation of crosslinking reactants in microspheres (adapted from ref. 109). (C) 3D microvascular networks filled with reactive crosslinkers (adapted from ref. 113). (D) Polymer matrix containing superparamagnetic nanoparticles (adapted from ref. 127).

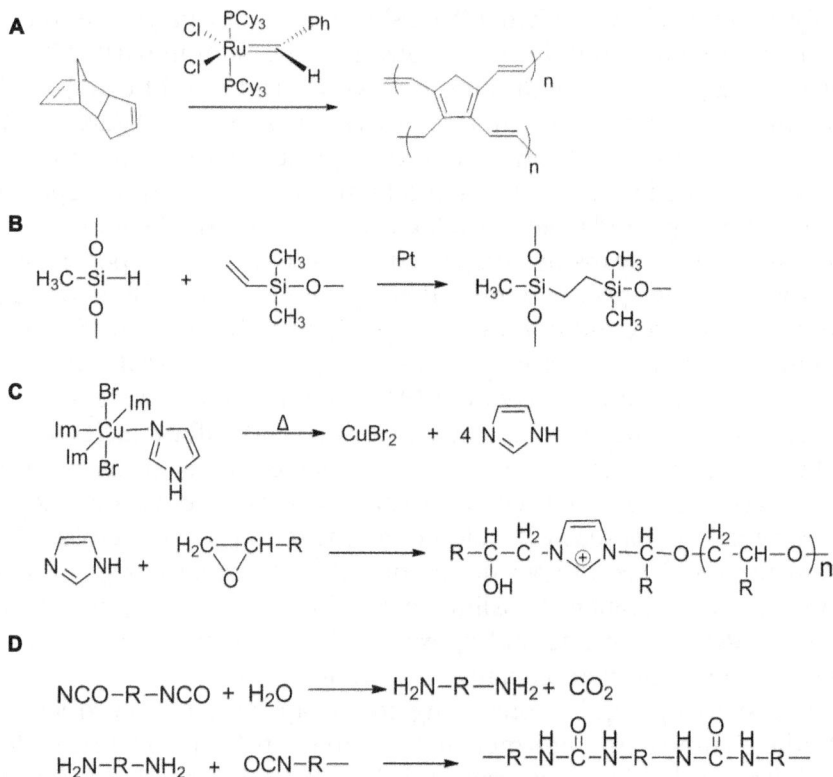

Figure 11.14 Reactions utilized in chemo-mechanical self-healing. (A) Ring-opening metathesis polymerization (ROMP) of dicyclopentadiene (DCPD) and ruthenium catalyst. (B) Cross-linking reactions of poly(dimethyl siloxane) (PDMS) copolymer and vinyl-terminated poly(dimethyl siloxane) (vinyl-PDMS) *via* Pt catalyst. (C) Epoxide/CuBr$_2$ 2-methylimidazole complex (CuBr$_2$(2-MeIM)$_4$) healing reaction. (D) Polymerization of isocyanate in the presence of water.

order to obtain better self-healing efficiency.[120,121] There is a certain degree of skepticism related to the practicality of the encapsulation concept attributed to the fact that introducing a liquid capsule or hollow fibers will weaken polymer matrix mechanical properties. In order to obtain a stable network towards water and air, the use of not easily attainable Ru catalyzed system was replaced with Sn catalyzed phase-separated droplets containing hydroxyl end-functionalized polydimethylsiloxane (HOPDMS) and polydiethoxysiloxane (PDES) dispersed in vinyl ester, where the catalyst dibutyltin dilaurate was encapsulated.[122] An extension of this concept was the development of a two capsule system, one consisting of vinyl terminated poly(dimethyl siloxane) (vinyl-PDMS) resin and Pt catalyst complexes, and the second containing a poly(dimethyl siloxane) (PDMS) co-polymer capable of crosslinking with vinyl-PDMS *via* Pt catalyst. The chemical reactions involve in this process are shown in Figure 11.14B. Also, due to useful adhesive properties epoxides have been utilized as healing agents in hollow fibers and Figure 11.14C illustrates epoxide/ $CuBr_2$ 2-methylimidazole complex ($CuBr_2(2\text{-MeIM})_4$) healing system containing microcapsules dispersed in epoxy resin.[123] Upon heating to 130 °C at the specific damaged site, 2-methylimidazole is released from the complex and initiate ring-opening polymerization of epoxide groups. Similarly, diglycidyl ether bisphenol A (DGEBA) epoxy monomer/ethyl phenylacetate (EPA) microcapsules were utilized, where the presence of dispersed scandium (III) triflate ($Sc(OTf)_3$) facilitated network repairs at 45 °C.[124] To achieve ambient temperature self-healable systems, epoxy/mercaptan reactions were used *via* double encapsulation approach.[125] Liquid amine encapsulated crosslinkers were also employed in epoxide reactions.[126] Taking advantage of moisture reactive isocyanate crosslinking reactions shown in Figure 11.14D. Besides ROMP, siloxane and epoxy-based chemistries, the isocyanate-based system also demonstrated self-healing attributes.[127]

If dispersing liquid containing microcapsules, hollow fibers, or fluidic channels in a polymer matrix shown in Figure 11.14A–C does not alter mechanical properties, this is perhaps a good choice for self-repairs. Numerous applications demand maintaining mechanical properties, toxicity, and release control.

Although chemo-mechanical approaches require no external energy source, serving as internal glue, they exhibit various drawbacks. Nevertheless, the concept alone stimulated innovative approaches to achieve self-healing properties. For example, the concept was further explored to directly embedding reactive particles into polymer network without capsulation or fibrous networks. Grain thermoplastic epoxy particles adhesive within glass–epoxy composite matrices melt upon heating, and have shown to repair damaged areas. The same concept was used by blending thermoplastic poly(bisphenol-A-*co*-epichlorohydrin) with epoxy resin at 80 °C to dissolve it into the thermosets. Further details concerning advantages and the limitations of these approaches have been extensively outlined in the literature.[25–27] Another extension of the encapsulation concept was suggested by trapping benzotriazole (BTA) anticorrosive agent inside silica/zirconia nanotubes dispersed in several crosslinked/polymerized stimuli-responsive monomers containing acrylic acid, isopropylacrylamide and disulfide functionalities to form a coating.[128]

Caution should be exercised when designing encapsulated materials in general, and polymers in particular. In an ongoing quest for self-healing materials, several studies published in the last couple years offered encapsulating hazardous reactive components. While it might be obvious to chemists that, for example, reactive isocyanates may be dangerous, it may not be as obvious to the end users. These issues are critical and represent significant technological challenge. Many reactive chemicals cannot be used for that purpose, yet there are reports on encapsulating such reactants as benzoxazines, iodonium bis(4-methylphenyl), hexafluorophosphates and others that are reluctantly used for the purpose of self-healing without precautionary measures. Perhaps the most appealing extension of encapsulation is remote self-healing.

11.4.2 Remote Self-healing

It is well known that polymers in a molten state will exhibit significantly enhanced chain mobility and interfacial diffusion. If interfacial diffusion is induced locally in a damage area of a polymer, it will facilitate repairs. Taking advantage of the fact that by incorporating superparamagnetic γ-Fe_2O_3 nanoparticles into a thermoplastic polymer network and applying remotely oscillating magnetic field, localized interfacial diffusion offered another method of repairing polymers.[129] This is depicted in Figure 11.15, in which γ-Fe_2O_3 nanoparticles oscillate at the frequency of the magnetic field, thus increasing the nanoparticle–polymer interface temperature. Generated localized melt flow permanently repair physically separated polymer interfaces and the process can be repeated multiple times. Taking advantage of selective microwave absorption generating localized heat, a composite system consisting of graphene layers dispersed in a polyurethane matrix also showed self-healing attributes.[130] In other studies the use of optically responsive materials was used in anticorrosion coatings to remotely trigger photo-induced release of healing agents, in which corrosion inhibitor benzotriazole was encapsulated in TiO_2 or SiO_2 particles.[131,132] Upon UV radiation, released of anticorrosive chemicals alone with self-healing occurred.

The use of selective wavelengths to initiate self-healing that converts electromagnetic radiation to other form of energy is another unique feature of this approach. Imbedded silver nanoparticles that exhibit surface plasma adsorption bands in the visible range of electromagnetic radiation inside the capsules[133] offered the advantage of photo-induced heating enhanced by plasma resonances from

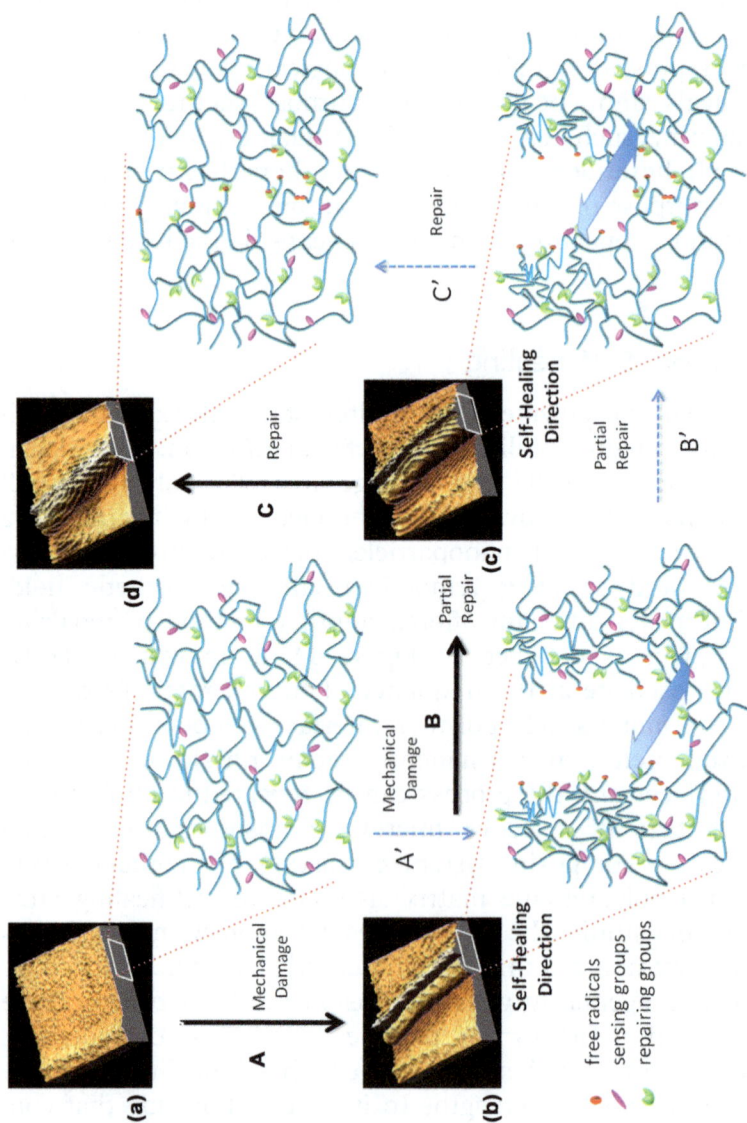

Figure 11.15 Schematic illustration of damage and self-healing process: a–d are the images of the damage repair process of modified polyurethane networks. The self-healing process is represented by a sequence of events A through C, and corresponds to a network repair process shown in cartoons A'–C'. Reproduced from ref. 2 with permission from The Royal Society of Chemistry.

metal nanoparticles. As a result, repairs were achieved. The use of noble metal nanoparticles offers new opportunities for the use of tunable energy sources capable of high efficiency localized conversions of electromagnetic radiation to heat.[134,135] Taking this concept further, nonlinear optical materials with plasmonic properties were developed. For example, gold colloids containing zinc phthalocyanines[136] that exhibit self-repairing photo-fragmented nanoparticles offered the advantage of remotely triggered responses.[137]

Although wide ranges of monomers and synthetic approaches have been proposed to construct re-mendable polymers, understating of molecular level processes governing self-healing behavior requires orchestrated efforts. At a molecular level, the balance between hydrophilic and hydrophobic interactions (if can be measured quantitatively) will play a key role in achieving phase separation and the ability to self-heal. Synthetic efforts, paralleled by modern analytical approaches in molecular imaging allowing monitoring localized transient events during self-healing combined with macroscopic physical network rearrangement analysis will enable limited molecular level understanding of these fascinating processes. The ultimate goal of the future studies should focus on the developments of new paths to achieve 'metabolic-like' self-replicating materials capable of dynamically adapting to environmental changes. Defined in biological systems as 'self-cannibalization' or autophagy, metabolism in materials can be viewed as self-healing by replacing 'outdated' degradation products or minute product side reactions. Because neither control nor elimination of side product reactions are trivial, one can envision that combining selected elements of supramolecular networks, covalent bonding, and recently discovered shape memory macromolecular segments will shape paths to the new generation of self-regulating materials. Within these SMP networks the simultaneous presence of signaling and responding groups induced by temperature, acidity–basicity, or electromagnetic radiation are obvious extensions. But perhaps the most intriguing aspect of future self-healing materials is the use of air components, such as carbon dioxide, water, and/or oxygen, in quantitative amounts that supporting living functions will revolutionize future technologies. Synthetic materials capable of responses to external or internal stimuli represent one of the most exciting and emerging areas of scientific interest and unexplored commercial applications. While there are many exciting challenges facing this field, there are a number of opportunities in design, synthesis, and engineering of stimuli-responsive polymeric systems and Mother Nature serves as a supplier of endless inspirations.

References

1. M. A. C. Stuart, W. T. Huck, J. Genzer, M. Müller, C. Ober and M. Stamm *et al.*, Emerging applications of stimuli-responsive polymer materials, *Nat. Mater.*, 2010, **9**(2), 101–113.
2. Y. Yang and M. W. Urban, Self-healing polymeric materials, *Chem. Soc. Rev.*, 2013, **42**(17), 7446–7467.
3. G. Hultquist, B. Tveten, E. Hörnlund, M. Limbäck and R. Haugsrud, Self-repairing metal oxides, *Oxid. Met.*, 2001, **56**(3), 313–346.
4. S. Thomas, N. Birbilis, M. Venkatraman and I. Cole, Self-repairing oxides to protect zinc: Review, discussion and prospects, *Corros. Sci.*, 2013, **69**, 11–22.
5. C. Dry and W. McMillan, Three-part methylmethacrylate adhesive system as an internal delivery system for smart responsive concrete, *Smart Mater. Struct.*, 1996, **5**(3), 297.
6. C. M. Dry, Self-repairing, reinforced matrix materials, *US Pat.*, 5 660 624, 1997.
7. T. W. Duerig, K. Melton and D. Stöckel, *Engineering Aspects of Shape Memory Alloys*, Butterworth-Heinemann, 2013.
8. M. Fremond and S. Miyazaki, *Shape Memory Alloys*, Springer, 2014.
9. K. Otsuka and C. M. Wayman, *Shape Memory Materials*, Cambridge University Press, 1999.
10. X. Chen, M. A. Dam, K. Ono, A. Mal, H. Shen and S. R. Nutt *et al.*, A Thermally Re-mendable Cross-Linked Polymeric Material, *Science*, 2002, **295**(5560), 1698–1702.
11. X. Chen, F. Wudl, A. K. Mal, H. Shen and S. R. Nutt, New Thermally Remendable Highly Cross-Linked Polymeric Materials, *Macromolecules*, 2003, **36**(6), 1802–1807.
12. N. Bai, K. Saito and G. P. Simon, Synthesis of a diamine cross-linker containing Diels–Alder adducts to produce self-healing thermosetting epoxy polymer from a widely used epoxy monomer, *Polym. Chem.*, 2013, **4**(3), 724–730.
13. C. Zeng, H. Seino, J. Ren, K. Hatanaka and N. Yoshie, Bio-Based Furan Polymers with Self-Healing Ability, *Macromolecules*, 2013, **46**(5), 1794–1802.
14. Y.-L. Liu and Y.-W. Chen, Thermally Reversible Cross-Linked Polyamides with High Toughness and Self-Repairing Ability from Maleimide- and Furan-Functionalized Aromatic Polyamides, *Macromol. Chem. Phys.*, 2007, **208**(2), 224–232.
15. M. J. Barthel, T. Rudolph, A. Teichler, R. M. Paulus, J. Vitz and S. Hoeppener *et al.*, Self-Healing Materials via Reversible

Crosslinking of Poly(ethylene oxide)-Block-Poly(furfuryl glycidyl ether) (PEO-b-PFGE) Block Copolymer Films, *Adv. Funct. Mater.*, 2013, **23**(39), 4921–4932.

16. K. K. Oehlenschlaeger, J. O. Mueller, J. Brandt, S. Hilf, A. Lederer and M. Wilhelm *et al.*, Adaptable Hetero Diels–Alder Networks for Fast Self-Healing under Mild Conditions, *Adv. Mater.*, 2014, **26**(21), 3561–3566.

17. C.-M. Chung, Y.-S. Roh, S.-Y. Cho and J.-G. Kim, Crack Healing in Polymeric Materials via Photochemical $[2 + 2]$ Cycloaddition, *Chem. Mater.*, 2004, **16**(21), 3982–3984.

18. J. Ling, M. Z. Rong and M. Q. Zhang, Photo-stimulated self-healing polyurethane containing dihydroxyl coumarin derivatives, *Polymer*, 2012, **53**(13), 2691–2698.

19. P. Froimowicz, H. Frey and K. Landfester, Towards the Generation of Self-Healing Materials by Means of a Reversible Photo-induced Approach, *Macromol. Rapid Commun.*, 2011, **32**(5), 468–473.

20. A. S. Fawcett and M. A. Brook, Thermoplastic Silicone Elastomers through Self-Association of Pendant Coumarin Groups, *Macromolecules*, 2014, **47**(5), 1656–1663.

21. T. Ono, T. Nobori and J.-M. Lehn, Dynamic polymer blends-component recombination between neat dynamic covalent polymers at room temperature, *Chem. Commun.*, 2005, **12**, 1522–1524.

22. G. Deng, C. Tang, F. Li, H. Jiang and Y. Chen, Covalent Cross-Linked Polymer Gels with Reversible Sol – Gel Transition and Self-Healing Properties, *Macromolecules*, 2010, **43**(3), 1191–1194.

23. W. W. Cleland, Dithiothreitol, a New Protective Reagent for SH Groups*, *Biochemistry*, 1964, **3**(4), 480–482.

24. A. V. Tobolsky, W. J. MacKnight and M. Takahashi, Relaxation of Disulfide and Tetrasulfide Polymers, *J. Phys. Chem.*, 1964, **68**(4), 787–790.

25. B. Adhikari, D. De and S. Maiti, Reclamation and recycling of waste rubber, *Prog. Polym. Sci.*, 2000, **25**(7), 909–948.

26. J. Canadell, H. Goossens and B. Klumperman, Self-Healing Materials Based on Disulfide Links, *Macromolecules*, 2011, **44**(8), 2536–2541.

27. J. A. Yoon, J. Kamada, K. Koynov, J. Mohin, R. Nicolaÿ and Y. Zhang *et al.*, Self-Healing Polymer Films Based on Thiol-Disulfide Exchange Reactions and Self-Healing Kinetics Measured Using Atomic Force Microscopy, *Macromolecules*, 2011, **45**(1), 142–149.

28. U. Lafont, H. van Zeijl and S. van der Zwaag, Influence of Cross-linkers on the Cohesive and Adhesive Self-Healing Ability of Polysulfide-Based Thermosets, *ACS Appl. Mater. Interfaces*, 2012, **4**(11), 6280–6288.

29. M. AbdolahZadeh, A. C. C. Esteves, S. van der Zwaag and S. J. Garcia, Healable dual organic–inorganic crosslinked sol–gel based polymers: Crosslinking density and tetrasulfide content effect, *J. Polym. Sci., Part A: Polym. Chem.*, 2014, **52**(14), 1953–1961.

30. P. Zheng and T. J. McCarthy, A Surprise from 1954: Siloxane Equilibration Is a Simple, Robust, and Obvious Polymer Self-Healing Mechanism, *J. Am. Chem. Soc.*, 2012, **134**(4), 2024–2027.

31. H. Ying, Y. Zhang and J. Cheng, Dynamic urea bond for the design of reversible and self-healing polymers, *Nat. Commun.*, 2014, 5.

32. Y. Amamoto, J. Kamada, H. Otsuka, A. Takahara and K. Matyjaszewski, Repeatable Photoinduced Self-Healing of Co-valently Cross-Linked Polymers through Reshuffling of Trithiocarbonate Units, *Angew. Chem.*, 2011, **123**(7), 1698–1701.

33. Y. Amamoto, H. Otsuka, A. Takahara and K. Matyjaszewski, Self-Healing of Covalently Cross-Linked Polymers by Reshuffling Thiuram Disulfide Moieties in Air under Visible Light, *Adv. Mater.*, 2012, **24**(29), 3975–3980.

34. L. M. García-Con, M. J. Whitcombe, E. V. Piletska and S. A. Piletsky, A Sulfur·Sulfur Cross-Linked Polymer Synthesized from a Polymerizable Dithiocarbamate as a Source of Dormant Radicals, *Angew. Chem., Int. Ed.*, 2010, **49**(24), 4075–4078.

35. K. Imato, M. Nishihara, T. Kanehara, Y. Amamoto, A. Takahara and H. Otsuka, Self-Healing of Chemical Gels Cross-Linked by Diarylbibenzofuranone-Based Trigger-Free Dynamic Covalent Bonds at Room Temperature, *Angew. Chem., Int. Ed.*, 2012, **51**(5), 1138–1142.

36. C. Yuan, M. Z. Rong, M. Q. Zhang, Z. P. Zhang and Y. C. Yuan, Self-Healing of Polymers via Synchronous Covalent Bond Fission/Radical Recombination, *Chem. Mater.*, 2011, **23**(22), 5076–5081.

37. Z. P. Zhang, M. Z. Rong, M. Q. Zhang and C. Yuan, Alkoxyamine with reduced homolysis temperature and its application in re-peated autonomous self-healing of stiff polymers, *Polym. Chem.*, 2013, **4**(17), 4648–4654.

38. C. Yuan, M. Z. Rong and M. Q. Zhang, Self-healing polyurethane elastomer with thermally reversible alkoxyamines as cross-linkages, *Polymer*, 2014, **55**(7), 1782–1791.

39. M. K. Beyer, The mechanical strength of a covalent bond calculated by density functional theory, *J. Chem. Phys.*, 2000, **112**(17), 7307–7312.

40. B. Ghosh and M. W. Urban, Self-repairing oxetane-substituted chitosan polyurethane networks, *Science*, 2009, **323**(5920), 1458–1460.

41. B. Ghosh, K. V. Chellappan and M. W. Urban, UV-initiated self-healing of oxolane–chitosan–polyurethane (OXO–CHI–PUR) networks, *J. Mater. Chem.*, 2012, **22**(31), 16104–16113.

42. B. Ghosh, K. V. Chellappan and M. W. Urban, Self-healing inside a scratch of oxetane-substituted chitosan-polyurethane (OXE-CHI-PUR) networks, *J. Mater. Chem.*, 2011, **21**(38), 14473–14486.

43. Y. Yang and M. W. Urban, Self-Repairable Polyurethane Networks by Atmospheric Carbon Dioxide and Water, *Angew. Chem., Int. Ed.*, 2014, **45**, 12142–12147.

44. S. Telitel, Y. Amamoto, J. Poly, F. Morlet-Savary, O. Soppera and J. Lalevee *et al.*, Introduction of self-healing properties into covalent polymer networks via the photodissociation of alkoxyamine junctions, *Polym. Chem.*, 2014, **5**(3), 921–930.

45. J.-M. Lehn, Supramolecular Chemistry—Scope and Perspectives Molecules, Supermolecules, and Molecular Devices (Nobel Lecture), *Angew. Chem., Int. Ed. Engl.*, 1988, **27**(1), 89–112.

46. P. Cordier, F. Tournilhac, C. Soulié-Ziakovic and L. Leibler, Self-healing and thermoreversible rubber from supramolecular assembly, *Nature*, 2008, **451**(7181), 977–980.

47. F. Herbst, S. Seiffert and W. H. Binder, Dynamic supramolecular poly (isobutylene) s for self-healing materials, *Polym. Chem.*, 2012, **3**(11), 3084–3092.

48. G. M. L. van Gemert, J. W. Peeters, S. H. M. Söntjens, H. M. Janssen and A. W. Bosman, Self-Healing Supramolecular Polymers In Action, *Macromol. Chem. Phys.*, 2012, **213**(2), 234–242.

49. Y. Chen, A. M. Kushner, G. A. Williams and Z. Guan, Multiphase design of autonomic self-healing thermoplastic elastomers, *Nat. Chem.*, 2012, **4**(6), 467–472.

50. D. Montarnal, F. Tournilhac, M. Hidalgo, J.-L. Couturier and L. Leibler, Versatile one-pot synthesis of supramolecular plastics and self-healing rubbers, *J. Am. Chem. Soc.*, 2009, **131**(23), 7966–7967.

51. R. Zhang, T. Yan, B.-D. Lechner, K. Schröter, Y. Liang and B. Li *et al.*, Heterogeneity, Segmental and Hydrogen Bond Dynamics, and Aging of Supramolecular Self-Healing Rubber, *Macromolecules*, 2013, **46**(5), 1841–1850.

52. F. Maes, D. Montarnal, S. Cantournet, F. Tournilhac, L. Corté and L. Leibler, Activation and deactivation of self-healing in supramolecular rubbers, *Soft Matter*, 2012, **8**(5), 1681–1687.

53. C. Hilger, R. Stadler, L. Liane and dL Freitas, Multiphase thermoplastic elastomers by combination of covalent and association chain structures: 2. Small-strain dynamic mechanical properties, *Polymer*, 1990, **31**(5), 818–823.

54. C. Hilger and R. Stadler, Cooperative structure formation by combination of covalent and association chain polymers: 4. Designing functional groups for supramolecular structure formation, *Polymer*, 1991, **32**(17), 3244–3249.

55. F. Herbst, K. Schröter, I. Gunkel, S. Gröger, T. Thurn-Albrecht and J. Balbach *et al.*, Aggregation and Chain Dynamics in Supramolecular Polymers by Dynamic Rheology: Cluster Formation and Self-Aggregation, *Macromolecules*, 2010, **43**(23), 10006–10016.

56. F. H. Beijer, R. P. Sijbesma, H. Kooijman, A. L. Spek and E. Meijer, Strong dimerization of ureidopyrimidones via quadruple hydrogen bonding, *J. Am. Chem. Soc.*, 1998, **120**(27), 6761–6769.

57. S. H. Söntjens, R. P. Sijbesma, M. H. van Genderen and E. Meijer, Stability and lifetime of quadruply hydrogen bonded 2-ureido-4 [1 H]-pyrimidinone dimers, *J. Am. Chem. Soc.*, 2000, **122**(31), 7487–7493.

58. K. E. Feldman, M. J. Kade, T. F. de Greef, E. Meijer, E. J. Kramer and C. J. Hawker, Polymers with multiple hydrogen-bonded end groups and their blends, *Macromolecules*, 2008, **41**(13), 4694–4700.

59. K. Yamauchi, J. R. Lizotte, D. M. Hercules, M. J. Vergne and T. E. Long, Combinations of Microphase Separation and Terminal Multiple Hydrogen Bonding in Novel Macromolecules, *J. Am. Chem. Soc.*, 2002, **124**(29), 8599–8604.

60. K. Yamauchi, J. R. Lizotte and T. E. Long, Thermoreversible Poly(alkyl acrylates) Consisting of Self-Complementary Multiple Hydrogen Bonding, *Macromolecules*, 2003, **36**(4), 1083–1088.

61. J. Cui and A. del Campo, Multivalent H-bonds for self-healing hydrogels, *Chem. Commun.*, 2012, **48**(74), 9302–9304.

62. J. Hentschel, A. M. Kushner, J. Ziller and Z. Guan, Self-Healing Supramolecular Block Copolymers, *Angew. Chem., Int. Ed.*, 2012, **51**(42), 10561–10565.

63. Y. Chen and Z. Guan, Multivalent hydrogen bonding block copolymers self-assemble into strong and tough self-healing materials, *Chem. Commun.*, 2014, **50**(74), 10868–10870.

64. I. Manners, *Synthetic Metal-containing Polymers*, John Wiley & Sons, 2006.

65. G. R. Whittell, M. D. Hager, U. S. Schubert and I. Manners, Functional soft materials from metallopolymers and metallo-supramolecular polymers, *Nat. Mater.*, 2011, **10**(3), 176–188.
66. C. G. Hardy, J. Zhang, Y. Yan, L. Ren and C. Tang, Metallopo-lymers with transition metals in the side-chain by living and controlled polymerization techniques, *Prog. Polym. Sci.*, 2014, **10**(39), 1742–1796.
67. K. A. Williams, A. J. Boydston and C. W. Bielawski, Towards electrically conductive, self-healing materials, *J. R. Soc., Interface*, 2007, **4**(13), 359–362.
68. M. Burnworth, L. Tang, J. R. Kumpfer, A. J. Duncan, F. L. Beyer and G. L. Fiore *et al.*, Optically healable supramolecular poly-mers, *Nature*, 2011, **472**(7343), 334–337.
69. S. Bode, L. Zedler, F. H. Schacher, B. Dietzek, M. Schmitt and J. Popp *et al.*, Self-Healing Polymer Coatings Based on Cross-linked Metallosupramolecular Copolymers, *Adv. Mater.*, 2013, **25**(11), 1634–1638.
70. N. Holten-Andersen, M. J. Harrington, H. Birkedal, B. P. Lee, P. B. Messersmith and K. Y. C. Lee *et al.*, pH-induced metal-ligand cross-links inspired by mussel yield self-healing polymer networks with near-covalent elastic moduli, *Proc. Natl. Acad. Sci.*, 2011, **108**(7), 2651–2655.
71. H. Ceylan, M. Urel, T. S. Erkal, A. B. Tekinay, A. Dana and M. O. Guler, Mussel Inspired Dynamic Cross-Linking of Self-Healing Peptide Nanofiber Network, *Adv. Funct. Mater.*, 2013, **23**(16), 2081–2090.
72. Z. Wang and M. W. Urban, Facile UV-healable polyethylenimine–copper (C 2 H 5 N–Cu) supramolecular polymer networks, *Polym. Chem.*, 2013, **4**(18), 4897–4901.
73. J. F. Gohy, B. G. Lohmeijer and U. S. Schubert, Reversible Metallo-Supramolecular Block Copolymer Micelles Contain-ing a Soft Core, *Macromol. Rapid Commun.*, 2002, **23**(9), 555–560.
74. A. Wild, A. Winter, F. Schlütter and U. S. Schubert, Advances in the field of π-conjugated 2, 2′: 6′, 2″-terpyridines, *Chem. Soc. Rev.*, 2011, **40**(3), 1459–1511.
75. E. C. Constable, 2, 2′: 6′, 2″-Terpyridines: From chemical ob-scurity to common supramolecular motifs, *Chem. Soc. Rev.*, 2007, **36**(2), 246–253.
76. E. Vaccaro and J. H. Waite, Yield and post-yield behavior of mussel byssal thread: a self-healing biomolecular material, *Bio-macromolecules*, 2001, **2**(3), 906–911.

77. B. P. Lee, J. L. Dalsin and P. B. Messersmith, Synthesis and gelation of DOPA-modified poly (ethylene glycol) hydrogels, *Biomacromolecules*, 2002, **3**(5), 1038–1047.

78. M. Krogsgaard, M. A. Behrens, J. S. Pedersen and H. Birkedal, Self-healing mussel-inspired multi-pH-responsive hydrogels, *Biomacromolecules*, 2013, **14**(2), 297–301.

79. L. He, D. E. Fullenkamp, J. G. Rivera and P. B. Messersmith, pH responsive self-healing hydrogels formed by boronate–catechol complexation, *Chem. Commun.*, 2011, **47**(26), 7497–7499.

80. M. Vatankhah-Varnoosfaderani, S. Hashmi, A. GhavamiNejad and F. J. Stadler, Rapid self-healing and triple stimuli responsiveness of a supramolecular polymer gel based on boron–catechol interactions in a novel water-soluble mussel-inspired copolymer, *Polym. Chem.*, 2014, **5**(2), 512–523.

81. F. R. Kersey, D. M. Loveless and S. L. Craig, A hybrid polymer gel with controlled rates of cross-link rupture and self-repair, *J. R. Soc., Interface*, 2007, **4**(13), 373–380.

82. Z. Wang, Y. Yang, R. Burtovyy, I. Luzinov and M. W. Urban, UV-induced self-repairing polydimethylsiloxane–polyurethane (PDMS–PUR) and polyethylene glycol–polyurethane (PEG–PUR) Cu-catalyzed networks, *J. Mater. Chem. A*, 2014, **2**(37), 15527–15534.

83. S. Burattini, H. M. Colquhoun, B. W. Greenland and W. Hayes, A novel self-healing supramolecular polymer system, *Faraday Discuss.*, 2009, **143**, 251–264.

84. S. Burattini, H. M. Colquhoun, J. D. Fox, D. Friedmann, B. W. Greenland and P. J. F. Harris *et al.*, A self-repairing, supramolecular polymer system: healability as a consequence of donor-acceptor [small pi]-[small pi] stacking interactions, *Chem. Commun.*, 2009, **44**, 6717–6719.

85. S. Burattini, B. W. Greenland, D. H. Merino, W. Weng, J. Seppala and H. M. Colquhoun *et al.*, A healable supramolecular polymer blend based on aromatic $\pi - \pi$ stacking and hydrogen-bonding interactions, *J. Am. Chem. Soc.*, 2010, **132**(34), 12051–12058.

86. S. Burattini, B. W. Greenland, W. Hayes, M. E. Mackay, S. J. Rowan and H. M. Colquhoun, A Supramolecular Polymer Based on Tweezer-Type $\pi - \pi$ Stacking Interactions: Molecular Design for Healability and Enhanced Toughness, *Chem. Mater.*, 2010, **23**(1), 6–8.

87. Z. Xu, J. Peng, N. Yan, H. Yu, S. Zhang and K. Liu *et al.*, Simple design but marvelous performances: molecular gels of superior strength and self-healing properties, *Soft Matter*, 2013, **9**(4), 1091–1099.

88. A. Eisenberg, B. Hird and R. Moore, A new multiplet-cluster model for the morphology of random ionomers, *Macromolecules*, 1990, **23**(18), 4098–4107.

89. S. J. Kalista Jr., *Self-healing of Thermoplastic poly (ethylene-co-methacrylic acid) Copolymers Following Projectile Puncture*, Virginia Polytechnic Institute and State University, Blacksburg, USA, 2003.

90. S. J. Kalista and T. C. Ward, Thermal characteristics of the self-healing response in poly (ethylene-co-methacrylic acid) copolymers, *J. R. Soc., Interface*, 2007, **4**(13), 405–411.

91. R. J. Varley, S. Shen and S. van der Zwaag, The effect of cluster plasticisation on the self healing behaviour of ionomers, *Polymer*, 2010, **51**(3), 679–686.

92. A. Reisch, E. Roger, T. Phoeung, C. Antheaume, C. Orthlieb and F. Boulmedais *et al.*, On the Benefits of Rubbing Salt in the Cut: Self-Healing of Saloplastic PAA/PAH Compact Polyelectrolyte Complexes, *Adv. Mater.*, 2014, **26**(16), 2547–2551.

93. S. J. Kalista, J. R. Pflug and R. J. Varley, Effect of ionic content on ballistic self-healing in EMAA copolymers and ionomers, *Polym. Chem.*, 2013, **4**(18), 4910–4926.

94. A. B. South and L. A. Lyon, Autonomic Self-Healing of Hydrogel Thin Films, *Angew. Chem.*, 2010, **122**(4), 779–783.

95. M. J. Serpe, C. D. Jones and L. A. Lyon, Layer-by-layer deposition of thermoresponsive microgel thin films, *Langmuir*, 2003, **19**(21), 8759–8764.

96. X. Wang, F. Liu, X. Zheng and J. Sun, Water-Enabled Self-Healing of Polyelectrolyte Multilayer Coatings, *Angew. Chem., Int. Ed.*, 2011, **50**(48), 11378–11381.

97. Y. Huang, P. G. Lawrence and Y. Lapitsky, Self-Assembly of Stiff, Adhesive and Self-Healing Gels from Common Polyelectrolytes, *Langmuir*, 2014, **30**(26), 7771–7777.

98. B. K. Ahn, D. W. Lee, J. N. Israelachvili and J. H. Waite, Surface-initiated self-healing of polymers in aqueous media, *Nat. Mater.*, 2014, **13**(9), 867–872.

99. A. Izquierdo, S. Ono, J.-C. Voegel, P. Schaaf and G. Decher, Dipping versus spraying: exploring the deposition conditions for speeding up layer-by-layer assembly, *Langmuir*, 2005, **21**(16), 7558–7567.

100. E. V. Skorb and D. V. Andreeva, Layer-by-Layer approaches for formation of smart self-healing materials, *Polym. Chem.*, 2013, **4**(18), 4834–4845.

101. H. Lodish, *Molecular cell biology*, Macmillan, 2008.

102. C. A. Schalley, Supramolecular chemistry goes gas phase: the mass spectrometric examination of noncovalent interactions in host–guest chemistry and molecular recognition, *Int. J. Mass Spectrom.*, 2000, **194**(1), 11–39.

103. M. D. Pluth and K. N. Raymond, Reversible guest exchange mechanisms in supramolecular host–guest assemblies, *Chem. Soc. Rev.*, 2007, **36**(2), 161–171.

104. M. Nakahata, Y. Takashima, H. Yamaguchi and A. Harada, Redox-responsive self-healing materials formed from host–guest polymers, *Nat. Commun.*, 2011, **2**, 511.

105. T. Kakuta, Y. Takashima, M. Nakahata, M. Otsubo, H. Yamaguchi and A. Harada, Preorganized Hydrogel: Self-Healing Properties of Supramolecular Hydrogels Formed by Polymerization of Host–Guest-Monomers that Contain Cyclo-dextrins and Hydrophobic Guest Groups, *Adv. Mater.*, 2013, **25**(20), 2849–2853.

106. E. A. Appel, F. Biedermann, U. Rauwald, S. T. Jones, J. M. Zayed and O. A. Scherman, Supramolecular Cross-Linked Networks via Host – Guest Complexation with Cucurbit [8] uril, *J. Am. Chem. Soc.*, 2010, **132**(40), 14251–14260.

107. R. Wool and K. O'connor, A theory crack healing in polymers, *J. Appl. Phys.*, 1981, **52**(10), 5953–5963.

108. P.-G. de Gennes, Reptation of a polymer chain in the presence of fixed obstacles, *J. Chem. Phys.*, 1971, **55**(2), 572–579.

109. C. Dry, Procedures developed for self-repair of polymer matrix composite materials, *Compos. Struct.*, 1996, **35**(3), 263–269.

110. M. Motuku, U. Vaidya and G. Janowski, Parametric studies on self-repairing approaches for resin infused composites subjected to low velocity impact, *Smart Mater. Struct.*, 1999, **8**(5), 623.

111. S. Bleay, C. Loader, V. Hawyes, L. Humberstone and P. Curtis, A smart repair system for polymer matrix composites, *Composites, Part A*, 2001, **32**(12), 1767–1776.

112. J. W. Pang and I. P. Bond, A hollow fibre reinforced polymer composite encompassing self-healing and enhanced damage visibility, *Compos. Sci. Technol.*, 2005, **65**(11), 1791–1799.

113. R. Trask, G. Williams and I. Bond, Bioinspired self-healing of advanced composite structures using hollow glass fibres, *J. R. Soc., Interface*, 2007, **4**(13), 363–371.

114. S. R. White, N. Sottos, P. Geubelle, J. Moore, M. R. Kessler and S. Sriram *et al.*, Autonomic healing of polymer composites, *Nature*, 2001, **409**(6822), 794–797.

115. S. Kim, S. Lorente and A. Bejan, Vascularized materials: tree-shaped flow architectures matched canopy to canopy, *J. Appl. Phys.*, 2006, **100**(6), 063525.

116. H. Williams, R. Trask and I. Bond, Self-healing composite sandwich structures, *Smart Mater. Struct.*, 2007, **16**(4), 1198.

117. H. Williams, R. Trask, A. Knights, E. Williams and I. Bond, Biomimetic reliability strategies for self-healing vascular networks in engineering materials, *J. R. Soc., Interface*, 2008, **5**(24), 735–747.

118. C. J. Hansen, W. Wu, K. S. Toohey, N. R. Sottos, S. R. White and J. A. Lewis, Self-Healing Materials with Interpenetrating Microvascular Networks, *Adv. Mater.*, 2009, **21**(41), 4143–4147.

119. J. D. Rule and J. S. Moore, ROMP Reactivity of endo- and exo-Dicyclopentadiene, *Macromolecules*, 2002, **35**(21), 7878–7882.

120. X. Liu, X. Sheng, J. K. Lee, M. R. Kessler and J. S. Kim, Rheokinetic evaluation of self-healing agents polymerized by Grubbs catalyst embedded in various thermosetting systems, *Compos. Sci. Technol.*, 2009, **69**(13), 2102–2107.

121. G. O. Wilson, M. M. Caruso, N. T. Reimer, S. R. White, N. R. Sottos and J. S. Moore, Evaluation of ruthenium catalysts for ring-opening metathesis polymerization-based self-healing applications, *Chem. Mater.*, 2008, **20**(10), 3288–3297.

122. S. H. Cho, H. M. Andersson, S. R. White, N. R. Sottos and P. V. Braun, Polydimethylsiloxane-Based Self-Healing Materials, *Adv. Mater.*, 2006, **18**(8), 997–1000.

123. T. Yin, M. Z. Rong, M. Q. Zhang and G. C. Yang, Self-healing epoxy composites–preparation and effect of the healant consisting of microencapsulated epoxy and latent curing agent, *Compos. Sci. Technol.*, 2007, **67**(2), 201–212.

124. T. S. Coope, U. F. Mayer, D. F. Wass, R. S. Trask and I. P. Bond, Self-Healing of an Epoxy Resin Using Scandium (III) Triflate as a Catalytic Curing Agent, *Adv. Funct. Mater.*, 2011, **21**(24), 4624–4631.

125. Y. C. Yuan, M. Z. Rong, M. Q. Zhang, J. Chen, G. C. Yang and X. M. Li, Self-healing polymeric materials using epoxy/mercaptan as the healant, *Macromolecules*, 2008, **41**(14), 5197–5202.

126. D. A. McIlroy, B. J. Blaiszik, M. M. Caruso, S. R. White, J. S. Moore and N. R. Sottos, Microencapsulation of a reactive liquid-phase amine for self-healing epoxy composites, *Macromolecules*, 2010, **43**(4), 1855–1859.

127. M. Huang and J. Yang, Facile microencapsulation of HDI for self-healing anticorrosion coatings, *J. Mater. Chem.*, 2011, **21**(30), 11123–11130.
128. G. L. Li, Z. Zheng, H. Möhwald and D. G. Shchukin, Silica/Polymer Double-Walled Hybrid Nanotubes: Synthesis and Application as Stimuli-Responsive Nanocontainers in Self-Healing Coatings, *ACS nano*, 2013, **7**(3), 2470–2478.
129. C. C. Corten and M. W. Urban, Repairing polymers using oscillating magnetic field, *Adv. Mater.*, 2009, **21**(48), 5011–5015.
130. L. Huang, N. Yi, Y. Wu, Y. Zhang, Q. Zhang and Y. Huang *et al.*, Multichannel and Repeatable Self-Healing of Mechanical Enhanced Graphene-Thermoplastic Polyurethane Composites, *Adv. Mater.*, 2013, **25**(15), 2224–2228.
131. E. V. Skorb, D. V. Sviridov, H. Möhwald and D. G. Shchukin, Light responsive protective coatings, *Chem. Commun.*, 2009, **40**, 6041–6043.
132. E. V. Skorb, A. G. Skirtach, D. V. Sviridov, D. G. Shchukin and H. Mohwald, Laser-controllable coatings for corrosion protection, *ACS nano*, 2009, **3**(7), 1753–1760.
133. I. Alessandri, Writing, Self-Healing, and Self-Erasing on Conductive Pressure-Sensitive Adhesives, *Small*, 2010, **6**(15), 1679–1685.
134. M. B. Cortie and A. M. McDonagh, Synthesis and optical properties of hybrid and alloy plasmonic nanoparticles, *Chem. Rev.*, 2011, **111**(6), 3713–3735.
135. V. Giannini, A. I. Fernandez-Dominguez, S. C. Heck and S. A. Maier, Plasmonic nanoantennas: fundamentals and their use in controlling the radiative properties of nanoemitters, *Chem. Rev.*, 2011, **111**(6), 3888–3912.
136. V. Amendola, D. Dini, S. Polizzi, J. Shen, K. M. Kadish and M. J. Calvete *et al.*, Self-healing of gold nanoparticles in the presence of zinc phthalocyanines and their very efficient nonlinear absorption performances, *J. Phys. Chem. C*, 2009, **113**(20), 8688–8695.
137. V. Amendola and M. Meneghetti, Advances in self-healing optical materials, *J. Mater. Chem.*, 2012, **22**(47), 24501–24508.

12 Shape Memory Materials

12.1 Introduction

A change of materials shape as a result of external stimuli is referred to as induced shape memory. The external stimuli can be temperature changes, or exposure to electromagnetic radiation or electromagnetic fields. The shape memory is unrelated to responsiveness of individual polymer chains or macromolecular segments, but instead is viewed as a global phenomenon attributed to a combination of polymer morphology changes due to processing conditions capable of 'programming' or reprograming desirable shapes. The process of programing and recovery is schematically depicted in Figure 12.1, where first a polymer is conventionally processed to obtain a permanent shape. Upon deformation to a specified and desirable shape, a temporary shape is fixed. This process is referred to as programing and typically consists of exposure to desired temperature, deformation to desirable shape, and cooling. Now that the permanent shape is stored, a specimen exhibits the temporary shape. When such a programmed shape memory polymer is heated above a transition temperature T_{trans}, the shape memory effect (SME) is induced, leading to the recovery of the stored permanent shape. This is the recovery stage. In terms of shape memory materials (SMMs), there are two classes: shape memory alloys (SMAs) and shape memory polymers (SMPs).

Stimuli-Responsive Materials: From Molecules to Nature Mimicking Materials Design
By Marek W. Urban
© Marek W. Urban, 2016
Published by the Royal Society of Chemistry, www.rsc.org

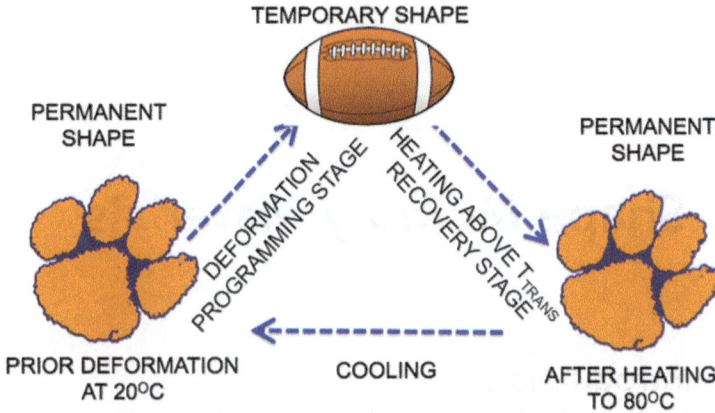

Figure 12.1 Simplistic diagram depicting the thermally induced one-way shape memory effect. The permanent shape is transferred to the temporary shape during the programming stage, but upon heating to the transition temperature T_{trans} the recovery to permanent shape occurs.

12.2 Shape Memory Alloys (SMAs)

SMAs are a group of metallic alloys that are capable of returning to their original shape or size upon being subjected to a memorization process between two transformation phases. Typically, the transformations occur upon exposure to temperature or magnetic field. The fundamental application of these materials is quite appealing: upon deformation by an external force, a material will contract to recover to its original shape upon external stimuli. However, it should be kept in mind that there is a distinct difference between SMPs and stimuli-responsive shape changing polymers. For example, thermally responsive polymers (discussed in Chapter 3) simply respond to temperature changes manifested by shape/size changes, but do not exhibit shape memory attributes.

The first solid phase transformation in SMAs was discovered in 1932,[1] on gold–cadmium (Au–Cd) alloys that were plastically deformed upon cooling, but returned to the original configuration upon heating. Other alloys, including copper–zinc (Cu–Zn), copper–tin (Cu–Sn),[2] In–Tl and Cu–Al–Ni showed similar behavior.[3,4] These early discoveries attracted scientific and technological interest, but applications were limited due to their complexity and unattractive properties. However, as technological advances were made, the demand for SMAs for engineering and technical applications continued to increase in many commercial fields. These included many consumer products and industrial applications, structures and composites, automotive, aerospace, mini actuators, microelectromechanical systems (MEMS), and biomedical applications. In general, iron-based and copper-based SMAs (for example, Fe–Mn–Si, Cu–Zn–Al, Cu–Al–Ni) are low cost and commercially available, but their instability and brittleness make these materials less desirable, and thus the preference is to use Ni–Ti based SMAs. Excellent reviews on this topic are available,[5,6] which summarize numerous applications of SMAs. Figure 12.2 illustrates the two examples of automotive and airspace applications, where numerous sensing components may serve in a variety of ways to safely operate and use cars or aircraft. Of course there are many futuristic aspects of these components, but it will not be long before, for example, a retractable car roof will be on the market, just like rain-activated windshield wipers.

As polymers and composites replace metal alloys, there are many opportunities for the development of polymer-based sensing materials. SMAs have been exploited and used in a diverse range of commercial robotic systems, especially as micro-actuators or artificial muscles. Aside from the challenges leading to the increase of intelligence, other challenges in using SMAs in robotic applications are the enhanced performance and miniaturization/weight. Again, polymeric materials and composites will dominate this field. There are also endless existing and potential SMA applications in the biomedical field, which usually requires high corrosion resistance, biocompatibility, and non-magnetic properties that may lead to the replication of certain human tissues and bones capable of responding to the human body temperature changes. Again, the challenges are precision and reliable miniature instruments to achieve accurate positioning and functioning for complex medical treatments and surgical procedures.

Figure 12.3 illustrates various phase transition that typically occur in SMAs.[5,7-9] Typically, SMAs can exist in two different phases with three different crystal structures: twinned martensite, detwinned martensite and austenite, and six possible transformations. The austenite structure is stable at high temperature, whereas the martensite structure is stable at lower temperatures. Upon heating a SMA, the transformation from martensite to the austenite phase occurs. There are two characteristic temperatures that are important: the austenite-start-temperature (A_s) is the temperature at which the

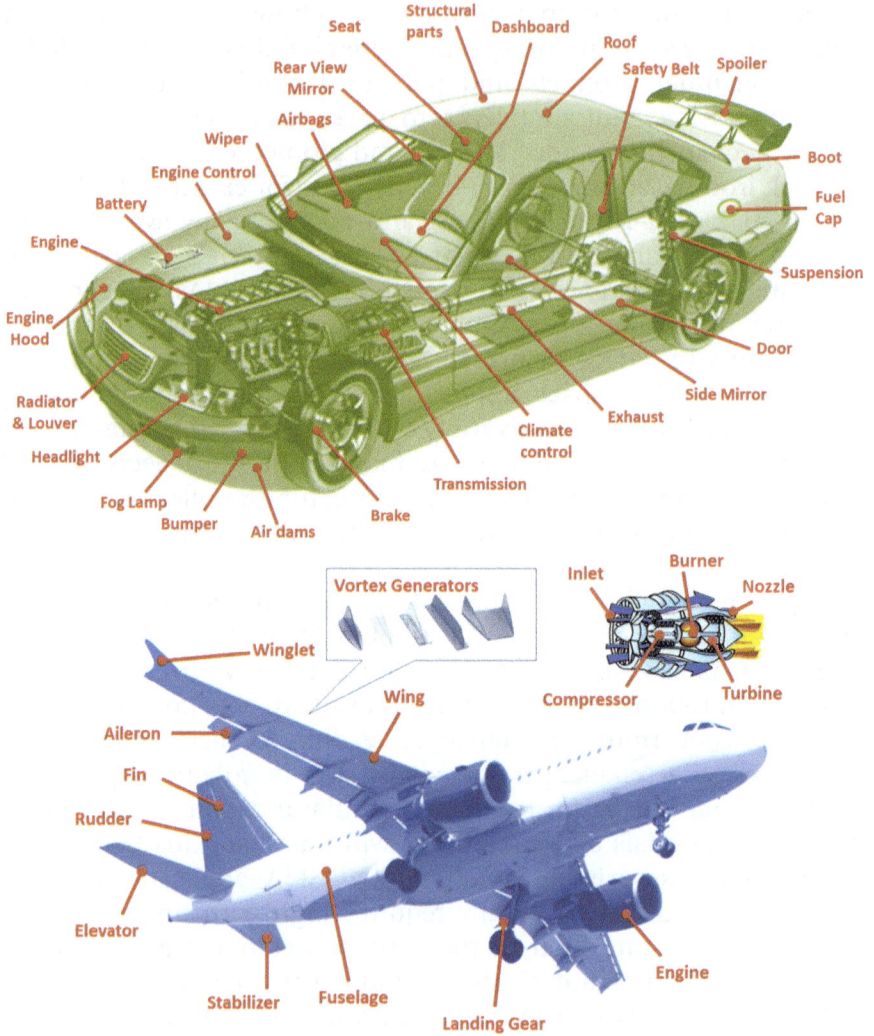

Figure 12.2   Existing and potential applications of SMAs in the automotive and airspace industries.
Adapted and reprinted from ref. 5. Copyright (2014) with permission from Elsevier.

transformation begins and the austenite-finish-temperature (A_f) at which the process is complete. Once a SMA is heated beyond A_s it begins to contract and transform into the austenite structure, resulting in the recovery to its original shape.

The highest temperature at which martensite can no longer be stress induced is referred to as M_d, and above this temperature the SMA is permanently deformed.[10] The shape change effects

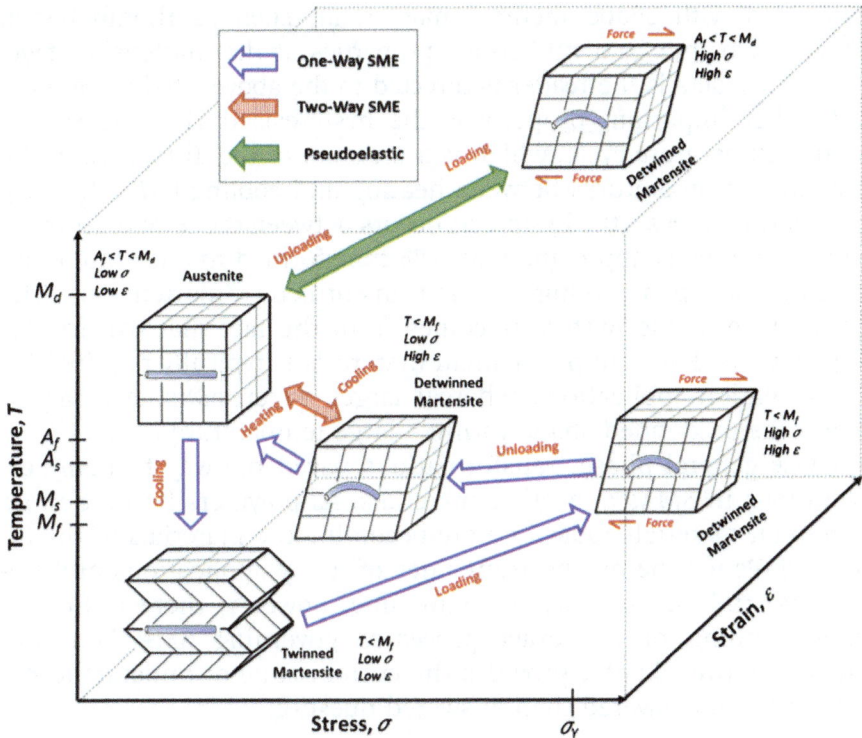

Figure 12.3 Schematic diagram of SMA phases and crystal structures that develop during each transition.
Adapted from ref. 6 to 9 and reproduced from ref. 5, copyright © 2013 Elsevier Ltd. Published by Elsevier Ltd. All rights reserved.

responsible for the SME and pseudoelasticity (or superelasticity) are often classified into three categories:

1. One-way shape memory effect (1-SME), defined as the process during which a material retains a deformed state after the removal of an external force, but recovers to its original shape upon heating.
2. Two-way shape memory effect (2-SME) or reversible SME, during which, in addition to the 1-SME, an alloy material remembers its shape at both high and low temperatures.
3. Pseudoelasticity (PE) or superelasticity (SE), during which an alloy SMA reverts to its original shape after applying mechanical loading at temperatures between A_f and M_d, without the need for thermal activation.

A typical SME is a diffusionless solid phase transition between martensitic and austenitic crystal structures.[11,12] Other transformations associated with shape memory may occur, such as rhombohedral (R-phase), bainite and rubber-like properties at the martensite stage. For further details the reader is directed to the above cited literature.

Another important property is the hysteresis during the stress–strain temperature cycle, which is a measure of the difference in the transition temperatures between heating and cooling ($\Delta T = A_f - M_s$). In a typical process, the hysteresis occurs between the temperatures at which an alloy is approximately 50% transformed to austenite upon heating and upon cooling 50% is transformed to martensite.[13] The importance of the hysteresis comes from the fact that will impact applications. For example, a small hysteresis is typically required for fast actuation applications, whereas larger hysteresis is necessary to retain the predefined shape within a large temperature range.

These general phase transformations paved the way to many explorations in SMAs and other materials. As polymers become more dominant materials today and are becoming the materials of the future, SMPs will be on the front lines of many applications and numerous studies. At this point and time, however, there is limited understanding of the exact processes governing SMP behavior. Therefore, this chapter provides the current state of understanding, with more unanswered than answered questions.

12.3   Shape Memory Polymers (SMPs)

Similarly to SMAs, SMPs have been of interest since the 1940s, with the first patent in methacrylic ester resins used in dental applications.[14] Two decades later, heat-shrinkable polyethylenes[15] and polyethylene-silicones[16] were developed, and today another wave of interest has arrived: multi-stage SMPs. Many recent developments in SMMs are driven by applications, and what is lacking is the fundamental understanding of precise molecular events responsible for this fascinating phenomenon. Thus, in general terms, SMPs are polymeric materials that are able to 'memorize' a permanent shape and maintain it under appropriate conditions. However, when triggered by stimuli such as heat or light, a transformation of the temporary shape to the memorized permanent shape will take place. Because temperature is often used as an external stimulus, Figure 12.4 depicts the basic concept involved in the SMPs.[13,17] While the interested reader is directed to a few excellent review articles describing specific polymer systems,[17–19] here we will focus on the fundamental physical and chemical principles governing this behavior.

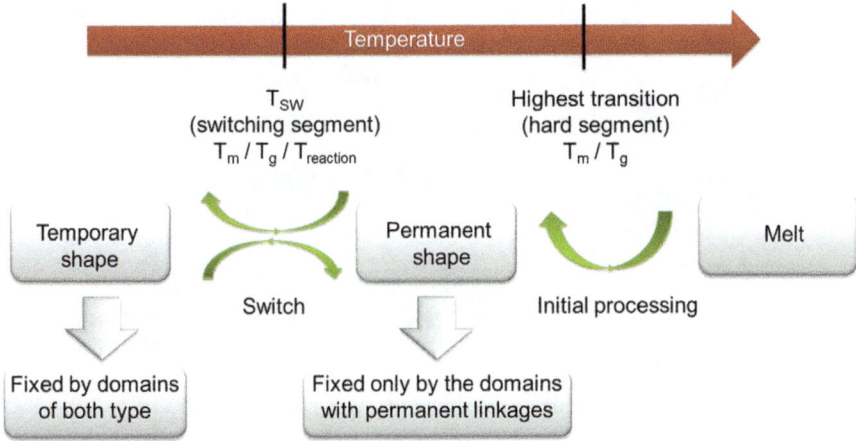

Figure 12.4 Schematic diagram describing the principles of SMPs. Reprinted from ref. 19. Copyright (2015) with permission from Elsevier.

Figure 12.4 illustrates temperature as a trigger of the SMP behavior, and schematically illustrates the process of designing a SMP, which may include, but is not limited to, an initial preparation step, such as extrusion, spinning, pressing or other processing techniques, and subsequent programming by means of applying mechanical force. This stage is required to establish the permanent shape above the T_{sw}, but the force should be applied below T_{sw}. While in principle this process can be repeated many times, it should be kept in mind that during this stage the thermal exposure above the T_{sw} will result in transformation back to the permanent shape.

Typical SMPs possess at least two 'phases': one represents a stable adoptable network capable of adopting shape changes, whereas the other one will respond to an external trigger. The stable phase facilitates mechanical integrity to a polymer network, thus retaining the original shape. This is schematically depicted in Figure 12.5. If deformation occurs, this phase is responsible for the shape recovery. The stability can be facilitated by the introduction of chemical crosslinks, incorporation of crystalline phases, or interpenetrating networks. The second phase temporarily fixes the temporary shape by covalent or non-covalent reversible bonding ($T_{reaction}$), or by crystallization (or a melting transition that results in the shape recovery; T_m), and by a glass transition temperature (T_g), or liquid crystalline transiton (T_{LC}). While the former can be accomplished by incorporating well-documented reversible covalent bonding, such as Diels–Alder, redox reactions, or supramolecular interactions,

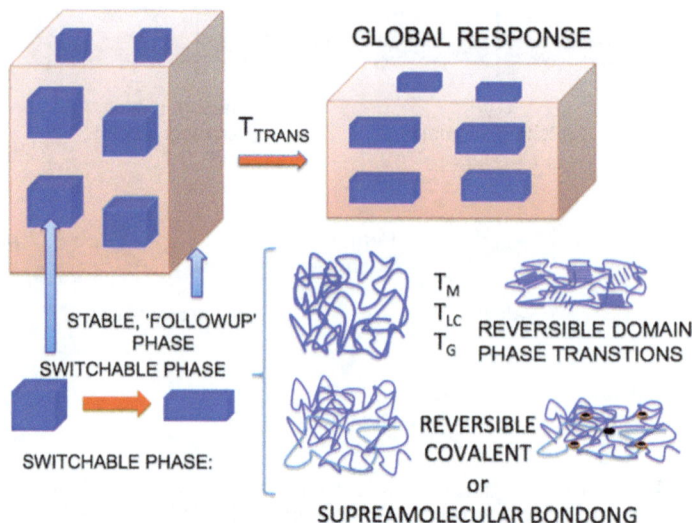

GLOBAL RESPONSE

T_{TRANS}

STABLE, 'FOLLOWUP' PHASE

SWITCHABLE PHASE

SWITCHABLE PHASE:

T_M
T_{LC} REVERSIBLE DOMAIN
T_G PHASE TRANSTIONS

REVERSIBLE
COVALENT
or
SUPREAMOLECULAR BONDONG

Figure 12.5   Schematic diagram of SMP responses to temperature. Reversible phase transitions resulting in shape changes can be induced by a melting temperature, liquid crystalline transition temperature, or glass transition temperature of a switchable phase in thermoplastic polymers. Reversible covalent or supramolecular bonding will be responsible for reversible thermosetting polymers, or in polymer networks with supramolecular components.

the latter will be driven by a physical segmental network or other macromolecular rearrangements. The construction of shape SMPs requires an initial preparation step, which may be extrusion, spinning or pressing. This step is essential in determining the permanent shape above the 'switch temperature' T_{sw}. The subsequent programming of the temporary shape requires an external mechanical force to be applied to the material below T_{sw}. In cases where the formed device is exposed to temperatures above T_{sw}, the transformation will occur, taking back a given material shape. Although this step presumably can be repeated many times, the fatigue factor will determine the actual reversibility outcomes. One major drawback of the simple SME approach described above is the lack of reversibility of shape changes. Therefore, once the permanent shape is recovered, a new programming will be required to rebuild the temporary shape.

Although the most explored T_{trans} in SMPs is based on T_m and T_g transitions, the T_{LC} based transitions have not been as widely developed. Table 12.1A summarizes selected examples of SMPs relying on T_M, whereas part B lists T_g-based materials.

As illustrated in Table 12.1A, the majority of T_M-type SMPs are based on polyolefins, which also include those mentioned above: polyethylene in heat-shrinkable polymers, polyethers or polyesters (poly(ε-caprolactone)). One of the attractive features of these materials is the presence of a low melting temperature phase in the vicinity of a crystalline 'hard phase', facilitating the elevated switching temperature. On the other hand, peroxide crosslinking of low density PE (LDPE) leads to a heat-induced shrinkage.[20] Ultra-high molecular weight PE (UHMWPE) can be used in fiber SMPs with a T_{trans} of ~150 °C.[21] Notably, polyolefins can also be used in the triple SMP (discussed in the next section) triggered by infrared radiation.[22]

Natural rubber offers another popular material in the category of SMPs. The effect is based on strain-induced crystallization[23–25] which can be triggered by temperature, application of transversal stresses or solvent vapors.[26] Another potential approach is the use of fatty acids, as demonstrated in the blends of natural rubber and stearic acid. The latter facilitates the SME by melting/crystallization of the fatty acid.[27]

Polyethers feature typically low melting temperatures, thus making them promising candidates as the switching blocks in SMPs. For example, one approach is to incorporate a polyethylene oxide (PEO) block segment into a multiblock copolymer containing poly(ethylene terephthalate) (PET) hard blocks.[28] As one would anticipate, a higher content of the soft block will facilitate a higher switching transition temperature, attributed to more efficient crystallization of the soft block. Also, PEO blocks can be incorporated into polyurethane (PU),[29] crosslinked polymethacrylate networks[30] and poly(p-dioxanone)-poly(ethylene glycol) networks.[31] One can combine interpenetrating networks of linear and star-shaped PEOs with methylmethacrylates.[32] Among many combinations of copolymers, aliphatic polyesters with low melting temperature are quite attractive because they can be used with the soft poly(ε-caprolactone) (PCL) blocks. There are many examples of SMPs, including polyurethanes,[33–36] polyamides and poly-aramides, serving as hard blocks in combination with PCL,[37,38] in particular with switching temperatures in the physiological temperature region.[39]

Polymers with T_g above room temperature can be used for T_g-based SMPs. Table 12.1B summarizes selected examples of T_g-based SMPs. Because all polymers possess a T_g, one could argue that all polymers exhibit SME. Indeed, this may be the case, but this does not necessarily imply that all polymers exhibit desirable responses that would make them SMPs. Because the T_g itself is a non-linear second-order phase transition, in comparison to the T_M-based SMPs, the T_g-based

Table 12.1 A Selected examples of melting/crystallization-based SMP and their transitions. B Selected examples of glass transition-based SMP. C Selected examples of LCP utilized in SMP applications.

Material	Hard phase	Soft phase	Phase transition temperature (°C)	Ref.
Part A. T_M-based SMP				
Epoxy	Epoxy	Jeff amine	30–93	40
Natural rubber	Chemical crosslinks	Isoprene	~30–50	41
Polyurethane	PU	Copolyester	~70	42
Polyester	Nanocrystals	Polyester	45	43
Polymethacrylate	Crosslinked			
LDPE	Crosslinks	LDPE	60 to 100	20
Ethylene-1-octene copolymers	Crosslinks	Ethylene-1-octene copolymers	60 to 100	44
UHMWPE	Physical entanglements	PE	145	21
PEO-PU	PU	PEO	60	27, 45
Natural rubber	Crosslinks	Polyisoprene/stearic acid	75	27
PEO-PET multiblock copolymer	PET	PEO	45 to 55	46
PU	PU	Castor oil derived polyol	60	36
PMMA-PEG IPN	PMMA	PEG	50 °C	32
PCL-PU	PU	PCL	43–60	39
PCL-PIBMD	PIBMD	PCL	60	47
PCL-Methacrylate network	Network	OCL	30–50	48
PCL-PA	PA	PCL	65	37
Radiation crosslinked PCL	Crosslinked	PCL	55	49
PCL-SBS blend	SBS	PCL	55	39

Part B. T_G-based SMP

Material				Ref.
Epoxy	Epoxy	Jeff amine	31–93	40
Crosslinked natural rubber	Crosslinks	Polyisoprene	~30–50	41
Polyurethane	PU	Copolyester	70	42
Polyurethane	Crystalline PU	Amorphous PU	30	50
Polyurethane	Crosslinks	PU	55	51, 52
Polyester	Nanocrystals	Polyester	45 °C	43
Polymethacrylate network	Crosslinks	Low T_g methacrylate	n.d.	53
Polymethacrylate network	Crosslinks	PEGDMA	56–92 (T_g)	54
Polymethacrylate network	Crosslinks	Low T_g methacrylate	65	55
Polyimide	Crosslinks	Polyimide	220	56
Polyaspartimide-urea	Crosslinks	Polyaspartimide-urea	150	54
PEEK	Crystalline domains	Amorphous domains	>180	57

Material	Principle	Switching temperature T_{TRAN} (°C)	Ref.
Part C			
LCE	Transition of nematic LC	160	68
LCE	Transition of smectic LC	90	68
xLCE	Transition of nematic LC	90	66
LCE	Transition of nematic LC	75	67
PCO	Crystallization under stress	~60	69
PCL	Crystallization under stress	~65	70
Polymer network	Crystallization under stress	40 and 71	71
Polymer network	Reversible crystallization/melting	27 and 38	72
Poly(octylene adipate)	Reversible crystallization/melting	5 and 38	73

SMPs exhibit relatively slow shape recovery. Thus, applications in which quick responses are desired are very limited. SMPs based on the epoxy resin DGEBA (diglycidyl ether bisphenol A) and varying content of curing agent DDM (4,4-diaminodiphenyl methane) can be produced and the T_g can be tuned from 45 to 145 °C by controlling the degree of cure.[58] Another example is epoxidized natural rubber crosslinked with zinc diacrylate in an oxa-Michael addition reaction,[41] in which enhanced crosslinking density resulted in higher switching temperatures. Considering the broad range of properties epoxies offer, the take-home lesson from many epoxy-based SMP studies is that these materials exhibit a relatively broad range of transition temperatures and for higher T_g, the transitions are much sharper.[59] Polyurethanes represent perhaps the most diversified group of SMPs. They offer T_M- as well as T_g-based SMMs. They can also form copolyester–polyurethane SMPs based on star-shaped oligo[(*rac*-lactide)-*co*-glycolide] soft segments.[42] Also, thermoplastic PU that exhibits a SME can be obtained by copolymerization of a soft block of 1,3-butanediol and hexamethylene diisocyanate and a hard block of 4,4′-bis-(6-hydroxyhexoxy)biphenyl and toluene-2,4-diisocyanate.[50] When C=C bond containing monomers are introduced to thermoplastic PU, crosslinking can be achieved, thus facilitating the control of crosslinking density.[51,52] Polyesters based on polyols and sebacic acid containing cellulosic nanocrystals also exhibit SME properties.[43] The majority of (meth)acrylate-based covalently crosslinked SMP networks are based on the glass transition of the soft phase,[53,54,60,61] offering several bioapplications attributed to biocompatibility[55] or biodegradability. The methacrylate network structure can be also utilized in shape memory hydrogels.[55] As illustrated in Table 12.1B, the majority of T_g-based SMPs exhibit transition temperatures below 100 °C. However, for many applications, the challenge lies in the high-temperature SMPs; for example, a low crosslink density polyimide features switching temperatures at $\sim$220 °C.[56,62] Notably, polyimides with <1% crosslinks exhibit a few second shape recovery >99% and similar behavior was reported in a polyaspartimide–urea system.[63]

   What we have described so far is a one-way SME, in which the permanent shape is recovered after heating. However, a temporary shape can only be obtained by mechanical deformation, which offers the ability of switchability between two different shapes upon an external stimulus without the need for additional mechanical deformation.[64] Under these circumstances the reversibility of transitions can be achieved. From the application point of view, this is highly desirable for many biomedical applications including artificial muscles, or reversible technologically critical processes, such as on-demand adhesion, and many others. Thus, in designing these SMPs it is necessary to consider reversible contraction/extension behavior. One group of materials that may facilitate these attributes are liquid crystalline elastomers (LCEs). One can take advantage of the transition between the anisotropic and the isotropic phase, which will lead to a contraction of the polymer and, in contrast to a one-way SMP, LCEs will expand again usually in a fairly uniform fashion. In contrast to a one-way SMP, the polymer will expand again, when the temperature is below the transition temperature. The uniform alignment of the mesogens is important to achieve the desirable responsiveness. Using a glass-forming nematic network, a liquid crystalline monomer can be polymerized by acyclic diene metathesis (ADMET) followed by crosslinking, resulting in a two-way SMP with the nematic-to-isotropic phase transition near 160 °C.[65]

Also, polymer networks incorporating smectic liquid crystalline monomers have been used[66] to overcome the limitations caused by the required macroscopic orientation.[67] In these experiments a LCE with exchangeable covalent bonds containing epoxide with carboxylic acid groups was utilized. Reversibility and ease of processing can be accomplished by the presence of the LC moieties. Table 12.1C summarizes selected example of LCEs used in SMPs.

In view of these considerations one can envisage that the two or more temporary phases in the presence of the permanent 'follow-up' phase will lead to multiple SMEs. Figure 12.6a illustrates that a SMP can be programmed to fix one temporary shape, but recovery to its permanent shape will occur upon applying a stimuli.[74] Under these conditions, the applied force required to fix a given shape will determine the shape. This is also referred to as a programing step. Because this process requires two shape changes (temporary and fixed), it is a dual shape memory process[75] driven by a combination of the reversible switching of network structural components and has been known for a long time. This behavior is often observed in chemically or physically crosslinked polymers as well as amorphous or crystalline polymers.[17,18,76–78] The programming illustrated in Figure 12.6 is external and typically achieved by applying directional forces. Consequently, the same material (in principle, if one neglects a fatigue factor) can be reprogrammed multiple times.[79] This is shown in Figure 12.6b. The unique feature of this phenomenon is that a variety of shape fixing pathways can be chosen and reprogrammed as desired. To illustrate the SME depicted in Figure 12.6b, let us consider two polyester-based materials with melting temperatures of 64 °C

(a)

(b)

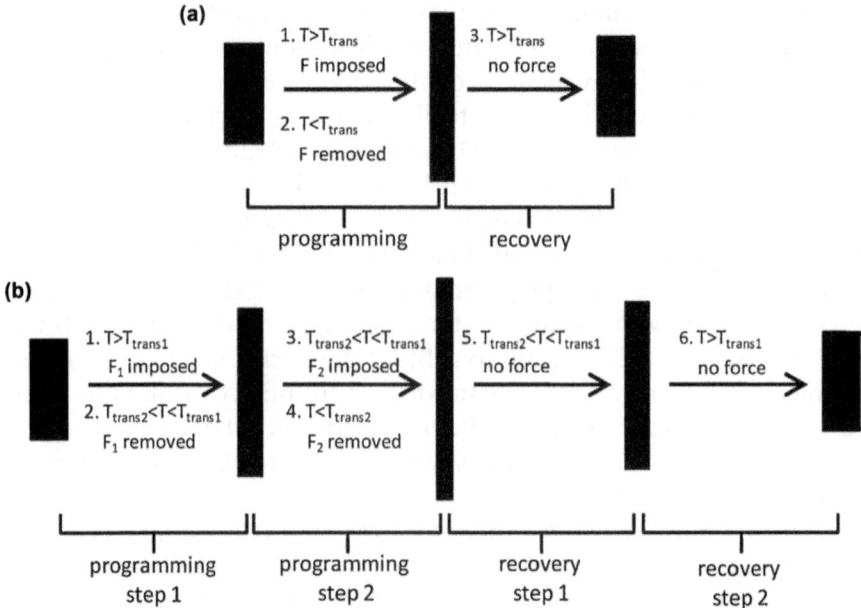

Figure 12.6   Schematic illustration of (a) dual SMEs; and (b) triple SMEs. Reproduced from ref. 74 with permission from Elsevier.

$(T_{m,high})$ and 34 °C $(T_{m,low})$.[64] These materials were synthesized by reacting a polyesterurethane (PEU) network with poly(ω-pentadeca-lactone) [PPDL] and PCL segments. Following the diagram in Figure 12.6b, these materials exhibit a two-way SME. Initially, un-perturbed PEU (permanent shape C) is deformed at $T_{reset} > T_{m,high}$ by the application of an external force. Upon removal of the force de-formation was fixed at a $T_{low} < T_{m,low}$. Upon stress removal, a shape B was generated, which completed the programming step. Again, the programming step is a typical shape memory cycle. At this point, PEU was reheated to a T_{high}, meeting this condition: $T_{m,low} < T_{high} < T_{m,high}$. As a result, shape A was produced. The intriguing feature of this ex-ercise is that PEU exhibits a reversible shape transformation between shapes A and B between temperatures T_{high} and T_{low}. The uniqueness of this behavior is that the shape changes occurred without using external stress; in essence, upon heating to T_{reset} PEU can recover to shape C. In accordance with the process depicted in Figure 12.6b, reprogramming can be applied again to develop different shapes.

The structural rearrangements responsible for this behavior are depicted in Figure 12.7. At T_{reset}, PEU exhibits the original shape Cm, and network macromolecular segments adopt random confirmations.

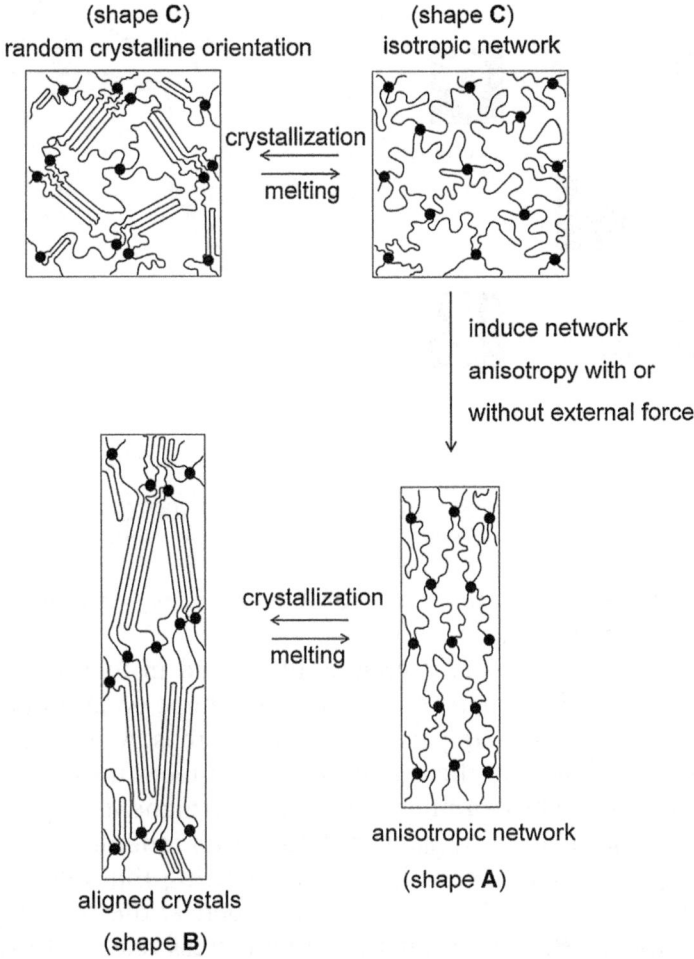

Figure 12.7 Schematic diagram of molecular mechanism 2W-SME of a crosslinked crystalline polymer with an isotropic or anisotropic network morphology.
Reprinted from ref. 74. Copyright (2015) with permission from Elsevier.

Upon application of force and deformation during the programming step, the chain conformations associated with both crystalline phases undergo shape changes, adopting shape A. Exposure to T_{high} only partially removes PCL chain orientation, but the orientation of PPDL chains remains, thus resulting in the development of anisotropy within PCL domains without the use of external forces. In essence, PCL segments are oriented due to constraints imposed by deformed PPDL segments. If the development of network anisotropy is responsible for the reversible shape shifting, it would imply that

(a)

(b)

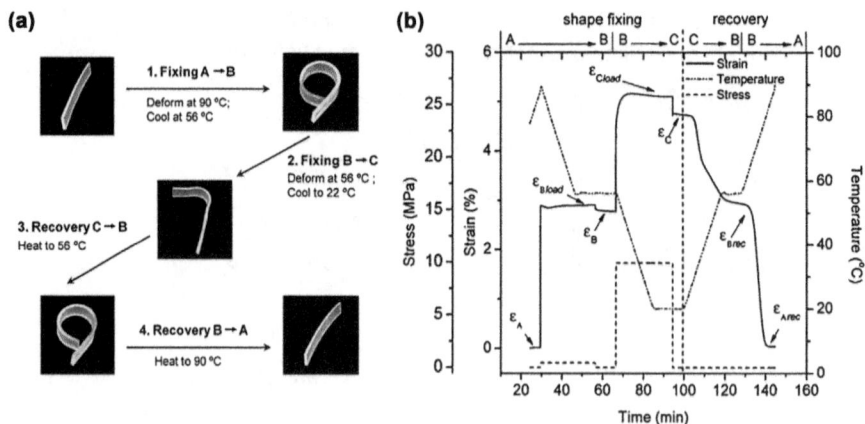

Figure 12.8 Triple-shape memory cycles: (a) visual demonstration; and (b) quantitative thermo-mechanical cycle.
Reproduced with permission from ref. 80. Copyright © 2009 WILEY-VCH Verlag GmbH & Co. KGaA, Weinheim.

internal stress, which is a measurable quantity, can be used to quantify this behavior.

A quantitative triple-shape cycle is illustrated in Figure 12.8.[80] In comparison to a dual-shape cycle, the triple-shape cycle contains two programmable shape fixing and two recovery steps. A prerequisite for a polymer is to have two thermal transition temperatures, $T_{trans,B}$ and $T_{trans,A}$. During the programming stage, the permanent shape A (εA) is deformed to εB load at T_{high}. The deformation should occur above both T_{trans}, followed by cooling to T_{mid}, meeting the following condition: $T_{trans,B} < T_{mid} < T_{trans,A}$. As a consequence, the resulting temporary shape B (εB) will be generated upon stress release. During the second programming step, the shape B can be further deformed to εC load, assuming the second temporary shape C (εC), which is fixed upon cooling to $T_{low} < T_{trans,B}$ and subsequent removal of the second force. The shape recovery can be conducted under a stress-free environment, but heating shape C from T_{low} to T_{mid} recovers the shape B ($\varepsilon Brec$). As heating continues to T_{high} the shape A ($\varepsilon Arec$) is recovered.

Although SMM in general, and SMPs in particular, offer many unique properties, the field is relatively new and further studies are needed to establish fundamental principles governing the SME and, equally important, be able to quantify this behavior. There are excellent review articles that offer many applications and the interested reader is encouraged to explore these.[15,75,79] The remaining section of this chapter will outline challenges and opportunities of SMM used in the context of self-healing.

As pointed out above, a thermally induced SME has been well known for different classes of polymers,[81] but the ability to memorize a permanent shape that can substantially differ from their mechanically induced temporary shape offers a unique opportunity to use these polymers in self-healing composites.[82] At some point, introduced as the 'close-then-heal' concept,[83] shape memory offers an interesting response of a polymer due to external impacts. A flow of loose chain ends and low molecular weight components to compensate for the access of surface energy at the bottom of a scratch facilitates the self-healing process depicted in Figure 12.9A. Shape memory thermosetting blends containing thermoplastic particles offered an unprecedented aid in achieving self-healing attributes. Figure 12.9B depicts the concept in which a SMP is depicted as a spring, which upon damage, is being compressed. But shortly after, it decompresses, bringing two surfaces together. This concept was initially used in SMA wires, which upon mechanical damage exerted a contractual force bringing both surface sides together.[84] Although this approach offered an excellent concept with somewhat limited practical solutions recognized by the necessity of local pre-orientation of wires or fibers, its extension was the utilization of SMPs, which opened new possibilities in this field. One of the attractive

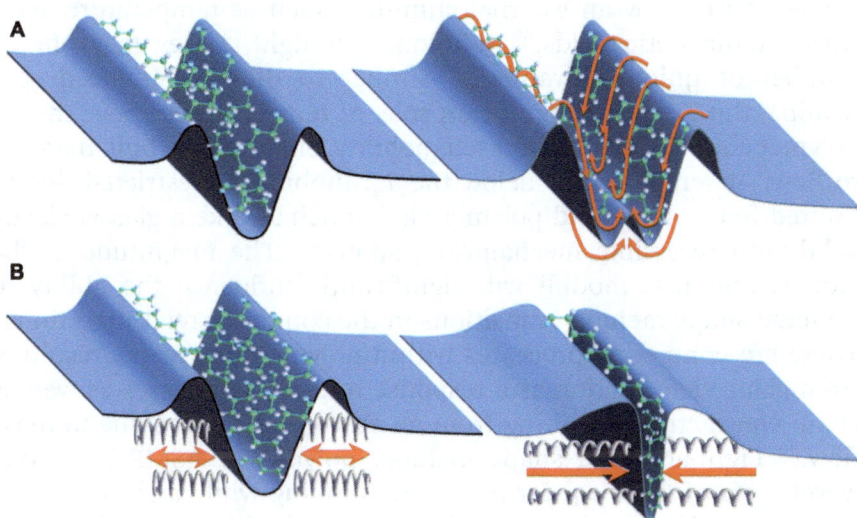

Figure 12.9   (A) A flow of loose chain ends and low molecular weight components to compensate for the access of surface energy at the bottom of a scratch facilitates self-healing process. (B) SMPs facilitate initial closure of a scratch, followed by self-healing reactions.

approaches is the use of a poly(ε-caprolactone) (PCL) based molecular composite system that consists of a two-component blend composed of a thiolene crosslinked PCL network (n-PCL) and a high molecular weight linear PCL (l-PCL) interpenetrating the crosslinked network.[85] The presence of n-PCL secures full recovery of large plastic deformations upon heating due to damage at room temperature below the shape memory transition temperature. The shape memory n-PCL facilitates closing of cracks resulting from mechanical damage, whereas mobile l-PCL chains eliminate cracks and restore mechanical strength by tackifying the crack by diffusion. The advantage of this approach is simultaneous shape recovery and melt/flow of the fibers to rebound the crack upon exposure to elevated temperatures. In essence, the use of macromolecular 'springs' in a form of SMP offers an outstanding opportunity for bringing surfaces together that are typically not easily reformable, especially for polymers with high T_g.

While the presence of stable reactive groups during self-healing reactions is one aspect of self-healing, another aspect is physical network remodeling. If there are polymer network components that 'remember' their shape before damage, their presence may be advantageous in designing self-healing networks. The SMPs represent this group of mechanically responsive materials that have the ability to return from a temporarily altered physical state to its original shape. Induced by an external stimulus, such as temperature, electrical or magnetic fields,[86] solution,[87] or light,[88] they may offer a number of unique advantages. Similar to diffusion-induced self-healing, the T_g plays a significant role. At temperatures above the T_g, polymer chains display increased mobility, enabling the bulk material to flow. At temperatures below the T_g, mobility is restricted due to limited free volume and polymer chains behave like a glassy, elastic solid with reversible mechanical responses. The magnitude of the storage and loss moduli will significantly influence the ability to undergo shape memory transitions in the context of relaxation times. There are competing processes within polymeric networks resulting from chain deformation as a response to external stresses as well as chain contractions resulting from an entropic response due to mass flow.[86] Light-activated shape memory polymers (LASMPs) use one wavelength of light (ν_1) for photo-crosslinking while a second wavelength (ν_2) of light cleaves the photo-crosslinked bonds, which helps the material to reversibly switch between an elastomer and a rigid polymer by changing T_g. Crystallinity may also play a significant role. For example, damage to semi-crystalline thermosetting polymers at temperatures below the melting temperature (T_m) can be repaired by heating above T_m, upon which the original shape will be recovered due to the presence of covalent crosslinks.

The primary advantage of using SMPs in self-healing materials is driven by an opportunity to close a crack, thus enhancing healing efficiency. Upon stimulus leading to self-healing, shape memory components can exert a recovery force to minimize the size of the crack, and aid the repair process *via* self-healing agents blended into the network. Embedding SMA wires into self-healing epoxy with manually injected or microencapsulated healing agents may also offer certain advantages. If aligned correctly, the wires contract at or above the shape change transformation temperature, thus physically decreasing the crack dimension. If alignment is not correct, the effect will be counterproductive. Perhaps the most pragmatic examples include blending of shape memory thermosetting polymers with small amounts of thermoplastics, where the former forces closure of the crack, and the latter repairs the damage by diffusion, entanglements and/or polymerization. Examples include a polystyrene-based shape memory thermosetting polymer blended with 6% by volume copolyester thermoplastic particles, which is able to repair at 150 °C for 20 min,[83] and a crosslinked poly(ε-caprolactone) (n-PCL) network interpenetrated with linear poly(ε-caprolactone) (l-PCL) which can be repaired by heating above T_m.[89] Recently, shape-memory assisted self-healing films were prepared using electro-spun thermoplastic poly(ε-caprolactone) fibers which were randomly distributed in a shape-memory epoxy matrix, featuring a flow of fibers upon heating to rebond the crack besides the SME. The advantage of this approach is simultaneous shape recovery of the epoxy matrix and melt/flow of the fibers to rebond the crack upon exposure to elevated temperatures. If one places this approach in the context of the previous studies that examined the crack healing phenomenon and concluded that when two surfaces approach each other, macromolecular segments rearrange upon physical contact, followed by 'wetting' and inter-diffusion, and mixing, bringing reversible or irreversible self-healing reactions inside a crack will not only facilitate chemical self-healing, but will allow the control of kinetics of chemical and physical aspects of self-healing.

One question is how to measure molecular level events responsible for self-healing, which usually occur inside a scratch. Experimentally, this can be accomplished using imaging spectroscopic tools (Raman, IR), which are often easier to control than following a mass flow during self-healing events. The fundamental question to address is how it is that localized individual reactions are capable of generating cascading responses leading to self-healing? If we consider chain rupture, depicted in Figure 12.10A/A′ – an outcome of mechanical damage – this process will create n isolated reactive ends (free radicals or other reactive groups). The repair process will begin longitudinally along the bottom of a scratch, depicted as an interactive process of connecting the ends, which will be completed when all reacting ends form N pairs. However during this process, after a given time, a percolation transition, *i.e.* the minimum threshold of connected ends, will occur when a finite number of ends are connected longitudinally. This is shown in Figure 12.10B/B′. The reactions responsible for repair and mass flow are initiated along the longitudinal direction, pointed out by an arrow, as illustrated by the enclosed images. The driving forces are high surface energies in the smaller curvature areas at the bottom of a scratch facilitating closer physical proximity of the reactive groups and higher segmental mobility. Simple geometrical considerations suggest that, in contrast to terminal –CH_3 represented by sp^3 hybridization, CH and CH_2 free radicals adopt highly directional sp and sp^2 hybridizations. The distances between the free radicals to initiate coupling are roughly 2.35–3.44 Å and typically

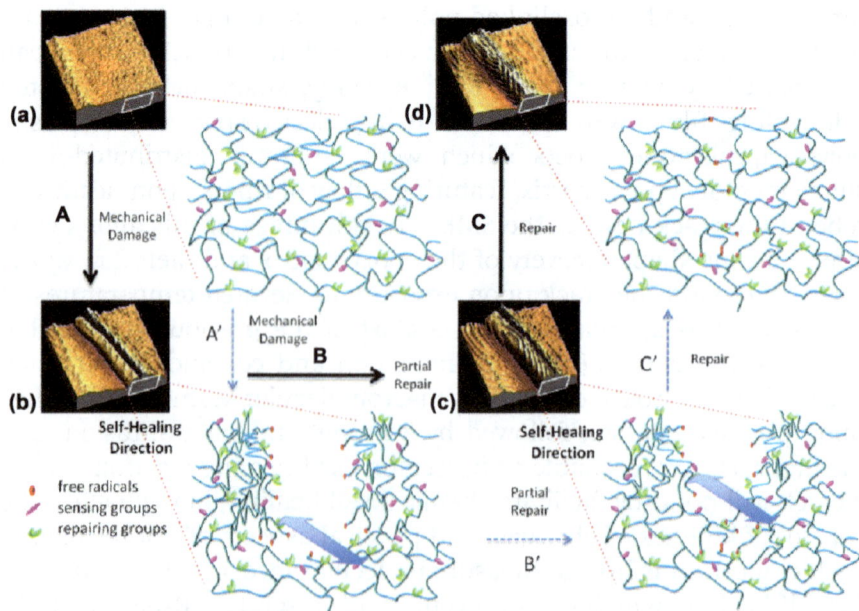

Figure 12.10   Schematic illustration of the self-healing process: a–d are the images of the damage repair process of modified polyurethane networks (adapted from ref. 45). The self-healing process is represented by a sequence of events A through C, and corresponds to a network repair process shown in cartoons A′–C′. Reproduced from Ref. 91 with permission from The Royal Society of Chemistry.

requires ~ 20 kcal mol^{-1}. When coupling occurs, the bonding distances will become significantly smaller, ranging from ~ 0.95–1.60 Å, which requires ~ 10 kcal mol^{-1}.[90] It is therefore reasonable to expect that reactive groups on one side of the surface/interface inside a scratch will exhibit stronger affinity towards reactants with their counterparts 'across-the-isle' on the other surface. One can envisage that the large-scale connectivity of a damaged polymer matrix begins with one-at-a-time longitudinal to a scratch formation of nodes, and the percolation transition will occur when a finite fraction of chain ends are connected in the longitudinal direction (Figure 12.10B′/C′). As self-healing progresses, at some point the critical value will be reached. Below this point, it is anticipated that the power law governed distribution will be obeyed, but self-healing will continue linearly. But when the transition points occur earlier, 'explosive' self-healing of a non-linear nature may take place, such as observed in the networks that require minimal remodeling. For delayed transitions, spontaneous transitions may occur at the later stages, and there is growing evidence that exposure to electromagnetic radiation may cause delayed, but more pronounced transitions at the later stages of self-healing. Experimentally determined morphology changes inside scratches of OXE–CHI–PUR networks indeed showed that self-healing is initiated from the bottom of the damage and progresses to the surface, as shown in Figure 12.10a–d.[91,92] These observations can be confirmed by atomic force microscopy experiments on disulfide polymer systems in which self-healing proceeds in a 'zip-up' fashion, starting from the bottom and edge of the damage up to the surface. Remodeling of neighboring chains as illustrated in Figure 12.10C/C′ depicts the final stages of the process, during which the flow from the side banks of damage-induced surfaces due to liquid-like properties supplies the majority of reactive groups, which are consumed resulting in bottom-up self-repair. This is somewhat counter-intuitive, but this directional altitudinal growth should be clearly differentiated from a lateral closure due to elastic recovery of deformed crosslinked or entangled polymer networks and has little to do with self-healing.

References

1. A. Ölander, An electrochemical investigation of solid cadmium-gold alloys, *J. Am. Chem. Soc.*, 1932, **54**, 3819–3833.
2. A. B. Greninger and V. G. Mooradian, Strain transformation in metastable beta copper-zinc and beta copper-Ti alloys, *Trans. Metall. Soc. AIME*, 1938, **128**, 337–369.

3. G. Kurdjumov and L. Khandros, First reports of the thermoelastic behaviour of the martensitic phase of Au-Cd alloys, *Dokl. Akad. Nauk SSSR*, 1949, **66**, 211–213.

4. L. Chang and T. Read, Plastic deformation and diffusionless phase changes in metals - The gold-cadmium beta-phase, *Trans. Am. Inst. Min. Metall. Eng.*, 1951, **191**, 47–52.

5. J. M. Jani, M. Leary, A. Subic and M. A. Gibson, A review of shape memory alloy research, applications and opportunities, *Mater. Des.*, 2014, **56**, 1078–1113.

6. L. Sun, W. M. Huang, Z. Ding, Y. Zhao, C. C. Wang, H. Purnawali and C. Tang, Stimulus-responsive shape memory materials: a review, *Mater. Des.*, 2012, **33**, 577–640.

7. I. Mihálcz, Fundamental characteristics and design method for nickel-titanium shape memory alloy, *Mech. Eng.*, 2001, **45**, 75–86.

8. L. Sun and W. Huang, Nature of the multistage transformation in shape memory alloys upon heating, *Met. Sci. Heat Treat.*, 2009, **51**, 573–578.

9. D. Hartl and D. Lagoudas, *Thermomechanical Characterization of Shape Memory Alloy Materials*, Springer, 2008.

10. T. Duerig and A. Pelton, *Materials Properties Handbook: Titanium Alloys*, 1994, pp. 1035–1048.

11. H. Funakubo and J. Kennedy, Shape memory alloys, *Gordon and Breach, xii + 275, 15×22 cm, Illustrated*, 1987.

12. L. Delaey, Diffusionless transformations, *Materials Science and Technology*, 1991.

13. T. Tadaki, K. Otsuka and K. Shimizu, Shape memory alloys, *Annu. Rev. Mater. Sci.*, 1988, **18**, 25–45.

14. L. B. Vernon and H. M. Vernon, Process of manufacturing articles of thermoplastic synthetic resins, *US Pat.*, 2 234 993, 1941.

15. J. J. Hitov, W. C. Rainer, E. M. Redding, A. W. Sloan and W. D. Stewart, Irradiated, crosslinked polyethylene copolymer, *US Pat.*, 3097150 A, 1964.

16. R. J. Perrone, Silicone-rubber, polyethylene composition; heat shrinkable articles made therefrom and process therefor, *US Pat.*, 3326869 A, 1967.

17. C. Liu, H. Qin and P. Mather, Review of progress in shape-memory polymers, *J. Mater. Chem.*, 2007, **17**, 1543–1558.

18. A. Lendlein and S. Kelch, Shape-memory polymers, *Angew. Chem., Int. Ed.*, 2002, **41**, 2034–2057.

19. M. D. Hager, S. Bode, C. Weber and U. S. Schubert, Shape memory polymers: past, present and future developments, *Prog. Polym. Sci.*, 2015, **49**, 3–33.

20. J. Morshedian, H. Khonakdar, M. Mehrabzadeh and H. Eslami, Preparation and properties of heat-shrinkable cross-linked low-density polyethylene, *Adv. Polym. Technol.*, 2003, **22**, 112–119.

21. A. Maksimkin, S. Kaloshkin, M. Zadorozhnyy and V. Tcherdyntsev, Comparison of shape memory effect in UHMWPE for bulk and fiber state, *J. Alloys Compd.*, 2014, **586**, S214–S217.

22. J. Zhao, M. Chen, X. Wang, X. Zhao, Z. Wang, Z.-M. Dang, L. Ma, G.-H. Hu and F. Chen, Triple shape memory effects of cross-linked polyethylene/polypropylene blends with cocontinuous architecture, *ACS Appl. Mater. Interfaces*, 2013, **5**, 5550–5556.

23. F. Katzenberg, B. Heuwers and J. C. Tiller, Superheated rubber for cold storage, *Adv. Mater.*, 2011, **23**, 1909–1911.

24. B. Heuwers, D. Quitmann, F. Katzenberg and J. C. Tiller, Stress-induced melting of crystals in natural rubber: a new way to tailor the transition temperature of shape memory polymers, *Macromol. Rapid Commun.*, 2012, **33**, 1517–1522.

25. B. Heuwers, A. Beckel, A. Krieger, F. Katzenberg and J. C. Tiller, Shape-memory natural rubber: an exceptional material for strain and energy storage, *Macromol. Chem. Phys.*, 2013, **214**, 912–923.

26. D. Quitmann, N. Gushterov, G. Sadowski, F. Katzenberg and J. C. Tiller, Solvent-sensitive reversible stress-response of shape memory natural rubber, *ACS Appl. Mater. Interfaces*, 2013, **5**, 3504–3507.

27. N. R. Brostowitz, R. Weiss and K. A. Cavicchi, Facile fabrication of a shape memory polymer by swelling cross-linked natural rubber with stearic acid, *ACS Macro Lett.*, 2014, **3**, 374–377.

28. X. Luo, X. Zhang, M. Wang, D. Ma, M. Xu and F. Li, Thermally stimulated shape-memory behavior of ethylene oxide-ethylene terephthalate segmented copolymer, *J. Appl. Polym. Sci.*, 1997, **64**, 2433–2440.

29. B. C. Chun, T. K. Cho, M. H. Chong and Y.-C. Chung, Structure–property relationship of shape memory polyurethane cross-linked by a polyethyleneglycol spacer between polyurethane chains, *J. Mater. Sci.*, 2007, **42**, 9045–9056.

30. T. Liu, J. Li, Y. Pan, Z. Zheng, X. Ding and Y. Peng, A new approach to shape memory polymer: design and preparation of poly (methyl methacrylate) composites in the presence of star poly (ethylene glycol), *Soft Matter*, 2011, **7**, 1641–1643.

31. Y. Niu, P. Zhang, J. Zhang, L. Xiao, K. Yang and Y. Wang, Poly(p-dioxanone)–poly(ethylene glycol) network: synthesis, characterization, and its shape memory effect, *Polym. Chem.*, 2012, **3**, 2508–2516.

32. X. Li, T. Liu, Y. Wang, Y. Pan, Z. Zheng, X. Ding and Y. Peng, Shape memory behavior and mechanism of poly(methyl methacrylate) polymer networks in the presence of star poly(ethylene glycol), *RSC Adv.*, 2014, **4**, 19273–19282.
33. J. Hu, Z. Yang, L. Yeung, F. Ji and Y. Liu, Crosslinked polyurethanes with shape memory properties, *Polym. Int.*, 2005, **54**, 854–859.
34. B. Yang, W. Huang, C. Li, C. Lee and L. Li, On the effects of moisture in a polyurethane shape memory polymer, *Smart Mater. Struct.*, 2004, **13**, 191.
35. L. Peponi, I. Navarro-Baena, A. Sonseca, E. Gimenez, A. Marcos-Fernandez and J. M. Kenny, Synthesis and characterization of PCL–PLLA polyurethane with shape memory behavior, *Eur. Polym. J.*, 2013, **49**, 893–903.
36. A. Saralegi, E. J. Foster, C. Weder, A. Eceiza and M. A. Corcuera, Thermoplastic shape-memory polyurethanes based on natural oils, *Smart Mater. Struct.*, 2014, **23**, 025033.
37. H. Y. Lee, H. M. Jeong, J. S. Lee and B. K. Kim, Study on the shape memory polyamides. Synthesis and thermomechanical properties of polycaprolactone-polyamide block copolymer, *Polym. J.*, 2000, **32**, 23–28.
38. G. Zhu, G. Liang, Q. Xu and Q. Yu, Shape-memory effects of radiation crosslinked poly (ε-caprolactone), *J. Appl. Polym. Sci.*, 2003, **90**, 1589–1595.
39. P. Ping, W. Wang, X. Chen and X. Jing, Poly (ε-caprolactone) polyurethane and its shape-memory property, *Biomacromolecules*, 2005, **6**, 587–592.
40. I. A. Rousseau and T. Xie, Shape memory epoxy: composition, structure, properties and shape memory performances, *J. Mater. Chem.*, 2010, **20**, 3431–3441.
41. T. Lin, S. Ma, Y. Lu and B. Guo, New design of shape memory polymers based on natural rubber crosslinked via oxa-Michael reaction, *ACS Appl. Mater. Interfaces*, 2014, **6**, 5695–5703.
42. A. Alteheld, Y. Feng, S. Kelch and A. Lendlein, Biodegradable, amorphous copolyester-urethane networks having shape-memory properties, *Angew. Chem., Int. Ed.*, 2005, **44**, 1188–1192.
43. Á. Sonseca, S. Camarero-Espinosa, L. Peponi, C. Weder, E. J. Foster, J. M. Kenny and E. Giménez, Mechanical and shape-memory properties of poly (mannitol sebacate)/cellulose nanocrystal nanocomposites, *J. Polym. Sci., Part A: Polym. Chem.*, 2014, **52**, 3123–3133.

44. I. S. Kolesov, K. Kratz, A. Lendlein and H.-J. Radusch, Kinetics and dynamics of thermally-induced shape-memory behavior of crosslinked short-chain branched polyethylenes, *Polymer*, 2009, **50**, 5490–5498.

45. Q. Meng and J. Hu, A poly(ethylene glycol)-based smart phase change material, *Sol. Energy Mater. Sol. Cells*, 2008, **92**, 1260–1268.

46. F. Li, W. Zhu, X. Zhang, C. Zhao and M. Xu, Shape memory effect of ethylene–vinyl acetate copolymers, *J. Appl. Polym. Sci.*, 1999, **71**, 1063–1070.

47. Y. Feng, M. Behl, S. Kelch and A. Lendlein, Biodegradable multiblock copolymers based on oligodepsipeptides with shape-memory properties, *Macromol. Biosci.*, 2009, **9**, 45–54.

48. A. Lendlein, A. M. Schmidt, M. Schroeter and R. Langer, Shape-memory polymer networks from oligo (ε-caprolactone) dimethacrylates, *J. Polym. Sci., Part A: Polym. Chem.*, 2005, **43**, 1369–1381.

49. H. Zhang, H. Wang, W. Zhong and Q. Du, A novel type of shape memory polymer blend and the shape memory mechanism, *Polymer*, 2009, **50**, 1596–1601.

50. H. M. Jeong, S. Y. Lee and B. K. Kim, Shape memory polyurethane containing amorphous reversible phase, *J. Mater. Sci.*, 2000, **35**, 1579–1583.

51. K. Hearon, L. D. Nash, B. L. Volk, T. Ware, J. P. Lewicki, W. E. Voit, T. S. Wilson and D. J. Maitland, Electron beam crosslinked polyurethane shape memory polymers with tunable mechanical properties, *Macromol. Chem. Phys.*, 2013, **214**, 1258–1272.

52. K. Hearon, C. J. Besset, A. T. Lonnecker, T. Ware, W. E. Voit, T. S. Wilson, K. L. Wooley and D. J. Maitland, A structural approach to establishing a platform chemistry for the tunable, bulk electron beam crosslinking of shape memory polymer systems, *Macromolecules*, 2013, **46**, 8905–8916.

53. D. L. Safranski and K. Gall, Effect of chemical structure and crosslinking density on the thermo-mechanical properties and toughness of (meth)acrylate shape memory polymer networks, *Polymer*, 2008, **49**, 4446–4455.

54. C. M. Yakacki, R. Shandas, D. Safranski, A. M. Ortega, K. Sassaman and K. Gall, Strong, tailored, biocompatible shape-memory polymer networks, *Adv. Funct. Mater.*, 2008, **18**, 2428–2435.

55. J. Hao and R. Weiss, Mechanically tough, thermally activated shape memory hydrogels, *ACS Macro Lett.*, 2013, **2**, 86–89.

56. H. Koerner, R. J. Strong, M. L. Smith, D. H. Wang, L.-S. Tan, K. M. Lee, T. J. White and R. A. Vaia, Polymer design for high temperature shape memory: low crosslink density polyimides, *Polymer*, 2013, **54**, 391–402.

57. X. L. Wu, W. M. Huang, Z. Ding, H. X. Tan, W. G. Yang and K. Y. Sun, Characterization of the thermoresponsive shape-memory effect in poly(ether ether ketone) (PEEK), *J. Appl. Polym. Sci.*, 2014, **131**, 34844–39850.

58. Y. Liu, C. Han, H. Tan and X. Du, Thermal, mechanical and shape memory properties of shape memory epoxy resin, *Mater. Sci. Eng., A*, 2010, **527**, 2510–2514.

59. K. S. Kumar, R. Biju and C. R. Nair, Progress in shape memory epoxy resins, *React. Funct. Polym.*, 2013, **73**, 421–430.

60. L. Song, W. Hu, G. Wang, G. Niu, H. Zhang, H. Cao, K. Wang, H. Yang and S. Zhu, Tailored (meth)acrylate shape-memory polymer networks for ophthalmic applications, *Macromol. Biosci.*, 2010, **10**, 1194–1202.

61. N.-Y. Choi and A. Lendlein, Degradable shape-memory polymer networks from oligo[(L-lactide)-ran-glycolide] dimethacrylates, *Soft Matter*, 2007, **3**, 901–909.

62. M. Yoonessi, Y. Shi, D. A. Scheiman, M. Lebron-Colon, D. M. Tigelaar, R. Weiss and M. A. Meador, Graphene polyimide nanocomposites; thermal, mechanical, and high-temperature shape memory effects, *ACS Nano*, 2012, **6**, 7644–7655.

63. J. Shumaker, A. McClung and J. Baur, Synthesis of high temperature polyaspartimide-urea based shape memory polymers, *Polymer*, 2012, **53**, 4637–4642.

64. M. Behl, K. Kratz, J. Zotzmann, U. Nöchel and A. Lendlein, Reversible bidirectional shape-memory polymers, *Adv. Mater.*, 2013, **25**, 4466–4469.

65. H. Qin and P. T. Mather, Combined one-way and two-way shape memory in a glass-forming nematic network, *Macromolecules*, 2008, **42**, 273–280.

66. K. Hiraoka, N. Tagawa and K. Baba, Shape-memory effect controlled by the crosslinking topology in uniaxially-deformed smectic elastomers, *Macromol. Chem. Phys.*, 2008, **209**, 298–307.

67. Z. Pei, Y. Yang, Q. Chen, E. M. Terentjev, Y. Wei and Y. Ji, Mouldable liquid-crystalline elastomer actuators with exchangeable covalent bonds, *Nat. Mater.*, 2014, **13**, 36–41.

68. H. Finkelmann, H. J. Kock and G. Rehage, Investigations on liquid crystalline polysiloxanes 3. Liquid crystalline elastomers—a

new type of liquid crystalline material, *Makromol. Chem., Rapid Commun.*, 1981, **2**, 317–322.

69. A. Agrawal, T. Yun, S. L. Pesek, W. G. Chapman and R. Verduzco, Shape-responsive liquid crystal elastomer bilayers, *Soft Matter*, 2014, **10**, 1411–1415.

70. T. Chung, A. Romo-Uribe and P. T. Mather, Two-way reversible shape memory in a semicrystalline network, *Macromolecules*, 2008, **41**, 184–192.

71. S. Pandini, S. Passera, M. Messori, K. Paderni, M. Toselli, A. Gianoncelli, E. Bontempi and T. Ricco, Two-way reversible shape memory behaviour of crosslinked poly (ε-caprolactone), *Polymer*, 2012, **53**, 1915–1924.

72. X. Zhang, Q. Zhou, H. Liu and H. Liu, UV light induced plasticization and light activated shape memory of spiropyran doped ethylene-vinyl acetate copolymers, *Soft Matter*, 2014, **10**, 3748–3754.

73. J. Zotzmann, M. Behl, D. Hofmann and A. Lendlein, Reversible triple-shape effect of polymer networks containing poly-pentadecalactone-and poly(ε-caprolactone)-segments, *Adv. Mater.*, 2010, **22**, 3424–3429.

74. Q. Zhao, H. J. Qi and T. Xie, Recent progress in shape memory polymer: new behavior, enabling materials, and mechanistic understanding, *Prog. Polym. Sci.*, 2015, **49**, 89–95.

75. T. Xie, Recent advances in polymer shape memory, *Polymer*, 2011, **52**, 4985–5000.

76. W. Wagermaier, K. Kratz, M. Heuchel and A. Lendlein, Characterization methods for shape-memory polymers, in *Shape-memory Polymers*, Springer, 2010, pp. 97–145.

77. P. T. Mather, X. Luo and I. A. Rousseau, Shape memory polymer research, *Annu. Rev. Mater. Res.*, 2009, **39**, 445–471.

78. J. Hu, Y. Zhu, H. Huang and J. Lu, Recent advances in shape-memory polymers: structure, mechanism, functionality, modeling and applications, *Prog. Polym. Sci.*, 2012, **37**, 1720–1763.

79. T. Xie and I. A. Rousseau, Facile tailoring of thermal transition temperatures of epoxy shape memory polymers, *Polymer*, 2009, **50**, 1852–1856.

80. T. Xie, X. Xiao and Y. T. Cheng, Revealing triple-shape memory effect by polymer bilayers, *Macromol. Rapid Commun.*, 2009, **30**, 1823–1827.

81. A. Charlesby, *Atomic Radiation and Polymers: International Series of Monographs on Radiation Effects in Materials*, Elsevier, 2013.

82. R. Zirbs, A. Lassenberger, I. Vonderhaid, S. Kurzhals and E. Reimhult, Melt-grafting for the synthesis of core–shell nanoparticles with ultra-high dispersant density, *Nanoscale*, 2015, 7, 11216–11225.
83. J. Nji and G. Li, A biomimic shape memory polymer based self-healing particulate composite, *Polymer*, 2010, 51, 6021–6029.
84. E. L. Kirkby, J. D. Rule, V. J. Michaud, N. R. Sottos, S. R. White and J. A. E. Månson, Embedded shape-memory alloy wires for improved performance of self-healing polymers, *Adv. Funct. Mater.*, 2008, 18, 2253–2260.
85. X. Luo and P. T. Mather, Shape memory assisted self-healing coating, *ACS Macro Lett.*, 2013, 2, 152–156.
86. R. Mohr, K. Kratz, T. Weigel, M. Lucka-Gabor, M. Moneke and A. Lendlein, Initiation of shape-memory effect by inductive heating of magnetic nanoparticles in thermoplastic polymers, *Proc. Natl. Acad. Sci. U. S. A.*, 2006, 103, 3540–3545.
87. H. Lu, Y. Liu, J. Leng and S. Du, Comment on "Water-driven programmable polyurethane shape memory polymer: Demonstration and mechanism", *Appl. Phys. Lett.*, 2010, 97, 056101.
88. A. Lendlein, H. Jiang, O. Jünger and R. Langer, Light-induced shape-memory polymers, *Nature*, 2005, 434, 879–882.
89. E. D. Rodriguez, X. Luo and P. T. Mather, Linear/network poly(ε-caprolactone) blends exhibiting shape memory assisted self-healing (SMASH), *ACS Appl. Mater. Interfaces*, 2011, 3, 152–161.
90. F. T. Wang, L. Chen, C. J. Tian, Y. Meng, Z. G. Wang, R. Q. Zhang, M. X. Jin, P. Zhang and D. J. Ding, Interactions between free radicals and a graphene fragment: physical versus chemical bonding, charge transfer, and deformation, *J. Comput. Chem.*, 2011, 32, 3264–3268.
91. B. Ghosh, K. V. Chellappan and M. W. Urban, Self-healing inside a scratch of oxetane-substituted chitosan-polyurethane (OXE-CHI-PUR) networks, *J. Mater. Chem.*, 2011, 21, 14473–14486.
92. B. Ghosh, K. V. Chellappan and M. W. Urban, UV-initiated self-healing of oxolane–chitosan–polyurethane (OXO–CHI–PUR) networks, *J. Mater. Chem.*, 2012, 22, 16104–16113.

Subject Index